The Engineering Ethics Decision Matrix
This is an example of one particular Engineering Ethics Decision Matrix

Options → NSPE Canons ↓	Go along with the decision	Appeal to higher management	Quit your job	Write your state representative	Call a newspaper reporter
Hold paramount the safety, health and welfare of the public.					
Perform services only in the area of your competence					
Issue public statements only in an objective and truthful manner					
Act for each employer or client as faithful agents or trustees					
Avoid deceptive acts					
Conduct themselves honorably ...					

Exploring Engineering
An Introduction to Engineering and Design

Fourth Edition

Exploring Engineering
An Introduction to Engineering and Design

Fourth Edition

Philip Kosky

Robert Balmer

William Keat

George Wise

ELSEVIER

AMSTERDAM • BOSTON • HEIDELBERG • LONDON
NEW YORK • OXFORD • PARIS • SAN DIEGO
SAN FRANCISCO • SINGAPORE • SYDNEY • TOKYO
Academic Press is an imprint of Elsevier

Academic Press is an imprint of Elsevier
125, London Wall, EC2Y 5AS, UK
525 B Street, Suite 1800, San Diego, CA 92101–4495, USA
225 Wyman Street, Waltham, MA 02451, USA
The Boulevard, Langford Lane, Kidlington, Oxford OX5 1GB, UK

Notices

Knowledge and best practice in this field are constantly changing. As new research and experience
broaden our understanding, changes in research methods, professional practices, or medical treatment
may become necessary.

Practitioners and researchers must always rely on their own experience and knowledge in evaluating
and using any information, methods, compounds, or experiments described herein. In using such
information or methods they should be mindful of their own safety and the safety of others, including
parties for whom they have a professional responsibility.

To the fullest extent of the law, neither the Publisher nor the authors, contributors, or editors, assume
any liability for any injury and/or damage to persons or property as a matter of products liability,
negligence or otherwise, or from any use or operation of any methods, products, instructions, or ideas
contained in the material herein.

Library of Congress Cataloging-in-Publication Data
A catalog record for this book is available from the Library of Congress

British Library Cataloguing in Publication Data
A catalogue record for this book is available from the British Library

ISBN: **978-0-12-801242-0**

For information on all Academic Press publications
visit our website at http://store.elsevier.com/

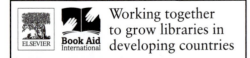

Working together
to grow libraries in
developing countries

www.elsevier.com • www.bookaid.org

"… it is engineering that changes the world …"

Isaac Asimov, *Isaac Asimov's Book of Science and Nature Quotations* (New York: Simon and Schuster, 1970)

"… [Engineering] … is the art of doing that well with one dollar that any bungler can do with two …"

Arthur Wellington, *Economic Theory of the Location of Railways*, 2nd ed. (New York: Wiley, 1887)

"The engineer has been, and is, a maker of history."

James Kip Finch (1883–1967), an American engineer and educator.

"Engineering is not merely … analysis; engineering is not merely the possession of the capacity to get elegant solutions to non-existent engineering problems; engineering is practicing the art of the organizing forces of technological change … Engineers operate at the interface between science and society."

Gordon Stanley Brown in Bert Scalzo, et al., *Database Benchmarking: Practical Methods for Oracle & SQL Server* (2007), 37.

"Scientists investigate that which already is; Engineers create that which has never been."

Albert Einstein (1879–1955), German-born American physicist who developed the special and general theories of relativity and the photoelectric effect, the latter for which he received the Nobel Prize for Physics in 1921.

"Scientists dream about doing great things. Engineers do them."

James A. Michener (1907–1997), U.S. novelist and short-story writer.

"A great pleasure in life is doing what people say you cannot do."

Walter Bagehot (1826–1877), English economist, political journalist, and critic.

"I often say that when you can measure what you are speaking about, and express it in numbers, you know something about it; but when you cannot measure it, when you cannot express it in numbers, your knowledge is of a meagre and unsatisfactory kind."

Lord Kelvin (William Thomson, 1824–1907), English mathematical physicist.

"Whenever you look at a piece of work and you think the fellow was crazy, then you want to pay some attention to that. One of you is likely to be, and you had better find out which one it is. It makes an awful lot of difference."

Charles Franklin Kettering (1876–1958), U.S. engineer and inventor.

"… the explosion of knowledge, the global economy, and the way engineers will work will reflect an ongoing evolution that began to gain momentum a decade ago."

Educating the Engineer of 2020, National Academy of Engineering, October 2005

Contents

Preface

Engineers have made remarkable innovations during the twentieth century. The National Academy of Engineering (NAE) recently identified the top 20 engineering achievements of the twentieth century, achievements that "shaped a century and changed the world":

1. Electrification—to supply our homes and businesses with electricity
2. Automobile—for leisure and commercial transportation
3. Airplane—for rapidly moving people and goods around the world
4. Water supply and distribution—to supply clean, germ-free water to every home
5. Electronics—to provide electronic control of machines and consumer products
6. Radio and television—for entertainment and commercial uses
7. Agricultural mechanization—to increase the efficiency of food production
8. Computers—a revolution in the way people work and communicate
9. Telephone—for rapid personal and commercial communication
10. Air conditioning and refrigeration—to increase the quality of life
11. Highways—to speed transportation of people and goods across the land
12. Spacecraft—to begin our exploration of limitless space
13. Internet—a cultural evolution of the way people interact
14. Imaging—to improve healthcare
15. Household appliances—to allow women to enter the workplace
16. Health technologies—to improve the quality of life
17. Petroleum and petrochemical technologies—to power transportation systems
18. Laser and fiber optics—to improve measurement and communication systems
19. Nuclear technologies—to tap a new natural energy source
20. High-performance materials—to create safer, lighter, and better products

However, engineering students are less interested in what was or what is than in what *will be*. Young men and women exploring engineering as a career are excited about the future—*their future*—and about the engineering challenges 10-20 years from now when they are in the spring and summer of their careers. In the words of the four-time Stanley Cup winner and Hockey Hall of Fame member Wayne Gretzky,

> *I skate to where the puck is going to be, not where it's been.*

The National Academy of Engineering also proposed the following 14 Grand Challenges for Engineering in the twenty-first century. In our fourth edition of this text, we have chosen to highlight material that engages these topics because they represent the future of engineering creativity.

1. Make solar energy economical
2. Provide energy from fusion
3. Develop carbon sequestration methods
4. Manage the nitrogen cycle
5. Provide access to clean water
6. Restore and improve urban infrastructure
7. Advance health informatics
8. Engineer better medicines
9. Reverse-engineer the brain

10. Prevent nuclear terror
11. Secure cyberspace
12. Enhance virtual reality
13. Advance personalized learning
14. Engineer the tools of scientific discovery

The twenty-first century will be filled with many exciting challenges for engineers, architects, physicians, sociologists, and politicians. Figure 1 illustrates an enhanced set of future challenges as envisioned by Joseph Bordogna, deputy director and chief operating officer of the National Science Foundation.[1]

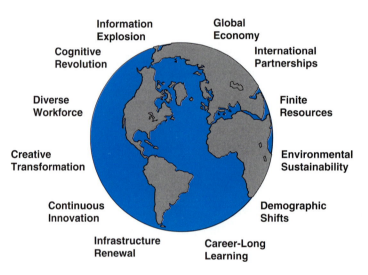

FIGURE 1 Future trajectories in science, engineering, and technology.

1 THE STRUCTURE OF THIS TEXT

The first and second editions of this text were organized around the theme of twenty-first century engineering based on the technologies to be found in a modern automobile because we thought that incoming students could identify with the subject of cars, but we were somewhat disappointed at the superficiality of that knowledge.[2] Several of our reviewers thought the text appeared to be a text for automotive engineering, which was not our intention. For that reason, this fourth edition has been expanded and reorganized quite differently. This expansion has taken the fourth edition into new territory with new subjects such as industrial engineering and aeronautical engineering as well as deeper into nuclear engineering than was the third edition. Overall, the coverage of engineering specialties is now unusually broad so that the text gives the student the possibility to discover what appeals to them while still presented at an elementary level—and what does not.

---

[1]http://www.nsf.gov/news/speeches/bordogna/jb98_nrl/sld001.htm.
[2]Ask your class what fuel is used to heat their houses in winter. Our experience is that few recognize that *any* fuel is used or that it is necessary.

No question but there is too much material to be completed in any one term or semester class. The text is 30 chapters long enough we consider to be sufficiently comprehensive to cover the modern world of engineering. In an ideal world, this is enough for students to choose in some detail which area of engineering is their forte. But realities of class sizes and professorial preferences will limit the areas to be taught.

Nor was it our intention to have students cover all of this book. Our conception is to give a basic introduction to many engineering methods, to apply principles and methods that have been learned in the first part of this book to areas of technology covered in one or more specialized chapters, and, finally, to present a design challenge to the students. Accordingly, this text is divided into three cohesive parts: Part 1, Lead-On; Part 2, Minds-On; and Part 3, Hands-On.

We tried to provide an exciting introduction to the engineering profession. Between its covers, you will find material on classical engineering fields as well as introductory material for emerging twenty-first century engineering fields.

- **Part 1**, which we call **Lead-On**, emphasizes what we consider the fundamentals that thread through virtually all branches of engineering. There are just five chapters in Part 1: "What Engineers Do," "Elements of Engineering Analysis," "Force and Motion," "Energy," and "Engineering Economics." Most chapters in Part 1 are organized around just one or two principles, have several worked examples, and include exercises with several levels of difficulty. Occasionally, answers are given to selected exercises to encourage students to work toward self-proficiency. Solutions are available online for instructors at the publisher's Web site.

  We recommend all the first five chapters for all first-year engineering courses.

  These chapters describe how to organize and solve engineering problems. Spreadsheets are introduced early and repeatedly throughout the text. We include an eclectic number of engineering tools unexplored by most comparable texts. We favor the use of rigorously structured answers to problems that, if followed, not only contribute to a solution for engineering problems but also leave an audit trail that can be followed later, should the need arise. We include how to address engineering ethics problems using a matrix approach, the importance of units and dimensions, and the use of tabular methods for multifaceted problems.

- **Part 2**, which we call **Minds-On**, covers introductory material explicitly from the following well-established engineering disciplines: aeronautical engineering, chemical engineering, civil engineering, computer engineering, electrical engineering, industrial engineering, manufacturing engineering, materials engineering, mechanical engineering, and nuclear engineering. There is some minor overlap because some engineering fields *do* overlap. In addition, three more chapters are devoted to emerging engineering fields: bioengineering, electrochemical engineering, and "green" energy engineering. The topics covered in each chapter are kept to a level commiserate with the background of first-year students. Our recommendation is to select appropriate discipline-specific chapters. Roughly speaking, first-year students are able to absorb an amount about equal to Part 1 in Part 2, but clearly it's up to the teaching faculty to cherry-pick which chapters best suit their specific needs. Instructors should expect to find shortcuts in methodology that might pain purists; nevertheless, we tried to be accurate as to basic principles.

- **Part 3** provides the content for a design studio and is associated with the design of multifaceted engineering *systems*. It should be covered in full if time permits. This **Hands-On** section is just as essential and challenging as the engineering aspects covered in Parts 1 and 2, and for most students, *it is a lot more fun*. Few things are more satisfying than seeing a machine, an electronic device, or a computer program you designed and built, doing exactly what you intended it to do. Such initial successes may sound simple, but they provide the basis of a rigorous system that will enable an engineering graduate, as part of a team of engineers, to achieve the even greater satisfaction in designing a system that can provide new means of transportation, information access, medical care, energy supply and can change for the better lives of people around the world.

Our experience is that any chapter can be sufficiently covered in about 2 h of lecture class time per week and that the students can complete the rest of the chapter unaided. Chapter 2 covers much of what is basic and common to all areas of engineering such as effective diagrams, variables, units, number systems, significant figures, and spreadsheets and might thus take 2 weeks to complete.

On the other hand, the design studio needs up to 3 contiguous laboratory hours per week to do it justice. It culminates in a team-orientated competition. Typically, student teams build a small model "device" that has wheels, walks, or floats; that may be wireless or autonomous; and so forth. Students then compete head-to-head against other teams from the course with the same design goals plus an offensive and defensive strategy to overcome all the other teams in the competition. Our experience is that this is highly motivating for the students.

Given the necessary breaks for recitation sections, testing, and a final examination, a typical class will cover all the chapters of Part 1. In Part 2, faculties choose which subjects they want to cover.

The approach taken in this first- or second-year text is unique, in part, because of the atypical character of authorship. Two of the authors have *industrial* backgrounds mostly at the GE Research Center in Niskayuna, New York. Together, they bring a working knowledge of what is core to a practicing engineer. The other two authors followed more traditional academic paths and have the appropriate academic experience and credentials to draw upon. We believe the synergy of the combined authorship provides a fresh perspective for beginning engineering education. Specifically, though elementary in coverage, this textbook parallels the combined authors' wide experience that engineering is not a "spectator sport." We therefore do not duck the introduction of relatively advanced topics in this otherwise elementary text. Here are some of the nonstandard approaches to familiar engineering topics:

1. We introduce spreadsheets early in the text, and almost every chapter of Part 1 and Part 2 has one or more spreadsheet exercises.

2. We try to rigorously enforce the use of appropriate significant figures throughout the text. For example, we always try to differentiate between 60. and 60 (notice the decimal point or its absence). We recognize it often appears to be clumsy to write numbers such as 6.00×10^1, but we do so to discourage bad habits, such as electronic calculator answers to undeserved significant figures.

3. Except for answers to engineering ethics problems, which have their own formalism, our exercise solutions use a rigorous format using a simple mnemonic **Need-Know-How-Solve** to discourage the student who thinks he or she knows the answer and writes the wrong one down (or even the correct one!). This too can appear to be clumsy in usage, but it is invaluable in training a young engineer to leave an audit trail of his or her methods, a good basic work habit of practicing engineers.

4. We recognize that the Engineering English unit system of lbf, lbm, and g_c will be used throughout the careers of many, if not most, of today's young engineers. A clear methodology is used to develop it and to use it, so we can avoid the terrible results of a factor of 32.2 that should or shouldn't be there!

5. Conservation principles, particularly energy and mass, are introduced early in the text as well as emphasis on the use of control boundaries that focus on the essential problem at hand.

6. The use of matrices/tables is a powerful tool, in the hands of both students and qualified engineers. We developed a number of tabular methods for many kinds of problems. Methods based on tables are also fundamental to design principles as taught in the design studio section of the book.

7. We emphasize the power of electrical switches as vital elements of computer design and their mathematical logic analogues.

8. We developed a simple solution method for standard one-dimensional kinematics problems using a visual-geometric technique of speed-time graphs rather than applying the standard equations by rote.

We believe this is a useful visual way to deal with multielement kinematics problems. Of course, we also quoted, but not developed, the standard kinematics equations because they are derived in every introductory physics textbook and their use does not increase basic understanding of kinematics *per se*.

9. The design methodology in the design studio is presented in a stepwise manner to lead student and instructor through a hands-on design project.

10. Pacing of hands-on projects is accomplished through design milestones. These are general time-tested project assignments that we believe are the most powerful tool in getting a freshman design course to work well.

11. The design examples were selected from past student projects, ranging from the freshman to the senior year, to appeal to and be readily grasped by the beginning engineering student. In one chapter, we present a couple of typical first-year design projects and follow the evolution of one team's design from clarification of the task to detailed design.

12. The culmination of the hands-on design studio is a head-to-head team competition, and it is recommended that all first-year engineering courses based on this text should strive to include it.

13. This textbook does not use the calculus *per se*, but a new chapter in the fourth edition is on the theory of flight. Conformal mapping is avoided, but complex numbers are at the root of the theory. The chapter may appear "too mathematical" for a beginning aeronautical engineering, but it contains nothing but the manipulations of imaginary numbers that can be accomplished by advanced high school seniors.

14. The Accreditation Board for Engineering and Technology (ABET) sets curriculum criteria[3] that require students to have "an understanding of professional and ethical responsibility." To avoid creating this unintentional contrast between ethics and engineering, we introduce a new pedagogical tool the **Engineering Ethics Decision Matrix**. The rows of the matrix are the canons of engineering ethics and the columns are possible ways to resolve the problem. Each box of the matrix must be filled with a very brief answer to the question, "Does this one particular solution meet this one particular canon?" This is a structured approach that will bring discipline to this subject for first-year engineers. Each chapter in Parts 1 and 2 has ethics problems pertinent to that particular chapter. We believe that it is more useful to infuse ethics continually during the term, rather than as a single arbitrarily inserted lecture.

PGK, RTB, WDK, GW
Union College, Schenectady, New York
University of Wisconsin—Milwaukee, Milwaukee, Wisconsin

---

[3]According to ABET, engineering programs must demonstrate that students attain an ability to (a) apply the knowledge of mathematics, science, and engineering; (b) design and conduct experiments and analyze data; (c) design a system, component, or process within economic, environmental, social, political, ethical, health-safety, manufacturability, and sustainability constraints; (d) function on multidisciplinary teams; (e) identify, formulate, and solve engineering problems; (f) understand professional and ethical responsibility; (g) communicate effectively; (h) understand engineering solutions in a global, economic, environmental, and societal context; (i) engage in life-long learning; (j) gain a knowledge of contemporary issues; and (k) apply modern engineering tools to engineering practice.

A companion website for this textbook is available at http://booksite.elsevier.com/9780128012420

It has resources, including time management and study skills information, links to unit conversion programs, and practice exercises with some solutions.

For instructors, a solution manual, design contest material, and Power Point lecture slides are available by registering at www.textbooks.elsevier.com. It contains worked solutions to every exercise using the Need-Know-How-Solve paradigm as developed in this text.

Acknowledgments

This textbook is for beginning engineering students, but it comprehensively covers introductory material for almost all disciplines in engineering. Therefore, we want to thank many colleagues proficient in those subjects and who gave us some of their precious time.

We acknowledge help, suggestions, and advice from several Union College colleagues and especially from coteachers for the Union freshman engineering course: Professors Brad Bruno, James Hedrick, Thomas Jewell, John Spinelli, Cherrice Travers, and Frank Wicks. Professors Bruno and Wicks also reviewed chapters on aeronautical and nuclear engineering, respectively. In addition, we have received advice and a chapter review from Professor Nicholas Krouglicof of Memorial University, Newfoundland, Canada. Other assistance came from Dr. John Rogers, Mechanical Engineering Division, West Point, and Dr. Andrew Wolfe, Civil Engineering Technology, SUNY Institute of Technology, Utica.

We would like to thank the following faculty, who provided feedback on the revision plans:

Aaron Budge, Minnesota State University
Mauro Caputi, Hofstra University
Kelly Crittenden, Louisiana Technical University
Brian DeJong, Central Michigan University
Mebougna Drabo, Alabama A&M University
Darin Gray, University of Southern California
Michael Gregg, Virginia Polytechnic Institute
Daniel Guino, Ohio University
Jerry Hamann, University of Wyoming
Robert Krchnavek, Rowan University
Steven McIntosh, University of Virginia
Francelina Neto, California State Polytechnic University
Jin Park, Minnesota State University
James Riddell, Baker College

In addition, we also thank these instructors, who reviewed chapter drafts:

Stephanie Adams, Virginia Tech
Stacy Birmingham, Grove City College
Mebougna Drabo, Alabama A&M University
Darin Gray, University of Southern California
Ken Manning, SUNY Adirondack
Samuel Morton, James Madison University

The competition-based hands-on approach to teaching design was inspired by Professor Michael C. Larson, Mechanical Engineering Department, Tulane University, New Orleans, Louisiana, and by Mr. Daniel Retajczyk, then a graduate student at Clarkson University, New York.

Mr. Craig Ferguson (Computer Science/Mechanical Engineering, Union College) developed the student design for the "A Bridge Too Far" example in Part 3.

Finally, we thank Mr. Dan Beller (University of Wisconsin-Milwaukee), for his assistance in developing the material on drawing and sketching.

Many of the graphic illustrations were produced by Ted Balmer at March Twenty Productions (http://marchtwenty.com).

Lead-On

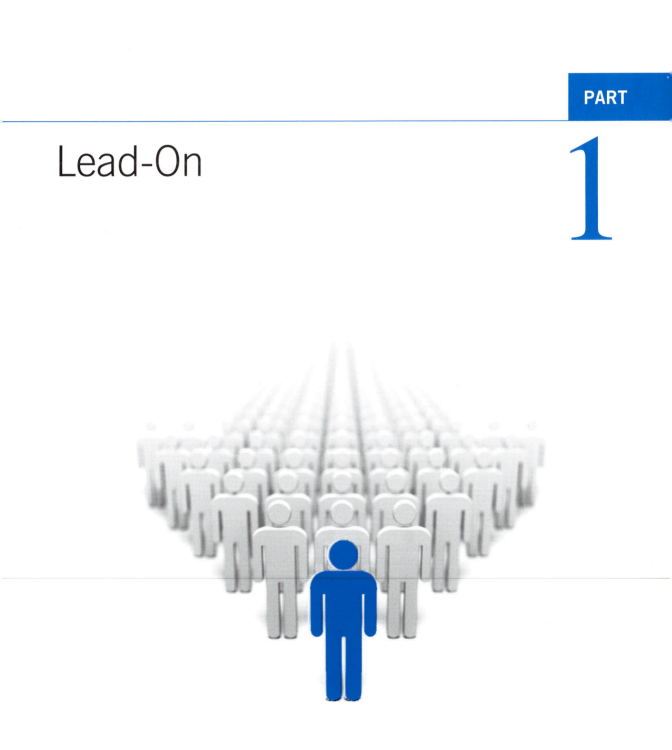

What Engineers Do

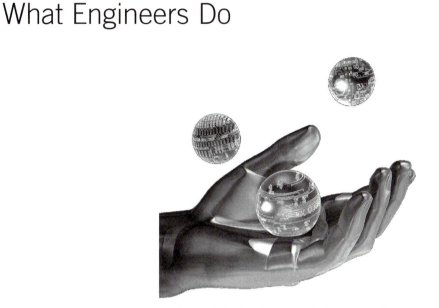

Source: iStockphoto.com/Antonis Papantoniou.

1.1 INTRODUCTION

What is an engineer, and what does he or she do? You can get a good answer to this question by just looking at the word itself. The word *engine* comes from the Latin *ingenerare*, meaning "to create." About 2000 years ago, the Latin word *ingenium* (the product of genius) was used to describe the design of a new machine. Soon after, the word *ingen* was used to describe all machines. In English, *ingen* was spelled "engine," and people who designed creative things were known as "engine-ers." In French, German, and Spanish today, the word for *engineer* is *ingenieur*, and in Italian, it is *ingegnere*.

What does the word engineer mean?

Answer: *The word engineer refers to someone who is a creative, ingenious person who finds solutions practical problems.*

Today the word *engineer* refers to people who use creative design and analysis processes that incorporate energy, materials, motion, and information to serve human needs in innovative ways. Engineers express knowledge in the form of variables, numbers, and units. There are many kinds of engineers, but all share the ideas and methods introduced in this book.

1.2 WHAT IS *ENGINEERING?*

The late scientist and science fiction writer Isaac Asimov once said that "Science can amuse and fascinate us all but it is engineering that changes the world."[1] Almost everything you see around you has been touched by an engineer. Engineers are creative people who use mathematics, scientific principles, material properties, and computer methods to design new products and to solve human problems. Engineers can and do just about anything, designing and building roads, bridges, cars, planes, space stations, cell phones, computers, medical equipment, and so forth.

Engineers can be classified according to the kind of work they do—administration, construction, consulting, design, development, teaching, planning (also called *applications engineers*), production, research, sales, service, and test engineers. Because engineering deals with the world around us, the number of engineering disciplines is very large. Table 1.1 lists some of the many engineering fields.

Table 1.1 A Few of the Many Engineering Fields Available Today

Aerospace/aeronautical	Ceramic	Electrical	Mechanical	Petroleum
Agricultural	Chemical	Environmental	Metallurgical	Sanitary
Architectural	Civil	Geological	Mining	Systems
Automotive	Computer	Manufacturing	Nuclear	Textile
Biomedical	Ecological	Marine	Ocean	Transportation

1.3 WHAT DO ENGINEERS DO?

Most engineers specialize in a specific field of engineering. The following list contains information on a few of the engineering fields in the Federal Government's Standard Occupational Classification (SOC) system.[2] Note that some of the engineering fields may have several subdivisions. For example, civil engineering includes structural and transportation engineering, and materials engineering includes ceramic, metallurgical, and polymer engineering.

- An **aeronautical, or aerospace, engineer**[3] applies scientific and technological principles to research, design, develop, maintain, and test the performance of civil and military aircraft, missiles, weapons systems, satellites, and space vehicles. They also work on the different components that make up these aircraft and systems.

- **Biomedical engineers** develop devices and procedures that solve medical and health-related problems by combining biology and medicine with engineering principles. Many biomedical engineers develop and evaluate systems and products such as artificial organs, instrumentation, and health management and care delivery systems.

- **Chemical engineers** apply the principles of chemistry to solve problems involving the production or use of chemicals and other products. They design equipment and processes for biotechnical use, chemical

---

[1]Isaac Asimov's *Book of Science and Nature Quotations*, (New York: Simon & Schuster, 1970).
[2]Abstracted from the Bureau of Labor Statistics (http://www.bls.gov/oco/ocos027.htm).
[3]www.prospects.ac.uk

manufacturing, plan, and test methods of manufacturing products and treating byproducts, and supervise production.

- **Civil engineers** design and supervise the construction of roads, buildings, airports, tunnels, dams, bridges, and water supply and sewage systems. Civil engineering is one of the oldest engineering disciplines[4] and encompasses many specialties. The major ones are structural, water resources, construction, transportation, and geotechnical engineering.

- **Computer engineers** research, design, develop, test, and oversee the manufacture and installation of computer hardware, including computer chips, circuit boards, computer systems, and related equipment, such as keyboards, routers, and printers. Computer engineers may also design and develop the software systems that control computers.

- **Electrical engineers** design, develop, test, and supervise the manufacture of electrical equipment. Some of this equipment includes electric motors; machinery controls, lighting, and wiring in buildings; radar and navigation systems; communications systems; and power generation, control, and transmission devices used by electric utilities.

- **Environmental engineers** use the principles of biology and chemistry to develop solutions to environmental problems. They are involved in water and air pollution control, recycling, waste disposal, and public health issues. Environmental engineers conduct hazardous-waste management studies in which they evaluate the significance of the hazard, and develop regulations to prevent mishaps.

- **Industrial and manufacturing engineers** determine the most effective ways to use the basic items of production—people, machines, materials, information, and energy—to make a product or provide a service. They are concerned with increasing productivity through the management of people, methods of business organization, and technology. These engineers study product requirements and then design manufacturing systems to meet those requirements.

- **Materials engineers** are involved in the development, processing, and testing of the materials used to create a range of products, from computer chips and aircraft wings to golf clubs and snow skis. They work with metals, ceramics, plastics, semiconductors, and composites to create new materials that meet certain mechanical, electrical, and chemical requirements.

- **Mechanical engineers** research, design, develop, manufacture, and test all types of mechanical devices. Mechanical engineering is one of the broadest engineering disciplines. Mechanical engineers work on power-producing machines such as electric generators, internal combustion engines, and steam and gas turbines; they also work on power-using machines such as refrigeration and air-conditioning equipment, machine tools, material-handling systems, and robots.

- **Nuclear engineers** research and develop the processes, instruments, and systems used to derive benefits from nuclear energy and radiation. They design, develop, monitor, and operate nuclear plants to generate power. They may work on the nuclear fuel cycle—the production, handling, and use of nuclear fuel and the safe disposal of nuclear waste.

You can find more about what today's engineers do within their specialties by searching the Internet. Here are some of the engineering societies[5] that represent different engineering fields: AIAA (aeronautical engineering),

---

[4]The oldest type of engineering is Military Engineering. Civil engineers are specifically called *civil engi*neers to distinguish them from *military* engineers. The word *civil* is a contraction of the word *civilian*.

[5]A typical engineering society has several functions. They define the core disciplines needed for membership and advocate for them. They also define codes and standards for their discipline, provide further educational courses, and offer a code of engineering ethics customized for that particular profession. Canadian engineering societies basically follow a similar nomenclature as do others worldwide.

AIChE (chemical engineers), ANS (nuclear engineering), ASCE (civil engineers), ASME (mechanical engineers), ASTM (materials and testing engineers), BMES (biomedical engineering), IEEE (electrical engineers), and many others.

Unsurprisingly you will discover that the basic college engineering courses have much in common with all engineering disciplines. They cover scientific principles, application of logical problem solving processes, principles of design, value of teamwork, and engineering ethics. If you are considering an engineering career, we highly recommend you consult web resources to refine your understanding of the various fields of engineering.

1.4 WHERE DO ENGINEERS WORK?

Most engineers work in office buildings, laboratories, or industrial plants. Others may spend time outdoors at construction sites and oil and gas exploration and production sites where they monitor or direct operations or solve onsite problems. Some engineers travel extensively to plants or worksites here and abroad. Many engineers work a nominal 40 h week. At times, deadlines or design standards may bring extra pressure to a job, requiring engineers to work longer hours.

Engineers usually work in teams. Sometimes, the team has only two or three engineers, but in large companies, engineering teams can have hundreds of people working on a single project (the design and manufacture of a large aircraft, for example). Engineers are responsible for communicating, planning, designing, manufacturing, and testing among other duties.

Engineers are capable of designing the processes and equipment needed for a project, and sometimes, that involves inventing new technologies. Engineers must also test their work carefully before it is used by trying to anticipate all of the things that could go wrong and make sure that their products perform safely and effectively.

More than 1.2 million engineers work in the United States today, making engineering the nation's second-largest profession. According to the 2014 survey by the National Association of Colleges and Employers, engineering majors have 8 of the 10 the highest baccalaureate degree starting salaries averaging $67,480 per year.

An engineering degree also opens doors to other careers. Engineering graduates can move into other professions, such as medicine, law, and business, where their engineering problem solving ability is a valuable asset. A list[6] is available of famous engineers who became American Presidents, Nobel Prize winners, astronauts, corporate presidents, entertainers, inventors, and scientists.

In the United States, distinguished engineers may be elected to the National Academy of Engineering (NAE); it is the highest national honor for engineers. In many countries, there are parallel organizations (e.g., The Royal Academy of Engineering in the United Kingdom).

1.5 WHAT IS ENGINEERING TECHNOLOGY?

The following definition of engineering technology was established by the Technology Accreditation Commission of ABET, Inc. (Accreditation Board for Engineering and Technology) and was approved by the Engineering Technology Council of the American Society for Engineering Education.

> *Engineering technology is the profession in which a knowledge of mathematics and natural sciences gained by higher education, experience, and practice is devoted primarily to the implementation and extension of existing technology for the benefit of humanity.*

Engineering technologists work closely with engineers in coordinating people, material, and machinery to achieve the specific goals of a particular project. The engineering technologist is often responsible for design and development.

---

[6]See http://www.sinc.sunysb.edu/Stu/hnaseer/interest.htm.

Many engineering technicians work in quality control, inspecting products and processes, conducting tests, or collecting data. In manufacturing, they may assist in product design, development, or production.

There is a wide range of options when it comes to educational preparation in engineering technology. Most employers prefer to hire engineering technologists with at least a 2 year associate degree in engineering technology. Some universities offer 2 year associate degrees, others offer 3 or 4 year BS degree programs, and some offer both types of degrees.

1.6 WHAT MAKES A "GOOD" ENGINEER?

This is actually a difficult question to answer because the knowledge and skills required to be an engineer (i.e., to create ingenious solutions) is a moving target. The factors that will lead to your career success are not the same as they were 20 years ago (and never will be). In this book, we illustrate the key characteristics of a successful twenty-first century engineer by exploring the multidisciplinary creative engineering process required to produce "good" competitive products for the twenty-first century.

So just what *does* the twenty-first century hold for the young engineer? It will be characterized by the *convergence* of many technologies and engineering systems. The products of today and tomorrow will be "smarter." The incorporation of computers, sensors, controls, modern alloys, and plastics are as important as continuing expertise in the traditional engineering disciplines. This book is also intended to appeal to a number of aspects of modern engineering subdisciplines.

1.7 WHAT THIS BOOK COVERS

In your mind what makes a "good" consumer product, say, an automobile? If you were in the market to purchase one, you might want one that has high performance and good gas mileage and is roomy, safe, and stylish. Or you might describe it in categories like new or used; sedan, sports car, or SUV; two doors or four doors. Or, maybe, you would be interested in only the price tag.

As a consumer making a decision about purchasing a car, it is enough to use the preceding words, categories, and questions to reach a decision. But engineers think differently. They design and analyze, and consequently, they must have a different set of words, categories, and questions. To design and analyze, engineers ask precise questions that can be answered with **variables**, **numbers**, and **units**. They do it to produce a safe and reliable product. From this point of view, an automobile is an engineer's answer to the question "What is a good way to move people safely and reliably?"

The purpose of this book is to introduce you to the engineering profession. It does so by introducing you to the way engineers think, ask, and answer questions like these: What makes an automobile—or a computer, or an airplane, or a washing machine, or a bridge, or a prosthetic limb, or an oil refinery, or a space satellite—*good*?

We use the automobile as an example at this point strictly for convenience. Presumably, you have, or at least think you have, some idea of what constitutes an automobile. But, it no more or no less expresses the essence of engineering than would an example based on a computer, an airplane, a washing machine, a bridge, a prosthetic limb, an oil refinery, or a space satellite. In each case, the essence of the example would focus on the creative use of energy, materials, motion, and information to serve human needs, so a more detail-oriented engineer might answer our original question like this:

> *A good 21st-century automobile employs stored energy (on the order of 100 million joules), complex materials (on the order of 1000. kilograms [about one ton] of steel, aluminum, glass, and plastics), and information (on the order of millions of bits processed every second)so that it is capable of high speed (on the order of 40 meters/second ≈ 90 mph), low cost (a few tens of a cent per mile), low pollution (a few grams of pollutants per mile), and high safety.*

That is a long and multidimensional answer, but an engineer would be unapologetic. Engineering is *inherently* multidimensional and multidisciplinary. It needs to be multidimensional to create compromises among conflicting criteria, and it needs to be multidisciplinary to understand the technical impact of the compromises. Making a car heavier, for example, might make it safer, but it would also be less fuel efficient. Engineers often deal with such competing factors. They break down general issues into concrete questions. They then answer those questions with design variables, units, and numbers.

Engineering is not a spectator sport! It is a *hands-on* and *minds-on* activity. In this book, you will be asked to participate in a "Design Studio." This is the part of the course that is "hands on"—and, it is *fun!* But, you will still learn the principles of good design practice (irrespective of your intended engineering major), and you will have to integrate skills learned in construction, electrical circuits, logic, and computers in building a device (the "device" could be a car, robot, boat, bridge, or anything else appropriate to your course). It will have to compete against similar devices built by other young engineers also in your class and whose motivation may be to stop your device from succeeding in achieving the same goals! You will learn how to organize data and the vital importance of good communication skills. You will present your ideas and your designs orally and in written format. In the Design Studio, you will design and build increasingly complex engineering systems, starting with the tallest tower made from a single sheet of paper and ending with a controlled device combining many parts into a system aimed at achieving complex goals.

As a start to the "minds-on" portion of the book, can you mentally take apart and put back together an imaginary automobile or toaster, or computer or bicycle? Instead of using wrenches and screwdrivers, your tools are mental and computerized tools for engineering thought.

Example 1.1

Figure 1.1 shows a generic car with numbered parts. Without cheating from the footnote, can you fill in the correct number corresponding to the parts in each of the blanks.[7]

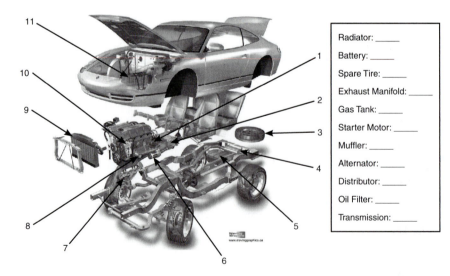

Radiator: _____	
Battery: _____	
Spare Tire: _____	
Exhaust Manifold: _____	
Gas Tank: _____	
Starter Motor: _____	
Muffler: _____	
Alternator: _____	
Distributor: _____	
Oil Filter: _____	
Transmission: _____	

FIGURE 1.1 An exploded view of a modern automobile. Source: © Moving Graphics.

---

[7]Answer: 1—distributor, 2—transmission, 3—spare tire, 4—muffler, 5—gas tank, 6—starter motor, 7—exhaust manifold, 8—oil filter, 9—radiator, 10—alternator, 11—battery.

As visually appealing as Figure 1.1 is, an engineer would consider it inadequate because it fails to express the functional connections among the various parts. Expressing in visual form, the elements and relationships involved in a problem is a crucial tool of engineering, called a **conceptual sketch**. A first step in an engineer's approach to a problem is to draw a conceptual sketch of the problem. Artistic talent is not an issue nor is graphic accuracy. The engineer's conceptual sketch may not look at all like the thing it portrays. Rather, it is intended to (1) help the engineer identify the elements in a problem, (2) see how groups of elements are connected together to form subsystems, and (3) understand how all those subsystems work together to create a working system.

Example 1.2

On a piece of paper, draw a conceptual sketch of what happens when you push on the pedal of a bicycle. Before you begin here are some questions you should think about:

- What are the key components that connect the pedal to the wheel?
- Which ones are connected to each other?
- How does doing something to one of the components affect the others?
- What do those connections and changes have to do with accomplishing the task of accelerating the bicycle?

Solution

Figure 1.2a shows what your sketch probably looks like. But this is just the final form of the bicycle; it does not give much insight into what was needed to design and to build it. It's the utterly unphysical representation in Figure 1.2b that will clarify the functions needed to design it, "Form *follows* function" means first it has to work.

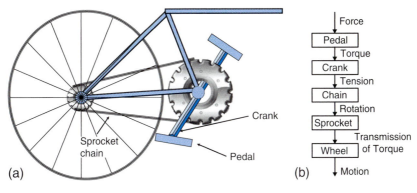

FIGURE 1.2 Bicycle transmission.

For any engineering concept, many different conceptual sketches are possible. You are encouraged to draw conceptual sketches of each of the key points in the learning sections in this book.

1.8 PERSONAL AND PROFESSIONAL ETHICS

The word "ethics" can have several meanings. It can mean philosophical ethics where reason is used to define the limits of right and wrong behavior in society or a profession. It can also mean a person's own ethical behavior relative to others. What are *personal* ethics and what do they have to do with engineering?

Personal ethics are the standards of human behavior that individuals of different cultures have constructed to make moral judgments about personal or group situations. Ethical principles developed as people reflected on the intentions and consequences of their acts. Naturally, they vary over time and from culture to culture, resulting in conflict when what is acceptable in one culture is not in another. For example, the notion of privacy in U.S. culture is very strong, and a desk is considered an extension of that privacy, whereas in another culture, such as Japan, office space is open and one's desk would be considered public domain.

Suppose you are a passenger in a car driven by a close friend. The friend is exceeding the speed limit and has an accident. There are no witnesses, and his lawyer tells you that if you testify that your friend was not exceeding the speed limit, it will save him from a jail sentence. What do you do?

Lying is more accepted in cultures that stress human relationships, but it is less accepted in cultures that stress laws. People in cultures that emphasize human relationships would most likely lie to protect the relationship, whereas people in cultures that put greater value on laws would lie less in order to obey the law.

How do you reconcile a belief in certain moral absolutes such as "I will not kill anyone" with the reality that, in some circumstances (e.g., war), it might be necessary to endanger or kill innocent people for the greater good? This issue gets particularly difficult if one denies tolerance to other faiths, yet the prevailing morality that most of us would describe as "good" is to extend tolerance to others.

1.8.1 The Five Cornerstones of Ethical Behavior

Here are some examples of codes of personal ethics. At this point, you might want to compare your own personal code of ethics with the ones listed here.[8]

1. Do what you say you will do.
2. Never divulge information given to you in confidence.
3. Accept responsibility for your mistakes.
4. Never become involved in a lie.
5. Never accept gifts that compromise your ability to perform in the best interests of your organization.

1.8.2 Top 10 Questions You Should Ask Yourself When Making An Ethical Decision[9]

10. Could the decision become habit forming? *If so, don't do it.*
9. Is it legal? *If it isn't, don't do it.*
8. Is it safe? *If it isn't, don't do it.*
7. Is it the right thing to do? *If it isn't, don't do it.*
6. Will this stand the test of public scrutiny? *If it won't, don't do it.*
5. If something terrible happened, could I defend my actions? *If you can't, don't do it.*
4. Is it just, balanced, and fair? *If it isn't, don't do it.*
3. How will it make me feel about myself? *If it feels lousy, don't do it.*
2. Does this choice lead to the greatest good for the greatest number? *If it doesn't, don't do it.*

And the number 1 question you should ask yourself when making an ethical decision:

1. Would I do this in front of my mother? *If you wouldn't, don't do it.*

---

[8]Manske, F.A., Jr., *Secrets of Effective Leadership*, (Columbia, SC: Leadership Education and Development, Inc., 1999).
[9]From: http://www.cs.bgsu.edu/maner/heuristics/1990Taylor.htm.

1.9 WHAT ARE *PROFESSIONAL* ETHICS?

A professional code of ethics has the goal of ensuring that a profession serves the legitimate goals of *all* its constituencies: self, employer, profession, and public. The code protects the members of the profession from some undesired consequences of competition (for example, the pressure to cut corners to save money) while leaving the members of the profession free to benefit from the desirable consequences of competition (for example, invention and innovation).

Having a code of ethics enables an engineer to resist the pressure to produce substandard work by saying, "As a professional, I cannot ethically put business concerns ahead of professional ethics." It also enables the engineer to similarly resist pressure to allow concerns such as personal desires, greed, ideology, religion, or politics to override professional ethics.

1.9.1 National Society of Professional Engineers (NSPE) Code of Ethics for Engineers

Engineering is an important and learned profession. As members of this profession, engineers are expected to exhibit the highest standards of honesty and integrity. Engineering has a direct and vital impact on the quality of life for all people. Accordingly, the services provided by engineers require honesty, impartiality, fairness, and equity and must be dedicated to the protection of the public health, safety, and welfare. Engineers must perform under a standard of professional behavior that requires adherence to the highest principles of ethical conduct.[10]

1.9.2 Fundamental Canons[11]

Engineers, in the fulfillment of their professional duties, shall

- hold paramount the safety, health, and welfare of the public;
- perform services only in areas of their competence;
- issue public statements only in an objective and truthful manner;
- act for each employer or client as faithful agents or trustees;
- avoid deceptive acts;
- conduct themselves honorably, responsibly, ethically, and lawfully so as to enhance the honor, reputation, and usefulness of the profession.

Example 1.3 An Ethical Situation

The following scenario is a common situation faced by engineering students. Read it and discuss how you would respond. What are your ethical responsibilities?

You and your roommate are both enrolled in the same engineering class. Your roommate spent the weekend partying and did not do the homework that is due on Monday. You did your homework, and your roommate asks to see it. You are afraid he/she will just copy it and turn it in as his/her own work. What are you ethically obligated to do?

---

[10]See http://www.nspe.org/ethics/eh1-code.asp.
[11]Canons were originally church laws; the word has come to mean rules of acceptable behavior for specific groups.

a. Show your roommate the homework?
b. Show the homework but ask your roommate not to copy it?
c. Show the homework and tell the roommate that if the homework is copied, you will tell the professor?
d. Refuse to show the homework?
e. Refuse to show the homework but offer to spend time tutoring the roommate?

Solution

For the purposes of this course, the answer to an ethics question consists of appropriately *applying a code of ethics*. In this example, the five cornerstones of ethical behavior are used, as they are familiar to you in one form or another.

Let us see which of the five cornerstones apply here:

1. **Do what you say you will do.** If the teacher has made it clear that this is an individual assignment, then by participating in the assignment, you have implicitly agreed to keep your individual effort private. Allowing one's homework to be copied means going back on this implicit promise. This implies that answer d or e, "Refuse to show the homework" is at least part of the right answer.
2. **Never divulge information given to you in confidence.** Again, homework is implicitly a confidential communication between an individual student and a teacher. By solving the problem, you have created a confidential communication with the teacher. This is more support for choice d or e.
3. **Accept responsibility for your mistakes.** Sharing your homework enables your roommate to evade this standard. Being an accomplice in the violation of standards by others is itself an ethical violation. This is further support for choice d or e.
4. **Never become involved in a lie.** Allowing your homework to be copied is participating in a lie: that the work the roommate turns in is his or her own work. This further supports choice d or e.
5. **Never accept gifts that compromise your ability to perform in the best interests of your organization.** Since the roommate has not offered anything in exchange for the help, this standard appears not to apply in this case.

Four of the five cornerstones endorse choice d or e, refuse to show the homework, while the fifth cornerstone is silent. These results indicate that your ethical obligation under this particular code of personal ethics is to refuse to show your homework.

Many people find the five cornerstones to be incomplete because they lack a canon common to most of the world's ethical codes: the Golden Rule.[12] Including the Golden Rule would create the additional obligation to show some empathy for your roommate's plight, just as you would hope to receive such empathy if you were in a similar situation. This suggests the appropriateness of choice e, offering to tutor the roommate in doing the homework. In much the same way, in subsequent exercises, you may feel the need to supplement the Code of Ethics for Engineers with elements from your own personal code of ethics. However, this must not take the form of *replacing* an element in the Code of Ethics for Engineers with a personal preference.

In subsequent chapters, the NSPE Code of Ethics for Engineers is used, but this does not constitute an endorsement of the code or any other particular code for personal ethics. Use of the NSPE Code of Ethics for Engineers in subsequent answers, by contrast, *does* constitute a reminder that you must accept that code in your professional dealings if you want to be a professional engineer.

---

[12]There are many versions of the Golden Rule in the world's major religions. Here's one attributed to Confucius: "Do not do to others what you would not like yourself."

1.10 ENGINEERING ETHICS DECISION MATRIX

To avoid creating an unintentional contrast between ethics and engineering, you will be asked to focus on a particular tool: the **engineering ethics decision matrix**. This tool presents a simple way of applying the canons of engineering ethics and further to see the spectrum of responses that might apply in a given situation. In particular, it should give you pause not to accept the first simple do/do not response that comes to you.

In Table 1.2, the rows of the matrix are the canons of engineering ethics (here, the NSPE set) and the columns are possible ways to resolve the problem. (You should add additional ones as they occur to you.) Each box of the matrix must be filled with a very brief answer to the question: "Does this one particular solution meet this particular canon?" Like other engineering tools, the ethical decision matrix is a way to divide-and-conquer a problem, rather than trying to address all its dimensions simultaneously.

Table 1.2 The Engineering Ethics Decision Matrix

Options → NSPE Canons↓	Go Along with the Decision	Appeal to Higher Management	Quit Your Job	Write Your State Representative	Call a Newspaper Reporter
Hold paramount the safety, health, and welfare of the public					
Perform services only in the area of your competence					
Issue public statements only in an objective and truthful manner					
Act for each employer or client as faithful agents or trustees					
Avoid deceptive acts					
Conduct themselves honorably					

Example 1.4

You are a civil engineer on a team designing a bridge for a state government. Your team submits what you believe to be the best design by all criteria, at a cost that is within the limits originally set. However, some months later, the state undergoes a budget crisis and cuts your funds. Your supervisor, also a qualified civil engineer, makes design changes to achieve cost reduction that he/she believes will not compromise the safety of the bridge. You are not so sure, though you cannot conclusively demonstrate a safety hazard. You request that a new safety analysis be done. Your supervisor denies your request on the grounds of time and limited budget. What do you do?

Solution

Table 1.3 shows a typical set of student responses. How would *you* fill out this table?

Table 1.3 Student Responses to the Ethical Scenario

Options → Canons ↓	Go Along with the Decision	Appeal to Higher Management	Quit Your Job	Write Your State Representative	Call a Newspaper Reporter
Hold paramount the safety, health, and welfare of the public	No Total assent may put public at risk	Maybe Addresses risk, but boss may bury issue	No If you just quit, risks less likely to be addressed	Yes Potential risk will be put before public	Yes Potential risk will be put before public
Perform services only in the area of your competence	Yes You are not a safety expert	Yes Though not a safety expert, you are competent to surface an issue	Maybe	No You are not an expert in government relations	No You are not an expert in press relations
Issue public statements only in an objective and truthful manner	No Silence may seem untruthful assent	Maybe You are publicly silent but have registered dissent	No Quitting to avoid the issue is being untruthful	Maybe Your personal involvement may hurt your objectivity	No The press is likely to sensationalize what is as yet only a potential issue
Act for each employer or client as faithful agents or trustees	Yes As an agent, you are expected to follow orders	Yes As an agent, you are expected to alert management to potential problems	Maybe Quitting a job is not bad faith	No As an agent or trustee, you may not make internal matters public without higher approval	No As an agent or trustee, you may not make internal matters public without higher approval
Avoid deceptive acts	No Assent to something you disagree with is deceptive	Yes You honestly reveal your disagreement	No Quitting to avoid responsibility is deceptive	Yes You honestly reveal your disagreement	Yes You honestly reveal your disagreement
Conduct themselves honorably	No Deceptive assent dishonors the profession	Yes Honorable dissent is in accord with obligations	Maybe	Yes Honorable dissent is in accord with obligations	Maybe Might be publicity seeking, not honorable dissent
Totals	Yes=2; no=4; maybe=0	Yes=4; no=0; maybe=2	Yes=0; no=3; maybe=3	Yes=3; no=2; maybe=1	Yes=2; no=3; maybe=1

Notice the multidimensional character of these answers. Here's one way to make some sense of your answer. Total the yes's and the no's in each column (ignore "maybe"). By this criterion, you should appeal to higher management, which of course might still ignore you. But that is the first action you should consider even though your boss may strenuously disagree with you. You have a powerful ally in the engineering ethics decision matrix to persuade others to your point of view. Some engineering ethics decision matrices have just one overwhelming criterion that negates all other ethical responses on your part; if so, you must follow that path—but usually the engineering ethics decision matrix has multiple conflicting factors. All you should expect from the matrix is that it will stimulate most or all of the relevant terms you should consider and help you avoid immediately accepting the first thought that entered your head.

1.11 WHAT SHOULD YOU EXPECT FROM THIS BOOK?

The old joke goes something like this: A year ago, I couldn't even spell *injuneer*, but now I *are* one! Well, you will *not* be an engineer at the end of this course, and if anything, you will learn at least that much. On the other hand, if you pay attention, you will learn the following:

1. Engineering is based on well-founded fundamental principles grounded in physics, chemistry, mathematics, and in logic, to name just a few skills.
2. Its most general principles include (a) definition of a force unit, (b) conservation of energy, (c) conservation of mass, and (d) the use of system control boundaries.
3. Engineering problems are multidisciplinary in approach, and the lines among each subdiscipline blur.
4. Engineering success is often based on successful teamwork.
5. The ability to carry out an introductory analysis in several engineering disciplines should be based on fundamental principles. It often depends on
   a. Identifying the basic steps in the design process.
   b. Applying those basic steps to simple designs.
   c. Completing a successful team design project.
6. You require sound thinking skills as well as practical hands-on skills.
7. The Design Studio will teach you that you also need writing and oral presentation skills.
8. No project is complete without reporting what you have accomplished. Therefore, you need to demonstrate effective communication skills.
9. Computer skills are essential to answer many kinds of practical engineering problems.
10. Engineering skills can be intellectually rewarding as well as demanding.
11. You should come away with some idea of what is meant by each subdiscipline of engineering and, for those who will continue to seek an engineering career, some idea of which of these subdisciplines most appeals to you.
12. We offer a practical way to ask if your behavior is ethical according to well-established engineering ethical canons. If you always act in concordance, no matter the short-term temptations not to, you *will* come out ahead.

SUMMARY

Engineering is about changing the world by creating new solutions to society's problems. This text covers introductions to most engineering subdisciplines: aeronautical engineering, bioengineering, chemical engineering, civil engineering, computer engineering, electrical engineering, electrochemical engineering, environmental engineering (a.k.a. green energy engineering), industrial engineering, manufacturing engineering, materials engineering, mechanical engineering, nuclear engineering as well as engineering economics, and design engineering.

What is common to the branches of engineering is their use of fundamental ideas involving **variables, numbers,** and **units** and the creative use of energy, materials, motion, and information. Engineering is both hands-on and minds-on. The hands-on activity for this book is the Design Studio, in which good design practices are used to construct a "device" to compete against similar devices built by other students. You will learn how to keep a log book and how to protect your designs. You will use conceptual sketches to advance your designs.

EXERCISES

1. Draw a conceptual sketch of your computer. Identify the keyboard, screen, power source, and information storage devices using arrows and labels.

2. Draw a conceptual sketch of an incandescent light bulb. Identify all the components using arrows and numbers as in Figure 1.1.

3. Draw a conceptual sketch of a ballpoint pen. Identify all the components with arrows and labels as in Figure 1.2B.

4. Check this exploded view of a table. Identify and label all the components.

5. Check this exploded view of a box. Identify and label all the components. Unless otherwise directed in the problems that follow, if a code of ethics is required, use that of the National Society of Professional Engineers (NSPE) Code of Ethics (http://www.nspe.org/Ethics/CodeofEthics/index.html). Format as in Figure 1.2B.

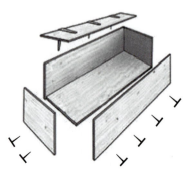

6. Repeat Example 1.3 using the NSPE Code of Ethics for Engineers. Solve using the engineering ethics matrix.

7. Repeat Example 1.4 using the five cornerstones of ethical behavior. Solve using the engineering ethics matrix.

8. It is the last semester of your senior year and you are anxious to get an exciting electrical engineering position in a major company. You accept a position from company A early in the recruiting process but continue to interview hoping for a better offer. Then, your dream job offer comes along from company B. More salary, better company, more options for advancement, it is just what you have been looking for. What should you do?
   a. Just don't show up for work at company A.
   b. Send a letter to A retracting your job acceptance with them.
   c. Ask company B to contact company A and tell them you won't be working for them.
   d. Reject the offer from company B and work for company A anyway.

9. A company purchased an expensive computer program for your summer job with them. The license agreement states that you can make a backup copy, but you can use the program on only one computer at a time. Your senior design course professor would like you to use the program for your senior design project. What should you do?
   a. Give the program to your professor and let him/her worry about the consequences.
   b. Copy the program and use it because no one will know.
   c. Ask your supervisor at the company that purchased the program if you can use it at school on your senior project.
   d. Ask your professor to contact the company and ask for permission to use the program at school.

10. You are attending a regional conference along with five other students from your institution. The night before the group is scheduled to return to campus, one of the students is arrested for public intoxication and is jailed. Neither he nor the other students have enough cash for bail, and he doesn't want his parents to know. He asks you to lend him the organization's emergency cash so that he doesn't have to spend the night in jail; he'll repay you as soon as his parents send money. What should you do?
   a. Lend him the money, as his parents are wealthy and you know he can repay it quickly.
   b. Tell him to contact his parents now and ask for help.
   c. Give him the money but ask him to write and sign a confessional note to repay it.
   d. Tell him to call a lawyer as it's not your problem.

11. You are testing motorcycle helmets manufactured by a variety of your competitors. Your company has developed an inexpensive helmet with a liner that will withstand multiple impacts but is less effective on the initial impact than your competitor's. The vice president for sales is anxious to get this new helmet on the market and is threatening to fire you if you do not release it to the manufacturing division. What should you do?
   a. Follow the vice president's orders, as he/she will ultimately be responsible for the decision.
   b. Call a newspaper to "blow the whistle" on the unsafe company policies.
   c. Refuse to release the product as unsafe and take your chances on being fired.
   d. Stall the vice president while you look for a job at a different company.

12. Paul Ledbetter is employed at Bluestone Ltd. as a manufacturing engineer. He regularly meets with vendors who offer to supply Bluestone with needed services and parts. Paul discovers that one of the vendors, Duncan Mackey, like Paul, is an avid golfer. They begin comparing notes about their favorite golf courses. Paul says he's always wanted to play at the Cherry Orchard Country Club, but since it is a private club, he's never

had the opportunity. Duncan says he's been a member there for several years and that he's sure he can arrange a guest visit for Paul. What should Paul do?[13]

a. Paul should accept the invitation since he has always wanted to play there.

b. Paul should reject the invitation, as it might adversely affect his business relationship with Duncan.

c. Paul should ask Duncan to nominate him for membership in the club.

d. Paul should ask his supervisor if it's OK to accept Duncan's invitation.

13. Some American companies have refused to promote women into positions of high authority in their international operations in Asia, the Middle East, and South America. Their rationale is that business will be hurt because some foreign customers do not wish to deal with women. It might be contended that this practice is justified out of respect for the customs of countries that discourage women from entering business and the professions.

    Some people feel that such practices are wrong and that gender should not to be used in formulating job qualification. Further, they believe that customer preferences should not justify gender discrimination. Present and defend your views on whether or not this discrimination is justified.

14. Marvin Johnson is an environmental engineer for one of several local plants whose water discharges flow into a lake in a flourishing tourist area. Included in Marvin's responsibilities is the monitoring of water and air discharges at his plant and the periodic preparation of reports to be submitted to the Department of Natural Resources.

    Marvin just prepared a report that indicates that the level of pollution in the plant's water discharge slightly exceeds the legal limitations. However, there is little reason to believe that this excessive amount poses any danger to people in the area; at worst, it will endanger a small number of fish. On the other hand, solving the problem will cost the plant more than $200,000.

    Marvin's supervisor says the excess should be regarded as a mere "technicality" and he asks Marvin to "adjust" the data so that the plant appears to be in compliance. He explains: "We can't afford the $200,000. It would set us behind our competitors. Besides the bad publicity we'd get, it might scare off some of tourist industry." How do you think Marvin should respond to Edgar's request?

15. Derek Evans used to work for a small computer firm that specializes in developing software for management tasks. Derek was a primary contributor in designing an innovative software system for customer services. This software system is essentially the "lifeblood" of the firm. The small computer firm never asked Derek to sign an agreement that software designed during his employment there becomes the property of the company. However, his new employer did.

    Derek is now working for a much larger computer firm. Derek's job is in the customer service area, and he spends most of his time on the telephone talking with customers having systems problems. This requires him to cross-reference large amounts of information. It now occurs to him that by making a few minor alterations in the innovative software system he helped design at the small computer firm, the task of cross-referencing can be greatly simplified.

    On Friday, Derek decides he will come in early Monday morning to make the adaptation. However, on Saturday evening, he attends a party with two of his old friends, you and Horace Jones. Since it has been some time since you have seen each other, you spend some time discussing what you have been doing recently. Derek mentions his plan to adapt the software system on Monday. Horace asks, "Isn't that unethical? That system is really the property of your previous employer." "But," Derek replies, "I'm just trying to

---

[13]Extracted from *Teaching Engineering Ethics, A Case Study Approach*, Michael S. Pritchard, editor, Center for the Study of Ethics in Society Western Michigan University: http://ethics.tamu.edu/pritchar/golfing.htm.

make my work more efficient. I'm not selling the system to anyone, or anything like that. It's just for my use—and, after all, I did help design it. Besides, it's not exactly the same system—I've made a few changes." What should be done about this situation?[14]

16. Jan, a professional engineer on unpaid leave, is a part-time graduate student at a small private university and is enrolled in a research class for credit taught by Dimanro, a mechanical engineering professor at the university. Part of the research being performed by Jan involves the use of an innovative geothermal technology.

   The university is in the process of enlarging its facilities, and Dimanro, a member of the university's building committee, has responsibility for developing a request for proposal (RFP) to solicit interested engineering firms. Dimanro plans to incorporate an application of the geothermal technology into the RFP. Dimanro asks Jan to serve as a paid consultant to the university's building committee in developing the RFP and reviewing proposals. Jan's employer will not be submitting a proposal and is not averse to having Jan work on the RFP and proposal reviews. Jan agrees to serve as a paid consultant.

   Is it a conflict of interest for Jan to be enrolled in a class for credit at the university and at the same time serve as a consultant to the university?[15]

FINAL THOUGHTS[16]

A Calvin and Hobbes comic strip nicely illustrates the importance of thinking ahead in engineering and ethical issues. As they are cascading down a treacherous hill in Calvin's wagon, they discuss their circumstance:

Calvin. Ever notice how decisions make chain reactions?

Hobbes. How so?

Calvin. Well, each decision we make determines the range of choices we'll face next. Take this fork in the road for instance. Which way should we go? Arbitrarily I choose left. Now, as a direct result of that decision, we're faced with another choice: Should we jump this ledge or ride along the side of it? If we hadn't turned left at the fork, this new choice would never have come up.

Hobbes. I note with some dismay, you've chosen to jump the ledge.

Calvin. Right. And *that* decision will give us *new* choices.

Hobbes. Like, should we bail out or die in the landing?

Calvin. Exactly. Our first decision created a chain reaction of decisions. Let's jump.

After crash-landing in a shallow pond, Calvin philosophizes: "See? If you don't make each decision carefully, you never know *where* you'll end up. That's an important lesson we should learn some time." Hobbes replies, "I wish we could talk about these things without the visual aids." Hobbes might prefer that they talk through a case study or two before venturing with Calvin into engineering practice.

---

[14]Adapted from: http://ethics.tamu.edu/pritchar/property.htm.

[15]Adapted from NSPE Board of Ethical Review Case No. 91-5.

[16]This section is from Michael S. Pritchard, Copyright 1992: Center for the Study of Ethics in Society, http://ethics.tamu.edu/pritchar/an-intro.htm.

Elements of Engineering Analysis

2.1 INTRODUCTION

How is an engineering problem posed so it can be analyzed? Often, there is an initial sketch that is a useful indication of how things may evolve. Often, these sketches are rough and are intended as a way station until thoughts crystallize in an engineer's brain. Sometimes these sketches become drawings, drawings that obey a number of graphical rules that make spatial ideas clear. But many physical problems of interest to engineers are modeled by mathematical analysis alone, so they may or may not be accompanied by any sketches. In the following chapters, you will learn about a few models accompanied by either or both mathematical models or by mechanical models or, most likely, by both.

All these models and the analysis methods used to construct them share some key elements. One of them, **numerical value**, is familiar to you. Answering a numerical question requires coming up with the right number. But in engineering that is only one part of answering such a question. This chapter introduces other core elements of engineering analysis: engineering drawing and sketching, **engineering variables**, **dimensions**, engineering **units of measurement**, **significant figures**, and **spreadsheet analysis** as well as a fail-safe method of dealing with units and dimensions.

We also offer a formalized method using a self-prompting mnemonic, **need-know-how-solve**, to set up problems. It accomplishes two things: (1) it leaves an "audit trail" of how you solved the problem and (2) it provides a methodology that sorts out and simplifies the problem in a formal way.

The essential idea to take away from this chapter is that arriving at the right numerical value in performing an analysis or solving a problem is only one step in the engineer's task. The result of an engineering calculation must involve the appropriate variables; it must be expressed in the appropriate units; it must express the numerical value (with the appropriate number of digits, or significant figures); and it must be accompanied by an explicit method so that others can understand and evaluate the merits and defects of your analysis or solution.

In addition to the preceding concepts, modern engineers have computerized tools at their fingertips that were unavailable just a generation ago. Because these tools can significantly enhance an engineer's productivity, it is necessary for the beginning engineer to learn them as soon as possible in his or her career. Today, all written reports and presentations are prepared on a computer. But there is another comprehensive computer tool that all engineers use: spreadsheets. This tool is another computer language that the engineer must master. We study it in this chapter, so you can soon get some early practice in the use of this tool. We use spreadsheet techniques throughout this book in the relevant exercise sections.

2.2 ENGINEERING DRAWING AND SKETCHING

What are "engineering drawings" and why do we engineers need them? Engineering drawings are used to *communicate* design ideas and technical information to engineers and other professionals throughout the design process. An engineering drawing represents a complex three-dimensional object on a two-dimensional piece of paper or computer screen by a process called *projection*. The most common types of engineering drawing projections are shown in Table 2.1.

Table 2.1 Types of Drawing Projections

Drawing Type	Example
Isometric—The isometric projection is the basis for the typical three-dimensional engineering sketch. The three axes of the isometric drawing form mutual 120 degrees angles with each other. Circles appear as ellipses in isometric drawings. Isometric grids are a convenient aid in sketching isometric drawings with both straight edges and circular features.	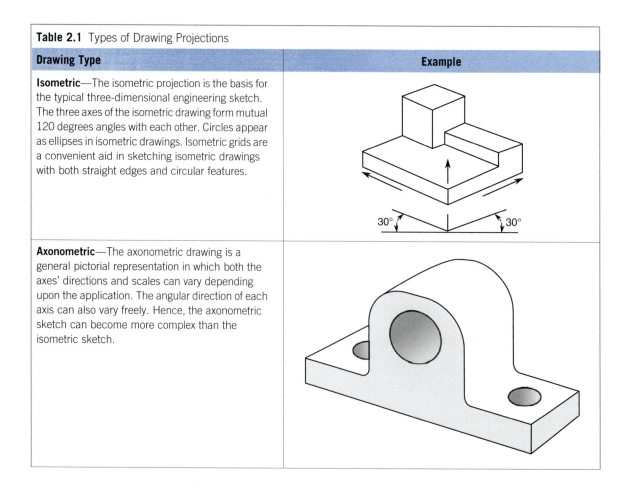
Axonometric—The axonometric drawing is a general pictorial representation in which both the axes' directions and scales can vary depending upon the application. The angular direction of each axis can also vary freely. Hence, the axonometric sketch can become more complex than the isometric sketch.	

Table 2.1 Types of Drawing Projections—cont'd	
Drawing Type	**Example**
Oblique—The three axes of an oblique drawing are drawn horizontal, vertical, and at a receding angle that can vary from 30 to 60 degrees. The main advantage of an oblique drawing is that circles that are parallel to the front plane of the projection are drawn true size and shape.	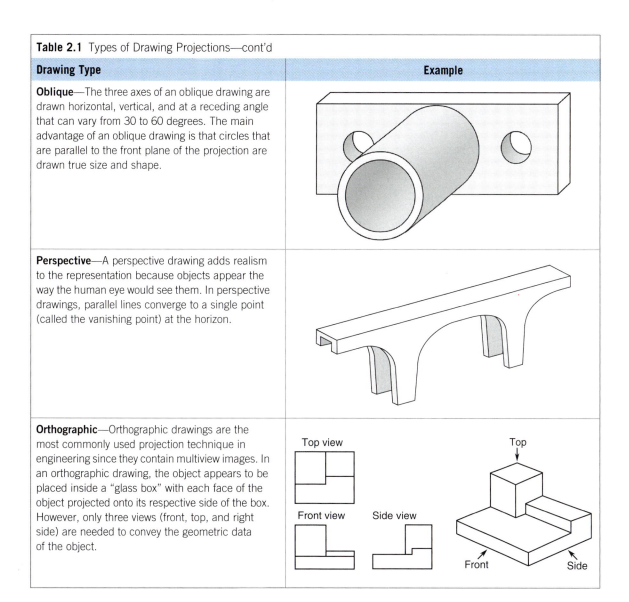
Perspective—A perspective drawing adds realism to the representation because objects appear the way the human eye would see them. In perspective drawings, parallel lines converge to a single point (called the vanishing point) at the horizon.	
Orthographic—Orthographic drawings are the most commonly used projection technique in engineering since they contain multiview images. In an orthographic drawing, the object appears to be placed inside a "glass box" with each face of the object projected onto its respective side of the box. However, only three views (front, top, and right side) are needed to convey the geometric data of the object.	

2.2.1 Drawing Scale and Dimensioning

Quite often, you will not be able to create a drawing at its actual size. The scale of a drawing is represented as the ratio of the size on the drawing to the size in reality. So if a drawing has a scale of 1:2, the drawing is ½ the actual object size.

When you add dimensions to a drawing, put in only as many dimensions necessary for a person to understand or manufacture the object. Repeatedly measuring from one point to another on the object will lead to confusion and inaccuracies. It is better to measure from one side of the object to various relevant points (center of a hole, another side, etc.). Try to place the dimensions in a way that will help a machinist create the object as in Figure 2.1.

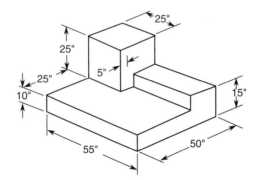

FIGURE 2.1 A properly dimensioned isometric engineering drawing.

2.2.2 Computer-Aided Design (CAD) Software

Today, engineering drawings are produced on computers with Computer-Aided Design (CAD) software such as SolidWorks, AutoCAD, Pro/ENGINEER, and so forth.[1] Every engineering program now includes training in one or more of these sophisticated computer programs. Not only you can accurately represent a three-dimensional image of an object, but also you can carry out complex engineering stress-strain, fluid flow, and heat transfer analysis for various loading conditions on the object (Figure 2.2). The resulting computer files can then be downloaded to a rapid prototyping machine that will make the object to scale.

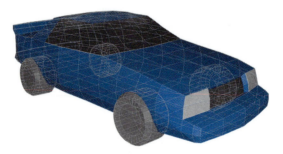

FIGURE 2.2 A 3-D CAD file. http://www.lasercamm.com/rapid.htm

2.2.3 Engineering Sketches

An engineering sketch is a free-hand drawing used to get ideas on paper quickly. A sketch does not need to be made to an exact scale, but it should be roughly proportional to the actual object.

How you prepare a sketch depends on with whom you are communicating. A pictorial sketch works best when communicating design concepts with other engineers. When communicating technical details to a technician, a multiview (orthographic) sketch would be the appropriate choice (Figure 2.3).

---

[1]SketchUp is a basic 3-D CAD program that is available for free at http://www.sketchup.com.

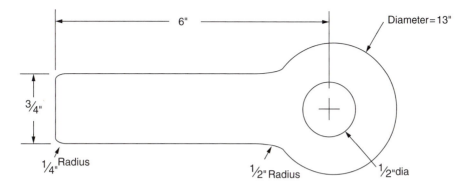

FIGURE 2.3 A typical engineering sketch.

Just because a sketch is a "free-hand" drawing, it should still be relatively neat. However, sketching does not require any real "artistic" skill. Use a soft pencil (e.g., 0.5 mm HB) to achieve a range of tone from light guide-lines that will not show on a photocopy to very dark grey lines that will photocopy black. Lines drawn quickly and confidently look better, but sharp lines tend to magnify errors, whereas fuzzy lines produce sketches where objects "appear" more as they should. The skills you will need to acquire are (1) sketching straight lines and (2) sketching arcs.

Straight lines are easy to sketch if you keep your eye on where you want to draw to, not on the end of the pencil. Sketch horizontal lines away from your body (otherwise you tend to have an up-curve at their end). Sketch vertical lines downward, otherwise they will tend to curve in the direction of the hand holding the pencil. For inclined lines, turn the paper until it's either horizontal or vertical.

Lightly mark points at each end of the intended line. Move your pencil between the points a couple of times, with the pencil point just clear of the paper, to "teach yourself to follow the route." Draw several light guidelines between the points and then go over the "best line" more firmly.

With practice, sketching arcs and circles is relatively easy. To sketch a complete circle, start by sketching a square box that will enclose the circle. Mark the centers of the sides of the square and then sketch arcs that are tangent to the sides of the square. For larger diameters, you can add diagonals to the square and by marking the circle's radius on the diagonals to get additional points to join with a smooth curve.

Do not try to draw the circle in one motion; instead, draw it as a series of connected arcs. To draw an ellipse or oval, sketch it inside a rectangular box. For more complex curves, join dots along the curve.

Lastly, even though a sketch is defined as a free-hand drawing, here are some common "aids" to help you sketch:

- Graph paper—helps to maintain proportion and keep straight lines.
- Scraps of paper—good for transferring distances
- Coins—make good circle templates
- Credit cards—make nice straight edges

2.3 ENGINEERING VARIABLES

Engineers typically seek answers to such questions as "How hot will this get?" "How heavy will it be?" and "What's the voltage?" Each of these questions involves a variable, a precisely defined quantity describing an aspect of nature. What an engineering calculation does is different from what a pure mathematical calculation might do. The latter usually focuses on the final numerical answer as the end product of an analysis. For example, $\pi = 3.1415926$ is a legitimate answer to the question "What is the value of π?" Similarly, the question "How hot?" is answered using the variable "temperature." The question "How heavy?" uses the variable "weight," "What voltage?" uses the variable "electric potential," and so forth.

For our purposes, variables almost always are defined in terms of measurements made with familiar instruments, such as thermometers, rulers, and clocks. Speed, for example, is defined as a ruler measurement of distance, divided by a clock measurement of time. This makes possible what a great engineer and scientist, William Thomson, Lord Kelvin (1824–1907), described as the essence of scientific and engineering knowledge.

> *I often say that when you can measure what you are speaking about, and express it in numbers, you know something about it; but when you cannot measure it, when you cannot express it in numbers, your knowledge is of a meager and unsatisfactory kind: it may be the beginning of knowledge, but you have scarcely, in your thoughts, advanced to the stage of science, whatever the matter may be.*

But, expressing something in numbers is only the beginning of engineering knowledge. In addition to variables based on measurements and expressed as numbers, achieving Lord Kelvin's aspiration requires a second key element of engineering analysis: units.

2.4 ENGINEERING UNITS OF MEASUREMENT

What if you are stopped by the highway patrol on a Canadian highway and get a ticket saying you were driving at "100"? You would probably guess that the variable involved is speed. But it would also be of interest to know if the claim was that you were traveling at 100 *miles* per hour or 100 *kilometers* per hour, knowing that 100 kph is only 62 mph. Units can and do make a difference!

While the fundamental laws of nature are independent of the system of units we use with them, in engineering and in the sciences, a calculated quantity always has two parts: the numerical value *and* its associated units, if any.[2] Therefore, the result of any engineering calculation must always be correct in two separate categories: It must have the correct numerical value, and it must have the correct units.

Units are a way of quantifying the underlying concept of dimensions. *Dimensions* are the fundamental quantities we perceive, such as mass, length, and time. Units provide us with a numerical scale, whereby we can carry out a measurement of a quantity in some dimension. The units are established quite arbitrarily and are codified by civil law or cultural custom. How the dimension of length ends up being measured in units of feet or meters has nothing to do with any physical law. It is solely dependent on the creativity and ingenuity of people. Therefore, the basic tenets of units systems are often grounded in the complex roots of past civilizations and cultures.

2.4.1 The SI Unit System

The international standard in units is the SI system or, officially, the International System of Units (in French, Le Système International d'Unités). It is the standard of modern science and technology and is based on MKS units (meter, kilogram, second). The fundamental units in the SI system are:

The meter (m), the fundamental unit of length.
The second (s), the fundamental unit of time.
The kilogram (kg), the fundamental unit of mass.

---

[2]Some engineering quantities legitimately have no associated units; for instance, a ratio of like quantities.

The degree kelvin (K), the fundamental unit of temperature.

The mole (mol), the fundamental unit of quantity of particles.

The ampere (A), the fundamental unit of electric current.

Table 2.2 illustrates a variety of SI units that were all derived from proper names of scientists who made discoveries in each of the fields in which these units are used.

Table 2.2 Some SI Units and Their Abbreviations		
Ampere (A)	Henry (H)	Pascal (Pa)
Becquerel (Bq)	Hertz (Hz)	Siemens (S)
Celsius (°C)	Joule (J)	Tesla (T)
Coulomb (C)	Kelvin (K)	Volt (V)
Farad (F)	Newton (N)	Watt (W)
Gray (Gy)	Ohm (Ω)	Weber (Wb)

Here are some rules regarding unit names.

- All unit names are written without capitalization (unless they appear at the beginning of a sentence), regardless of whether they were derived from proper names.
- When the unit is to be abbreviated, the abbreviation is capitalized if the unit was derived from a proper name.
- Unit abbreviations use two letters *only* when necessary to prevent them from being confused with other established unit abbreviations[3] (e.g., Wb for the magnetic field unit weber[4] to distinguish it from the more common W, the watt unit of power) or to express prefixes (e.g., kW for kilowatt).
- A unit abbreviation is never pluralized, whereas the unit's name may be pluralized. For example, kilograms is abbreviated as kg, *not* kgs; newtons as N, *not* Ns; and the correct abbreviation of seconds is s, neither sec nor secs.
- Unit name abbreviations are never written with a terminal period unless they appear at the end of a sentence.
- All other units whose names were not derived from the names of historically important people are both written and abbreviated with lowercase letters; for example, meter (m), kilogram (kg), and second (s).

In examples that follow in this text, we introduce a *fail-safe* method that will always allow you to develop the correct units. We will follow the numerical part of the question with square brackets [...] enclosing the unit conversions that are needed. For example, knowing there are 12 inches in 1 foot would produce the units conversion factor [12 in/ft], so if we wanted to convert 12.7 ft^2 to in^2, we would write 12.7 [ft^2] [12 in/ft]^2 = 12.7 × 144 [ft^2][in/ft]^2 = 1830 in^2.

---

[3]Non-SI unit systems do not generally follow this simple rule. For example, the English length unit, foot, could be abbreviated f rather than ft. However, ft is well established within society, and changing it at this time would only cause confusion.

[4]Pronounced veyber as in German.

Example 2.1

The height of horses is still measured in the old unit of "hands." How many feet high is a horse that is 13 hands tall?

The conversion of old units to more familiar ones is of course known:

Unit	Equivalent	[Conversion]
2 hands	1 span	½ [span/hand]
2 spans	1 cubit	½ [cubit/span]
2 cubits	3 feet.	$^3/_2$ [feet/cubit]

The height of the horse is 13 [hands] × ½ [span/hand] × ½ [cubit/span] × $^3/_2$ [foot/cubit] = 13 × ½ × ½ × $^3/_2$ [feet] = 4.9 ft.

While this may seem ponderous in this example, in examples that are more complicated *it is essential* to follow this methodology. We will attempt to be consistent in presenting solutions and follow this technique throughout this text.

There are many SI units pertaining to different quantities being measured and multiples thereof (Tables 2.3 and 2.4, respectively). Some of the rationale for the fundamental units will become clearer as we proceed.

Table 2.3 has value beyond merely listing these units: it relates the unit's name to the fundamental MKS units, that is, the fact that a frequency is expressed in hertz may not be as useful as the fact that a hertz is nothing but the name of an inverse second, s^{-1}. In Table 2.4, multiples of these quantities are arranged in factors of 1000 for convenience for very large and very small multiples thereof.[5]

Table 2.3 Some Derived SI Units

Quantity	Name	Symbol	Formula	Fundamental Units
Frequency	Hertz	Hz	1/s	s^{-1}
Force	Newton	N	kg m/s^2	m kg s^{-2}
Energy	Joule	J	N m	m^2 kg s^{-2}
Power	Watt	W	J/s	m^2 kg s^{-3}
Electric charge	Coulomb	C	A s	A s
Electric potential	Volt	V	W/A	m^2 kg s^{-3} A^{-1}
Electric resistance	Ohm	Ω	V/A	m^2 kg s^{-3} A^{-2}
Electric capacitance	Farad	F	C/V	m^{-2} kg^{-1} s^4 A^2

---

[5]In recent years, a new subcategory of materials and technology known as *nanotechnology* has arisen; it is so called because it deals with materials whose characteristic size is in the nanometer or 10^{-9} m range.

Table 2.4 SI Unit Prefixes

Multiples	Prefixes	Symbols	Submultiples	Prefixes	Symbols
10^{18}	exa	E	10^{-1}	deci	d
10^{15}	peta	P	10^{-2}	centi	c
10^{12}	tera	T	10^{-3}	milli	m
10^{9}	giga	G	10^{-6}	micro	μ
10^{6}	mega	M	10^{-9}	nano	n
10^{3}	kilo	k	10^{-12}	pico	p
10^{2}	hecto	h	10^{-15}	femto	f
10^{1}	deka	da	10^{-18}	atto	a

2.5 SIGNIFICANT FIGURES

Having defined your variable and specified its units, you now calculate its value. Your calculator will obediently spew out that value to as many digits as its display will hold. But, how many of those digits really matter? How many of those digits actually contribute toward achieving the purpose of engineering, which is to design useful objects and systems and to understand, to predict, and to control their function in useful ways? This question introduces into engineering analysis a concept you have possibly seen in your high school science or mathematics courses: the concept of significant figures.

Even the greatest scientists have howling errors by quoting more significant figures than were justified. Newton wrote that "the mass of matter in the Moon will be to the mass of matter in the Earth as 1 to 39.788" (*Principia*, Book 3, proposition 37, problem 18). Since the ratio of the mass of the Earth to the mass of the Moon is actually $M_e/M_m = 81.300588$, it is clear that Newton had gone wrong somewhere. His value of M_e/M_m to five significant figures was completely unjustified.[6]

The use of the proper number of significant figures in experimental work is an important part of the experimentation process. Reporting a measurement of 10 meters, 10. meters (notice the period after the zero), 10.0 meters, or 10.00 meters implies something about how accurately the measurement was made. The implication of 10 meters as written is that the accuracy of our measuring rule is of the order of ± 10 meters. However, 10. meters implies the measurement was good to ± 1 meter. Likewise, 10.0 meters implies accuracy to 0.1 meter and 10.00 meters to 0.01 meters, and so on, a convention we try to maintain henceforth in this book. Unless they are integers, numbers such as 1, 30, 100 each have just one significant figure.

The concept of significant figures arises because arithmetic alone does not increase the accuracy of a measured quantity.[7] If arithmetic is applied indiscriminately, it might actually decrease the accuracy of the result. The use of significant figures is a method to avoid such blunders as $10/6 = 1.666666667$ (as easily obtained on many electronic calculators), whereas the strict answer is 2. (Since 6 and 10 are apparently known only to 1

[6]D. W. Hughes, "Measuring the Moon's Mass," *Observatory*, 122 (April 2002), 62.
[7]The averaging of a number of repeated measurements of the same quantity might seem to violate this statement, but it allows increased confidence only in the interval in which the averaged number will lie.

significant figure, and if they represent real physical measurements, they apparently have not been measured to the implied accuracy of the arithmetical operation that produced 1.666666667.) In this sense, physical numbers differ from pure mathematical numbers.

"Exact" numbers (numbers such as in 1 foot $=12$ inches, or numbers that come from counting, or in definitions such as diameter $=2 \times$ radius) have no uncertainty and can be assumed to have an infinite number of significant figures. Thus, they ordinarily do not limit the number of significant figures in a calculation.

Definition

A significant figure[8] is any one of the digits 1, 2, 3, 4, 5, 6, 7, 8, 9, and 0. Note that zero is a significant figure except when it is used simply to fix the decimal point or to fill the places of unknown or discarded digits.

The number 234 has three significant figures, and the number 7305 has four significant figures, since the zero within the number is a legitimate significant digit. But leading zeroes before a decimal point are not significant. Thus, the number 0.000452 has three significant figures (4, 5, and 2), the leading zeroes (including the first one before the decimal point) being place markers rather than significant figures.

How about trailing zeroes? For example, the number 12,300 is indeed twelve thousand and three hundred, but we can't tell without additional information whether the trailing zeroes represent the precision[9] of the number or merely its magnitude. If the number 12,300 was precise only to ± 100, it has just three significant figures. If it were truly precise to ± 1, then all five figures are significant. To convey unequivocally which ending zeroes of a number are significant, it should be written as 1.2×10^4 if it has only two significant figures, as 1.23×10^4 if it has three, as 1.230×10^4 if it has four, and as 1.2300×10^4 (or 12300.—note the decimal point here) if it has five.

The identification of the number of significant figures associated with a measurement comes only through knowledge of how the measurement was carried out. For example, if we measure the diameter of a shaft with a ruler, the result might be 3.5 inches (two significant figures), but if it is measured with a digital micrometer, Figure 2.4, it might be 3.512 inches (four significant figures).

FIGURE 2.4 Digital micrometer.

Engineering calculations often deal with numbers having unequal numbers of significant figures. A number of logically defensible rules have been developed for various computations. These rules are actually the result of strict mathematical understanding of the propagation of errors due to arithmetic operations such as addition, subtraction, multiplication, and division. Rule 1 is for addition and subtraction.

---

[8]There are many good websites on the Internet that deal with this concept. The following sites have self-tests that you can use to check yourself http://science.widener.edu/svb/tutorial/sigfigures.html.
[9]The word *precision* has a specific meaning: it refers to how many times using independent measurements we can *reproduce* the number. If we throw many darts at a dartboard and they all cluster in the double three ring, they are precise but, if we were aiming at the bull's eye, they were not accurate!

Rule 1

The sum or difference of two values should contain no significant figures further to the right of the decimal place than occurs in the least precise number in the operation.

For example, $113.2 + 1.43 = 114.63$, which must now be rounded to 114.6. The least precise number in this operation is 113.2 (having only one place to the right of the decimal point), so the final result can have no more than one place to the right of the decimal point. Similarly, $113.2 - 1.43 = 111.77$ must now be rounded to 111.8. This is vitally important when subtracting two numbers of similar magnitudes, since their difference may be much less significant than the two numbers that were subtracted. For example, $113.212 - 113.0 = 0.2$ has only one significant figure even though the "measured" numbers each had four or more significant figures.

There is a corresponding rule for multiplication and division of figures.

Rule 2

The product or quotient should contain no more significant figures than are contained by the term with the least number of significant figures used in the operation.

For example, $(113.2) \times (1.43) = 161.876$, which must now be rounded to 162, and $113.2/1.43 = 79.16$, which must now be rounded to 79.2 because 1.43 contains the least number of significant figures (i.e., three).

Finally, Rule 3 is a rule for "rounding" numbers up or down.[10]

Rule 3

When the discarded part of the number is 0, 1, 2, 3, or 4, the next remaining digit should not be changed. When the discarded part of the number is 5, 6, 7, 8, or 9, then the next remaining digit should be increased by 1.

For example, if we were to round 113.2 to three significant figures, it would be 113. If we were to round it further to two significant figures, it would be 110, and if we were to round it to one significant figure, it would be 100 with the trailing zeroes representing placeholders only. As another example, 116.876 rounded to five significant figures is 116.88, which further rounded to four significant figures is 116.9, which further rounded to three significant figures is 117. As another example, 1.55 rounds to 1.6, but 1.54 rounds to 1.5.

Example 2.2

    a. Determine the number of significant figures in 2.2900×10^7
    b. Determine the number of significant figures in 4.00×10^{-3}
    c. Determine the number of significant figures in 480
    d. What is the value of $345.678 - 345.912$
    e. What is the value of $345.678 - 345.9$

Solution

    a. The number of significant figures in 2.2900×10^7 is **5**
    b. The number of significant figures in 4.00×10^{-3} is **3**
    c. The number of significant figures in 480 is **2**
    d. The value of $345.678 - 345.912 = -\mathbf{0.234}$
    e. The value of $345.678 - 345.9 = -0.222 = -\mathbf{0.2}$

---

[10]There is another round-off rule corresponding to Rule 3: the so-called Bankers' Rule was used before computers to check long columns of numbers. When the discarded part of the number is exactly 5 followed only by zeros (or nothing), then the previous digit should be rounded up if it is an odd number, but it remains unchanged if it is an even number. It was meant to average out any rounding bias in adding the columns.

2.6 THE "NEED-KNOW-HOW-SOLVE" METHOD

We suggest the **need-know-how-solve** method for setting up and solving problems. It has the powerful benefit of enabling you to attack complicated problems in a systematic manner, thereby making success more likely. It may seem cumbersome in simple problems, but for those with an eye to their final course grades, it has an additional benefit of ensuring at least partial credit when a miscalculated number results in an incorrect answer. It also tracks an essential element of modern engineering practice by providing an "**audit trail**" by which past errors can be traced to their source, easing subsequent corrections. Finally, the need-know-how-solve method is a self-contained mnemonic device that should also help direct your thought processes.

We have previously emphasized the need to find the proper variables, to express them in a consistent set of units, and to express the answer to the appropriate number of significant figures. But, what guarantees do we have that the answer, even if correctly expressed in this way, is actually the correct solution to a problem or a useful result that accurately reflects or predicts the performance of an actual object or system in the real world? The correctness of an answer can be no better than the correctness of the methods and assumptions used to obtain it. Typically, the more systematic your method, the more likely you are to avoid errors. In addition, the best way to ensure the correctness of a method is to submit it to evaluation and criticism. The use of an explicit method leaves an audit trail for your colleagues and customers (and for your future boss!).

In the need-know-how-solve method, the **need** is the variable or variables for which you are solving. It is the very first thing you should write down.[11] The **know** is the quantities that are known, either through explicit statement of the problem or through your background knowledge. Write these down next. They may be graphical sketches or schematic figures as well as numbers or principles. The **how** is the method you will use to solve the problem, typically expressed in an equation, although rough sketches or graphs may also be included. The how also includes the assumptions you make to solve the problem. Write it out in sentence or symbolic form before applying it. Only then, with need, know, and how explicitly laid out, should you proceed to **solve** the problem. Often, it is also necessary to discuss the implications of your solution.

The pitfall this method guards against is the normal human temptation to look at a problem, think you know the answer, and simply write it down. But, rather than gamble with this hit and miss method, the need-know-how-solve scheme gives a logical development that shows your thought processes.

Many of the problems you solve in this book involve specific equations. But, do not be deceived into thinking that this method involves the blind plugging of numbers into equations if you do not understand where those equations came from. The essence of the method is to realize that, once you have correctly defined the variable you need, you know a lot more about the problem than you thought you did. With the aid of that knowledge, there is a method (how) to solve the problem to an appropriate level of accuracy. Sometimes this method requires no equations at all.

It bears emphasizing: the need-know-how-solve process is guaranteed to get you a better grade—even if the final answer is the same one that got an F when all you bothered to write down was the final answer!

When applying the method, it may be especially helpful, as part of the how step, to sketch the situation described in the problem. Many people are discouraged from sketching by lack of artistic talent. Do not let this stop you. As the next example illustrates, even the crudest sketch can be illuminating.

---

[11]In some sense, there is even a prior step before need: read the question and then reread the question. Few things are more frustrating than solving the wrong problem!

Example 2.3

A Texan wants to purchase the largest fenced-in square ranch she can afford. She has exactly $320,000 available for the purchase. Fencing costs exactly $10,000 a mile, and land costs exactly $100,000 a square mile. How large a ranch, as measured by the length of one side of the square, can she buy?

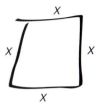

Need: The length of a side of the largest square of land the Texan can buy.

Know: Fencing costs $10,000 a mile, and land costs $100,000 a square mile. Our Texan has $320,000 to invest.

How: Let the unknown length $=x$ miles. It may not be immediately obvious how to write an equation to find x. So sketch the ranch.

From the crude picture (and that is generally all you need), it is immediately obvious that the length of the fence surrounding the ranch is $4x$, and the area of the ranch is x^2.

So, the cost of the ranch is

$$\text{Length of fence [miles]} \times 10,000 \text{ [\$/mile]} + \text{area of ranch [square miles]} \times$$

$$100,000 \text{ [\$/sq. mi]} = \text{total cost of ranch} = \$320,000$$

or

$$4x \text{ [miles]} \times 10,000 \text{ [\$/mile]} + x^2 \text{ [sq. miles]} \times 100,000 \text{ [\$/sq. mi]} = \$320,000$$

Solve: $(4x) \times (10,000) + x^{2.} \times (100,000) = 320,000$

Therefore, $10x^2 + 4x - 32 = 0$, which is a quadratic equation whose solutions are

$$x = +1.6 \text{ and } x = -2.0 \text{ (each to two significant figures)}$$

So there are two solutions, $x = 1.6$ miles and $x = -2$ miles. Here one must apply a bit of knowledge so obvious that, although known, was not listed in the Know section: that the length of the side of the ranch must be greater than zero (if for no other reason than to have room to put the cattle!). This yields the final answer. **The ranch has 1.6 miles on a side.** (It's *always* worth checking if this is correct: $1.6^2 \times \$100,000 + 4 \times 1.6 \times \$10,000 = \$3.2 \times 10^5$, which is arithmetically OK.)

Example 2.4

Consider the following problem: How many barbershops are there in the city of Schenectady (population about 60,000 people)? Your first reaction may be "I haven't got a clue." (You may want to try the need-know–how-solve method on this problem for yourself before looking at what follows.)

Need: The number of barbershops.

Know: There are about 60,000 people in the city of Schenectady, of whom about half are male. Assume the average male gets about 10 haircuts a year. A barber can probably do one haircut every half an hour, or about 16 in the course of an 8 hour day. There are about three barbers in a typical barbershop.

How: The number of haircuts given by the barbers must be equal to the number of haircuts received by the customers. So, if we calculate the number of haircuts per day received by all those 30,000 males, we can find the number of barbers needed to give those haircuts. Then, we can calculate the number of shops needed to hold those barbers. Assume that barbershops are open 300 days per year.

Solve: 30,000 males require about 10 [haircuts/male-year] × 30,000 [males] = 300,000 haircuts/year.

On a per day basis, this is about 300,000 [haircuts/year] × [1 year/300 days] = 1000 haircuts/day. This requires: 1000 [haircuts/day] × [1 barber-day/16 haircuts] = 62.5 barbers. At three barbers per shop, this means 62.5 [barbers] × [1 shop/3 barbers] = 20.8 shops.

So the solution is **20 barbershops** (since surely not more than one of the digits is significant).

Looking in a recent yellow pages directory and counting the number of barbershops in Schenectady gives the result 23 barbershops. So we are within about 10 percent, which is a fortuitously good answer, given the roughness of our estimates and the many likely sources of error.

Notice also in this problem how the method of carrying the units in square brackets [..] helps your analysis of the problem and directs your thinking.

Haircuts and barbershops and Texas ranches are not among the variables or units used by engineers. But, applying common sense as well as equations is a crucial component of engineering analysis as demonstrated in the next example.

Example 2.5

A 2.00 m steel wire is suspended from a hook in the ceiling by a mass of 10.0 kg that is tied to its lower end; the wire stretches by 15.0 mm under this load. If the same mass is used to stretch a 4.00 m piece of the same steel wire, how much will it stretch?

Need: Stretch = ____ mm for a 4.00 m piece of wire.

Know: 10.0 kg mass stretches a 2.00 m wire by 15.0 mm.

How: We need to deduce a possible law for extending a wire under load. Without experimentation, we cannot know if our "theoretical law" is correct, but a mixture of common sense and dimensional analysis can yield a plausible relationship. A longer wire should stretch further than a shorter one if otherwise equivalent (a larger mass presumably also stretches the wire further).

Solve: A *plausible* model is therefore the extension x which is proportional to the unstretched length of wire L (all other things such as the stretching force and the wire's cross section being equal).

Then, under a fixed load, $x \propto L$ and the ratio between the two cases is

$$\frac{x_2}{x_1} = \frac{L_2}{L_1}.$$

Therefore, the new extension $x_2 = 15.0 \times 4.00/2.00$ [mm][m/m] = **30.0 mm**.

The proportionality "law" derived in this manner is not guaranteed to be correct. To guarantee that it is physically correct, we need to either understand the underlying principles of wire stretching or have good experimental data relating wire stretching to the wire's length. In fact, it is correct and is part of a physical law with the name *Hooke's Law*.

In some problems, the need-know-how-solve method is decidedly clumsy in execution; nevertheless, it is recommended that you use the basic method and learn how to modify it to suit the peculiarities of the problem at hand. You will soon find there is ambiguity among the *know*, *how*, and even *solve* components of the method. Don't mind this: the relevant information can go into one or the other baskets, just as long as it acts as a jog in the direction you need to solve the problem.

The elements of engineering analysis introduced in this chapter provide systematic ways of applying sense, both common and uncommon, to help meet engineering challenges.

2.7 SPREADSHEET ANALYSIS

Unfortunately, even though the need-know-how-solve method gets you to an understanding of the answer, the subsequent mathematics may be too difficult to solve on a piece of paper, you may need multiple cases, or graphical solutions, or there may be other complications that prevent you from immediately writing down the answer you are seeking. Today, computers come to your aid; an experienced engineer on a laptop computer can obtain solutions to problems that were daunting a generation ago and might have required a room full of people cranking out small pieces of mathematical puzzles by hand. In addition, spreadsheets do a superior job of displaying data— no small advantage if you want a busy person, perhaps your customer or your manager, to pay attention to the data.

Spreadsheets allow you to distinguish among three disparate concepts—***data***, ***information***, and ***knowledge***. First, what are data[12]?

What we call *data* is just a jumble of facts or numbers such as 4, 3, 1, 1, 6, or a set of colors such as red, violet, green, blue, orange, indigo, and yellow. They become "information" when you sort them out in some way such as 3, 1, 4, 1, 6, or red, orange, yellow, green, blue, indigo, and violet. They respectively become "knowledge" when the numerical sequence is interpreted as 3.1416 (i.e., π), and the color sequence is recognized as the visible light spectrum arranged as dispersed by a prism (mnemonic ROYGBIV).

Spreadsheets were invented at the Harvard Business School in 1979 by a student[13] who was apparently bored with repetitive hand calculations. He and a fellow entrepreneur made the first spreadsheet tool, VisiCalc. It was originally intended simply to take the drudgery out of relatively mundane business ledgers. Not only do modern spreadsheets take the drudgery out of the computations, but they are also wonderful at displaying data in neat and tidy ways that can be real aids in understanding the displayed data.

The spreadsheet concept became one of the first so-called killer-apps—an application tool that no self-respecting personal computer could do without! Lotus 123 (now part of IBM) soon followed VisiCalc, and that was soon followed by Microsoft® Excel. In this book, we explicitly use Excel, but the principles you learn here are readily transferred to IBM's Lotus 123 or to Corel's Quattro®. VisiCalc has faded from common use.

Until VisiCalc, computers had tedious formatting requirements to display their results and typically as much time and energy was expended on getting results from the computer in an understandable way as was spent computing the results in the first place! What made spreadsheets interesting was the visual way they could handle large arrays of numbers; these numbers were identified by their position in these arrays by reading across their rows and down their columns. Interactions among these numbers were rapidly enhanced by providing sophisticated and complex mathematical functions that would manipulate the numbers in these arrays. Further, these arrays could be graphed in many ways so that an explicit crafted and individualized output was easily produced. Gone were most of the arcane skills in formatting that mainframe computers had demanded, and since it was visual, silly mistakes instantly stood out like sore thumbs and could be expeditiously corrected. No wonder spreadsheets were one of the killer applications that drove the sales of PCs as indispensable tools for engineering, science, and business.

Figure 2.5 is a reproduction of an Excel spreadsheet with some guidelines as to its key parameters. It is navigated by the intersection of rows and columns. The first cell is thus A1. Active cells, such as E10 in the example, are boxed in a bold border. You can scroll across and down the spreadsheet using the horizontal and vertical scroll bars, respectively. Note there is even a provision to have multiple pages of spreadsheets that can be thought of as a 3D spreadsheet. Finally, the various icons and menu items that appear above and below this particular spreadsheet are idiosyncratic and can be changed at will to reflect your own preferences chosen among hundreds of different capabilities.

---

[12]Singular form *datum*.
[13]http://www.bricklin.com/visicalc.htm.

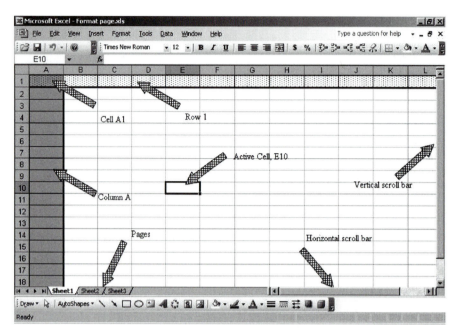

FIGURE 2.5 A typical spreadsheet.

2.7.1 Cell Addressing Modes

The window that you view is just a small fraction of the whole spreadsheet; the remaining virtual spreadsheet has thousands of rows and hundreds of columns of available space as well as many interactive sheets per workbook, enabling "3D addressing." Movement within a window is by the scroll bars either across or down. Movement to virtual positions can also be effected by moving the cursor to the boundary you wish to extend. Types of data that can be in a cell include:

Alphanumeric (text)
   1. Headings
   2. Labels
Numeric
   1. Numbers: integers and floating point (meaning with decimals or exponents)
   2. References to the contents of other cells (using the cell address)
Equations
Formulae such as $=B1+B4/H13$ add the contents of cell B1 to the result of dividing the contents of cell B4 by the contents of cell H13.
Predefined functions
   1. Sum
   2. Average
   3. Max
   4. Min

Many other functions: trigonometric, exponential, logical, statistical, etc.

You will need to explicitly tell the computer what data type is to be stored in a cell.

The default cell entry is text or plain numbers; text is automatically left justified, and numbers are right justified. (You can change these using the justifying pictographs in the heading rows.)

If you put a single quotation mark (') at the beginning of your cell entry, that means left-justified alphanumeric to follow. If you put an equal sign (=) at the beginning of your cell entry, that means equation or numeric to follow. If you put a $+$ or $-$ at the beginning of your cell entry, that means numeric to follow.[14]

Example 2.6

FleetsR'Us owns a fleet of rental cars; it wants to offer its customers an inclusive all-in-one-contract that takes account of the cost of the gasoline used. To estimate the cost and, therefore, pricing to their customers, they do a survey of the typical renter's journey and determine how many miles are driven by a selection of typical drivers. The raw data are given in Table 2.5.

Table 2.5 Input to Spreadsheet for Example 2.6

Renter	City miles	Suburban miles	Highway miles
Davis	16	28	79
Graham	10	31	112
Washington	4	7	158
Meyers	22	61	87
Richardson	12	56	198
Thomas	5	22	124
Williams	4	14	142

Using a spreadsheet, give the total miles in each category of driving by each driver and the average miles driven in each category. Show the algorithm used in the spreadsheet.

Need: Driver mileage and average miles driven in each segment of renter's journey.

Know: Mileage data in Table 2.5.

How: Total miles for each driver is the sum of the individual miles driven, and the average mileage in each category of driving is the total miles divided by the number of drivers.

Solve: Use a spreadsheet to input the data and the indicated mathematical operations (Table 2.6).

The sheet title heading conveniently appears in cell A1 (even though the actual input sprawls across adjacent cells). It helps in organizing the spreadsheet to entitle it descriptively so that it can be read by you (or others) later, when the purpose of the spreadsheet has faded from immediate memory. In addition to the preceding spreadsheet, descriptive labels for some columns are given in row 4 as A4:E4. (Read the colon to mean it includes all the cells in the rectangular area prescribed by the limits of the addresses, A4 and E4 in this case.) The labels may be simple, such as these are, or they may be informative, such as the units involved in the column (or row) that follows. Other descriptive labels that appear in this spreadsheet are A5:A11 and A13. All of these are **text** statements. The labels are the first steps in interpreting data as more than mere entries. Labels are there to codify what you want to arrange as information.

---

[14]But "+ name" means a specially defined macro (which is just a type of program).

Table 2.6 Output of Spreadsheet for Example 2.6

	A	B	C	D	E
1	Miles driven in various categories				
2					
3					
4	Renter	City miles	Suburban miles	Highway miles	Total miles
5	Davis	16	28	79	123
6	Graham	10	31	112	153
7	Washington	4	7	158	169
8	Meyers	22	61	87	170
9	Richardson	12	56	198	266
10	Thomas	5	22	124	151
11	Williams	4	14	142	160
12					
13	Average	10.4	31.3	128.6	170.3

The numerical input data for this example are given in the block B5:D11. The results of some arithmetical operations appear as row E5:E11 and as the block B13:E13. These data are the beginnings of information about the journeys. Note the variables are formatted to a reasonable number of significant figures. (The formatting structure appears under the headings "Format, Cells, Number," and it is highly recommended you to experiment with fonts, borders, alignment, and number formats.) A block of cells can be highlighted before each operation and the formatting applied en masse.

Let's first get average miles driven in each category. Those answers are shown in Table 2.6 in B13:D13. There's a quick bonus that we can also get the total miles driven by each driver, E5:E11 and the overall average in E13.

Table 2.7 shows the mathematical operations behind Table 2.6 and what's in each cell. You can get these results from the spreadsheet in Table 2.6 by using a *simultaneous* control key (^) and a tilde (~). This operation is written in shorthand as ^~ .

Table 2.7 Expanded Spreadsheet from Table 2.6 (Using ^~)

	A	B	C	D	E
1	Miles driven				
2					
3					

Table 2.7 Expanded Spreadsheet from Table 2.6 (Using ˆ~)—cont'd

	A	B	C	D	E
4	Renter	City miles	Suburban miles	Highway miles	Total miles
5	Davis	16	28	79	=SUM(B5:D5)
6	Graham	10	31	112	=SUM(B6:D6)
7	Washington	4	7	158	=SUM(B7:D7)
8	Meyers	22	61	87	=SUM(B8:D8)
9	Richardson	12	56	198	=SUM(B9:D9)
10	Thomas	5	22	124	=SUM(B10:D10)
11	Williams	4	14	142	=SUM(B11:D11)
12					
13	Average	=AVERAGE (B5:B11)	=AVERAGE (C5:C11)	=AVERAGE (D5:D11)	=AVERAGE (E5:E11)

Note the actual arithmetical operations now appear in cells E5:E11 and B13:E13. We declared them as such by inputting the mathematical operation after an " = " sign. The mathematical operations are the sum and the average, both with the specified ranges. Finally, notice in this view that the individual formatting for each cell's arithmetic has been removed and the actual input data as typed are revealed, while the long statement that appeared in cell A1 has now been truncated.

FleetsR'Us now realizes that the gas mileage in each segment of the journeys is different; the length of the journey is less important than the gallons used in each segment. It also realizes that gas mileage is a moving target; vehicles can get both more and less fuel efficient (a large SUV vs. a compact hybrid vehicle, for example). It needs to keep its options open and be able to update the spreadsheet when the market changes. Spreadsheets have a very powerful way of showing you how to do this. They introduce a powerful new tool: **absolute and relative cell addressing.** An absolute address is preceded by a "$" sign for either or both rows and columns; relative addressing is the default mode.

Example 2.7
FleetsR'Us wants to add the gas mileage to its spreadsheet in a flexible manner, so it can be updated later if needed. It wants to now know the gallons used per trip and the average used by their typical drivers. Here are the miles per gallon (mpg) by journey segment:

	City	Suburbs	Highway
mpg	12.	18.	26.

Need: Average gallons per journey segment and average per driver.
Know: Mileage data from Example 2.6 plus mpg just given.
How: Divide mileage by mpg to get gallons used in each segment of the journey.
Solve: Table 2.8 shows several ways we can program our spreadsheet. (We just look at the spreadsheet mathematics using our ˆ~ method. Be assured that each cell *will* give the correct answer.)

Table 2.8 Constants and Absolute Addressing

	A	B	C	D	E
4	**Renter**	**City gallons**	**Suburban gallons**	**Highway gallons**	**Total gallons**
5	Davis	=16/12	=28/18	=79/26	=SUM(B5:D5)
6	Graham	=10/B17	=31/$C$17	=112/D$17	=SUM(B6:D6)
7	Washington	=4/B17	=7/$C$17	=158/D$17	=SUM(B7:D7)
8	Meyers	=22/B17	=61/$C$17	=87/D$17	=SUM(B8:D8)
9	Richardson	=12/B17	=56/$C$17	=198/$D17	=SUM(B9:D9)
10	Thomas	=5/B17	=22/$C$17	=124/$D17	=SUM(B10:D10)
11	Williams	=4/B17	=14/$C$17	=142/$D17	=SUM(B11:D11)
12					
13	Average	=AVERAGE(B5:B11)	=AVERAGE(C5:C11)	=AVERAGE(D5:D11)	=AVERAGE(E5:E11)
14					
15					
16		**City**	**Suburbs**	**Highway**	
17	**mpg**	12	18	26	

What, if anything, is wrong with these solutions? To start with, cells B5:D5 do exactly what they are asked to do, but they are *inflexible* should we later want to enter different mpg, since we would have to hand input them to every affected cell (here just 3, but possibly 10,000 or more).

Table 2.8 shows that we can use a "constant," such as the contents of cell B17, to divide each cell in column B; then, when we wish to again change the mpg, all we do is change the number in the cell B17, and it will be updated wherever it is used. This is a powerful improvement over entering the data in individual cells.

We can also change the weightings of the cells to reflect mpg in columns C and D. We use a slightly different method. In fact, we fix the contents of the cell C17 in a tricky way, using the address $C$17. Such an address, for reasons seen shortly, is called *absolute addressing*. It is used here to modify cells C6:C11; we can also use either D$17 or $D17 as a constant (called a *mixed* mode of relative and absolute addressing) to modify the cells in D6:D11. What these do is *fix* the reference to the row or number operated on by the "$" sign. What does this mean and why do it? The importance of these operations is best explained by copying and pasting a section of the spreadsheet as in the next example.

Example 2.8

FleetsR'Us wants to copy part of its spreadsheet to another area on the spreadsheet to make it clearer for someone to read. Assume the origin is no longer cell A1 and shift the whole spreadsheet to a new origin H1. Analyze what you see and report whether the translocated spreadsheet is correct or not; if not, explain what is wrong.

Need: Copied spreadsheet.

Know: You can copy and paste across the spreadsheet form.

How: Use "Edit, Copy, Paste" commands.

Solve: Suppose you want to copy the table to another part of the spreadsheet, all you do is highlight the block of information A1:E13 that you want to move, go to "Edit," and then "Copy."[15] Move your cursor to the corner cell you want to move to, say, cell H1 and then go to "Edit," "Paste." Observe the effects after this is done in Table 2.9.

Table 2.9 Contents of Cells after Shifting Location

	H	I	J	K	L
4	Renter	City gallons	Suburban gallons	Highway gallons	Total gallons
5	Davis	=16/12	=28/18	=79/26	=SUM(I5:K5)
6	Graham	=10/I17	=31/$C$17	=112/K$17	=SUM(I6:K6)
7	Washington	=4/I17	=7/$C$17	=158/K$17	=SUM(I7:K7)
8	Meyers	=22/I17	=61/$C$17	=87/K$17	=SUM(I8:K8)
9	Richardson	=12/I17	=56/$C$17	=198/$D17	=SUM(I9:K9)
10	Thomas	=5/I17	=22/$C$17	=124/$D17	=SUM(I10:K10)
11	Williams	=4/I17	=14/$C$17	=142/$D17	=SUM(I11:K11)
12					
13	Average	=AVERAGE(I5:I11)	=AVERAGE(J5:J11)	=AVERAGE(K5:K11)	=AVERAGE(L5:L11)

We did not get all the expected results: instead of modifying cells to the correct answers, cells I6:I11 and K6:K8 will give the wrong answers. (Other divisions by zero also occur, since the spreadsheet's mathematical functions of sum and average also reference cells with one or more zero divisions.) What has happened? Use the toggle ^~ again and observe the results in Table 2.10.

Of course, the manually transplanted input cells I5:K5 are correct, since they call for no information outside of the particular cells, but the entries to I6:I11 are divided, not by B17, but by I17. In other words, when we translocated the cells, we also shifted B17 "relative" to the move (i.e., cell B17 is now displaced to I17 as is cell A1 to H1). And cell I17 (not shown) contains a default empty cell value of zero—hence, the apparent disaster that the gallons used in city driving are now all infinite!

Notice that cells J6:J11 are multiplied by the contents of cell $C$17 as desired, since we used the "$" designator for both row and column. Cell $C$17 has not translocated relative to the move; its address is therefore "absolute."

The sundry results of the translocated highway mileage column K are explained by the "mixed" mode of addressing used there. Cells K6:K8 now have been divided by K$17, since the column designator translocated from D to K (again, just as column A did to H). Since we fixed the row designator at $17, it did not translocate to a new row. But, the cell K$17 contains a default value of zero, and hence, the cells K6:K8, all of which contain this as a divisor, are each infinite.

However, cells K9:K11 gave the correct answer because they were divided by $D17. The $ fixed the column designator at column $D. We did not choose to change the rows when we copied the block of cells from A1 to H1, and the divisor

---

[15]There are alternate ways of copying and pasting using "smart tags"; however, the method suggested above is conceptually easier for a beginner in spreadsheet manipulations.

Table 2.10 Contents of Cells after Shifting Location

	H	I	J	K	L
1	**Miles driven in various categories**				
2					
3					
4	**Renter**	**City gallons**	**Suburban gallons**	**Highway gallons**	**Total gallons**
5	Davis	1.3	1.6	3.0	5.9
6	Graham	#DIV/0!	1.7	#DIV/0!	#DIV/0!
7	Washington	#DIV/0!	0.4	#DIV/0!	#DIV/0!
8	Meyers	#DIV/0!	3.4	#DIV/0!	#DIV/0!
9	Richardso	#DIV/0!	3.1	7.6	#DIV/0!
10	Thomas	#DIV/0!	1.2	4.8	#DIV/0!
11	Williams	#DIV/0!	0.8	5.5	#DIV/0!
12					
13	Average	#DIV/0!	1.7	#DIV/0!	#DIV/0!

$D17 still refers to the original mpg in cell $D17. We simply took advantage of our simple horizontal move from A1 to H1 in that it did not change the row locations. Had we also changed the row location—say, by pasting the block at A1:E13 to H2 instead of H1—the rest of the highway gallons calculations would also have been in error due to an imputed mpg call to cell $D18.

Sometimes we want to use relative, absolute, or mixed modes. It depends on what we want to do. With experience, your ability to do this will improve. One nice feature of spreadsheets is that your mistakes are usually embarrassingly and immediately clear. If, for example, you forget to fix an absolute cell address and add, multiply, divide, exponentiate, and so on by some unintended cell, the worksheet will complain as the target cell picks up whatever cell values were in the inadvertent cells.

Example 2.9

Correct Example 2.8 using relative and absolute addressing modes.

  Need: Corrected spreadsheet.
  Know: How to use different cell addressing modes.
  How: Make sure that no cell addresses are translocated to undesired cells.
  Solve: Had all the cells B5:D11 row by row been divided by their respective absolute constants, $B$17:$D$17, the final translocated result would have been as initially desired—see Table 2.11.

Table 2.11 Results of Spreadsheet That Used Absolute Cell Constants

	H	I	J	K	L
1	**Miles driven in various categories**				
2					
3					
4	**Renter**	**City gallons**	**Suburban gallons**	**Highway gallons**	**Total gallons**
5	Davis	1.33	1.56	3.04	5.93
6	Graham	0.833	1.72	4.31	6.86
7	Washington	0.333	0.389	6.08	6.80
8	Meyers	1.83	3.39	3.35	8.57
9	Richardson	1.00	3.11	7.62	11.7
10	Thomas	0.417	1.22	4.77	6.41
11	Williams	0.333	0.778	5.46	6.57
12					
13	Average	0.869	1.74	4.95	7.55

2.7.2 Graphing in Spreadsheets

One of the big advances of spreadsheeting is that we can display our results in many ways that can visually convey a great deal of information. In a real sense, most of our "knowledge" is accessible only then, since the raw data, even if arranged as logical information in neat tables, may still be difficult to interpret.

We confine ourselves for the moment to simple Cartesian graphing, but you will soon recognize the pattern to follow if you want to use other modes of display.

Example 2.10

SpeedsR'Us wants to sell after-market booster kits for cars. They equipped a car with a booster kit and found that it had the performance on a level road shown in Table 2.12.

Table 2.12 Performance spreadsheet for a modified car

Acceleration time (seconds, s)	Car speed (miles/hour, mph)
0.0	0.00
1.0	18.6
2.0	32.1

(Continued)

Table 2.12 Performance spreadsheet for a modified car—cont'd

Acceleration time (seconds, s)	Car speed (miles/hour, mph)
3.0	46.5
4.0	58.2
5.0	78.6
6.0	86.9
7.0	96.9
8.0	103

The car manufacturer says that the *un*modified car speed characteristic on a level road obeys the following equation: $mph = 136.5(1 - e^{-0.158t})$ with t in seconds. Is the modified car faster to 100 mph than the standard car?

Need: To compare the tabular data to the equation given.
Know: A spreadsheet provides multiple ways of looking at data.
How: Graph the tabular data and compare that data to the theoretical equation.
Solve: Set up the spreadsheet in Table 2.13.

Table 2.13 Setup for Graphing Function

	A	B	C	D
1	Compare measured car speed to manufacturer's specification			
2				
3				
4	Speed in mph = $136.51(1 - e^{-0.158t})$ with t in seconds			
5				
6	Time, s	Mph, actual	Mph, theory	
7	0	0	=136.5*(1 − EXP(−0.158*A7))	
8	=A7+1	18.6	=136.5*(1 − EXP(−0.158*A8))	
9	=A8+1	32.1	=136.5*(1 − EXP(−0.158*A9))	
10	=A9+1	46.5	=136.5*(1 − EXP(−0.158*A10))	
11	=A10+1	58.2	=136.5*(1 − EXP(−0.158*A11))	
12	=A11+1	78.6	=136.5*(1 − EXP(−0.158*A12))	
13	=A12+1	86.9	=136.5*(1 − EXP(−0.158*A13))	
14	=A13+1	96.9	=136.5*(1 − EXP(−0.158*A14))	
15	=A14+1	102.9	=136.5*(1 − EXP(−0.158*A15))	

Start with a spreadsheet title as in cell A1. Make it descriptive. The cell labels in A6:C6 should be helpful as to what the immediate columns under them mean. Of course, the numbers generated as A7:C15 are just raw data—not easy to interpret by merely looking at them. In columns A, B, and C are the data you want to plot, that is, the value of the variable "mph" given as input data and calculated by the manufacturer's equation.

You don't have to enter too much to get a lot of mileage.[16] In cells A7:B7, enter both 0, and in C7, enter "136.5*(1 − exp(−0.158*A7))." The result in C7 is zero, since $1 - e^0 = 1 - 1$ is zero. Now in A8, write $= A7 + 1$. The value that appears in A8 is 1, since A7 is zero. Now copy A8 to A9:A15, then copy C7 to C8:C15, and you will have the manufacturer's function defined corresponding to each time of measurement.

You might want to graph your results, A7:A15 vs. B7:B15 and also A7:A15 vs. C7:C15. No sweat: just highlight A7:C15 and then go to "Insert, Chart." Then tell the spreadsheet what kind of graph you want. You want a "xy *(Scatter)*" plot. ("Scatter" is just a regular Cartesian graph!) You will see two sets of points corresponding to each column of speed data. Then, hit "Next" and fill in your preferred titles. Optionally, toggle off the "Legend" box. You can finish by pasting the graph on the current worksheet (i.e., the working page in the spreadsheet). If you want to, you can highlight either series of data and format them to discrete points or to a continuous line.

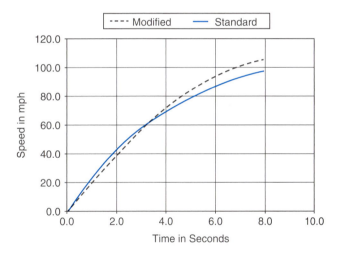

FIGURE 2.6 SpeedsR'Us modification.

You should get the graph in Figure 2.6 for your troubles. Spreadsheets leave you plenty of flexibility if you use your imagination. On the basis of Figure 2.6, has the speed modification achieved its goal?

SUMMARY

To summarize this chapter, the results of an engineering analysis must be correct in *four* ways: it must involve the appropriate **variables**; it must be expressed in the appropriate **units**; it must express the correct **numerical value**; and it must have the appropriate number of **significant figures**.

Sketching the object you are analyzing can be helpful in focusing on the project at hand. Some **sketches** are just that—crude unscaleable drawings done with a pencil and paper. But some are more formal and properly

---

[16]As previously noted, you can short-circuit much of this discussion by taking advantage of "smart tags"; these are pull-downs on the cells that automatically copy or increment adjacent cells. See the Help menu for your particular spreadsheet to see if this usage is supported.

called "**engineering drawings**." Fortunately, these days you do not need to be a qualified draftsperson. There are computerized apps known as **Computer-Aided Design (CAD)** that are easy to use and will accomplish much of what a professional draftsperson would have labored weeks over just a few years ago.

A problem-solving method that guarantees you a path to the answer to engineering solutions is a valuable addition to an engineer's tool kit; we recommend the self-prompting mnemonic approach we call the **need-know-how-solve** method. It allows for a systematic approach to engineering problems. We suggest you couple this method with the units method, [..]. Between these two methods, you should be able to reason your way through even arcane problems that appear impossible at first glance. In addition, the systematic approach means you leave an auditing trail that can be used in your absence to check on, extend, or even to correct your solution.

Many solutions to engineering problems require either or both complex analysis and repetitive solutions for many cases; still others require graphical representation. In other cases, we may be confronted with jumbles of apparently uncorrelated **data**. These situations are where spreadsheet analysis stands out. **Spreadsheeting** allows you to organize these data and sort them, so you have some understanding of the gross **information** involved. Then you can interpret the data to gain **knowledge** from the model you developed. The heart of this method is to recognize that cells in a spreadsheet which may be manipulated either in an **absolute** sense or in a **relative** sense. The former allows for facile repetition of algebraic and arithmetical operations using **constants**, and the latter allows one to easily extend tables of data, even while carrying their embedded mathematical operations.

This section on spreadsheeting merely skims the surface of the very powerful tool of spreadsheet analysis. With time, both as a beginning engineer and as a practicing engineer, you will discover many of the additional features of this tool. Used with skill and precision, this will become one of the most useful and generic skills that you will acquire. The sooner you make use of it, the sooner you acquire skill. You will find practice problems in virtually every chapter of this text that are susceptible to spreadsheet analysis.

Together, with an appreciation of the significance of engineering numbers and dimensional analysis, the contents of this chapter are fundamental to what an engineer does and how successful he/she will be in pursuing an engineering career.

EXERCISES

To help get you in the habit of applying these elements, in the following exercises, you will be graded on use of *all* of the elements we discussed in this chapter. Make sure you are reporting the solution to the proper number of significant figures.

Some of these exercises can be done using hand calculators; alternatively, you can also use a spreadsheet even for simple problems. *If you choose a spreadsheet, then for all your spreadsheet exercises for this course, you should go to "File," "Print setup," "Sheet," and then tick the boxes for "Gridlines" and "Row and columns headings"* so that your final cell locations are printed when you submit the answers to these problems. *You will lose points otherwise*, since it is nearly impossible to grade answers without the row and column indicators!

You should use the ***need-know-how-solve*** method in setting up all appropriate problems.

1. Sketch the isometric plate below and add the following dimensions: Height=5.0 cm, Length=40.0 cm, Width=20.0 cm. The diameter of the hole is 10.0 cm, and it is centered 15.0 cm from the front.

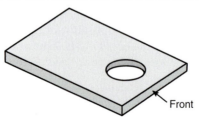
Front

2. Sketch and complete the orthographic drawing below.

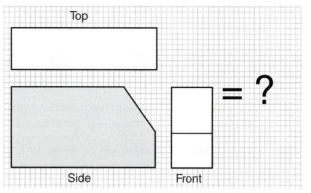

3. Sketch the top, left, and front views of the object below.

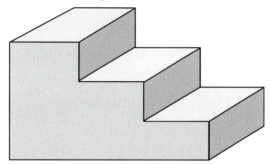

4. If a U.S. gallon[17] has a volume of 0.134 ft^3 and a human mouth has a volume of 0.900 in^3, then how many mouthfuls of water are required to fill a 5.00 gallon can? (**A: 1.29 $\times$ 10^3 mouthfuls.**)

5. Identify whether you would perform the following unit conversions by definition, conversion factors, geometry, or scientific law.
   a. How many square miles in a square kilometer?
   b. How many microfarads in a farad?
   c. What is the weight on Earth in N of an object with a mass of 10 kg?
   d. How many square miles on the surface of the Earth?

6. An acre was originally defined as the amount of area that an oxen team could plow in a day.[18] Suppose a team could plow 0.040 hectare per day, where a hectare is 10^4 m^2. There are 1609 meters in a mile. How many acres are there in a square mile?

7. There are 39 inches in a meter. What is the area in the SI system of the skin of a spherical orange that is 4.0 inches in diameter? (**A: 3.3 $\times$ 10^{-2} m^2.**)

8. There are 39 inches in a meter. What is the volume in the Engineering English system of a spherical apple that is 10. (note the decimal point here) cm in diameter?

---

[17]Distinguish U.S. gallons from the old English measure of Imperial gallons; 1 Imp. gallon$=$1.20 U.S. gallons.
[18]The furlong or "furrow-long" was the distance of 220 yards that the oxen could plow in a day times a width of one "chain" or 22 yards. Multiplied together they defined the area of an acre. The usage goes back to feudal times.

9. Suppose the ranch in Example 2.3 was a circle instead of a square. Using the same financial information ($320,000 available funds, $10,000 a mile for fence, and $100,000 per square mile land cost), what would be the diameter of the ranch?

10. The great physicist Enrico Fermi used to test the problem-solving ability of his students at the University of Chicago by giving them the following problem: How many piano tuners are there in the city of Chicago? (Assume the population of Chicago is 5 million people.) (**Plausible answers 50–250 tuners**[19])

11. Of all the rectangles that have an area of one square meter, what are the dimensions (length and width) of the one that has the smallest perimeter? Solve by graphing on a spreadsheet.

12. Suppose you want to make a cylindrical can to hold 0.01 m^3 of soup. The sheet steel for the can costs $0.01/m^2. It costs $0.02/m to seal circular pieces to the top and bottom of the can and along the seam. What is the cost of the cylindrical can that is least expensive to make? (**A: 3 cents/can.**)

13. Suppose the mass used in Example 2.5 was increased from 10.0 kg to 20.0 kg, and the wire stretched by twice as much. If the 20.0 kg mass was then used to stretch a 4.00 m piece of the same steel wire, how much would it axially stretch?

14. Use a spreadsheet analysis to determine the total miles driven by each driver and the average miles driven in each category for the following car renters:

Renter	City miles	Suburban miles	Highway miles
Geske	35	57	93
Pollack	27	11	275
Loth	14	43	159
Sommerfeld	12	31	305
Thunes	22	16	132
Lu	5.0	21	417

15. Using the renter mileage information given in the previous exercise and the miles per gallon information given in Example 2.7, determine the average gallons per journey segment and average per driver for this set of drivers.

16. Using the technique introduced in Example 2.10, create a spreadsheet graph of the following data for the median annual salaries in dollars for engineers based on years of experience, supervisory responsibility, and level of education.[20]

	Number of Years after BS Degree					
	0	5	10	15	25	35
Nonsupervisory						
B.S.	$55,341	$63,649	$73,162	$80,207	$85,116	$92,745
M.S.	—	$79,875	$86,868	$90,134	$97,463	$110,289
Ph.D.	—	—	$91,352	$98,053	$108,747	$122,886

---

[19]See http://www.grc.nasa.gov/WWW/K12/Numbers/Math/Mathematical_Thinking/fermis_piano_ tuner.htmS.
[20]These data are from the 2007 report of the Engineering Workforce Commission of the American Association of Engineering Societies (AAES).

Supervisory						
B.S.	—	—	$72,632	$80,739	$108,747	$122,886
M.S.	—	$99,367	$109,450	$110,360	$113,91	$117,146
Ph.D.	—	—	—	$110,877	$132,800	$147,517

Exercises 17–19 involve the following situation: Suppose that the weight of the gasoline in lbf in a car's gas tank equaled the weight of the car in lbf (lbs force) to the 2/3 power (i.e., if G=gasoline weight in lbf, and W=car weight in lbf, then $G=W^{2/3}$). Assume that gasoline weighs 8.0 lbf/gal, and gas mileage, measured in miles/gallon or mpg, varies with weight according to the empirical formula

Gas mileage in mpg = (84,500 mpg · lbf) ÷ (Car's weight in lbf) − 3. mpg

17. What is the fuel usage in miles per gallon of a 3.00×10^3 lbf car?

18. What is the heaviest car that can achieve a range of 600. miles? (**A: 3.7×10³ lbf.**)

19. Suppose the formula for weight of the gas was $G=W^b$, where b can be varied in the range of 0.50–0.75. Graph the range of a 3.69×10^3 lbm car as a function of b.

Exercises 20–23: These exercises are concerned with bungee jumping as displayed in the figure. At full stretch, the elastic rope of original length L stretches to $L+x$. For a person whose weight is W lbf and a cord with a stiffness K lbf/ft, the extension x is given by the following formula:

$$x = \frac{W}{K} + \sqrt{\frac{W^2}{K^2} + \frac{2W \times L}{K}}$$

which can be written in spreadsheet script as $x = W/K + \mathrm{sqrt}(W^2/K^2 + 2*W*L/K)$.

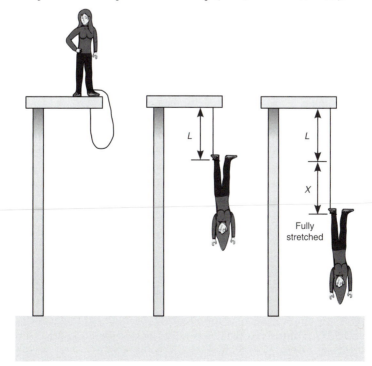

20. If the height of the cliff is 150.0 ft, $K = 6.25$ lbf/ft, $L = 40.0$ ft, and the person's weight is 150.0 lbf, will the person be able to bungee jump safely? Support your answer by giving the final value for length $= L + x$. (**A: 114 ft $<$ 150 ft, OK.**)

21. Americans are getting heavier. What's the jumper's weight limit for a 40.0 ft unstretched bungee with stiffness of $K = 6.25$ lbf/ft? Graph final length $L + x$ vs. W for weights from 100 lbf to 300 lbf in increments of 25 lbf. Print a warning if the jumper is too heavy for a 150 ft initial height. (**Hint:** Look up the application of the "IF" statement in your spreadsheet program.)

22. If the height of the parapet is 200.0 ft, the weight of the person is 150.0 lbf, and the unstretched length $L = 45.0$ ft, find a value of K that enables this person to stop exactly 5 feet above the ground. (**A: 2.60 lbf/ft.**)

23. By copying and pasting your spreadsheet from Exercise 22, find and plot the values of L needed (in ft) vs. W, weight of jumper (in lbf) for successful bungee jumps (coming to a stop 5 ft above the ground) for $K = 6.25$ lbf/ft and from a cliff of height 150.0 ft above the ground. The graph should cover weights from 100 lbf to 300 lbf in increments of 25 lbf. (**Hint:** The function "Goal seek" under "Tools" is one way to solve this exercise.)

Exercises 24–27: The fixed costs per year of operating an automobile is approximately 20.% of the initial price of the car. Therefore, the operating cost/mile $= 0.20$/yr $\times$ (purchase price of automobile)/(miles driven per year) + (price of gasoline/gallon) $\times$ (gallons used per mile). In the exercises that follow, assume that the automobile is driven 2.0×10^4 miles per year. Assume gasoline costs $5.00/gallon.

24. Estimate the operating cost per mile of an automobile with a price of $15,000 that gets 30.0 miles per gallon. (**A: $0.32/mile.**)

25. If one were to double the price of the automobile in Exercise 24, what would its gas mileage have to be to cost the same to operate per mile as the automobile in Exercise 24?

26. Suppose that the purchase price of automobiles varies with weight according to the formula that weight in lbf $\times$ $8.00, and gas mileage varies according to: mpg $= (84,500$ mile-lbf/gal$)/W - 3$. miles/gal.

Graph the cost per mile of operating a car as a function of the car's weight, in increments of 500.0 lbf from 2000.0 lbf to 5000.0 lbf. (**Partial A: 29 cents/mile for 2000 lbf cars and 76 cents/mile for 5000 lbf cars.**)

Exercises 27–29: In visiting stores, one finds the following prices for various things. Broccoli crowns cost $2.89 per pound. Soft drinks cost $2.00 per 2 liter bottle. (A liter is 0.001 m^3.) A new automobile weighs 2.50×10^3 lbf and costs 1.50×10^4. A dozen oranges, each of which is 0.06 m in diameter, costs $2.05. A 1.5 lb package of chicken thighs costs $5.35. A dictionary weighs 5.00 pounds and costs $20. A refrigerator weighs 200.0 lbf and costs $900. Assume that one cubic meter of any solid object or liquid weighs 1.00×10^4 N.

27. For the objects just listed, make a table and graph of the cost of objects in dollars as a function of their weight in newtons. It is suggested for this graph to use a *line* (not "scatter") graph with markers displayed at each data value. (Get rid of the unwanted line using the "Format series function.") The value of the line graph is that everything plotted is at the same horizontal displacement and is not dependent on its value.

28. What (perhaps surprising) simple generalization about the cost of things might one make based on the table and graph of Exercise 27? (**Hint:** Does the comparative lack of spread in price surprise you?)

29. Name a product or group of products that does *not* fit the generalization you made in Exercise 28 and add and label the point on the graph in Exercise 28. To get a better perspective, use a *log scale* for the y-axis, $/N, since the ordinate should be much larger than the coordinates of those points from Exercise 28.

30. An unnamed country has obtained the population of passenger cars on its roads as determined by 250 kg mass differences. These data are shown in the table that follows. You have to make these data clear to the undersecretary to that country's transport minister. Plot these data by two methods: (1) as a "pie chart" and (2) as a histogram to show the distribution in an effective manner.

	Car mass in kg	% of all cars
1	1000	12.1
2	1250	13.1
3	1500	15.4
4	1750	18.6
5	2000	14.8
6	2250	9.2
7	2500	7.5
8	2750	6.3
9	3000	2.0
10	3250	1.0

Exercises 31–33 deal with Hubbert's Peak,[21] a model of the supply and demand for oil. It looks at the amount of available oil and its rate of consumption to draw conclusions about continuing the current course of our oil-based economy.

31. Suppose the world originally had 3 trillion ($3. \times 10^{12}$) barrels of oil and exploration for oil began in 1850. Suppose 10.0% of the remaining *undiscovered* oil has been found in every quarter century since 1850. Call the *discovered*, but not yet consumed oil, *reserves*. Suppose oil consumption was 1.0×10^8 barrels in 1850, and further suppose oil consumption has grown by a factor of 5 in every quarter century since 1850. When will the oil start to run out? (That is, when will the reserves become negative?) Give your answer to the nearest 25 years and provide a spreadsheet showing reserves and consumption as a factor. (**A: 2025.**)

32. Suppose the world originally had 10 trillion (10.0×10^{12}) barrels of oil. Use the data of Exercise 31 to again predict when the oil will start to run out.

33. Repeat Exercise 32, but instead of assuming the exponential growth in consumption continuing unabated by a factor of 5 in every quarter century since 1850, curtail growth since 2000 and assume consumption has stayed constant since then. Again predict when the oil will start to run out.

34. Your friend tells you that the **need-know-how-solve** problem-solving method seems overly complicated. He just wants to find the answer to the problem in the quickest possible way—say, by finding some formula in the text and plugging numbers into it. What do you tell him?

---

[21]M. King Hubbert was a geologist with Shell Oil who, in the 1950s, correctly pointed out that the U.S. supply of oil was going to fall short of demand by the 1970s. His methods have since been applied to world oil production and, based on demand exceeding production, predict an ongoing oil supply crisis. See http://www.hubbertpeak.com/hubbert/.

    a. Go ahead and do whatever you want, then you'll flunk out, and I'll survive.
    b. Talk to the instructor and have her explain why this methodology works.
    c. Find someone who has used this method and ask to copy his homework.
    d. Explain why this technique will lead to a fail-safe method of getting the correct answer.

35. Calculate with the correct significant figures: (a) $100/(2.0 \times 10^2)$, (b) $1.0 \times 10^2/2.0 \times 10^2$. (**A: 0.5, 0.50.**)

36. Calculate with the correct significant figures: (a) 10/6, (b) 10.0/6, (c) 10/6.0, (d) 10./6.0, and (e) 10.0/6.00.

37. What is 2.68×10^8 minus 2.33×10^3 to the correct significant figures? (**A: 2.68×10^8.**)

38. A machinist has a sophisticated micrometer that can measure the diameter of a drill bit to 1/10,000 of an inch. What is the maximum number of significant figures that should be reported if the approximate diameter of the drill bit is (a) 0.0001 inches, (b) 0.1 inches, (c) 1 inch.

39. Round off to three significant places: (a) 1.53, (b) 15.345, (c) 16.67, (d) 102.04, (e) −124.7, and (f) 0.00123456.

40. Suppose you want to design a front door and doorway to fit snugly enough to keep out the drafts, yet still be easy to open. (You are not showing off precision carpentry here but merely designing a convenient ordinary door by standard methods.) The dimensions are to be given in inches. To how many significant figures would you specify the length and width of the door and doorway?[22] Assume a standard door is 30.0" by 81.0".

41. You are browsing the Internet and find some units conversion software that may be useful in this course. You would like to download the software on your PC at school and use it in this course. What do you do?
    a. Check with the Internet site to make sure this software is freeware for your use in this course.
    b. Just download the software and use it because no one will know.
    c. Download the software at home and bring it to school.
    d. Never use software found on the Internet.
    Suggested method: Apply the Fundamental NSPE Canons and fill in an Engineering Ethics Matrix:

42. On December 11, 1998, the Mars *Climate Orbiter* was launched on a 760 million mile journey to the Red Planet. On September 23, 1999, a final rocket firing was to put the spacecraft into orbit, but it disappeared. An investigation board concluded that NASA engineers failed to convert the rocket's thrust from pounds force to newtons (the unit used in the guidance software), causing the spacecraft to miss its intended 140–150 km altitude above Mars during orbit insertion, instead entering the Martian atmosphere at about 57 km. The spacecraft was then destroyed by atmospheric stresses and friction at this low altitude. As chief NASA engineer on this mission, how do you react to the national outcry for such a foolish mistake?
    a. Take all the blame yourself and resign.
    b. Find the person responsible, and fire, demote, or penalize that person.
    c. Make sure it doesn't happen again by conducting a software audit for specification compliance on all data transferred between working groups.
    d. Verify the consistent use of units throughout the spacecraft design and operations.
    Suggested method: Apply the Fundamental Canons and fill in an Engineering Ethics Matrix.

---

[22]The subject of tolerancing is important in mass manufacturing to ensure a proper fit with one part and another, since each part isl not exact, their combined tolerance determines how well, if at all, they fit together. It is the subject of significant statistical analysis. Applying the methods rigorously allowed the Japanese automotive industry to eclipse those of the rest of the world in terms of quality (see Chapter 10, "Manufacturing Engineering.").

43. You e-mail a classmate in this course for some information about a spreadsheet homework problem. In addition to answering your question, your classmate also attaches a spreadsheet solution to the homework. What do you do?

  a. Delete the spreadsheet without looking at it.
  b. Look at the spreadsheet to make sure your classmate did it correctly.
  c. Copy the spreadsheet into your homework and change the formatting so that it doesn't look like the original.
  d. E-mail the spreadsheet to all your friends so that they can have the solution, too.

44. Stephanie knew Adam, the environmental manager, would not be pleased with her report on the chemical spill. The data clearly indicated that the spill was large enough that regulations required it to be reported to the state. When Stephanie presented her report to Adam, he lost his temper. "A few gallons over the limit isn't worth the time it's going to take to fill out those damned forms. Go back to your desk and rework those numbers until it comes out right." What should Stephanie do?[23]

  a. Tell Adam that she will not knowingly violate state law and threaten to quit.
  b. Comply with Adam's request since he is in charge and will suffer any consequences.
  c. Send an anonymous report to the state documenting the violation.
  d. Go over Adam's head and speak to his supervisor about the problem.

---

[23] Abstracted from *Engineering Ethics: Concepts and Cases* at http://wadsworth.com/philosophy_d/templates/student_resources/0534605796_harris/cases/Cases.htm.

Force and Motion

3.1 INTRODUCTION

We have developed some visual and mathematical models that describe how to analyze the forces and motions of real objects, called **kinetics**. These are part of most engineering disciples—civil, mechanical, biomedical, and electrical, to name just a few. For example, electrical, computer, and control engineers need to understand the forces and motions of robots. Further, bioengineers study how the human body responds to forces and motions in normal and abnormal situations (such as in a fall). We also look at the movement of objects that does not involve either the object's mass or the forces that cause the motion. This subject is called **kinematics**, and we will conclude this chapter with an intuitive graphical method that simplifies how to understand and solve one-dimensional kinematic problems.

3.2 WHAT IS A FORCE?

The word *force* is used in many ways. The force of law, brute force, police force, driving force, nuclear force, electromagnetic force, and the force of gravity are all examples of its use. In engineering, a **force** is something that pushes or pulls and can cause an object to change speed, direction, or shape.[1]

---

[1]Engineering terms that are related to force include thrust, drag; shear, tensile, and compressive stress (these are forces per unit area) and torque.

Early philosophers, such as Aristotle and Archimedes, did not have a clear understanding of the relationship between force and motion. Erroneous theories existed until Sir Isaac Newton formulated his Laws of Motion. When he described the motion of objects using their inertia, he found that they obeyed three physical laws that form the basis for Newtonian mechanics.[2]

Newton's Laws of Motion can be summarized as follows:

1. **Newton's First Law.** The speed v of a body remains constant unless the body is acted upon by an external force F.
2. **Newton's Second Law.** The acceleration a of a body is parallel to and directly proportional to the net force F applied to it and inversely proportional to its mass m, or $a \propto F/m$.
3. **Newton's Third Law.** The mutual forces of action and reaction between two bodies are equal, opposite, and collinear.

3.3 NEWTON'S FIRST LAW

Newton's First Law of Motion (sometimes called the **law of inertia**) is

An object at rest stays at rest and an object in motion stays in motion with the same speed and in the same direction unless acted upon by an unbalanced force.

This can be written mathematically as:

$$\sum_{external} \text{Forces} = 0 \text{ implies that speed } v = \text{constant}$$

Consequently,

- An object at rest will stay at rest unless an unbalanced force acts upon it.
- An object in motion will not change its speed unless an unbalanced force acts upon it.

Newton's First Law directly connects inertia with the concept of relative speeds. Specifically, in systems where objects are moving with different speeds, it is impossible to determine which object is "in motion" and which object is "at rest." In other words, the Laws of Motion are the same in every inertial frame of reference.

Example 3.1

Neglecting aerodynamic drag, when you are traveling down the highway in an open convertible at a constant speed, you can throw a heavy ball straight up in the air and catch it just as you would if the car was not moving. But what does a person standing on the roadside see as you pass by?

Need:	What a stationary observer sees.
Know-How:	If you are traveling in a car at a constant speed, the Laws of Motion are the same as when the car is at rest.
Solve:	A person observing you from the roadside sees the ball follow a curve in the same direction as the motion of the car. From your perspective inside the car, the car and everything inside of it is at rest, it is the world outside of the car that is moving with a constant speed.

---

[2]In 1687, Newton published his three Laws of Motion in *Philosophiae Naturalis Principia Mathematica*.

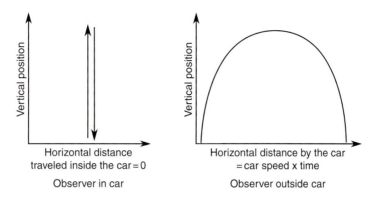

Observer in car
Horizontal distance
traveled inside the car = 0

Observer outside car
Horizontal distance by the car
= car speed x time

No experiment can tell whether it is the car that is at rest or the outside world that is at rest; the two situations are physically the same. Consequently, the concept of inertia applies to any constant speed motion (even when the speed is zero).

3.4 NEWTON'S SECOND LAW

Newton's Second Law can be stated as

*An object will accelerate if the unbalanced force on the object is not zero. The direction of the acceleration is the same as the direction of the unbalanced force. The magnitude of the acceleration is directly **proportional** to the unbalanced force, and inversely **proportional** to the mass of the object.*

We assumed you are familiar with **mass**, and you may indeed think you are, but it is not a trivial concept. As you will soon learn, mass is *not* weight. In a gravitational field, mass certainly produces weight, but the mass is present even where there is no gravity, such as in outer space. Mass is best considered as the quantity of matter; it is a property of the substance.

Central to any scientific set of units is the definition of **force**. You have probably had some prior introduction to this concept in high school. According to Newton's Second Law, force is *proportional* to mass and acceleration and can be written as

$$F \propto ma \tag{3.1}$$

where a is the **acceleration** of mass m. Acceleration is simply the rate of change of velocity (speed), so Newton's Second Law can be written for bodies of constant mass as $F \propto m(\Delta v/\Delta t)$ with Δv being the change in the speed and Δt being the change in the time during the acceleration period. To convert the proportionality in Newton's Second Law into an "equality," we need to introduce a **constant of proportionality**.

To phrase Equation (3.1) in terms of an equality, suppose there exists a set of units for which a force F_1 accelerates the mass m_1 by a_1. Then, Newton's Second Law can be written as

$$F_1 \propto m_1 a_1$$

If we now eliminate the proportionality by dividing Equation (3.1) by this equation, we obtain

$$\frac{F}{F_1} = \frac{m}{m_1}\frac{a}{a_1} \quad \text{or} \quad F = \left(\frac{F_1}{m_1 a_1}\right)ma \tag{3.2}$$

Clearly, the fact that the proportionality constant, $F_1/m_1 a_1$, is the ratio of a specific force to a specific mass and acceleration is very important to the calculations made with this equation. It has become customary to give the proportionality factor the unusual symbol $1/g_c$, where $g_c \equiv m_1 a_1 / F_1$. Then, Newton's Second Law becomes

$$F = \left(\frac{F_1}{m_1 a_1}\right) ma = \frac{ma}{\left(\frac{m_1 a_1}{F_1}\right)} = \frac{ma}{g_c} \tag{3.3}$$

We must choose both the magnitude and the dimensions of the proportionality constant g_c. Any consistent set of units satisfies this equation. With this degree of flexibility, it is easy to see how a large number of different unit systems have evolved.

Example 3.2

Suppose you define a mass unit called the **slug**, a rather ugly word for a mass that accelerates by "exactly" 1 ft/s² when a force of "exactly" 1 pound force (lbf) is applied to it. What is the weight of a mass of 5.00 slug?

Need: Weight of a mass of 5.00 slug.
Know: Weight is the force produced by the Earth's gravitational acceleration of 32.2 ft/s².
How: The first part of Equation (3.3) can be used here

$$F = \left(\frac{F_1}{m_1 a_1}\right) ma, \text{ in which } \frac{F_1}{m_1 a_1} = \frac{1}{1 \times 1} = 1 \left[\frac{\text{lbf} \cdot \text{s}^2}{\text{slug} \cdot \text{ft}}\right]$$

Therefore,

$$F = \left(\frac{F_1}{m_1 a_1}\right) mg = 1 \left[\frac{\text{lbf} \cdot \text{s}^2}{\text{slug} \cdot \text{ft}}\right] \times 5.00\,[\text{slug}] \times 32.2 \left[\frac{\text{ft}}{\text{s}^2}\right] = 161 \text{ lbf}$$

Note that with this definition of force,

$$g_c = \left(\frac{m_1 a_1}{F_1}\right) = 1 \left[\frac{\text{slug} \times \text{ft}}{\text{lbf} \times \text{s}^2}\right]$$

In fact, the slug system is still preferred in the United States in some engineering fields. Its obvious advantage is that Newton's Second Law constant of proportionality g_c is exactly 1, so Newton's Second Law can be written in the familiar form of $F = ma$, with the stipulation that mass m is in slugs and the acceleration is in ft/s². Its disadvantage is that you probably have little "feel" for the size of a mass in slugs.

The SI system is somewhat similar to the slug system with the choice for the constant of proportionality g_c being both easy and logical. In the SI system, the numerical value of the ratio $F_1/m_1 a_1$ was chosen to be exactly "1" and without (i.e., [0]) dimensions.[3] The unit of force must then simply be equivalent to the units of mass times the units of acceleration. It is convenient to give this set of units a unique new name (mainly to keep from having to write the combined mass times acceleration units for every force), and in the SI system, the force unit is called the **newton** (abbreviated N).

In other words, a force F_1 of 1 Newton (N) acting on a mass m_1 of exactly 1 kg causes the mass to accelerate a_1 at exactly 1 m/s². Thus, in the SI system, $g_c = m_1 a_1 / F_1 \equiv 1$ and Newton's Second Law can be written

$$F = ma \tag{3.4}$$

This simplification led to the familiar form of Newton's Law of Motion that most of you have seen before. Since the conversion factor g_c—that is, [(kg m/s²)/N]—is exactly unity, we have dropped it altogether.

---

[3]Read [0] to mean "dimensionless."

Example 3.3

What is the force in Newtons on a body of mass 102 g (0.102 kg) that is accelerated at the gravitational acceleration of 9.81 m/s^2?

Need: Force in Newtons on a mass of 0.102 kg accelerated at 9.81 m/s^2.
Know: Newton's Second Law applies using MKS units.
How: Equation (3.4) is the principle we use.
Solve: $F = ma = 0.102 \times 9.81 \text{[kg][m/s}^2] = 1.00 \text{[kg m/s}^2] = \mathbf{1.00 \ N}$.

These last two examples use an acceleration of special interest, which is that caused by Earth's gravity: $g = 32.2 \text{ ft/s}^2 = 9.81 \text{ m/s}^2$. Looking at Equation (3.4), the SI force acting on 1 kg mass due to gravity is not 1 N, but 9.81 N, a fact that causes distress to newcomers to the subject. Weight is thus just a special force—that due to gravity. In this sense, Equation (3.4) can be specialized for the acceleration of gravity to yield that force we call *weight*, W, by writing it as

$$W = mg/g_c \tag{3.5}$$

Example 3.4

What is the *weight* in Newtons of a mass of 0.102 kg?

Need: Weight in Newtons on a mass of 0.102 kg accelerated at 9.81 m/s^2.
Know: Newton's Second Law applies using MKS units.
How: Equation (3.5) with $g_c = 1$ (since we are using SI units).
Solve: The arithmetic is identical to that of Example 3.2—the force on 0.102 kg of mass (which is about 4 oz in common English units) is 1.00 N; in other words, *one Newton is just about the weight of a small apple here on Earth*! Perhaps, this will help you mentally imagine the magnitude of a force stated in Newtons.

In addition to SI units, a North American engineer must master at least one of the other systems that relate mass and force, one whose persistence in the United States is due more to custom than logic: the **Engineering English** system of units. In this system, the conversion factor g_c between force and mass × acceleration is *not* unity. Because of this, we must carry an explicit proportionality constant every time we use this unit system. In addition, because the proportionality factor is not unity, it requires the use of an explicit set of units as well.

The Engineering English system also evolved a rather unfortunate convention regarding both the pound (abbreviated lb) unit and the definition of force. It was decided that the name *pound* would be used *both* for mass *and* for weight (i.e., force). Since mass and force are distinctly different quantities, a modifier had to be added to the "pound" unit to distinguish which (mass or weight) is being used. This was solved by simply using the phrase **pound mass** and **pound force** with the associated abbreviations, lbm and lbf, respectively, to distinguish between them.

In the Engineering English system, it was decided that a "pound mass" should weigh a "pound force" at standard gravity, which accelerates the mass by 32.174 ft/s^2. This has the helpful convenience of allowing one the intuitive ability to understand immediately what is meant by, say, a force of 15 lbf. It would be the force you would experience if you picked up a rock of mass 15 lbm on the Earth's surface.

This convenience was accomplished by setting the ratio $(F_1/m_1 a_1) = 1/32.174$ [lbf s^2/lbm ft]. In other words, 1 lbf is defined as the force that will accelerate exactly 1 lbm by exactly 32.174 ft/s^2. The designers of the Engineering English system cleverly decided to define the numerical value of the proportionality constant as g_c, as

$$g_c \equiv 32.174 \frac{\text{lbm} \cdot \text{ft}}{\text{lbf} \cdot \text{s}^2} \tag{3.6}$$

The g_c symbolism was chosen because the numerical value (but *not* the dimensions) of g_c is the same as that of the acceleration in standard gravity in the English Engineering units system. This is awkward because it tends to make you think that g_c is the same as (i.e., equal to) the acceleration due to local gravity, g, *which it definitely is not.* The constant g_c is nothing more than a proportionality constant with dimensions of [mass × length/(force × time^2)]. Because the use of g_c is so widespread today in the United States and, because it is important that you are able to recognize the meaning of g_c when you see it elsewhere, it is used in the relevant equations in this course except when we are using the much more convenient (and universal) SI units. For example, in the English Engineering unit system, we henceforth write Newton's Second Law as

$$F = \frac{ma}{g_c} \tag{3.7}$$

The consequence of this choice for F_1, m_1, and a_1 as expressed by g_c is that you can easily calculate the force in lbf corresponding to an acceleration in ft/s^2 and a mass in lbm.

Example 3.5

What is the force necessary to accelerate a mass of 65.0 lbm at a rate of 15.0 ft/s^2?

Need: Force to accelerate mass of 65.0 lbm at 15.0 ft/s^2.

Know: Newton's Second Law in Engineering English units, in which $g_c = 32.2$ [lbm ft][lbf s^2].

How: Equation (3.7) is the principle used here:

$$F = \frac{ma}{g_c}$$

Solve: Since the problem is stated in Engineering English units, assume the answer is also required in these units.

$$F = \frac{ma}{g_c} = \frac{65.0 \times 15.0}{32.174} = [\text{lbm}] \left[\frac{\text{ft}}{\text{s}^2}\right] \left[\frac{\text{lbf} \times \text{s}^2}{\text{lbm} \times \text{ft}}\right] = 30.3 \text{ lbf}$$

Notice how the **units** as well as the value of g_c enter the problem; without g_c, our "force" would be in the nonsense units of lbm/ft/s^2 and our calculated numerical value in those nonsense units would be 975.

There are another series of consequences, too. In a subsequent chapter, you are introduced to the quantity **kinetic energy**. An engineer using the SI system defines kinetic energy as $\frac{1}{2}mv^2$ (remember, the v stands for speed). However, an engineer using the Engineering English system defines kinetic energy as $\frac{1}{2}(m/g_c)v^2$. The convenient mnemonic for all applications of the Engineering English system is

When you see a mass m in the Engineering English system, divide it by g_c.

Of course, it is logically more elegant to argue the units using the [..] bracket convention previously introduced. For example, if the grouping $\frac{1}{2}mv^2$ were expressed in Engineering English units, it would be

$[\text{lbm}] \times [\text{ft/s}]^2$ if g_c were ignored. This is a meaningless collection of units! But, using the g_c proportionality factor, the definition of *kinetic energy* now becomes

$$\frac{[\text{lbm}] \times [\text{ft/s}]^2}{[\text{lbm} \cdot \text{ft}]/[\text{lbf}/\text{s}^2]} = [\text{ft} \cdot \text{lbf}]$$

which is a legitimate unit of energy in the Engineering English unit system.

Understand that the force conversion constant g_c has the same value everywhere in the universe—that is, the value of 32.174 $\text{lbm} \cdot \text{ft}/(\text{lbf} \cdot \text{s}^2)$—even if its domain of acceptance is confined to the United States! In this regard, as previously stated, it should not be confused with the physical quantity g, the acceleration due to gravity, which has different numerical values at different locations (as well as different dimensions from g_c).

The concept of weight always has the notion of the *local* gravity associated with it. Weight thus varies with location—indeed only slightly over the face of the Earth but significantly on nonterrestrial bodies.

To restate what has been learned, the weight of a body of mass m on the surface of the Earth is the force on it due to the acceleration "g" due to gravity; that is,

$$\text{Weight} \quad W = \frac{mg}{g_c} \tag{3.8a}$$

where in the Engineering English units system, $g = 32.174$ ft/s^2 and $g_c = 32.174$ (lbm ft/lbf s^2), but in the SI units system, $g = 9.81$ m/s^2 and $g_c = 1$. In SI units, this is:

$$\text{Weight} \quad W = mg \tag{3.8b}$$

The weight of a body of mass m where the local gravity is g' is mg'/g_c. A person on the International Space Station experiences microgravity, a much smaller g than we do on Earth; and a person on the Moon experiences about $^1/_6$ of g compared to a person on the surface of the Earth. However, a person who is an engineer using the Engineering English system of units must use the same numerical value for g_c wherever in the universe he or she may be.

Example 3.6

  (a) What is the weight on Earth in Engineering English units[4] of a 10.0 lbm object?
  (b) What is the weight on Earth in SI units of a 10.0 kg mass?
  (c) What is the mass of a 10.0 lbm object on the Moon (assume the local $g = 1/6.00$ that of Earth')?
  (d) What is the weight of that 10.0 lbm object on the Moon?

Solution: This is a straightforward unit conversion problem
  (a) $W = mg/g_c$ in Engineering English units. Therefore,
     $W = 10.0 \times 32.2/32.2$ $[\text{lbm}][\text{ft/s}^2]/[\text{lbm} \cdot \text{ft/lbf} \cdot \text{s}^2] = 10.0$ lbf.
  (b) $W = mg/g_c$ where $g_c = 1$ in SI units. Therefore,
     $W = 10.0 \times 9.81$ $[\text{kg}][\text{m/s}^2] = 98.1$ $[\text{kgm/s}^2] = 98.1$ N.
     (See Table 2.2 for the definition of Newtons in terms of MKS fundamental units.)
  (c) Mass is a property of the material. Therefore, the object still has a mass of 10.0 lbm on Earth, on the Moon, or anywhere in the cosmos.
  (d) $W = mg/g_c$ in Engineering English units. On the Moon, $g = 32.2/6.00 = 5.37$ ft/s^2. Therefore, on the Moon, $W = 10 \times (5.37)/32.2$ $[\text{lbm}][\text{ft/s}^2]/[\text{lbm} \cdot \text{ft/lbf} \cdot \text{s}^2] = 1.67$ lbf.

---

[4]Generally, you should give your answers in the "natural" set of units suggested by the problem statement.

Until the mid-twentieth century, most English-speaking countries used one or more forms of the Engineering English units system. But, because of world trade pressures and the worldwide acceptance of the SI system, many engineering textbooks today present examples and homework problems in both the Engineering English and the SI unit systems. The United States is *slowly* converting to common use of the SI system. However, it appears likely that this conversion will take at least a significant fraction of your lifetime. So, to succeed as an engineer in the United States, you must learn the Engineering English system. Doing so will help you avoid future repetition of such disasters as NASA's embarrassing loss in 1999 of an expensive and scientifically important Mars Lander due to an improper conversion between Engineering English and SI units.[5]

Example 3.7

How much horizontal force is required to accelerate a 1000. kg car at 5.00 m/s²?

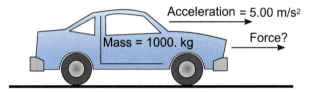

Need: The unbalanced force required to accelerate the car at 5.00 m/s².
Know: The mass of the car is 1,000. kg.
How: Newton's Second Law, in SI units.
Solve: Newton's Second Law relates an object's mass, the unbalanced force on it, and its acceleration: $F = ma/g_c$, so

$$F = ma/g_c = (1,000.\ \text{kg} \times 5.00/1)\ [\text{kg}]\left[\frac{\text{m}}{\text{s}^2}\right] = 5,000.\ \text{kg newtons}$$

3.5 NEWTON'S THIRD LAW

Newton's Third Law can be simply stated as:

For every action, there is an equal and opposite reaction.

For example, Newton's Third Law states that a force on object A that is due to the presence of a second object B is automatically accompanied by an equal and opposite force on object B due to the presence of object A. This means that there are no internal forces to the arrangement of the two objects that are unbalanced. That is, any action-and-reaction pair of forces between any two otherwise isolated objects does not cause the center of mass of these objects to accelerate. These objects can accelerate toward or away from each other, but their center of mass cannot accelerate (see Figure 3.1).

A common demonstration of Newton's Third Law, called *Newton's Cradle*, is shown in Figure 3.2. If one or more spheres is pulled away and then released, it strikes the next sphere in the series and comes to (nearly) a

---

[5]See http://www.cnn.com/TECH/space/9909/30/mars.metric for more information.

FIGURE 3.1 Relative acceleration of two objects.

complete stop. The impact force is transmitted through the group of spheres until an equal number of spheres on the opposite end are lifted by the reaction force. More than half the spheres can be set in motion. For example, three out of five spheres result in the central sphere swinging without any apparent interruption.

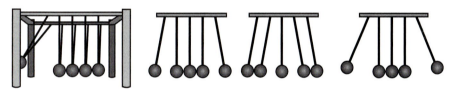

FIGURE 3.2 Newton's Cradle.

Example 3.8

The total weight of a new U.S. space transport (vehicle and launch rockets) at liftoff is 4.40×10^6 lbf and the liftoff thrust is 6.78×10^6 lbf. What is the unbalanced force on the shuttle at the moment of liftoff? What's the initial rocket's acceleration?

Need: The unbalanced force on the space shuttle at the moment of liftoff.

Know-How: A rocket launch is a classic example of Newton's Third Law. A chemical reaction in the rocket engine generates gases that exert a downward force (thrust) at the bottom of the engine. This is the **action**. A force equal and opposite to the thrust exerts an upward force on the rocket pushing it in the upward direction. This is the **reaction**.

Solve: Since for every action, there is an equal and opposite reaction, so the net unbalanced force on the shuttle assembly is

$$F_{\text{thrust}} - F_{\text{weight}} = F_{\text{unbalanced}} = 6.78 \times 10^6 \text{ lbf} - 4.40 \times 10^6 \text{ lbf} = \mathbf{2.38 \times 10^6 \text{ lbf}}$$

Note: The initial acceleration of this vehicle in "gees"[6] is

$$a_{\text{init}} = \frac{F_{\text{unbalanced}} g_c}{mg} [\text{lbf}] \left[\frac{\text{lbm} \cdot \text{ft}}{\text{lbf} \cdot \text{s}^2} \right] \left[\frac{\text{s}^2}{\text{lbm ft}} \right] = \frac{2.38 \times 10^6}{4.40 \times 10^6} \times \frac{32.2}{32.2} = 0.541$$

which is about one-half the standard acceleration due to gravity.

3.6 FREE-BODY DIAGRAMS

A force acts in a particular direction with an amount that depends on how strong the push or pull is. Because of these characteristics, forces have both direction and magnitude. This means that forces follow a different set of mathematical rules than quantities that do not have direction. For example, when determining what happens when two forces act on the same object, it is necessary to know both the magnitude and the direction of both forces to determine the resultant force.

Free-body diagrams can be used as a convenient way to keep track of forces acting on a system. Ideally, these diagrams are drawn with the angles and relative magnitudes of the forces so that graphical addition can be done to determine the resultant force (see Figure 3.3).

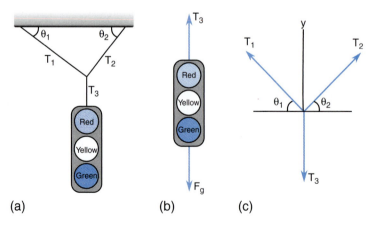

FIGURE 3.3 Free-body diagram of a traffic light suspension system.

[6]That is, in fractions or multiples of 32.2 ft/s².

When two forces act on an object, the *resultant* force can be determined by what is called the *parallelogram rule of addition*. The addition of two forces represented by sides of a parallelogram gives a resultant force that is equal in magnitude and direction to the diagonal of the parallelogram. For example, in Figure 3.3, the resultant force T_3 is equal to the vertical components of T_1 plus T_2, or

$$T_3 = T_1 \sin \theta_1 + T_2 \sin \theta_2$$

As well as being added, forces can also be resolved into independent components at right angles to each other. A force can be split into two perpendicular forces, one pointing in a specified x-direction and one pointing in a specified y-direction. For many problems, these are simply horizontal and vertical directions but not always. Resolving forces into x and y components is often an easier way to describe them than using magnitudes and directions.

Equilibrium occurs when the resultant force acting on an object is zero. There are two kinds of equilibrium: static and dynamic.

3.6.1 Static Equilibrium

The simplest case of static equilibrium occurs when two forces are equal in magnitude but opposite in direction. For example, an object on a level surface is pulled downward by the force of gravity (i.e., its weight). At the same time, a surface under the object resists the downward force with an equal upward force. This condition is one of zero net force and no acceleration.

Pushing horizontally on an object on a horizontal surface can result in a situation where the object does not move because the applied force is opposed by a force of static friction between the object and the surface. In this case, the static friction force balances the applied force resulting in no acceleration.

Static equilibrium between two forces is a common way of measuring the force of gravity (i.e., weight). This can be done using spring scales or balance scales, as shown in Figure 3.4.

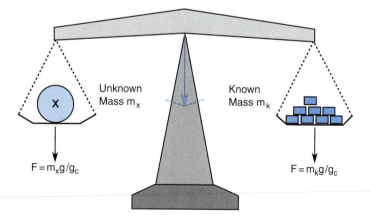

FIGURE 3.4 A balance scale free-body diagram illustrating static equilibrium.

3.6.2 Dynamic Equilibrium

The definition of *dynamic equilibrium* is when an object moves at a constant speed so that all forces on the object are balanced. A simple case of dynamic equilibrium occurs when a horizontal force is applied to an object causing it to move with a constant speed across a horizontal surface. In this case, the applied force is in the direction of motion, while a friction force in the opposite direction balances the applied force, producing no unbalanced force on the object. But, since the object is moving, it continues to move with a constant speed, as in Figure 3.5.

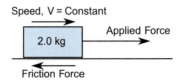

FIGURE 3.5 Dynamic equilibrium of a sliding object occurs when the applied force = friction force.

3.7 WHAT IS KINEMATICS?

Kinematics[7] is the study of how things move. Engineers need to understand the relationships among the important variables of distance, speed, acceleration, and time. These relationships are collectively called kinematics when they do not also involve the forces on objects or the inertia of the objects. Kinematics simply serves as a way to relate the effects of forces on a body to its subsequent motion without asking the questions of how we can achieve particular values of these variables.

Engineers are concerned with the kinematics variables because they are the basic variables used to understand the motion of objects, such as the motion of cars in traffic. In the simplest cases, they are related to each other by geometric methods—in more complicated situations by the methods of calculus.

3.7.1 Distance, Speed, and Acceleration

In this book, only motion in a single direction (often called *one-dimensional*) is considered. For convenience, think of positive **distance** as from left to right, and negative distance as from right to left, just as in Cartesian geometry. **Speed** is a variable commonly measured in miles per hour (mph) in the United States but in kilometers per hour (kph) in most of the rest of the world. But, for engineering design, the SI units of m/s are often more convenient. Speed is calculated by measuring the distance traveled and the time spent and dividing distance by time. In mathematical terms, speed $v=\Delta x/\Delta t$, with the uppercase Greek delta Δ symbol meaning "difference." In Cartesian geometry, this is the average slope of the line $x(t)$. In this case, speed means the final position less the initial position divided by the final time less the initial time (see Figure 3.6).

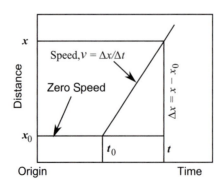

FIGURE 3.6 Constant speed is when $v=\Delta x/\Delta t=$ constant.

---

[7]Yes! "Kinematics" takes a singular verb.

Again, only one direction is considered. We use the symbol v (as in velocity) for speed. (If we used the symbol s, it might be confused for the symbol s for "seconds.") But there is an important differentiation between the concepts of *speed* and *velocity*. Speed is the *magnitude* of velocity. Velocity may have more than one component, which is so much in the y direction (as in Cartesian geometry), so much in the x direction, and so on. Of course, in our one-dimensional analyzes, speed and velocity are functionally synonymous. For acceleration, we use a, except for that due to gravity, when we use g (see Figure 3.6).

Adding a word to *speed* gets us to the phrase *speed up*. We have already met the speed up variable, **acceleration**. It is defined as change of speed per unit time or, more mathematically precise, as the rate of change of speed and is commonly measured in m/s^2 or ft/s^2. Please resist the temptation to express acceleration in such ugly and ultimately less useful hybrid forms as "miles per hour per second."

Finally, there is a fourth variable. It is mentioned only implicitly in the preceding description, but it is so important that, without formally introducing it, we have already used it. This is the variable **time**, measured in seconds or hours, as in miles per hour.

Strictly speaking, two of the preceding variables should be defined in two ways: in terms of *instantaneous* speed and acceleration and in terms of *average* speed and acceleration. These are important distinctions, as shown in Figure 3.7. An instantaneous variable, here the acceleration, is changing with the change in slope. Your physics courses enable you to understand how to deal with nonconstant acceleration, but in this introductory treatment, we can and will get along without nonconstant accelerations.

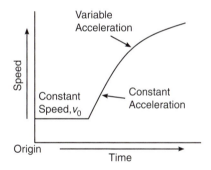

FIGURE 3.7 Acceleration is the local slope of the speed-time curve.

The only one of the four concepts of distance, speed, acceleration, and time that gives the typical student any difficulty is acceleration. So, before proceeding further, please read and understand the following example.

Example 3.9

A car enters an on-ramp traveling 15 miles per hour. It accelerates for 15 s. At the end of that time interval, it is traveling at 60.0 miles per hour. What is its average acceleration? How does that compare to g, the acceleration due to gravity?

Need: Acceleration = _____ m/s^2.

Know-How: Acceleration = (change in speed)/time.

Initial speed = 15 miles per hour = 22 ft/s = 6.7 m/s

Final speed = 60.0 miles per hour = 88.0 ft/s = 26.8 m/s

Solve: In Engineering English units, average acceleration = (88.0 - 22)/15 = 4.4 ft/s^2.

In MKS units, average acceleration = (26.8 − 6.7)/15 = 1.34 = 1.3 m/s^2 (in terms of a fraction of g, the gravity, the average acceleration here is 1.3/9.8 = **0.13 g**, so this is a mild acceleration).

3.7.2 The Speed versus Time Diagram

One could deal with problems of distance, speed, acceleration, and time using words and equations, but a much more intuitive tool makes possible an insightful description of motion problems. Frequently, it enables an engineer to solve the problem merely by inspecting a diagram. Formal calculation may be needed only to achieve the correct number of significant figures.

This versatile and essential tool is the speed versus time diagram (Figure 3.7). For brevity, we call our diagram the *v-t* diagram. The *v-t* diagram is just what it says. The horizontal axis represents time, with zero being the instant the situation being considered began. For each instant of time, the speed (or one-dimensional velocity) is plotted in the vertical direction.

Example 3.10

Plot the *v-t* diagram in SI units from Example 3.9.

Need: The *v-t* diagram describing the situation "a car enters an on-ramp traveling 15.0 miles per hour. It accelerates constantly for 15.0 s. At the end of that time interval, it is traveling at 60.0 miles per hour."

Know-How: First, convert all speeds to SI. Then, plot speed for $t = 0$ (when $v = 15.0$ mph $= 6.70$ m/s) and $t = 15.0$ s (when $v = 60.0$ mph $= 26.8$ m/s). Join the two points with a *straight line*. (What tells us the line is straight is the phrase *constant acceleration*.)

Solve:

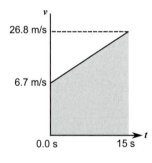

Why is this particular graph so important? There are two reasons.

1. The slope of the *v-t* graph measures acceleration.
2. The shaded area under the *v-t* graph measures distance traveled.

No other graph involving the four variables of interest summarizes so much information in so intuitive a manner. (If the concepts "slope" and "area" aren't intuitive to you, make them so. Review your high school Cartesian geometry, if necessary!)

You should recall *slope* being defined as the quotient of a vertical "rise" divided by a horizontal "run." For the standard Cartesian *x, y* diagram, the slope is $\Delta y/\Delta x$. That the slope of the *v-t* graph gives you acceleration is evident from considering dimensions. The "rise" of the graph has units of speed, m/s. The "run" of the graph has units of time, s. So slope $=$ rise/run $= \Delta v/\Delta t$ in units of [m/s]/[s] or m/s^2, valid dimensions of acceleration.

The second statement relating area under the curve to distance traveled is less obvious. In Example 3.10, the *average* speed over the 15.0 s period of acceleration is $(26.8 + 6.70)/2 = 16.8$ m/s, and so the vehicle will have covered $16.8 \times 15.0 = 250.$ m. Notice the shaded area is a trapezoid whose area is $0.5 \times$ (sum of parallel sides) × (perpendicular distance between them). It is identical to the calculation of distance covered. It can also be approached dimensionally. The "height" or ordinate has units of [m/s]. The horizontal axis has units of seconds. So length × height has units of [s × (m/s)] = [m], a unit of distance.

Those of you familiar with elementary calculus[8] will recognize that saying "the area under the v-t graph measures distance" is the same as saying "the integral of speed over time is distance." Saying that "the slope of the v-t graph measures acceleration" is the same as saying "the derivative of speed is acceleration." These statements are true even when the lines are curved, that is, for the case of nonconstant acceleration. In the general case, calculus is required to evaluate the slopes or areas under curves.

We, however, get along without calculus. Solving Examples 3.9 and 3.10 using the v-t diagram has done more than enable us to visualize that problem. It has provided a bonus. It has already answered a second question.

Example 3.11

What is the distance traveled in steadily accelerating from 15.0 mph to 60.0 mph in 15.0 s?

Need: Distance = _____ m.

Know-How: The distance is the area (shaded in the diagram for Example 3.10) beneath the v-t graph. Geometry tells us we can break the area of a trapezoid into a rectangle with length 15.0 s and height 6.70 m/s and a triangle with base 15.0 s and altitude (26.8 − 6.70) m/s.

Solve: Distance = Area = Area of rectangle + Area of triangle

$$= 6.70 \text{ [m/s]} \times 15.0 \text{ [s]} + (\tfrac{1}{2}) \times 15.0 \text{ [s]} \times (26.8 - 6.70) \text{ [m/s]}$$
$$= 100.5 \text{ m} + 150.75 \text{ m} = 251.2 \text{ m}$$

Therefore, **distance = 251 m** (there are only three significant figures here).

You may already know some formulae you could have used to solve this problem (e.g., $x = v_0 t + (\tfrac{1}{2})at^2$), but it is recommended that you use the v-t diagram on the problems and exercises of this chapter. Use of this tool develops a visual and intuitive appreciation of motion problems that cannot be obtained merely by manipulating equations. Formulae are useful once you fully understand the process, but the v-t diagram is the best way to achieve that understanding.

The answers to a number of other engineering problems are also contained in the v-t diagram. These range from the design of highway intersections to the design of a cannon that might be used to shoot payloads to the moon. Meanwhile, Example 3.12 shows the power of this method even for a relatively complex application.

Example 3.12

Provide a table and a graph of the length of a highway on-ramp as a function of acceleration ranging from 1.0 m/s² (about 0.1 g) to 10. m/s² (about 1 g) in increments of 1.0 m/s². Assume that the vehicle enters the on-ramp with a speed of 15. mph (6.7 m/s) and leaves the on-ramp at a speed of 60 mph (26.8 m/s).

Need: A table and a graph with entries for the length of on-ramp d in m vs. a m/s².

Know-How: The v-t diagram gives the relationships among a, d, and t. Let τ be the (unknown) time you spend on the on-ramp accelerating to speed.

---

[8]It's really easy with a little calculus; under a v-t curve an area $= \int_1^2 v \, dt$. Also, speed is related to distance by $v = dx/dt$ so that $x = \int_1^2 v \, dt$ is the distance traveled and is thus equal to the area under the speed curve.

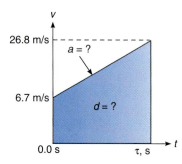

Solve: From the definition of acceleration as the slope of the v-t line

$$a = \frac{(26.8 - 6.7)}{\tau} \quad \text{or}$$

$$\tau = \frac{(26.8 - 6.7)}{a} [\text{m/s}][\text{s}^2/\text{m}] = \frac{20.1}{a} [\text{s}]$$

Also *d* is the area under the v-t curve, so

$$d = 6.7\tau + 1/2 \times (26.8 - 6.7)\tau [\text{m/s}][\text{s}] = 6.7\tau + 10.05\tau [\text{m}] = 16.75\tau.$$

Now, substitute for $\tau = 20.1/a$. Therefore, $d = 337/a$ in meters.

Take these relationships to a spreadsheet and prepare the table, and from the table, prepare graphs of *a* vs. τ and *d* vs. τ as shown below.

a, m/s²	d, m	τ, s
1.00	337	20.1
2.00	169	10.1
3.00	112	6.70
4.00	84	6.03
5.00	67	4.02
6.00	56	3.35
7.00	48	2.87
8.00	42	2.51
9.00	37	2.23
10.0	34	2.01

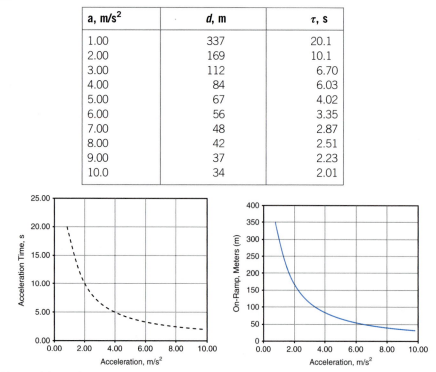

Notice this graph immediately tells us that the solution to the time and distance traveled while accelerating is quadratic, a fact that emerged automatically.

3.8 THE EQUATIONS OF ONE-DIMENSIONAL KINEMATICS

These equations are included only for completeness because we have circumscribed one-dimensional kinematics problems without needing these results. Of course, the student may use them if he/she wishes not to use the graphical method.

The key variables are

t, the time, typically measured in seconds (s)
x, the distance traveled, typically measured in meters (m)
v, the speed in one dimension, typically measured in m/s
a, the acceleration, typically measured in m/s^2

If the speed is constant at $v=v_0$ and the acceleration is zero ($a=0$), then the distance-time kinematic equation is

$$x = x_0 + v_0(t - t_0) \qquad (3.9)$$

If the acceleration is constant, then the distance-time kinematic equation is

$$x = x_0 + v_0(t - t_0) + 0.5a(t - t_0)^2 \qquad (3.10)$$

The speed-distance kinematic equation is

$$v^2 = v_0^2 + 2a(x - x_0) \qquad (3.11)$$

Notice some familiar forms from high school physics: if $x_0=0$ at $t_0=0$ in Equation (3.9), then $x=v_0 t$. Also, if $t_0=0$ and $x_0=v_0=0$, then Equation (3.10) gives $x=0.5at^2$. And, finally, if $x_0=0$ and $v_0=0$, then Equation (3.11) gives $v^2=2ax$.

More general relationships for nonconstant a can be derived using calculus, but we do not derive them here—all cases, constant acceleration or not, will be covered in your first-year physics classes.

SUMMARY

This chapter introduced the basic concepts of forces and kinematics. These concepts are necessary for many engineers as part of their fundamental knowledge.

The study of forces and their effects on inertial bodies was first formulated by Newton when he promulgated his three laws of mechanics. Newton's First Law deals with inertial bodies, which remain in their initial state unless forces act on them. The Second Law relates the acceleration of an inertial body to its mass and its acceleration, and the Third Law is the action/reaction law for interacting bodies.

The principal equation used by engineers is the **Newton's Second Law**. It is written as $F=ma$ in SI units of Newtons, kg, and s, and as $F=ma/g_c$ in English Engineering units of lbf, lbm, and s. In the latter system, $g_c=32.174$ lbm $\times$ ft(lbf $\times$ s^2).

Kinematics is the study of motion without regard to the forces that produce the motion. The relationships among **acceleration**, **speed**, and **time** are basically geometric (with some calculus needed in complicated cases, which we ignored). We presented kinematic relationships among distance, speed, and acceleration through a versatile tool, the **speed versus time (*v-t*) diagram**. These relationships are used in a wide variety of applications, and the subject of kinematics is integral to a number of engineering fields, such as mechanical, civil, control, and biomechanical engineering.

Free-body diagrams are used to analyze complex interactions on bodies subject to vectorial (i.e., multidirectional) forces.

Together, these concepts lie at the center of much of modern engineering. They allow engineers to design much of our modern world.

EXERCISES

1. Suppose the mass in Example 3.2 is 50.0 slugs. What would its weight be in lbf (pounds force)?

2. What would the 5.00 slug mass in Example 3.2 weigh on the Moon, where the acceleration of gravity is only one-sixth of that on Earth.

3. What would the force on the body in Example 3.3 be if its mass were 856 g?

4. What would the weight of the body in Example 3.4 be on the Moon, where the acceleration of gravity is just $g_{moon} = 1.64$ m/s^2?

5. What force would be necessary in Example 3.5 if the mass is 735 lbm?

6. What is the value and units of g_c in the Engineering English system on the Moon?

7. Acceleration is sometimes measured in g, where $1.0\ g = 9.8$ m/s^2. How many g's correspond to the steady acceleration of a car doing "zero to sixty"[9] in 10.0 s? (**A: 0.27 g.**)

8. What is your mass in kilograms divided by your weight in pounds? Do you have to step onto a scale to answer this question? How did you answer the question?

9. If power (measured in W, or watts) is defined as work (measured in J, or joules) performed per unit time (measured in s), work is defined as force (measured in N or Newtons) × distance (measured in m), and speed is defined as distance per unit time (measured in m/s), what is the power being exerted by a force of 1,000. N on a car traveling at 30. m/s. (Assume force and speed are in the same direction and treat all numbers as positive.) (**A: 3.0×10^4 W.**)

10. A rocket sled exerts 3.00×10^4 N of thrust and has a mass of 2.00×10^3 kg. What does it do "zero to sixty" in? How many g's (see Exercise 7) does it achieve?

11. A person pushes a crate on a frictionless surface with a force of 100. lbf. The crate accelerates at a rate of 3.0 feet per second^2. What is the mass of the crate in lbm? (**A: 1.07×10^3 lbm.**)

12. The force of gravity on the moon is one-sixth (i.e., 1/6.0) as strong as the force of gravity on Earth. An apple weighs 1.0 N on Earth. (a) What is the mass of the apple on the Moon in lbm? (b) What is the weight of the apple on the Moon, in lbf? (Conversion factor: 1.00 kg $= 2.20$ lbm.)

13. How many lbf does it take for a 4.0×10^3 lbm car to achieve zero to 60. mph in 10. s? (**A: 1.1×10^3 lbf.**)

14. Suppose a planet exerted a gravitational force at its surface that was 0.6 the gravitational force exerted by Earth. What is g_c on that planet?

It is strongly suggested you to use the *v-t* **diagram** for most of the following exercises.

15. A car is alone at a red light. When the light turns green, it starts at constant acceleration. After traveling 100. m, it is traveling at 15 m/s. What is its acceleration?

---

[9]In the language of the car enthusiast, a standard test to accelerate a car from a standing start to 60 mph is called "zero to sixty."

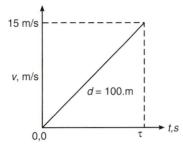

16. A car is alone at a red light. When the light turns green, it starts at constant acceleration. After traveling exactly one-eighth of a mile, it is traveling at 30. mph. What is its acceleration in m/s²?

17. An electric cart can accelerate from 0 to 60. mph in 15 s. The Olympic champion sprinter Usain Bolt can run 100.0 m in 9.69 s.[10] Which would win a 100.0 m race between this cart and the world champion sprinter? **(A: The sprinter wins by 16.7 m!)**

18. A car starts from a stop at a traffic light and accelerates at a rate of 4.0 m/s². Immediately on reaching a speed of 32 m/s, the driver sees that the next light ahead is red and instantly applies the brakes (reaction time = 0.00 s). The car decelerates at a constant rate and comes safely to a stop at the next light. The whole episode takes 15.0 s. How far does the car travel? (**Partial A:** See the following figure.)

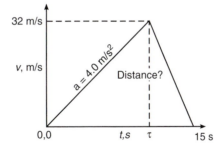

19. For the previous problem, supply a table and draw a graph showing the distance, speed, and acceleration of the car versus time.

20. Based on your experience as a driver or a passenger in a car, estimate the maximum deceleration achieved by putting maximum pressure on the brakes when traveling at 30 mph.

21. A car leaves a parking space from a standing stop to travel to a fast-food restaurant 950 m away. Along the journey, it has to stop after 325 m at a stop sign. It has a maximum acceleration of 3.0 m/s² and a maximum deceleration of −10. m/s². It never exceeds the legal speed limit of 15.0 m/s. What is the least possible time it can take until the car comes to a full stop in front of the fast-food restaurant? (**A: 70. s.**)

22. You are an engineer designing a traffic light. Assume a person can see a traffic light changing color from red to yellow, and it takes one second to respond to a change in color. Suppose the speed limit is 15.0 m/s. Your

[10]He has since improved this to a stunning 9.59 s. Go Usain!

goal is to enable drivers always to stop after seeing and responding to the yellow light with a maximum deceleration of -5.0 m/s^2. How long should the yellow light last?

23. You are a driver responding to the traffic light in the previous exercise. If it was correctly designed according to that problem, at what distance from the light should you be prepared to make your "to stop or not to stop" decision? Assume you are a safe driver who neither speeds up to get through the yellow light nor stops more suddenly than the deceleration rate of -5.0 m/s^2.

24. Suppose the deceleration of a car on a level off-ramp is -3.0 m/s^2. How long would the off-ramp have to be to allow a car to decelerate from 60. to 15. mph? (**A: 110 m.**)

25. An early proposal for space travel involved putting astronauts into a large artillery shell and shooting the shell from a large cannon.[11] Assume that the length of the cannon is 30. m and the speed needed by the shell to achieve orbit is 15,000 m/s. If the acceleration of the shell is constant and takes place only within the cannon, what is the acceleration of the shell in g's?

26. Suppose that a human body can withstand an acceleration of 5.0 g's, where 1 g is 9.8 m/s^2. How long would the cannon have to be in the previous exercise to keep the acceleration of the humans within safe limits? (**A: 4.6×10^6 m.**)

27. You wish to cover a 2.00 mile trip at an average of 30.0 mph. Unfortunately, because of traffic, you cover the first mile at just 15.0 mph. How fast must you cover the second mile to achieve your initial schedule?

28. You are an engineer with the responsibility for choosing the route for a new highway. You have narrowed the choice to two sites that meet all safety standards and economic criteria. (Call them Route A and Route B.) Route B is arguably slightly superior in terms of both safety and economics. However, Route B would pass by the site of an expensive new house recently built by your favorite niece, and the proximity to the highway would severely depress the value of the house. Since your niece's last name is different from yours and you have never mentioned her at the office, you would be unlikely to be "caught" if you chose Route A to save the value of her house. What do you do? (Use the Engineering Ethics Matrix.)
    (**a**) Choose Route A.
    (**b**) Choose Route B.
    (**c**) Ask to be relieved from the responsibility because of conflict of interest.

29. You are an engineer on a team designing a bridge for a state government. Your team submits what you believe to be the best design by all criteria, at a cost that is within the limits originally set. However, some months later, the state undergoes a budget crisis. Your supervisor, also a qualified engineer, makes design changes to achieve cost reductions that he believes will not compromise the safety of the bridge. You are not so sure, though you cannot conclusively demonstrate a safety hazard. You request that a new safety analysis be done. Your supervisor denies your request on the grounds of time and limited budget. What do you do? (Use the Engineering Ethics Matrix.)
    (**a**) Go along with the decision. You have expressed your concerns, and they have been considered.
    (**b**) Appeal the decision to a higher management level.
    (**c**) Quit your job.
    (**d**) Write your state representative.
    (**e**) Call a newspaper reporter and express your safety concerns.

---

[11] A number of military satellites have been delivered to space using specially adapted naval cannons from the Vandenberg Air Force Base in California.

Energy

4.1 INTRODUCTION

Is the world running out of energy? Where does the energy come from that provides the light by which you are reading these words? Does a car possess more energy when it is sitting in your driveway or 15 min later when it is traveling down the highway at 60 miles/h? In this chapter, you will learn how to answer these questions, as well as how to address the more quantitative ones involving energy that engineers of all types encounter in their work.

Energy is broadly defined as the capability to do *work*. An important point is that energy comes in several discrete kinds. These kinds are capable of conversion from one kind to another, but not necessarily at 100% efficiency. In the course of these conversions, the total *energy* is neither created nor destroyed, but rather *conserved*. Put succinctly, the total amount of energy in the universe remains constant. This is one of the most important principles you will have to master as an engineer. You will have to construct a useful system boundary known as a "control boundary" to monitor the flows of energy.

Using energy as a variable, engineers create models that help with a wide range of applications, such as deciding what fuel to use in an automobile, designing methods of protecting buildings from earthquakes or lightning strikes, or advising citizens as to the technical capability of a society to replace fossil fuels with renewable resources such as solar power, wind power, and biomass, or with controversial, but abundant energy sources such as uranium. But, in order to achieve these wide-ranging applications, engineers must be very specific how they understand and use the concept of energy.

Mechanical work is narrowly defined based on Newton's second law of motion because that introduced the notion of a force. In the simplest mathematical terms, work (symbol W) is defined as:

$$Work\,(W) = F \times d \tag{4.1}$$

in which a force F moves through a distance d, the distance and the force being parallel. To do 500 J of work, you could apply a force of 500 N of force through 1 m or 100 N through 5 m or infinitely in other ways. The work concept is broader than it appears from this straightforward definition and can be applied to a number of situations apparently far removed from this simple statement, while still inherently dependent on it.

Energy, E, has the same units as work, W. In an imaginary system, one might convert energy into work and then recover some amount of the original energy. But, overall, to do 500 J of work you must expend 500 J of energy.

Different kinds of energy are capable of conversion from one kind to another and, as such, are a core component of the responsibility of several kinds of engineers. Again, in the course of these conversions, energy is neither created nor destroyed, but conserved so that the total amount of energy in the universe remains constant.[1]

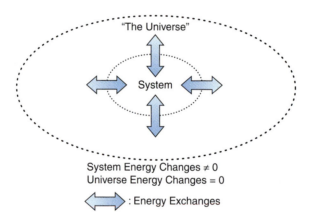

FIGURE 4.1 The Conservation of Energy

Figure 4.1 represents one of the most important principles you will have to master as an engineer. The engineer uses something that we can generically call a **system**. It might be an engine, a sailboat, a lawn mower, an electric kettle, a crane, or anything. It is that which is contained within the inner dotted boundary. It connects to the rest of the "universe" through whatever elements are interposed between the system and the universe. You will have to use engineering skills and know-how to construct a useful system boundary known as a *control boundary* such that you can monitor the flows of energy (perhaps of several kinds) across it. When you do this, the **law of conservation of energy** will enable you to determine the flows of energy necessary to maintain the balance.

When engineers use the term **energy conservation** they are neither offering exhortations to turn off the lights, nor are they opposing the use of gas-guzzling sport utility vehicles. Rather, they are expressing a scientific

---

[1]There is one caveat that cosmological physicists might add: radiant energy (which is traveling in the universe at the speed of light) may leak to beyond the observable edge of the expanding universe—something that is unlikely to disturb any engineers in the foreseeable future! For all practical purposes, the energy in the universe is indeed constant.

principle that is fundamental to engineering and that will remain in effect regardless of the energy policies that people and nations choose to follow. If you remember only three words from this chapter remember this: "**energy is conserved**." This is a very powerful tool in the hands of a capable engineer, since it allows the engineer to concentrate on the localized "system" of interest.

4.2 ENERGY IS THE ABILITY TO DO WORK

Understanding the variable "energy" begins by defining variables for the concepts of force and work. Force is a variable that may be thought of as a push or a pull, as measured with a spring balance or by the weight of an object in a gravitational field. As we have already seen, force has the units of *newtons* in the SI system and *pounds force* (lbf) in the Engineering English system. For example, at sea level, to lift a book of mass 1.0 kg requires a force of 9.8 N. In the Engineering English system, the force to lift this 1.0 kg book is 2.2 lbf (since 1.0 kg = 2.2 lbm, and the 2.2 lbm "weighs" 2.2 lbf at standard gravity).

The units of work are *joules* in the SI system, and *foot pounds force* (ft lbf) in the Engineering English system. A joule is exactly 1 N m or equivalently 1 kg m^2/s^2. For example, to raise a book with a mass of 1.00 kg (and thus weighing 9.8 N) from the surface of the Earth to a level 1.0 m above the surface of the Earth requires an amount of work that equals 9.8 [N] × 1.00 [m] = 9.8 [N m] = 9.8 J. In the Engineering English system, we have to account for the pesky g_c to calculate the amount of work as $(mg \times d)/g_c = 2.2 \times 32.2 \times 3.28/32.2$ [lbm] [ft/s^2] [ft] [lbf s^2/lbm ft] = 7.2 ft lbf (since 1.0 kg is 2.2 lbm and 1.00 m is 3.28 ft).

Energy may be stored in many objects, such as liquid gasoline, solid uranium, or a speeding train. Equally, the stored energy in a body can be released, such as by water running through a water turbine, heat from burning gasoline, or electrical current from a generator. More precisely, any object that does an amount of work consisting of some number of joules will see its energy content decrease by that same number of joules. Any object that has an amount of work done upon it by its surroundings will see its energy similarly increase.

Often, an engineer wants to know how fast that work was done or how rapidly the amount of energy possessed by an object changed. This is determined using the variable *power*, which is defined as the time rate of doing work or, equivalently, as the time rate of change of energy. It is measured in *watts* in SI, where 1 W is equal to 1 J/s. For example, if a person takes 2.0 s to lift a 1.0 kg book a height of 1.0 m above the surface of the Earth, that person is expending (9.8 J)/(2.0 s) = 4.9 W of power during those 2 s.

In the Engineering English system, power is normally expressed in units of ft lbf/s or in horsepower,[2] where 1 horsepower is 550 ft lbf/s. It is indeed roughly the power exerted by a working horse and roughly 5-10 times the power exerted by a person doing sustained physical labor such as shoveling or carrying.

A useful mnemonic connecting the SI and Engineering English systems is that 1 kW and 1 hp are the same order of magnitude (1 kW = 1.34 hp). Some other examples of quantities of energy are shown in Table 4.1.

In the past three centuries, a vast increase has occurred in the amount of power at the command of the ordinary person in the United States and other developed nations. In 1776, when Thomas Jefferson was writing the Declaration of Independence and James Watt was perfecting his steam engine, the average adult in the United States or in Britain could call on considerably less than 1 kW of power for some fraction of a day. Today, the average person (adults and children included) in the United States and other industrialized nations typically has at his or her command more than 10 kW (13.4 hp) every day of the year, every second, day, and night. This includes several kilowatts of electric power per person to say nothing of over 100 kW of automotive power for selected periods of the day. Other kilowatts per person serve us each day—for example, in the form of

---

[2]The Scottish engineer James Watt measured the rate of work that a good brewery horse could sustain as a way to sell his early steam engines.

Table 4.1 Typical Energy Magnitudes

Joules	Approximate Equivalent Energy Source
1	Kinetic energy of a small apple falling 1 m
10	Kinetic energy of a person swinging a baseball bat
10^2	Lighted match
10^3	Energy in 1 hamburger
10^4	Speeding bullet
10^5	Kinetic energy of a small car at 65 mph
10^6	A small meal
10^7	Lighting strike
10^8	One gallon of gasoline
10^9	Lighting strike
10^{10}	1 ton of oil
10^{11}	1 g of uranium in fission
10^{12}	1000 tons of TNT
10^{13}	Hiroshima size atomic bomb
10^{14}	Annihilation of 1 g of matter
10^{44}	Supernova explosion

the trucks that haul goods for us, the oil or gas that heats us in winter, the airplanes that transport us and our overnight deliveries, and the machinery that manufactures all these energy-using artifacts. So effectively each citizen of an industrialized nation continually has on call the power of hundreds of horses at his or her service for a full 24/7. At the same time, vast numbers of people elsewhere in the world are still in that eighteenth century less than one-horsepower state or, worse, a one-person-power state. Resolving this global power discrepancy in an environmentally acceptable manner is another one of the great engineering and social challenges of the twenty-first century.

Let us now turn from these general concepts of work, energy, and power to the specific kind of energy that scientists and engineers have defined and to the phenomena in nature and technology that these kinds of energy help engineers to quantify and to use in engineering models.

4.3 KINDS OF ENERGY

Energy comes in various forms. Energy due to motion is called kinetic energy (KE) and includes the first two major types of energy we will discuss, **translational kinetic energy**,[3] *TKE*, and, less obviously, *thermal energy*.

---

[3]In some cases, matter is not moving in a straight line but rather rotating around a central axis. In this case, the kinetic energy is called *rotational* kinetic energy, or RKE. It is not different in principle from TKE. It does, however, require more complicated mathematics, since many pieces of matter are moving in different directions and at different speeds. Due to these mathematical complications, RKE will not be the required knowledge in this book.

Energy due to position is called potential energy and includes **gravitational potential energy** (GPE) and, less obviously, **chemical energy**[4] and **electromagnetic energy**.[5] There are other important energy types such as Einstein's discovery that mass is a form of energy, as embodied in his famous equation $E = mc^2$ (see Chapter 15, "nuclear energy").

Translational kinetic energy (TKE) is the energy of mass in straight line motion. It is calculated by the formulae:

$$\text{TKE} = (\tfrac{1}{2})mv^2 \text{ (in SI units) or TKE} = (\tfrac{1}{2})mv^2/g_c \text{ (in English units)} \tag{4.2}$$

where m is mass and v is the speed. TKE is often assumed to be the only form of KE, and then it is simply called "kinetic energy" and then abbreviated "KE."

Note first that Equation (4.2) results in the correct SI units of joules, since speed is m/s, and mass is in kilograms, so $\tfrac{1}{2}mv^2$ has units [kg][m²/s²], which are the same units that make a joule. In Engineering English units, we have to divide by g_c to get the proper units of [ft lbf] from [lbm ft²/s²] [lbf s²/lbm ft].

Example 4.1

What is the TKE of an automobile with a mass of 1.00×10^3 kg traveling at a speed of 65. miles/h (29. m/s)?

Need: TKE of vehicle.
Know: Mass is 1.00×10^3 kg, speed is 29. m/s.
How: Apply Equation (4.2), $\text{TKE} = \tfrac{1}{2}mv^2$.
Solve: $\text{TKE} = \tfrac{1}{2} \times (1.00 \times 10^3) \text{ [kg]} \times 29.0^2 \text{ [m/s]}^2 = 423{,}410 \text{ kg m}^2/\text{s}^2 = 4.2341 \times 10^5 \text{ J}.$

Only two of these digits are significant (since the speed was stated to only two digits), so the answer is $4.2 \times 10^5 \text{ J} = 4.2 \times 10^2 \text{ kJ}$.

Anything that has mass and is moving in a straight line has TKE. Prominent examples of TKE in nature include the winds and the tides.

Example 4.2

Estimate the total KE of the wind on Earth. As an introduction to this problem, consider that the wind is a movement of air that is produced by temperature differences in the atmosphere. Since hot air is less dense than cold air, air heated by the sun at the equator rises until it reaches an altitude of about 6.0 miles (9.0 km) and then it spreads north and south. If the Earth did not rotate, this air would simply travel to the North and South Poles, cool down, and return to the equator along the surface of the Earth as wind. However, because the Earth rotates, the prevailing winds most of us see travel in a west-east rather than a south-north direction (in the northern hemisphere). In places where winds are strong and steady, it may make economic sense to install a windmill or wind turbine to capture that TKE.

---

[4]Unburned fuel such as a can of gasoline clearly has potential energy locked inside of it. Obviously this stretches the simple definition that potential energy is mass × height × g (see the section on GPE). Some part of this picture can be retained if you think of the electrons that surround the atoms of the fuel being in high energy states and endeavoring to reach lower or more stable states when being burnt in air.
[5]Electromagnetic energy depends on the potential energy of electrons to do work; hence voltage is also often called "potential."

Need: Total TKE of the wind in joules.

Know: The atmosphere is about 9.0 km thick (i.e., the height of Mt. Everest). The radius of the Earth is about 6.4 million meters. The surface area of a sphere is $4\pi R^2$, so the volume of an annular "shell" of thickness T around the Earth[6] is about $4\pi R^2 T$, since the thickness T is very small compared to the radius R. Air has a density of about 0.75 kg/m^3 (averaging from sea level to the top of Everest). This air is typically moving at about 10. m/s.

How: Find the mass of air in that 9.0 km thick shell around the Earth, and apply Equation (4.2), TKE$=\frac{1}{2}mv^2$.

Solve: Volume or air around the Earth is about $4\pi \times (6.4 \times 10^6)^2 \times 9000\,[\text{m}^2][\text{m}]=4.6 \times 10^{18}$ m^3. The mass of air around the Earth is, therefore, $0.75 \times 4.6 \times 10\,[\text{kg/m}^3][\text{m}^3]=3.5 \times 10^{18}$ kg.[7]

Therefore, TKE $\approx \frac{1}{2} \times (3.5 \times 10^{18}) \times 10^2\,[\text{kg}]\,[\text{m/s}]^2 = 1.7 \times 10^{20}$ J.

This is a number sufficiently large as to be meaningless to most people. But to an engineer it should inspire such questions as: Where does this energy come from? Where does it go? In fact, the total sun's energy reaching the Earth is about 1.4 kW for every m^2 of the Earth's surface. Of this, about 31% is reflected back into space. We thus receive about 1 kW/m^2 net solar radiation. Over the Earth this amounts to about 1.7×10^{17} kW or 1.7×10^{20} W.[8]

Since we receive this on the global average for 12 h every day, our beneficent sun delivers 7.3×10^{24} J/day; apparently, our estimate of the winds accounts for only $[1.7 \times 10^{20}/7.3 \times 10^{24}] \times 100 \approx 0.002\%$ of the daily received solar energy.

In addition, the sun is the ultimate source of the energy in the fossil fuels that power virtually all of our automobiles and about two-thirds of our electric power stations, as well as the source of the biomass that, in the form of food, powers us. Moreover, it drives the cycles of flowing or falling water that power hydroelectric generating plants.

The role that the winds or other solar energy sources might play in meeting the needs of those billions of people who still live at an eighteenth-century energy standard is still debated. The major problems are that this source of energy is diffuse, often unpredictable, and fluctuates daily and seasonally.

Thermal energy is often referred to as *heat*, and is a very special form of KE because it is the *random* motion of trillions and trillions of atoms and molecules that leads to the perception of temperature. Heat is simply the motion of things too small to see, an insight captured by the nineteenth-century German physicist Rudolf Clausius when he defined thermal energy as "the kind of motion we call heat."

There is a macroscopic analog to thermal energy that may be familiar to many of you. The "mosh pit" at a rock concert is an example of the jostling motion of many bodies expending KE, but not actually moving anywhere. It is the molecular amplification of this picture with trillions and trillions of atoms or molecules bouncing together that produces the net effect of what we call thermal energy.

Our analysis of the meaning of temperature will begin with the motion of a single particle of gas (either an atom or a molecule) in a box. We will point out the theoretical underpinnings of the famous "ideal-gas law" that you probably have seen before. The ideal-gas law for a fixed mass or equivalently, a fixed number of moles of gas is:

$$pV = NRT \tag{4.3}$$

---

[6]Since the volume of a sphere is $(4/3)\pi R^3$, then the volume of the atmosphere of thickness T is $(4/3)\pi(R+T)^3—(4/3)\pi R^3$, which is approximately $(4/3)\pi R^2 T$ if R is much greater than T. You can show this yourself be expanding the equation binomially and neglecting all terms containing T^2 and T^3.

[7]By measurement (see: Lide DR. Handbook of chemistry and physics. Boca Raton, FL: CRC; 1996:14-7), the atmospheric mass is about 5.1×10^{18} kg so we are low by about 50% with our crude estimate.

[8]And even this huge amount is only about 4×10^{-10} of the total energy generated by the sun.

where p is the pressure, V is the volume of the gas, T is its absolute temperature, and R is the gas constant per *mole* of gas,[9] and N is the number of moles of gas. This law is accurate for many real gases over a wide range of conditions.

We can use the ideal-gas law to analyze *pressure* (i.e., the force per unit area) from the point of view of the motion of a single gas atom in a box. The pressure on the walls of the box is the result of the atom repeatedly hitting the walls of the box and transferring its momentum to the enclosure's walls. If there are few collisions, the internal pressure on the box walls is small, but if there are many collisions, the pressure is larger. If the temperature rises, the atom has more speed and thus more momentum and consequently the pressure on the walls increases. Now we need to calculate the rate of change of momentum as our single atom hits a wall of the box.

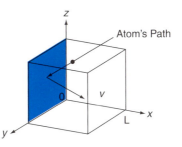

FIGURE 4.2 Cubical box of side L, volume V containing just one atom.

Assume that an atom is moving with speed v perpendicularly toward the box's wall (see Figure 4.2). On impact with a wall, the atom's momentum will be transferred to the wall of the box, and the resulting force on the wall is proportional to its rate of change of momentum, or:

$$F \propto mv/t \tag{4.4}$$

where the average time to reach the wall of the box is proportional to L/v where L is the size of the box. Thus, the *force* on the wall is $\propto mv^2/L$. The wall area for impact of atoms or molecules is L^2 and thus the pressure on this wall by our chosen particle is:

$$p \propto \frac{mv^2}{L^3} \propto \frac{mv^2}{V} \tag{4.5}$$

where V is the box's volume.

But there are millions of trillions of atoms (or molecules) in just 1 cm^3 of gas. Let this number of atoms be n. Thus, the total pressure on each of the box's walls is thus:

$$p \propto \frac{nmv^2}{V} \tag{4.6}$$

The "speed" of the atoms or molecules in this equation is that of the *fluctuations* in speed as they interact with each other. It is *not* a steady wind.

It is hard to comprehend just how large n really is. The next example will show you why.

---

[9] A *mole* of gas is its mass/molecular mass. For example, helium has a molecular mass of 4 kg/kg mole so that one kg mole of it has a mass of 4 kg, and 2 kg moles $=8$ kg, etc. For oxygen gas (a molecule with two oxygen atoms each of molecular mass 16 kg/kg mole), it has a molecular mass of 32 meaning 1 mole of it has a mass of 32 kg. These concepts are explained more fully in Chapter 6.

Example 4.3

If the number of atoms in 1 kg mole of a gas is 6.022×10^{26}, how many atoms of this gas occupy just 1.00 cm^3 at the standard atmospheric pressure of 1.00×10^5 N/m^2 and a temperature of 0 °C (273 K).

Need: $n =$ _____ atoms in 1.00 cm^3 of a gas at $p = 1.00 \times 10^5$ N/m^2 and $T = 273$ K.

$R = 8314$ l/(kg mole K) (R is the universal (molar) gas constant)

Know: 1 kg mole of the gas contains 6.022×10^{26} molecules.

How: Use the ideal-gas law $pV = NRT$ to find N (the number of moles), and then convert N, into n, the number of molecules.

Solve: $N = \dfrac{pV}{RT} = \dfrac{1.00 \times 10^5 \times 1.00}{8314 \times 273} \times [10^{-2}]^3 \left[\dfrac{N}{m^2} cm^3 \left(\dfrac{m}{cm}\right)^3 \dfrac{kg\,mole\,K}{J} \dfrac{1}{K}\right] = 4.41 \times 10^{-8}$ kg moles in 1.00 cm^3.

Finally, convert N into the number of atoms in 1 cm^3:

$$n = 4.41 \times 10^{-8} \times 6.022 \times 10^{26} \text{ [kg moles/cm}^3\text{][molecules/kg mole]} = 2.65 \times 10^{19} \text{ atoms.}$$

This is a very large number and there is no way of monitoring the motion of all of these gas atoms or molecules as they collide with each other and the walls in random directions and at random speeds. Thus, we use *averaging* techniques to describe for the behavior of large numbers of atoms or molecules.

Returning to our calculation of the pressure on the walls of a container, it is just the force per unit area, (F/A), and we can see that:

$$pV \propto nmv^2 \tag{4.7}$$

Note that nm is the total *mass* of all of the gas in the box.

Our two expressions for pV, the ideal-gas law $pV = NRT$ and our derived equation, $pV \propto nmv^2$ are very similar so that we can conclude that:

$$T \propto \tfrac{1}{2}mv^2 \tag{4.8}$$

The absolute temperature of a gas is thus proportional to the KE of the atoms or molecules, $\tfrac{1}{2}mv^2$ so when you feel hot it is because the KE of the molecules of air striking your body has increased.

It goes without saying that much more sophisticated models are available, but they use concepts well beyond this text. But, importantly, the reasoning above does contain the basic physics of this problem.

Example 4.4

If you feel hot at 35 °C and cold at 5.0 °C, what is the percent change in the average speed of molecules contacting your skin?

Hint: Change the temperatures into absolute quantities in degrees Kelvin by adding 273. (See the next page.)

Need: Percent change in average molecular speed between a cold gas (at 5.0 °C = 278 K) and a hot gas (at 35 °C = 308 K).

Know-how: From Equation (4.8) we have that $T \propto v^2$.

Solve: $\dfrac{v_H}{v_C} = \sqrt{\dfrac{T_H}{T_C}}$ in which the subscripts refer to hot and to cold.

Therefore, $\dfrac{v_H}{v_C} = \sqrt{\dfrac{308}{278}} = 1.05$ or the difference in the average air molecule's speed is only 5%.

The key principle here is that temperature is a result of the KE of atoms and molecules. If we want the temperature of a roast in the oven, or a piece of steel being welded, or a block of ice, we cannot measure all the speeds of all the molecules. In practice, we use thermometers and other kinds of devices that automatically average the speed of the atoms or molecules.

"Thermometry" is the technology of temperature measurement. People have always been able to experience the sensations of hot and cold, but the development of an accurate temperature measurement technology did not occur until the seventeenth century. Galileo is credited with constructing the first practical thermometer in about 1592. By the eighteenth century, more than 30 different temperature scales were in use. These scales were usually based on the use of two fixed standard temperatures with the distance between them divided into equally spaced degrees. Some of these early scales are shown in Table 4.2.[10]

Table 4.2 Early Temperature Scales

Inventor and Date	Fixed-Points
(1) Isaac Newton (1701)	Freezing water of (0°N) and human body heat (12°N)
(2) Daniel Fahrenheit (1724)	Freezing water (0 °F) and human body heat (96 °F)
(3) René Réaumur (1730)	Freezing water (0°Re) and boiling water (80°Re)
(4) Anders Celsius (1742)	Freezing water (0 °C) and boiling water (100 °C)

The 100 division Celsius temperature scale became very popular during the eighteenth and nineteenth centuries and was commonly known as the "centigrade" scale (from the Latin "centi" for "100" and "gradua" for "step") until 1948 when Celsius' name was formally attached to it and the name "centigrade" was officially dropped.

In 1848, William Thomson (Lord Kelvin) developed an absolute scale based on the Celsius degree size that now bears his name. Soon after Kelvin's scale, an absolute temperature scale based on the Fahrenheit degree size was developed and named after the Scottish engineer William Rankine. The relationship between the modern temperature scales is shown below.

$$T(°F) = (9/5)T(°C) + 32 = T(R) - 460$$

$$T(°C) = (5/9)[T(°F) - 32] = T(K) - 273$$

$$T(R) = (9/5)T(K) = (1.80)T(K) = T(°F) + 460$$

$$T(K) = (5/9)T(R) = T(R)/1.80 = T(°C) + 273$$

in which $T(R)$ and $T(K)$ (*without* the degree symbol on the R and K) represent the temperature in Rankin and Kelvin.

GPE is the energy acquired by an object by virtue of its position in a gravitational field—typically by being raised above the surface of the Earth. In SI units, it is calculated by the equation:

$$GPE = mgh \tag{4.9}$$

---

[10] 0 °F was the lowest temperature you could then reach with a mixture of ice and salt. The modern Fahrenheit scale uses the freezing point of water (32 °F) and the boiling point of water (212 °F) as its fixed-points. This change to more stable fixed-points resulted in changing the average body temperature reading from 96 °F on the old Fahrenheit scale to 98.6 °F on the new Fahrenheit scale.

where h is the height above some datum, usually the local ground level. In Engineering English units, this definition is modified by g_c to ensure that it comes out in ft lbf units:

$$GPE = \frac{mgh}{g_c} \tag{4.10}$$

Note that GPE only has meaning relative to a reference datum level. The choice of such a level is arbitrary but must be applied consistently throughout a problem or analysis.

Example 4.5

Wile E. Coyote holds an anvil of mass 100. lbm at the edge of a cliff, directly above the roadrunner who is standing 1000. ft below. Relative to the position of the roadrunner, what is the GPE of the anvil?

Need: GPE.

Know: The anvil has mass 100. lbm. The reference datum level, where the roadrunner is standing, can be chosen as 0. The height h of the anvil is 1000. ft referenced to roadrunner's datum. Because we are calculating in Engineering English units, we will also have to use g_c.

How: Equation (4.10) is in Engineering English units, $GPE = \frac{mgh}{g_c}$.

Solve: $GPE = \dfrac{100.0\,[\text{lbm}]\ 1000.0\,[\text{ft}]\ 32.17\,[\text{ft/s}^2]}{32.17\,[\text{lbm\,ft/lbf\,s}^2]} = 1.00 \times 10^5\,\text{ftlbf}$

correct to three significant figures (as is the least known variable, the mass of the anvil).

Electromagnetic energy (often just called "electricity") is a form of energy that is typically carried by electric charges[11] moving through wires, or electromagnetic waves moving through space. It will be the subject of Chapter 10, so we just touch on it briefly here.

Like all forms of energy and power, electromagnetic energy can be measured in joules, and electromagnetic power can be measured in watts. Consider the electromagnetic power used by an electrical device that is connected by wires to an electrical battery to create an electrical circuit,[12] as shown in Figure 4.3.

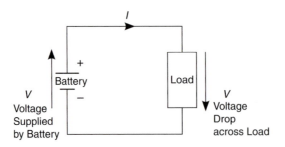

FIGURE 4.3 An electrical circuit.

---

[11]Specifically, negatively charged electrons.
[12]The word "circuit" has the idea of a *circle* built-in. Thus, an electrical circuit must start and finish in an unbroken loop. The physical reason is that the electrons as charge carriers are not consumed in the circuit.

An **electrical current**, represented by the letter I, flows into and out of the device. Electrical current is measured in units called *amperes* (symbol A), which is simply a measure of the number of electrons passing through any cross section of the wire every second. An electrical potential, represented by the letter V, can be measured at any point on the circuit. Electrical potential is measured in units called *volts* (V). It might be thought of as an "electrical pressure," originating in the battery that keeps the current flowing through the wire. The voltage has its most positive value at the terminal on the battery marked plus (+), and drops throughout the circuit, reaching its minimum value at the battery terminal marked minus (−). By measuring the voltages at any two points along the wire and subtracting to determine the difference between those two measured voltages, one can determine the voltage drop between those two points on a circuit. One typically determines the overall voltage drop by subtracting the voltage at the point nearer the minus terminal of the battery from the voltage at the point nearer the plus terminal of the battery so the voltage drop really can be thought of as a pressure "pushing" the current from the point of higher voltage to the point of lower voltage.

In mechanical systems, power, the rate of doing work, can be computed as the product of force × speed. From electrostatics, the force on a charge of Q coulombs[13] in a voltage gradient of V/d is $Q \times V/d$ (d is the distance over which the voltage changes from 0 to V)—hence, the *work* in moving the charge is $d \times QV/d = QV$. If the charge is moved along the voltage gradient in time t, the power to move the charge is VQ/t. We define the rate of movement of charge as the electric current, $I = Q/t$. If Q in coulombs, t in seconds, then I is in amperes. Further, if V is in volts, then the power is in watts. Simply put, what you need to remember is that the electrical power is expressed as:

$$\text{Electric Power}\,(P) = I \times V \qquad\qquad (4.11)$$

Example 4.6

A battery sustains a voltage drop of 3.0 V across a small lightbulb and produces a current of 0.1 A through the lightbulb. What is the power required by the lightbulb?

> **Need:** Power of lightbulb in watts (W).
> **Know:** Voltage across bulb is 3.0 V. Current through lightbulb $I = 0.1$ A.
> **How:**   Use Equation (4.11), $P = I \times V$, that is, power = voltage × current.
> **Solve:** $P = 3.0\,[\text{V}] \times 0.1\,[\text{A}] = 0.3$ W.

While the power into the lightbulb departs from the electric circuit, it does not disappear. Instead, some of it is radiated away from the bulb in the form of electromagnetic power (or, equivalently, in the form of massless particles called photons) as visible "light." Some of this electromagnetic energy is in the form of radiant heat. More energy is lost in lower grade heat by heating the lightbulb's local surroundings. Calculating the value of the power carried by these mechanisms will not be covered in this book.

Chemical energy is another form of potential energy in that it is determined by the relative distribution of electrons in the atoms that make up the structure of molecules. It is so important to our theme of twenty-first century engineering that we will devote an entire chapter to chemical energy. Suffice it here to say that it too is most conveniently measured in joules.

---

[13]The charge on a single electron is 1.6×10^{-19} coulombs.

4.4 ENERGY CONVERSION

Sunlight drives the winds. An anvil hoisted to a cliff top can be used to deliver KE to an unwary roadrunner below (though it is more likely to end up falling on the head of Wile E. Coyote himself!). Burning fossil fuel (ultimately a form of stored-up solar electromagnetic energy) results in the rotational KE (known there as "shaft work") of a turbine that is then converted back into electromagnetic energy in a generator to light up a city. All these occurrences suggest a second key fact about energy: its various kinds can be converted from one form to another.

Because all types of energy can be expressed in the same units, joules, this conversion can be expressed quantitatively in simple models. Even in Engineering English units, the number of conversion factors is relatively few.[14]

We mentioned earlier that the sun delivers more than 10^{24} J of energy daily to the Earth in the form of the electromagnetic solar energy we call sunlight. What happens to that electromagnetic energy? A discussion provides the clues that can be made into a model. Part of the energy heats the atmosphere and drives the winds (as previously noted) as well as driving the water cycle. Part of the solar energy that reaches the Earth is converted into chemical energy via photosynthesis in trees, grass, and agricultural crops. Part of it simply heats the Earth and the ocean. Much of it is reflected away back into deep space. One thing the model must answer is the following question: If all that energy is arriving from the sun every day, and much of it is converted to thermal energy, why is not the Earth accumulating more and more thermal energy and continuously getting hotter? (Try to answer that question yourself before reading the next paragraph.)

The fact that the temperature of the Earth is staying roughly constant requires another element in our simple model. Roughly speaking, every day the Earth must reflect or radiate away into space the same amount of energy that it receives from the sun. Should anything happen, either due to natural or human causes, to interfere with this energy conversion balance between absorption and radiation of solar energy, we will be in big trouble! The Earth might either cool-off (as it did in the various ice ages) or heat up—as it appears to be doing right now according to many scientists who believe that global warming is taking place. So understanding energy conversion is crucial to projecting the future of life on our planet.

We can perform a similar analysis of energy conversion on a technological system, such as an automobile, a house, a toaster, or anything we choose. In the case of an automobile, we can regard the initial source of the energy as the chemical energy pumped into the vehicle at the gas station. In the automobile's engine, that chemical energy is converted to the TKE of a piston in a cylinder. That TKE is then converted into rotational energy in the crankshaft, transmission, axles, and wheels. That rotational KE in turn provides the TKE of the automobile in its motion down the road. In this process, part of that energy—indeed, a substantial majority of the initial chemical energy—is transferred to the environment in the form of heat, either through the car's radiator and exhaust pipe or through road friction or air resistance.

---

[14]One useful conversion factor in the Engineering English system shows the relative magnitude of mechanical and thermal units: 1 Btu (British thermal unit) is the heat required to raise the temperature of 1 lbm of water by 1 °F; it is equivalent to 778 ft lbf of energy. In other words, if you dropped 1 lbm of water through 778 ft and all of the original GPE were converted to thermal energy, you would heat that water only by 1 °F. A useful conversion factor is 1 Btu is approximately equal to 1 kJ (actually, 1.000 Btu = 1.055 kJ).

Example 4.7

A gallon of gasoline can provide about 1.30×10^5 kJ of chemical energy. Based on Example 4.1, if all the chemical energy of a gallon of gasoline could be converted into the TKE of an automobile, how many gallons of gasoline would be equivalent to the vehicle's TKE if the automobile is traveling at 65.0 miles/h on a level highway?

Need: Gallons of gasoline to propel the automobile at 65. mph (29. m/s).

Know: TKE of vehicle $= 4.2 \times 10^2$ kJ (to two significant figures at 65. mph as per our previous calculation in Example 4.1). Energy content of gasoline is 1.30×10^5 kJ/gallon.

How: Set TKE of vehicle equal to chemical energy in fuel. Let $x =$ number of gallons needed to accomplish this.

Solve: $x \times 1.30 \times 10^5$ [gallons][kJ/gallon] $= 4.2 \times 10^2$ [kJ]. Therefore, $x = 0.0032$ gallons.

This is a tiny amount of gasoline. As we will see later, this is a misleadingly small amount of gasoline. In fact, most of the gasoline is *not* being converted into useful KE, but is mostly "wasted" elsewhere. The misleading nature of this calculation will become obvious as we proceed! The error is in the conditional statement "if all the chemical energy of a gallon of gasoline could be converted into the TKE of an automobile." The lesson to be learned here is to beware of your assumptions!

4.5 CONSERVATION OF ENERGY

It is possible in principle (though impossibly difficult in practice) to add up all the energy existing[15] in the universe at any moment and determine a grand total. This may sound like a theoretical claim of little practical use. However, engineers have developed a method of applying the fact that energy is conserved to practical problems while avoiding the inconvenience of trying to account for all the energy in the universe. This method is called "control boundary" analysis and indeed was schematically illustrated in Figure 4.1.

It consists simply of isolating the particular object or system under consideration, and making a simple model of the way that object or system exchanges energy with the rest of the universe. Making that model begins with a conceptual sketch of the object or system. You draw a dotted line representing the control boundary around the sketch to contain the item under analysis. Then draw arrows across the boundary representing specific types of exchanges of energy between the object or system and the rest of the universe. By limiting the energy exchange in this way, the principle of conservation of energy can be "imported" into the system under consideration to determine the results of various energy conversion processes that occur within the dotted line. Losses or gains are simply one form or another of energy that crosses that boundary.

As a simple example, consider your classroom as a closed system and draw an imaginary control boundary around it. Assume this boundary is impervious to energy flows so that what's inside in the classroom stays there. Now imagine you have a 1.0 kg book on your desk that is 1.0 m high. The book has GPE $= mgh = 1.0 \times 9.8 \times 1.0$ [kg][m/s^2][m] $= 9.8$ J with respect to the floor of your classroom. Suppose that book now falls to the floor, thus losing all of its GPE. Since the classroom boundaries are impervious to energy exchanges, where did that GPE go? Have we violated conservation of energy?

---

[15]In this case, we would also have to add in all the mass in the universe because cosmic events freely convert mass into energy and energy into mass via $E = mc^2$!

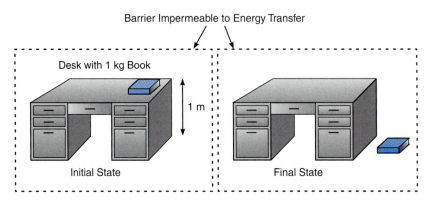

FIGURE 4.4 Where did the book's potential energy go?

The fact is that GPE is still trapped within the room, but not in its original form. What physically happens is: (1) the GPE of the falling book is converted to TKE; (2) when the book hit the floor, it sets up some sound waves in the floor material and in the surrounding air; and (3) eventually when all the transients died out, the forms of energy in (1) and (2) end up heating the room and its contents by exactly 9.8 J of energy (see Figure 4.4).

The principle of conservation of energy says energy is never lost, merely transformed into another form. This is one statement of energy conservation, which is also called the *first law of thermodynamics*. Ultimately all forms of energy degrade to heat; this is one statement of the *second law of thermodynamics*. Control boundary analysis is a very useful way to account for energy flows. This kind of a model captures enough of reality that it can be highly useful in engineering analysis and design, as Figure 4.5 shows.

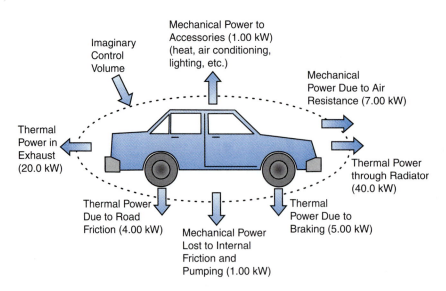

FIGURE 4.5 Control boundary analysis model of an automobile.

Figure 4.5 shows a control boundary analysis model as applied to an automobile. In words, the picture illustrates the following "energy accounting" of a typical automobile trip. As indicated by the arrows crossing the dotted line, an automobile traveling at 65. miles/h (29. m/s) transfers about 40.0 kW to the atmosphere in the form of thermal energy from the radiator, about 20.0 kW to the atmosphere in the form of heat and chemical energy out the exhaust, about 7.00 kW of mechanical power[16] to the atmosphere in overcoming air resistance, about 11.0 kW in frictional work (mainly the result of elastic compression and expansion of the rubber in the tires and applying the brakes), and about 2.00 kW of power for other purposes such as pumping and operating such accessories as heat, lights, and air conditioning. All of these contributions total about 80.0 kW.

Example 4.8

If gasoline contains 1.30×10^5 kJ/gallon, how many gallons of gasoline must be used per second to provide the power needed to sustain travel at 65. miles/h on a level road? From this, estimate the fuel economy in mpg (miles/gallon).

Need: Amount of gasoline consumed per second in gallons.

Know: An automobile traveling at 65. miles/h (29. m/s) transfers about 80.0 kW or 80.0 kJ/s (or 110. hp) into various forms.

How: Apply the principle of the conservation of energy. The energy lost by the car through the boundary of the control surface must come from somewhere. The only place it can come from within the car is by the decrease in the chemical energy of some of the gasoline. By the principle of conservation of energy, the decrease in chemical energy must equal the transfer out of the dotted lines of all the other types of energy.

Solve: Chemical energy needed per second = 80.0 kW or 80.0 kJ/s. Since gasoline contains about 1.30×10^5 kJ/gallon of chemical (i.e., potential) energy, this means $80.0/1.30 \times 10^5$ [kJ/s][gallon/kJ] $= 6.15 \times 10^{-4}$ gallons/s of gasoline are consumed.

Is this a reasonable answer? Note that a car going at 65. miles/h travels 65/3600. = 0.018 miles/s, and its fuel economy in miles/gallon is:

$$(0.018) \times 1/(6.15 \times 10^{-4}) \text{ [miles/s][s/gallon]} = 29. \text{ mpg}$$

This is a reasonable estimate given the roughness of our calculations.

Note a couple of implications of this example. First, one might, from an energy standpoint, view an automobile as a device for converting the chemical energy of fuel into thermal energy in the atmosphere. Second, operating an automobile requires a process for liberating 80.0 kJ of energy every second from liquid fuel. This process, called combustion, is covered in Chapter 7.

SUMMARY

The *principle of conservation of energy* can help us answer those questions that began this chapter. Is the world running out of energy? Drawing a dotted line around the Earth and applying *control boundary analysis* indicates that the amount of energy here on Earth is either remaining constant or if you believe that global warming is happening, increasing slightly (although still significantly). So the answer is no—the world certainly is not running of the total energy, although useful primary energy sources might be in decline.

---

[16]In the atmosphere, due to "friction" the displaced air will eventually slow, and the energy imparted to it will dissipate as an equivalent amount of thermal energy.

So where does the energy come from that provides the light by which you are reading these words? This gets to the real point that people are making when they assert we are in an "energy crisis." Though the amount of energy around us is constant, we are rapidly converting reserves of easily exploited chemical energy (which is actually a form of stored-up solar electromagnetic energy derived from past sunlight) into much less useful thermal energy. The light that you are reading this probably came from a lightbulb powered by a fossil-fuel-burning electric power plant. As you read these words, that light is being converted from electromagnetic energy into heat at room temperature—a form of energy of very little use for anything beyond keeping you warm.

Originally we asked, "Does a car possess more energy when it is sitting in your driveway or 15 min later when it is traveling down the highway at 65.0 miles/h?" Our control boundary analysis model of the automobile tells us that the automobile is continually transferring energy from its gas tank to the atmosphere. So it possesses less energy on the highway than it did 15 min earlier in the driveway, and until you pull into a gas station and fill up the gas tank, it will steadily decrease in the energy content within its dotted control boundary.

To sum up, an understanding of the concept of energy is one of the indispensable analytical tools of an engineer. That understanding begins with the following key ideas. *Energy is the capability to do work.* Energy comes in many kinds. These kinds are capable of conversion from one kind to another. In the course of these conversions, energy is neither created nor destroyed but, rather, conserved. This is a principle that an astute engineer can exploit to advantage in the analysis of complex systems using the concept of a control boundary that separates the universe into that which we are studying and the rest of the universe. Using this concept, we can directly calculate energy flows to and from the control boundary as an aid in analyzing its parts.

EXERCISES

Pay attention to the number of significant figures in your answers!
  Conversion factors: $1.00 \text{ J} = 0.738 \text{ ftl bf}$, $1.00 \text{ kg} = 2.20 \text{ lbm}$, and $g_c = 32.2 \text{ lbm ft/lbf s}^2$.

1. Determine the TKE of the automobile in Example 4.1 if its speed was reduced to 55. miles/h.

2. Determine the TKE in Engineering English units of the automobile in Example 4.1 if its mass was increased to 4.00×10^3 lbm.

3. Determine the TKE of the atmosphere in Example 4.2 if the average air speed increased to 15. m/s.

4. Repeat the calculation of Example 4.5 in Engineering English units. Check that your answers agree with the solution in Example 4.5 using the appropriate conversion factors.

5. What would the gravitational potential in SI units of the anvil in Example 4.5 be if its mass was 100. kg and the cliff was 1000. m high?

6. Determine the GPE of an 8.00×10^3 kg truck 30. m above the ground. (*A*: 2.4×10^6 J to two significant figures, since *h* is known only to two significant figures.)

7. A spring at ground level—that is, at height $= 0.00$ m—shoots a 0.80 kg ball upward with an initial KE of 245 J. Assume that all of the initial TKE is converted to GPE. How high will the ball rise (neglecting air resistance)?

8. Chunks of Earth orbital debris can have speeds of 2.3×10^4 miles/h. Determine the TKE of a 2.0×10^3 lbm chunk of this material in SI units. (A: 4.8×10^{10} J to two significant figures.)

9. An airplane with a mass of 1.50×10^4 kg is flying at a height of 1.35×10^3 m at a speed of 250.0 m/s. Which is larger—its TKE or its GPE with respect to the Earth's surface? (Support your answer with numerical evidence.) (A: TKE $= 4.69 \times 10^8$ J; GPE $= 1.99 \times 10^8$ J. Therefore, the TKE is *greater than* GPE.)

10. Determine the amount of gasoline required in Example 4.7 if the automobile was traveling at 55. miles/h.

11. Suppose the 1.00 kg book in Figure 4.4 fell from a height of 2.5 m. What would be the final energy of the classroom?

12. A vehicle of mass 1.50×10^4 kg is traveling on the ground with a TKE of 4.69×10^8 J. By means of a device that interacts with the surrounding air, it is able to convert 50% of the TKE into GPE. This energy conversion enables it to ascend vertically. To what height above the ground does it rise?

13. Aeronautical engineers have invented a device that achieves the conversion of kinetic to potential energy as described in Exercise 12. The device achieves this conversion with high efficiency. In other words, a high percentage of the TKE of motion is converted into vertical "lift" with little lost to horizontal "drag." What is the device called? (*Hint*: This is not rocket science.)

14. A hypervelocity launcher is an electromagnetic gun capable of shooting a projectile at very high speed. A Sandia National Laboratory hypervelocity launcher shoots a 1.50 g projectile that attains a speed of 14.0 km/s. How much electromagnetic energy must the gun convert into TKE to achieve this speed? Solve in SI. (A: 1.5×10^2 kJ.)

15. Solve Exercise 14 in Engineering English units. (Also check your answer by converting the final answer to Exercise 14 into Engineering English units.)

16. Micrometeoroids could strike the International Space Station with impact speeds of 19 km/s. What is the TKE of a 1.0 g micrometeoroid traveling at that speed? (A: 1.8×10^5 J.)

17. Suppose a spaceship is designed to withstand a micrometeoroid impact delivering a TKE of a million joules. Suppose that the most massive micrometeoroid it is likely to encounter in space has mass of 3 g. What is the maximum speed relative to the spaceship that the most massive micrometeorite can be traveling at for the spaceship to be able to withstand its impact?

18. A stiff 10.0 g ball is held directly above and in contact with a 600.0 g basketball and both are dropped from a height of 1.00 m. What is the *maximum* theoretical height to which the small ball can bounce?

19. What would be the power required by the lightbulb in Example 4.6 if it sustained a voltage drop of 120. V?

20. What would be the current in the lightbulb in Example 4.6 if it sustained a voltage drop of 120. V and required a power of 100. W?

21. An electric oven is heated by a circuit that consists of a heating element connected to a voltage source. The voltage source supplies a voltage of 110. V, which appears as a voltage drop of across the heating element. The resulting current through the heating element is 1.0 A. If the heating element is perfectly efficient at converting electric power into thermal power, what is the thermal power produced by the heating element? (A: 1.1×10^2 W to two significant figures.)

22. A truck starter motor must deliver 15 kW of power for a brief period. If the voltage of the motor is 12 V, what is the current through the starter motor while it is delivering that level of power?

23. A hybrid car is an automobile that achieves high fuel efficiency by using a combination of thermal energy and electrical energy for propulsion. One of the ways it achieves high fuel efficiency is by regenerative braking. That is, every time the car stops, the regenerative braking system converts part of the TKE of the car into electrical energy, which is stored in a battery. That stored energy can later be used to propel the car. The remaining part of the TKE is lost as heat. Draw a control surface diagram showing the energy conversions that take place when the hybrid car stops. (*A: See diagram.*)

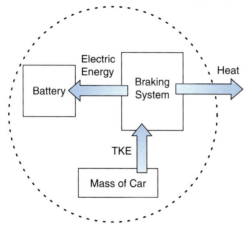

24. Suppose the car in Exercise 23 has a mass of 1000. kg and is traveling at 33.5 miles/h. As it comes to a stop, the regenerative braking system operates with 75% efficiency. How much energy per stop can the regenerative braking system store in the battery? Illustrate with a control boundary showing the energy flows.

25. Suppose the car in Exercises 23 and 24 has stored 1.00×10^2 megajoules (MJ) of energy in its battery. Suppose the electric propulsion system of the car can convert 90% of that energy into mechanical power. Suppose the car requires 30.0 kW of mechanical power to travel at 33.5 miles/h. How many miles can the car travel using the energy in its battery? (*A: 28 miles.*)

26. Determine the amount of gasoline consumed per second by the automobile in Example 4.8 if it was traveling at 41. m/s.

27. In order to maintain a speed v on a horizontal road a car must supply enough power to overcome air resistance. That required power goes up with increasing speed according to the formula:

$$P = Power\ in\ kW = K \times V^3$$

where v is the speed measured in miles/h and K is a constant of proportionality. Suppose it takes a measured 7.7 kW for a car to overcome air resistance alone at 30.0 mph.
a. What is the value of K in its appropriate units?
b. Using a spreadsheet, prepare a graph of power (kW on the y axis) as a function of speed (mph on the x axis) for speeds from 0 mph to 100 mph.

28. Review Exercises 20–23 in Chapter 2 concerning the dynamics (and consequent fate) of bungee jumpers. Draw a control surface around the jumper and cord. Show the various forms of energy possessed by

the jumper and cord, along with arrows showing the directions of energy conversion inside and across the control surface: (a) when the jumper is standing on the cliff top, (b) when the jumper is halfway down, and (c) when the cord brings the jumper to a safe stop.

29. After working for a company for several years, you feel you have discovered a more efficient energy conversion method that would save your company millions of dollars annually. Since you made this discovery as part of your daily job you take your idea to your supervisor, but he/she claims it is impractical and refuses to consider it further. You still feel it has merit and want to proceed. What do you do?
   a. You take your idea to another company to see if they will buy it.
   b. You contact a patent lawyer to initiate a patent search on your idea.
   c. You go over your boss's head and talk to his/her supervisor about your idea.
   d. You complain to your company's human resources office about having poor supervision.

30. Your course instructor claims that energy is not really conserved. He/she uses the example of a spring that is compressed and then tied with a nylon string. When the compressed spring is put into a jar of acid, the spring dissolves and the energy it contained is lost. How do you react?
   a. Ignore him/her and follow the established theories in the text.
   b. Go to the department chairperson and complain that the instructor is incompetent.
   c. Say nothing, but make detailed statements about the quality of the instructor on the course evaluation at the end of the term.
   d. Respectfully suggest that the energy in his/her spring example really is conserved.

Engineering Economics

Source: iStockphoto.com/Vasiliy Yakobchuk.

5.1 INTRODUCTION

This chapter is applicable to virtually *all* the fields of engineering since engineers design and make things that people buy. Many subdisciplines of engineering use this specialized form of economics, whether to financially embrace new products or to evaluate existing ones. In addition to industrial engineers, manufacturing engineers, mechanical engineers, chemical engineers, etc., all use engineering economics since these disciplines are often involved in the cost management of industrial products.

5.2 WHY IS ECONOMICS IMPORTANT?

We all make economic decisions every day. Can I afford that steak dinner, a new laptop computer, an iPhone, or should I get a hamburger, continue to use my old desktop computer, and keep the old cell phone I have used for years? Can I afford a new car? What about a mortgage for a new house? Engineers make similar decisions all the time to determine if what they are doing makes economic "sense." Unlike personal economic decisions, the criteria used by engineers are multidimensional and have several components.

An engineer should be familiar with several financial terms. These include **interest**, **time value of money**, **principal**, **present value**, **first cost**, and **future value** to name just a few. These terms are needed to assess the true cost of an investment, be that in buying a whole new factory or just buying an electric motor for an existing project such as manufacturing a vacuum cleaner. Other criteria used to evaluate the economics of a new investment include **break-even point (BEP)**, **process improvement (PI)**, and **return on investment (ROI)**. These terms represent but a few of many standard concepts used before an engineer expends large sums of money.

One's answer to almost all economic decisions made for personal use rests on **first cost**. **First cost** is what you pay for the item when you buy it. Few of us think in terms of the purchase's life cycle costs such as repairs, depreciation, and the cost of borrowing money when making a purchase.[1] But, if you are making a major business purchase, you better consider more than just the first cost. In the next section, we will concentrate on how the **cost of borrowing** may be as important as the cost of the item itself.

5.3 THE COST OF MONEY

Suppose you purchase a car for $10,000. Is that what it really costs you? Even the most elementary analysis suggests that this is not so. Because, if you had $10,000 to spend, you already have the cash *somewhere*. Suppose it was not kept under your mattress but in a bank in an interest-bearing account earning 5% **simple** interest. This means that each year you will have a deposit to your account of 5% of $10,000 or $500. Purchasing a car with that cash means that this income stream stops forever. If nothing else changes, your once $10,000 now forgoes the addition of $500 each and every year for a **future value** of infinity (should you live so long)!

So you see that money itself has a value, and a proper model of how to account for this value is important in an engineer's decision to buy a new car or anything else. It is very unusual to make cash purchases in a business (except for small "petty cash" purchases). If you want to make a significant purchase (a $10,000 car, for example), a business would probably borrow the money from a bank. Suppose the bank charges 5% simple interest on the loan, then once a year you owe the bank $500. With simple interest it's as if once a year you paid back the $10,000 and gave the bank $500 for the use of the $10,000, and then you immediately borrowed another $10,000 at the same interest for the next year. If you pay back the full amount after 5 years, you will pay the creditor $10,000 (the "**principal**," P)+$500/year × 5 years=$2500 (the "**interest**," I). Here you can see a finite time value of money—$2500 on a principal of $10,000.

But, if you were a potential creditor of a business, and knew that you would have to wait 5 years before seeing a penny back of the principal, wouldn't you also think that you have lent the business another $2500 spread over 5 years and on which you received nothing in return? The banker wants interest on this interest as well as on the original principal and so charges **compound** interest for the loan.

It works this way. Year 1 of the loan is on $10,000, year 2 is on $10,500, year 3 is on $11,025 (the interest being 5% of $10,500 in year 3, etc.), and so on until in year 5 you repay the creditor/banker the sum of

$$P = \$10,000 + I$$

in which

$$I = \$500 + \$525 + \$551 + \$579 + \$608 = \$2763$$

instead of just $I=\$2500$. Note each yearly interest is $(1+0.05)\times$ the previous year's balance. Table 5.1 summarizes this situation.

---

[1] Except possibly for a home mortgage because the monthly amounts are usually a large fraction of financial resources.

Table 5.1 Generalizations of Interest Accrual Equations

Type of Interest	Period	Beginning of Period	End of Period	Future Value, F
Simple[a]	1	P	$P + rP$	$P(1+r)$
	2	P	$P + rP$	$P(1+r) + rP = P(1+2r)$
	3	P	$P + rP$	$P(1+2r) + rP = P(1+3r)$
	...	...	...	...
	N	**P**	**P + rP**	**P(1 + Nr)**
Compound	1	P	$P(1+r)$	$P(1+r)$
	2	$P(1+r)$	$P(1+r) + rP(1+r)$	$P(1+r)^2$
	3	$P(1+r)^2$	$P(1+r)^2 + rP(1+r)^2$	$P(1+r)^3$
	...	...	...	...
	N	**$P(1+r)^{N-1}$**	**$P(1+r)^{N-1} + rP(1+r)^{N-1}$**	**$P(1+r)^{N}$**

[a] While we have also included a simple interest table, it is for pedagogical reasons only because it is easy to understand and not because it is used in practical engineering economics.

Table 5.1 defines the **future value**, F, of the **present value** of the principal, P. In other words, the future value is what today's principal is worth under some estimate of its future financial behavior. Virtually all business transactions use compound interest formula and other formulae derived from it.

Note: In the problems involving financing in this chapter, we will often ignore our rules concerning significant figures and conform to the language of the financial industry. We will normally express our calculations in whole dollars.[2]

Example 5.1

You wish to borrow $100,000 for 10 years at 5.0% annual interest. What is the difference in the cost of the loan if it is compounded yearly, monthly, or daily?

Need: Cost of borrowing $100,000 for 10 years at 5.0% under assumptions of 10 annual payment periods, 120 monthly periods, and 3650 daily periods, respectively.

Know-How: The formulae for compound interest from Table 5.1 is: $F = P(1+r)^N$.

Solve: (1) $N = 10$, $r = 0.050$, therefore

$$F = P(1+r)^N = \$100,000 \times (1+0.050)^{10} = \mathbf{\$162,889}.$$

(2) $N = 120$, $r = 0.050/12 = 0.00417$, therefore

$$F = P(1+r)^N = \$100,000 \times (1+0.00417)^{120} = \mathbf{\$164,701}.$$

(3) $N = 3650$, $r = 0.050/365 = 1.37 \times 10^{-4}$, therefore

$$F = P(1+r)^N = \$100,000 \times \left(1 + 1.37 \times 10^{-4}\right)^{3650} = \mathbf{\$164,866}.$$

[2] However, some transactions, such as small loans, are compounded to the penny.

Notice that the cost of the loan in Example 5.1 goes up when the interest is compounded more often. Note too that interest rates are nominally quoted in annual terms, but they must be adjusted to reflect the daily or monthly periods of compounding.

Example 5.2

Nuclear power plants cost billions of dollars. This reflects the need for the highest possible confidence in the construction of the reactor. A spending plan for a large generic nuclear power plant is shown in the figure below.

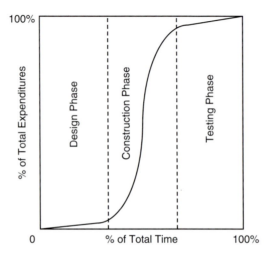

In a simplified case, each phase takes about 1/3 of the total time to build the power plant with the heaviest spending naturally occurring during the construction phase. A manufacturer of nuclear plants would typically get construction loans as needed to minimize the interest expense. Notice that significant loan costs are incurred during the final test phase since the construction money has already been spent, and therefore also borrowed, for activities that have yet to bring in any income.

Suppose the costs were $5.0 billion up to the beginning of the test phase for a 1000-MW nuclear power plant. Suppose further the whole construction period is 12 years and thus the final test phase takes 4 years. If the interest rate is 12.0% with quarterly compounding, how much does the testing phase cost in finance charges alone?

Need: Testing phase finances charges $=\$$_____ .

Know: Outstanding capital, $P=\$5.0$ billion. Period of testing$=4$ years. Costs are compounded over $N=4$ quarters/ year $\times 4$ years$=16$ quarters, and the interest rate $r=12.0\%/4=3.0\%=0.030$ per quarter.

How: Use the compound interest law from Table 5.1: $F=P(1+r)^N$.

Solve: $F=P(1+r)^N=\$5.0\times 10^9\times(1+0.030)^{16}=\$8.02$ billion. The finance charges during the testing phase are then $=F-P=8.0-5.0=$**$3.0 billion**.

The financial expenditures during the testing phase swamp the actual costs of testing itself by a considerable amount. For a project with a long construction time up front, the viability of the project may depend on financing charges more than any other single factor.

Rather than do the arithmetic on a calculator, any spreadsheet program you use almost certainly has many functions developed especially for financial analysis. Excel™ has the functions described in Table 5.2 (plus *many* other financial functions).

Table 5.2 Some Excel™ Financial Functions

Function	Use	Arguments
FV(*rate,nper, pmt,pv,type*)	Calculates future value	
PV(*rate,nper, pmt,fv,type*)	Calculates present value	• *rate*=*r* • *nper*=*N* • *pmt*=periodic payments of *P* and *r* • *pv*=*P*
NPER(*rate, pmt, pv, fv, type*)[a]	Calculates how long it will take to increase *PV* to *FV*	• *fv* is optional. It is the future value that you'd like the investment to be after all payments have been made. If this parameter is omitted, the NPER function will assume that *fv* is 0. • *type*=0 if the payment is at the start of the period. • *type*=1 if the payment is at the end of the period.

[a] *For example, suppose you deposit $10,000 in a savings account that earns an interest rate of 8%. To calculate how many years, it will take to double your investment, use NPER as follows: =NPER(0.08, 0, −10,000, 20,000, 0). This will return an answer of 9.01, which indicates that you can double your money in about 9 years.*

Example 5.3

Repeat Example 5.1 using the Excel (or equivalent) spreadsheet future value function, FV.

Need: Cost of borrowing $100,000 for 10 years at $r=5\%$ under assumptions of (1) $N=10$ payment periods, (2) $N=120$ payment periods, and (3) $N=3650$ payment periods.

Know-How: Use the FV function in Excel

Solve: FV(*rate, nper, pmt, pv, type*)

	A	B	C	D
1	*P*	r, %		
2	$100,000	5.00%		
3				
4	Case #	*N*	*r* per period	*FV*
5	1	10	5.00%	$162,889
6	2	120	0.42%	$164,701
7	3	3650	0.01%	$164,866

	A	B	C	D
1	*P*	r, %		
2	100000	0.05		
3				
4	Case #	*N*	*r* per period	*FV*
5	1	10	=$B$2	=FV(C5,B5,,-$A$2,1)
6	2	120	=B2/12	=FV(C6,B6,,-$A$2,1)
7	3	3650	=B2/365	=FV(C7,B7,,-$A$2,1)

Notice the convention that your principal (or, "investment") is a negative number (−$A$2), which is logical but confusing. If you enter it as negative number in $A$2, then you don't have to put a negative sign in front of $A$2 in the FV function. Also, notice that unless you are prepaying the owed principal or interest, you do not need to enter a number for *pmt* in the FV function. Just leave a blank space (or enter 0) for *pmt* so that, ", *pmt*" becomes " , ," as shown in the second spreadsheet.

Example 5.4

You want to reap $1,000,000 from an investment of $500,000. How many years will you have to wait if $r=7.50\%$?

Solve: Use the Excel spreadsheet function NPER(*rate, pmt, pv, fv, type*).

	A	B	C	D
12	P	r, %	FV	$1,000,000
13	$500,000	7.50%		
14				
15		N, years	r per period	
16		**9.58**	7.50%	

	A	B	C	D
12	P	r, %	FV	1000000
13	500000	0.075		
14				
15		N, years	r per period	
16		=NPER(B13, ,-A13, D12, 1)	=$B$13	

Thus you need to wait just over 9½ years to double your money.

To summarize what it really costs to make an engineering purchase, you need to include the present value (the principal) P and the interest charges on that amount. Clearly, you also have to add other costs such as labor costs, material costs, and overhead costs (such as the rent for your factory, the cost of power to keep the lights on, the cost of heat to keep the factory warm, and so on) to get the total costs.

5.4 WHEN IS AN INVESTMENT WORTH IT?

But how do you know if your project is worth the effort? In fact, most engineering businesses use one or more criteria to assess the value of a project. A major indicator occurs when you start to make a profit by selling enough of your widgets. This is called the "**Break-Even Point**" (**BEP**), and it is a standard measure of a project's value. It has a simple definition:

BEP occurs when the project has earned back what it took to make it.

Example 5.5

Assume the cost of producing a new product is $1,000,000. Then, the BEP occurs when net profit from the product reaches $1,000,000. Let's say the profit per widget is $1.00, and we're selling 1000/day.

Need: BEP=_____ years for a project costing $1,000,000 assuming you are selling 1000/day with a profit of $1.00 per widget.

Know-How: Equate cost to total money stream.

Solve: 1000 [widgets/day] × 1.00 [$/widget] × D [days] = $1,000,000. Solving for D gives:

$$D = 1000 \text{ days} = \textbf{2.74 years}.$$

In Example 5.5, it will take 2.74 years to reach the BEP. Is this good enough? It depends on the industry. Many companies would prefer a BEP of 18 months or less.

Example 5.6

One "long horizon" industry is the electric power industry. In Example 5.2, our 1000 MW nuclear power plant costs $8.0 billion, a very large sum of money. Yet its product sells for about 10 cents per kWh. What is its best possible[3] BEP after it starts to produce electricity?

Need: BEP=_____ years for a 1000-MW nuclear power plant that cost $8.0 billion and selling electrical energy for 10.0 cents/kWh = 0.10$/kWh.

Know: For t hours of operation, a 1000-MW nuclear power plant can produce a maximum of $1,000,000 \times t$ kWh of electricity.

How: Equate costs to total money stream produced by reactor.

Solve: $1,000,000 \times t \times 0.1$ [kWh][$/kWh] $= \$8.0 \times 10^9$ [$], or $t = 8.0 \times 10^4$ [$][h/$] or **BEP = 9.2 years**.

Is the BEP in Example 5.6 good enough? While nuclear power plants currently cost much more than fossil fuel power plants, the advantages of uranium over fossil fuel ultimately might increase as world oil and gas supplies dwindle and concerns about greenhouse gases grow

Example 5.7

An engineer proposes an improvement to an existing process. The cost required to make this **process improvement** (**PI**) is $100,000. Suppose the process makes 100,000 widgets/day. If the proposed PI saves one cent per unit ($0.01/unit), what is its BEP for the PI?

Need: BEP=_____ for a **PI** costing $100,000 assuming you are selling 100,000 widgets/day with a **PI** savings of $0.01/widget.

Know-How: You will save 100,000 [widgets/day] × 0.01 [$/widget] = $1000/day with the new and improved widget manufacture.

Solve: The **PI BEP** is given by how long to recover your $100,000 investment. This is 100,000/1000/[$][days]/[$] = 100 days = **3.3 months**.

---

[3]Ignoring other significant costs such as labor and uranium fuel, refueling outage, repairs, etc.

Is this PI BEP in Example 5.7 good enough? As a rule of thumb, a nominal BEP should be 18 months or less, and less than 12 months is preferable. In Example 5.5, the new widget is a marginal investment. But in Example 5.7, a PI BEP is a definite go.

One other financial term that is often used to determine if an engineering investment is satisfactory is the (annual) return on investment or "**ROI**." This is defined as

$$\text{ROI (in\%)} = \frac{\text{Annual return}}{\text{Cost of the investment}} \times 100$$

In the simplest case, if an investment of $500,000 produces an income of $40,000 per year, its ROI = $40,000/$500,000 = 0.08 = 8%. Many companies would not invest for such returns on their money. Many successful large companies operate with ROI's of 15% or more.

As Table 5.3 shows, there is a big difference among the ROIs of companies. You should not be surprised to see that Exxon Mobil has the highest ROI in this table due to its profits, and therefore the numerator of the ROI formula is high while the processing cost for the equipment for converting crude oil to refined products is small relative to the cost of the processed material.

Table 5.3 ROI of Several Large Chemical Companies

Company	ROI, Annual %
Dow Chemical	10.5
Exxon Mobil	22.4
DuPont	18.5
PPG Industries	20.2
Air Products	11.0
Eastman Chemical	10.9
W.R. Grace	9.8

Source: Chemical Engineering Progress, July 2, 2007

Example 5.8

Calculate the annual ROI for the PI in Example 5.5. Assume factory is operating 300 days per year.

Need: ROI annual % for a project costing $1,000,000 assuming you are selling 1000 widgets per day at a profit of $1.00 per widget for 300 days per year.

Know: Profit is 1000×1.00 [widgets/day][$/widget] = $1000/day

How: Compare annual profits to the investment cost.

Solve: Profit (i.e., "return") = 1000×300 [$/day][days/year] = $300,000 $/year.

Since the investment is $1,000,000, then the **ROI** = $300,000/$1,000,000 = 0.30 = **30%**.

Is the ROI in Example 5.8 good enough? Almost surely yes! This is a wonderful place to invest $1,000,000 (assuming all the assumptions and numbers are correct—such as the profit/widget, the number of days of operation/year, and the sales of widgets/day).

SUMMARY

This chapter was concerned with engineering economics, a discipline that is important to almost every branch of engineering, and particularly to industrial, manufacturing, or management engineering. Terms broached in this chapter include *compound interest*, *present value*, *principal*, and *future value*. All are concerned with the *time value of money* in making economic decisions. In some cases, the cost of financing new products may be the dominant cost in the product rather than the investment in hardware, materials, and labor. Further guidelines are given in how to assess whether a proposed project will be commercially viable. These are the *break-even point*, a measure of how long it takes to recover one's investments and *return on investment*, a ratio in percent, of the annual profit divided by the investment that produces it. Successful engineering businesses typically want to see a BEP of less than 18 months and an ROI greater than 15%.

EXERCISES

1. Find the cost of borrowing $100,000 in Example 5.1 if the interest is compounded hourly.

2. Rework Example 5.1 for a $15,000 loan at a 10% annual interest.

3. You need to borrow $12,000 to pay your tuition plus room and board. One bank offers to loan you the money for 10 years at 5.0% interest compounded annually. Another bank offers you the loan at 4.75% interest compounded monthly. Which bank has the better deal?

4. You are a big financial success, and you want to purchase the Remlab Company for $35 billion. You have $5 billion in cash but need to borrow the remaining $30 billion from your friendly banker. You banker says fine, I'll lend you the money for 10 years, but at 7.5% annual interest compounded quarterly. How much interest will you pay to the bank over the life of this loan?

5. What will be the testing phase finance charges in Example 5.2 if the nuclear power plant cost $10 billion and everything else stays the same?

6. It turns out that the testing phase in Example 5.2 takes 6 years instead of 4 years, and the entire project takes 14 years. What will the finance charges be if everything else remains the same?

7. Repeat Example 5.3 using the Excel spreadsheet (or equivalent) FV function to determine the cost of borrowing $100,000 for 10 years at 5% for 87,600 hourly payment periods.

8. Repeat Example 5.3 using the Excel spreadsheet (or equivalent) FV function to determine the cost of borrowing $100 billion for 10 years at 8% compounded quarterly.

9. Repeat Example 5.3 using the Excel spreadsheet (or equivalent) FV function to determine the cost of borrowing $3000 on a credit card for 1 year at 28% compounded monthly.

10. Use the NPER Excel function (or equivalent) in Example 5.4 to determine how long it will take to double your $500,000 investment if you earn 7.5% interest?

11. Use the NPER Excel function (or equivalent) in Example 5.4 to determine how long it will take to triple your $500,000 investment if you earn 11.3% interest?

12. Use the NPER Excel function (or equivalent) in Example 5.4 to determine how long it will take to increase an investment of $3500-4200 if you earn 3.5% interest?

13. How long will it take to double your money at an interest rate that is $\ll 100\%$ starting with the expression $F=P(1+r)^N$? **Hint[4]**: $\ln(1+r)\approx r$ if $r\ll1$. (**Answer**: Years to double the investment is $70\div r\%$, which is known as the **Rule of Seventy**.[5])

14. Compare the answer to Exercise 10 with the "Rule of Seventy" described in Exercise 13.

15. Suppose the profit on each widget sold in Example 5.5 was $10.00 instead of $1.00. What would the BEP be if everything else remained the same?

16. Suppose the new product in Example 5.5 has a profit of $5.00 per widget, but the sales projection is off by 50% and the company is only able to sell 500 per day. What is the new BEP?

17. You are a manufacturing engineer and have been asked to project the costs required to produce a new line of products for your company. After contacting the appropriate equipment manufacturers and consulting with management about labor and overhead costs, you determine that the manufacturing line will cost $8.3 million to get into operation. The marketing department tells you that the company can expect to see a profit of $37.50 per item produced and that they can sell 1500 items per day in big chain stores. You run into your boss at the coffee machine and he/she casually asks you what the BEP will be for your new line. What do you tell him/her?

18. The owners of the nuclear power plant in Example 5.6 want a BEP of 5 years. To do this, they hope to raise the price of the electricity they sell. How much do they need to sell the electricity for in order to meet the BEP?

19. In Example 5.6, the owners of the nuclear power plant decide to use substandard materials to reduce the cost of building and testing the power plant from $8.0 billion to $7.35 billion. If everything else remains the same (and no one finds out what they've done), what will be their new BEP?

20. Another "long horizon" industry is the commercial aircraft industry. A Boeing 747-8 Freighter costs $300 million. A commercial air freight company purchases this plane and keeps it in the air 12 hours per day for 280 days per year. If the air freight company makes a profit of $4500 per flight hour, how long before it breaks even with the cost of the plane?

21. In Example 5.7, it turns out that the new process does not save 1 cent per unit. It only saves 0.3 cents per unit. What is the new BEP, and is it worth the expense to make this improvement?

22. You are a manufacturing engineer and discover that you can purchase a machine that will replace a labor-intensive manual operation in your manufacturing line. The machine costs $1.5 million, but the labor savings costs are $7.80 per unit manufactured. Can you get a BEP less than 18 months if you make 8500 units per day?

---

[4]"ln" means the base of natural logarithms, $e=2.718$ ….
[5]Sometimes called the Rule of Seven.

23. An automobile manufacturer would like to add a new feature to their standard production model. The new feature will cost the manufacturer $8.67 per car to produce, and they expect to sell 100,000 cars per years. If the cars normally sell for $25,990 each, how much do they need to increase the price to reach a BEP in 18 months?

24. Calculate the annual ROI for the PI in Example 5.5 if the factory operates 300 days per year but only sells 500 widgets per day.

25. What is the ROI for the PI in Example 5.7 if the factory operates 300 days per year? Would you invest in this company?

Use the Engineering Ethics Matrix to analyze the following situations.

26. As a junior engineer in a small company, you are involved in designing a production line for a new product. The company CEO believes that this new product will make a great deal of money and wants to get it into production as soon as possible. Your immediate supervisor wants to please the CEO and get things moving, so he/she starts ordering equipment and hires contractors to expand the building. While glancing at a copy of the new project your boss gave to the CEO, you discover an error. The BEP is improbably low and the ROI is improbably high both by a factor of about 10. When you ask your immediate supervisor about this, you are told that it is none of your concern. What do you do?

27. You are the CEO of a large corporation, and a recent economic downturn requires you to eliminate all upcoming dividends to your company's stockholders. This is a highly confidential decision because it will negatively affect your stock price. Several of your wealthy neighbors own a considerable amount of your company stock. At a recent picnic in your back yard, one of them asks if she should sell some of your companies stock, considering how badly the stock market is doing in general. If you say "no," she could lose a considerable amount of money. What do you do?

Minds-On

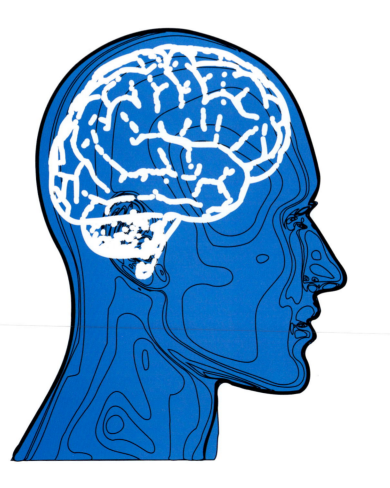

Aeronautical Engineering

6

Source: First Flight of the Wright Brothers, December 17, 1903 at Kitty Hawk, NC.

"It is better to be on the ground wishing you were flying, rather than up in the air wishing you were on the ground"

Anon[1].

6.1 INTRODUCTION

How does a fully laden Boeing 747 rise into the air despite weighing of about 450 tons? How can wings give rise to the forces that lift these 450 tons? How are those forces produced?

Practical powered flight started with the Wright Brothers' epic first flight.[2] Popular accounts of their achievement state that they designed and built their first aircraft in a bicycle shop and they used a trial and error process until they hit upon a successful design. There is some truth in this view, but they are also widely acknowledged for their many contributions to early aircraft design. Each brother had only a high school education but, as an example of their genius, they recognized that they needed some method of testing of their designs before building a final design. So they built their own wind tunnel in 1901 to test scale models of their aircraft designs. They used their wind tunnel to measure aerodynamic quantities such as "lift" and "drag" on wing designs, so their airplane at least had a chance to fly.

A large modern aircraft is the result of literally thousands of engineers and technicians working in discrete parts of its design and construction. There are many major technologies involved and, to mention a few, will certainly include mechanical engineers (on structural aspects of the aircraft), electrical engineers (traditional electrical engineering and also electronic systems), materials engineers (on structural aspects of the aircraft), aeronautical engineers (on the overall design including the theory of lift), and even possibly

---

[1]http://www.av8n.com/how/htm/4forces.html

[2]Scholars say Gustave Whitehead, a German immigrant, first flew his aircraft two years ahead of the Wrights. Based on new evidence, the aeronautical journal "Jane's All the World's Aircraft"—the bible of aviation—has officially recognized Whitehead as first in flight.

some mathematicians (on aerodynamics such as understanding shock waves, which is vital for supersonic flight). This chapter will focus on just one aspect of flight, how "lift" is generated since it is the core technology *sine qua non*. "Lift" is just one of the major forces acting over a plane in flight (Figure 6.1).

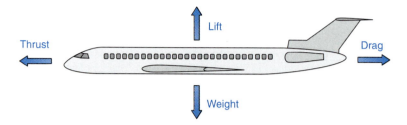

FIGURE 6.1 Forces on an airplane flying straight and level. Langleyflyingschool.com

An airplane in level flight and at constant altitude balances two major sets of forces, lift vs. weight and thrust vs. drag. Weight is an easy concept, e.g., imagine say a Boeing 787 jet liner sitting on the runway and weighing 250 tons.[3] "Lift" is the force needed to overcome gravity, and the "drag" is the penalty extracted as the plane speeds through the air. In steady level flight, the **lift = weight**. Since the wings of a Boeing 787 have an area of about 325 meters^2 (or 3500 square feet, which is more than the area of a tennis court), the magnitude of the lift is an amazing ~1 ton/m^2 of wing area. We shall examine how this is achieved.

In level flight, each of the two engines for a Boeing 787 provides enough thrust (about 267,000 N or 60,000 lbf), to speed this aircraft on its way at more than 500 mph. Unfortunately, the faster an aircraft flies, the more thrust is needed to overcome the air resistance. In steady level flight, the **drag = thrust**. Drag is mostly due to the change in momentum of the air flowing over the wings, the fuselage, and particularly the engines, which present bluff bodies to airflow at 500 mph. As the aircraft's speed increases, so does its drag, and the more fuel is uses. So an aircraft design is a tricky balance of several factors. It takes the combined expertise of many engineers to optimize these factors.

The Wright Brothers' empirical contributions needed a theoretical foundation. The engineer/scientist who filled that gap was born in Russia about 20 years before the Wright Brothers. His name was Nikolai Joukowski. Joukowski's contributions are vital to the understanding of flight.

The "Theory of Flight" is used in calculating **airfoil**[4] lift (the airfoil being the shape of the wing), so we need to start with some algebraic analysis. The topics we need to explore are airfoils and lift, imaginary numbers,[5] conformal mapping,[6] Joukowski airfoil theory, and the Kutta condition.

Don't panic! The required algebra is relatively simple, but it will lead to a very basic understanding of flight (which is especially important for those considering a career in aeronautical engineering). At this point, the student must be surprised to have seen that "imaginary numbers" is used in the theory of flight. Fortunately, the level required is only that learned by students in the 11th or 12th grade in high school. The student may have to remind himself/herself of that subject.

---

[3]The new Boeing 787 weighs roughly one half the weight of the older Boeing 747. Parts of the body and wings of the 787 are lighter than the corresponding parts of a 747 because it is made of a composite plastic-like material, not metal. The savings in weight leads to commensurate savings in fuel consumption.
[4]In British English "aerofoil".
[5]Imaginary numbers are based on $\sqrt{-1}$; they have been noted by mathematicians for 500 years but not fully appreciated until used by two mathematical giants, Euler in the 18th century and Gauss in the 19th century.
[6]The word "conformal" means mapping of one surface onto another surface so that all the angles between intersecting curves remain unchanged.

6.2 AIRFOILS AND LIFT

It's the wings that provide the lift by which an airplane defies gravity. You will convert something that is approximately parallel to the wing (thrust) into two discrete directions, one parallel to the wing (drag) and the second one that is perpendicular to the wing (lift). It's the shape and orientation of the wings that is important in producing lift.

The term "airfoil" is applied to a two-dimensional slice through the wing section. The flow across an infinitely extended airfoil is called two-dimensional or simply 2D flow, which is simply an approximation to a 3D wing—see Figure 6.2.

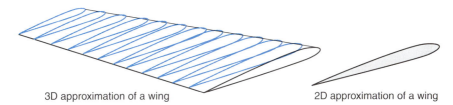

3D approximation of a wing   2D approximation of a wing

FIGURE 6.2 Three-dimensional model of wing approximated by 2D cross sections. http://www.pilot3d.com/Airfoil.htm, http://www.symscape.com/node/562

The lift is generated by the differences in air flow patterns over the top and bottom surfaces of the wing. The flow over the wing's upper surface will experience an increase in its kinetic energy and a corresponding decrease in its pressure thus producing a net vertical force on the wing. This is the simplest of many explanations of how the lift force is produced. At the leading edge of the wing, the incoming flow is split into two streams: one over the top surface of the wing and the other along the bottom. This flow can be accurately calculated by assuming the air is "frictionless".[7] We will use mathematics (specifically something called the Joukowski transform) to "morph" a simple 2D shape (a circle) into a 2D airfoil shape. The Joukowski transform will generate the profile of a wing and the air flow adjacent to it. But first we need to take a brief diversion into **imaginary numbers**.

6.3 THE ALGEBRA OF IMAGINARY NUMBERS

What is an imaginary number? An imaginary number is any real number multiplied by the imaginary unit $i = \sqrt{-1}$. Obviously, $i^2 = -1$, which is a real number that allows us to move between real and imaginary numbers, as explained below.

An imaginary number iy can be added to a real number x to form a **complex number** of the form $z = x + yi$, where x is called the *real part* and y is called the *imaginary part* of the complex number z (i.e., $x = \mathrm{Re}(z)$ and $y = \mathrm{Im}(z)$). For example, a complex number could be $z = 3.5 - i7.3$ or $z = 21.7 + i45.8$. You can plot these numbers in a "complex plane" called an **Argand diagram** (Figure 6.3).[8]

---

[7]Near the wing surfaces the airflows are dominated by the "viscosity" of the air, also called friction.
[8]Named after the Swiss mathematician Jean Robert Argand, it is a system of rectangular coordinates in which the complex number $x + iy$ is represented by the point whose coordinates are x and iy.

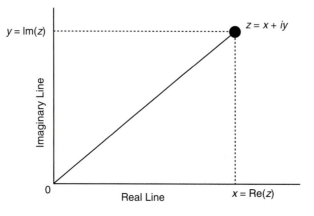

FIGURE 6.3 The complex or Argand plane for the point $z = x + iy$.

Example 6.1

Consider the complex number $A = 1/(x+iy)$. What are the real and imaginary parts of A in terms of x and y?

Need: Complex number in x, iy format for the given expression for A.

Know—How: Simple algebra and $i = \sqrt{-1}$ so that $i^2 = -1$.

Solve: $A = \dfrac{1}{x+iy} = \dfrac{x-iy}{(x-iy)(x+iy)}$

This gives: $A = \dfrac{x-iy}{x^2 - ixy + ixy - i^2 y^2} = \dfrac{x-iy}{x^2 - i^2 y^2} = \dfrac{x-iy}{x^2 + y^2}$

(Note that $x^2 - i^2 y^2 = x^2 - (-1)y^2 = x^2 + y^2$)

Thus the real part of A is

$$\mathbf{Re}(A) = \frac{x}{x^2 + y^2}$$

and the imaginary part of A is

$$\mathbf{Im}(A) = \frac{-y}{x^2 + y^2}$$

In Example 6.1, if $x = 7.3$ and $y = 4.6$, then:

$$\mathbf{Re}(A) = \frac{7.3}{7.3^2 + 4.6^2} = \frac{7.3}{74.45} = 0.098 \text{ and } \mathbf{Im}(A) = -\frac{4.6}{7.3^2 + 4.6^2} = -\frac{4.6}{74.45} = -0.062$$

then, $A = \dfrac{1}{(7.3 + i4.6)} = 0.098 - i0.062$. Note only the last form of this equation is in the standard form of a complex number.

Example 6.2

Let $A = B + iC$ be the complex number defined as $A = z + K^2/z$, where K^2 is a constant. Find the complex number A in standard format in terms of x and y.

Need: The complex number in x and iy format from $A = z + K^2/z$

Know: The real and imaginary parts of $1/z$ from Example 6.1

How: Add z to $K^2 \times$ previous result in Example 6.1

Solve: Add $\dfrac{K^2 x}{x^2 + y^2}$ and $\dfrac{-iK^2 y}{x^2 + y^2}$ to $z = x + iy$

Thus, $B = \mathrm{Re}(A) = x + \dfrac{K^2 x}{x^2 + y^2} = x\left(1 + \dfrac{K^2}{x^2 + y^2}\right)$

and $C = \mathrm{Im}(A) = y - \dfrac{K^2 y}{x^2 + y^2} = y\left(1 - \dfrac{K^2}{x^2 + y^2}\right)$

Hence $A = x\left(1 + \dfrac{K^2}{x^2 + y^2}\right) + iy\left(1 - \dfrac{K^2}{x^2 + y^2}\right)$

Note the appearance of $x^2 + y^2$ in the results of Example 6.2. This expression defines a circle of radius R about the origin (e.g., $R^2 = x^2 + y^2$). The arithmetic can be simplified by introducing dimensionless coordinates x/R and y/R which defines a *unit* circle described by the equation $x^2 + y^2 = 1$ (Figure 6.4).

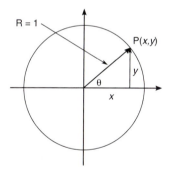

FIGURE 6.4 $x^2 + y^2 = 1$ for any point **P** (x, y) on the circumference of the unit circle.

Example 6.3

Simplify the complex number from Example 6.2 on the circumference of a unit circle.

Need: The complex number $A = B + iC = x\left(1 + \dfrac{K^2}{x^2 + y^2}\right) + iy\left(1 - \dfrac{K^2}{x^2 + y^2}\right)$ on a unit circle.

Know: The circumference of a unit circle lies on $x^2 + y^2 = 1$

How: Simplify the expression for A using $x^2 + y^2 = 1$

Solve: When $x^2 + y^2 = 1$, then $B = x\left(1 + \dfrac{K^2}{x^2 + y^2}\right) = x(1 + K^2)$ and $C = y\left(1 - \dfrac{K^2}{x^2 + y^2}\right) = y(1 - K^2)$

Thus, on the circumference of a unit circle, $A = x(1 + K^2) + iy(1 - K^2)$

All this algebra appears to have nothing to do with the theory of flight; however, the complex number $A = z + K^2/z$ is so important in the theory of flight it has been given a name, the **Joukowski Transform**. We will pursue why it is important after we have had a short look at **Conformal Mapping**.

6.4 CONFORMAL MAPPING[9]

Conformal mapping is the process of transforming a relationship from one Argand plane, $z = x + iy$, to another, $A = B + iC$, while preserving all the angles between the lines (Figure 6.5).

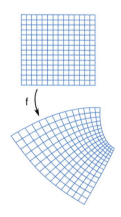

FIGURE 6.5 Example of a conformal mapping. http://math.fullerton.edu/mathews/c2003/ConformalMapDictionary.1.html

The heart of what we need to do is to have two Argand planes, one in (x, y) coordinates and one in (B, C) coordinates, and then *transform* a circular cylinder in the x, y Argand plane to an airfoil shape (i.e., a wing's cross section) in the B, C Argand plane. This is done using the Joukowski transform. Figure 6.6 summarizes what we are trying to do.

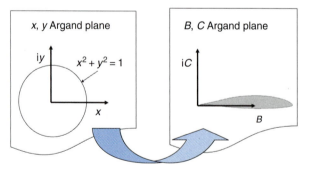

FIGURE 6.6 Conformal mapping.

[9]See http://www.grc.nasa.gov/WWW/k-12/airplane/map.html. This website simply describes conformal mapping and has an instructive and interactive model of resultant flows across airfoils.

Example 6.4

Using the Joukowski transform, $A = z + K^2/z = x(1+K^2) + iy(1-K^2)$ on a unit circle for which $y = \pm\sqrt{(1-x^2)}$. Plot the resulting graph in the $B = \text{Re}(A), C = \text{Im}(A)$ plane using a spreadsheet and explore the effects of varying the magnitude of K^2.

Need: A graph in the $B = \text{Re}(A), C = \text{Im}(A)$ plane of the transformation.

Know-How: Real and imaginary parts for A are already known from Example 6.3: $A = x(1+K^2) + iy(1-K^2)$, so $B = x(1+K^2)$ and $C = y(1-K^2)$, where $y = \pm\sqrt{(1-x^2)}$

Solve: See the spreadsheet solution in Table 6.1.

Table 6.1 Spreadsheet Solution for Example 6.4

The Joukowski Transform from the x, y plane to the B, C plane

$B = x(1+K^2)$				$K^2 = 0.980$		
Note range of x is from −1 to +1				$K = 0.990$		
$C = y(1-K^2)$ in which $y = +/- \text{SQRT}(1-x^2)$						
x	**$y = +\text{SQRT}(1-x^2)$**	**B pos root**	**C Pos root**	**$y = -\text{SQRT}(1-x^2)$**	**B Neg root**	**C Neg root**
−1.000	0.000	−1.98	0.000	0.000	−1.98	0.000
−0.950	0.312	−1.881	0.006	−0.312	−1.881	−0.006
−0.900	0.436	−1.782	0.009	−0.436	−1.782	−0.009
−0.850	0.527	−1.683	0.011	−0.527	−1.683	−0.011
−0.800	0.600	−1.584	0.012	−0.600	−1.584	−0.012
−0.750	0.661	−1.485	0.013	−0.661	−1.485	−0.013
−0.700	0.714	−1.386	0.014	−0.714	−1.386	−0.014
−0.650	0.760	−1.287	0.015	−0.760	−1.287	−0.015
−0.550	0.835	−1.089	0.017	−0.835	−1.089	−0.017

At $K^2 = 0.98$, we have generated a cross section of a symmetrical airfoil. Because this airfoil is symmetrical and if the air flow impinges parallel to the airfoil, there is no lift because the lift on the topside of the airfoil is exactly balanced by the negative lift on the underside of the airfoil. This simplified calculation is very useful since it shows the purpose of transforming from one Argand plane to another.

Expanded Spreadsheet from Table 6.1 (Using ^~)

x	$y = +\text{SQRT}(1-x^2)$	X pos root	Y = Pos root	$y = -\text{SQRT}(1-x^2)$	B Neg root	C Neg root
−1	=+SQRT(1−A12^2)	=A12*(1+$F$5)	=B12*(1−$F$5)	=−B12	=A12*(1+$F$5)	=E12*(1−$F$5)
=A12+0.05	=+SQRT(1−A13^2)	=A13*(1+$F$5)	=B13*(1−$F$5)	=−B13	=A13*(1+$F$5)	=E13*(1−$F$5)
=A13+0.05	=+SQRT(1−A14^2)	=A14*(1+$F$5)	=B14*(1−$F$5)	=−B14	=A14*(1+$F$5)	=E14*(1−$F$5)
=A14+0.05	=+SQRT(1−A15^2)	=A15*(1+$F$5)	=B15*(1−$F$5)	=−B15	=A15*(1+$F$5)	=E15*(1−$F$5)
=A15+0.05	=+SQRT(1−A16^2)	=A16*(1+$F$5)	=B16*(1−$F$5)	=−B16	=A16*(1+$F$5)	=E16*(1−$F$5)
=A16+0.05	=+SQRT(1−A17^2)	=A17*(1+$F$5)	=B17*(1−$F$5)	=−B17	=A17*(1+$F$5)	=E17*(1−$F$5)
=A17+0.05	=+SQRT(1−A18^2)	=A18*(1+$F$5)	=B18*(1−$F$5)	=−B18	=A18*(1+$F$5)	=E18*(1−$F$5)

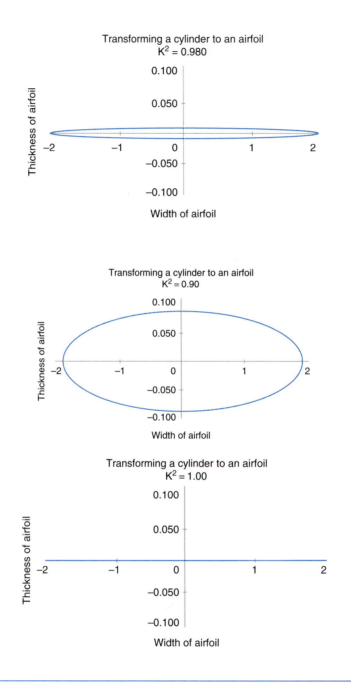

Notice that the airfoil in Example 6.4 gets thicker with smaller K^2. At $K^2 = 0.90$, there would be heavy drag on the airfoil. On the other hand, the maximum K is 1.00, and the resulting airfoil is totally flat and there is no lift.

6.5 THE JOUKOWSKI AIRFOIL THEORY

So far we have shown is how to transform a cylinder into an airfoil shape in Example 6.4, but you have yet to see how a simple flow over a circle transforms into the actual airflow. The calculated flow over an airfoil is shown in Figure 6.7.

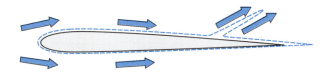

FIGURE 6.7 Airflow over a Joukowski airfoil.

Note that the flow has an obvious discontinuity at trailing edge of the airfoil. We will deal with this in the next section. So, ignoring it for a moment, see how the oncoming flow has split at the leading edge and is diverted around the airfoil, which therefore shows different flow patterns on its top and undersides. The airfoil's lift depends on these differences in flow patterns on the two sides of the airfoil.

6.6 THE KUTTA CONDITION

The flow separation as shown in Figure 6.7 should feel intuitively unreasonable because you should expect the flow to move smoothly off the trailing edge and not to back up against the local flow in that region. The discontinuity in Figure 6.7 can be dealt with by using the **Kutta** condition; it is named after a German applied mathematician who made important early contributions to the theory of flight. The effect is related to the **circulation**[10] of the flow which is denoted by an upper case Greek letter gamma (Γ). Circulation is, as the name implies, about the rotation of the flow. The Kutta condition corrects the apparent defect in airfoil flow by applying circulation[11] as follows:

> *"A body with a sharp trailing edge which is moving through a fluid will create about itself a circulation of sufficient strength to hold the rear stagnation point at the trailing edge."*

What this is saying is that flow in the close vicinity of an airfoil can be rotated until it is tangential to the trailing edge of an airfoil—see Figure 6.8.

FIGURE 6.8 The Kutta correction.

---

[10]Circulation is a mathematical operation that computes an integral of a flow with the parallel direction it covers. Circulation is calculable from the Kutta correction and the result can be used to compute the lift force of an airfoil.

[11]A.M. Kuethe and J.D. Schetzer, *Foundations of Aerodynamics*, Section 4.11 (2nd edition), John Wiley & Sons, Inc., New York, 1959.

The corresponding mathematics says that the amount of circulation you add by this rotation provides the lift required of the airfoil. The equation that is used to calculate the lift L' *per unit length of wingspan*[12] from the **circulation Γ** is:

$$L' = \rho V \Gamma \text{ in SI units} \tag{6.1}$$

$$L' = \frac{\rho V \Gamma}{g_C} \text{ in English engineering units} \tag{6.2}$$

in which the lift L' (lbf/ft of *wingspan* or N/m of *wingspan*) is equal to the product of the air density[13] ρ (lbm/ft^3 or kg/m^3), the airspeed V (ft/s or m/s), and the circulation Γ (ft^2/s or m^2/s). Equations (6.1) and (6.2) are known as the **Kutta-Joukowski theorem**. It is very worthwhile noting that the lift term L' is integrated around the contour of the wing and thus is a measure of the wing's total top and bottom of width.[14]

Example 6.5

A newly designed small airplane has a total mass of 2000. kg and a wingspan of 10.0 m. It is designed to fly at low altitude at a fixed speed of 150. km/h. At this speed and altitude, the circulation is $\Gamma = 50.0$ m^2/s and the air density is $\rho = 1.05$ kg/m^3. Is the calculated total lift, $L = L' \times wingspan$, sufficient given the plane's weight?

Need: Total lift on a small airplane of mass $m = 2000.$ kg
Know: Air density is 1.05 kg/m^3. Circulation $\Gamma = 50.0$ m^2/s and the acceleration of gravity is 9.81 m/s^2. Convert to SI units: 150. km/h = 41.7 m/s, and 2000. kg = 2000. × 9.81 [kg] [m/s^2] = 19,620 N.
How: Use the Kutta-Joukowski theorem in SI units.
Solve: $L' = \rho V \Gamma = 1.05 \times 41.7 \times 50.0$ [kg/m^3][m/s][m^2/s] = 2190 N/m

If the lift is constant over a wingspan of 10.0 m, the total lift is
$L = L' \times wingspan = 2190 \times 10.0$ [N/m][m] = 21,900 N
which is larger than 19,600 N, the weight of the airplane (to three significant figures).

The airplane in Example 6.5 will not need a wingspan quite as large as 10.0 m to maintain level flight. The designer might reduce or otherwise change the wingspan to save some fuel, or just keep it "in reserve."

Example 6.6

A jumbo jet with a take-off weight of 292. tons[15] has a wingspan of 200. ft and a wing area of 3.50×10^3 ft^2. At its cruising altitude, the air density is 0.022 lbm/ft^3 and its cruising speed is 567 mph. What is

(a) the lift per ft^2 of wing area and
(b) the circulation Γ needed to enable this design?

Need: The lift per ft^2 of wing area and the circulation Γ (lbf/ft) to keep a 292-ton jumbo jet flying at 567 mph, where the air density is 0.0.022 lbm/ft^3.

---

[12]Wingspan is the tip-to-tip distance along the wing.
[13]The air density varies with altitude. If you consider a column of air on top of your head, there is progressively less atmosphere weighing on you as you ascend because the weight of the column is reduced and so decreases the density. Contrariwise air density is increased as the temperature falls with altitude.
[14]This is important because it is obviously the total lift is dependent on the *area* of the wings, the more wing area the more lift. The area is implicit in L' being per unit of span but, less obviously, the circulation Γ implicitly contains the width of the wing.
[15]One (US) ton = 2000 lbm.

Know: Since 292. [tons] × 2000 [lbm][1/ton] = 584 × 10³ lbm and the weight of a mass of 1 lbm = 1 lbf on earth, the lift force is also 584 × 10³ lbf. Speed is 832 ft/s, wingspan = 200. ft, and the wing area = 3.50 × 10³ ft². In the English engineering system, $g_c = 32.2$ lbm ft/(lbf s²).

How: Lift/ft² comes directly from the geometry and circulation via the Kutta-Joukowski equation. Lift per foot of wingspan = $L' = (584 × 10³)/200$. [lbf]/[ft] = 2920 lbf/ft

Solve: **(a)** Lift/ft² of wing area = $(584 × 10³)/(3.50 × 10³)$ [lbf]/[ft²] = **167 lbf/ft².**

(b) The circulation can be calculated from the Kutta-Joukowski theorem. Since

$$L' = \frac{\rho V \Gamma}{g_C} = 2920 \, \text{lbf/ft}$$

then the circulation is

$$\Gamma = \frac{L' g_C}{\rho V} = \frac{2920 × 32.2 \, [\text{lbf/ft}][\text{lbm} × \text{ft}/(\text{lbf} × \text{s}^2)]}{0.022 × 832 \, [\text{lbm/ft}^3] × [\text{ft/s}]} = 5140 \, [\text{ft}^2/\text{s}]$$

6.7 SYMMETRIC AIRFOILS

By now you have realized that the circulation around an airfoil relates to its lift, the lift itself being generated by the flow differences between the flow on top of and the flow below an airfoil. What then happens in a symmetric airfoil with the same geometry top and bottom? Even more extreme is the airfoil shape designed in Example 6.3—it's elliptical. How can you get any lift?

6.7.1 The Angle of Attack

Figure 6.9 shows how a symmetric airfoil can achieve lift even in the absence of circulation.

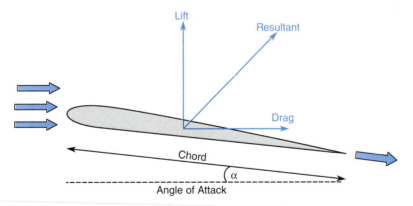

FIGURE 6.9 Lift on a symmetric airfoil. Minewiki.org

Basically, the lift arises due to the resolution of the components of forces on the wing. Have you ever put your hand out of the window of a moving car[16]? If your hand is perpendicular to the direction of motion, you feel a force pushing your hand back. If you rotate your hand parallel to the direction of motion, this is very little effect. But at a modest angle-of-attack, your hand feels both the resistance and the lift.

---

[16]You probably should not do this—it can be dangerous!

6.8 MAJOR FACTORS IN AIRCRAFT ECONOMY

For an airline to buy an airplane, there are several factors it must consider. A major item is the cost of aviation fuel.[17] The fuel is burnt in a turbine and converted to thrust. The thrust propels the aircraft and has to be sufficient to provide the desired speed. A slower aircraft might be more economical in fuel consumption, but the savings in fuel are offset by the reduced number of trips (and therefore a reduced number of tickets sold).

An engine's thrust can be increased by supplying more fuel to it. A variable in analyzing aircraft engine performance is its **specific fuel consumption**, abbreviated as **SFC**. It is defined as:

$$\text{SFC (in English units)} = (\text{lbm/hr of fuel})/(\text{lbf of thrust}) \qquad (6.3)$$

$$\text{SFC (in SI units)} = (\text{kg/s of fuel})/(\text{N of thrust}) \qquad (6.4)$$

There are oddly mixed units in both SFC definitions. Note that aviation fuel flow is measured in mass units and not in volume units,[18] since its volume will vary with temperature.

Example 6.7

At a cruising speed of 567 mph, a Boeing 787 consumed 2.73×10^4 US gallons of jet fuel (equivalent to 1.82×10^5 lbm) while flying 8000. miles. Its two engines delivered a total thrust of 2.16×10^4 lbf. What is the SFC of the engines?

Need: SFC in English engineering units for a Boeing 787 at cruising conditions. It uses 1.82×10^5 lbm of fuel for an 8000. mile flight. During the flight, its engines deliver 2.16×10^4 lbf of thrust.

Know—How: SFC = (lbm/h of fuel)/(lbf of thrust)

Solve: Time of flight is 8000./567 [miles]/[miles/h] = 14.1 h

Fuel consumption rate = $1.82 \times 10^5/14.1$ [lbm]/[h] = 1.29×10^4 lbm/h
Therefore, the **SFC** = $1.29 \times 10^4/2.16 \times 10^4$ = **0.598 (lbm/h of fuel)/(lbf of thrust)**.

Example 6.8

During a flight, the SFC of a Boeing 787 is 0.587 (lbm/h of fuel)/(lbf of thrust) and its engines produce a steady thrust of 2.16×10^4 lbf. Its cruising lift-to-drag force ratio is 15.0. What is

(a) the fuel consumption rate in lbm/h, and
(b) the weight and mass of the aircraft (including fuel and passengers)?

Need: Fuel consumption rate when the thrust = 2.16×10^4 lbf and the weight and mass of the aircraft when the total lift-to-drag force ratio is 15.0.

Know: SFC of this aircraft is 0.587 (lbm/h)/(lbf of thrust). In steady level flight, the drag must be equal to the thrust.

How: Fuel consumption rate can be calculated from the SFC, and lift = weight.

Solve: (a) The fuel consumption rate is

---

[17] Jet fuel is essentially 50/50 diesel fuel and gasoline mixture; as of the time of writing the price of jet aviation fuel averages about $6/US gallon.

[18] A jet ran out of fuel at 41,000 feet over central Canada because of incorrect conversion of mass and volume units when it was being refueled before takeoff! Amazingly the pilots were able to land without any physical injuries to the passengers and crew by gliding to a distant airfield. en.wikipedia.org/wiki/Gimli_Glider.

$SFC \times Thrust = 0.587\,[lbm/h]/[lbf] \times 2.16 \times 10^4\,[lbf] = \mathbf{1.27 \times 10^4\ lbm/h}$

(b) The total lift-to-drag force ratio is 15.0, and since the drag and thrust must be equal, then the lift = (lift-to-drag force ratio)(drag) = (lift-to-drag force ratio)(thrust) = $15.0 \times 2.16 \times 10^4$ lbf = 3.24×10^5 lbf.

In steady cruising, the lift = the **weight** $= \mathbf{3.24 \times 10^5\ lbf}$
and the mass is: **mass = Weight** $\times$ (g/g$_c$) $= \mathbf{3.24 \times 10^5\ lbm}$

SUMMARY

An aircraft flies because of **lift** generated by flow of air across its wings. It is the difference in flow between the upper and lower surfaces of the wing that generates the lift. The wings are airfoils directing the flow. In the simplest model of lift, an airfoil is mapped conformally between two complex planes or **Argand diagrams**. The **mapping of** $A = z + K^2/z$ predicts the shape of the **Joukowski airfoil** and the air flow around; it is the key step in understanding the origins of lift.

A difficulty that soon appears is that flow over the Joukowski airfoil's trailing edge detaches in a retrograde direction; it must be corrected by inserting circulation around the wing to force it to detach smoothly from the trailing edge (known as the **Kutta correction**). This correction leads to an equation for lift per unit of wingspan, $L' = \rho V \Gamma$ in which ρ is the air density, V is the aircraft's speed and Γ is the circulation to correct the Kutta condition.

Total lift is equal to weight in a cruising aircraft in level flight. The engine's thrust equates to the air drag, again while cruising in level flight.

Various ratios are used in understanding an aircraft's attributes. Important ones include the total lift to drag ratio, the **specific fuel consumption, SFC**, and others related to fuel consumption and costs.

EXERCISES

1. Complete the following table, where $i = \sqrt{-1}$.

i^{-3}	i^{-2}	i^{-1}	i^0	i^1	i^2	i^3	i^4	i^5	i^6
i			1	i	-1	$-i$			-1

2. Expand $(1+i)^2 = ?$ (**Answer:** $2i$)

3. Carry out the multiplication $(3+2i)(1+7i) = ?$ (**Answer:** $-11+23i$)

4. What is $(a+bi)(a-bi) = ?$ (**Answer:** $a^2 + b^2$)

5. The absolute value of a complex number is defined to be the distance from the origin to the number in the Argand plane using the Pythagorean Theorem. What is the absolute value of $z = x + iy$,? (**Answer:** the absolute value of z is $\sqrt{x^2 + y^2}$)

6. Simplify: $\dfrac{4-i}{7+i}$. **Answer:** $\dfrac{27}{50} - \dfrac{11i}{50}$

7. Simplify: $\dfrac{(3+2i)^2}{(-1+3i)}$.

8. What are the real and imaginary parts of $A = 1/z^2$? Write your answer in the form $A = B + iC = \text{Re}(A) + i\text{Im}(A)$.

9. Look up Euler's/de Moivre's Theorem in your high school mathematics text book (or the Internet). It says that $e^{i\theta} = \cos\theta + i\sin\theta$, where θ is the rotational angle in the complex plane.[19] Draw lines corresponding to $\theta = 0$, $\pi/4$, $\pi/2$, $3\pi/4$, and π. What are their x, y values on the surface of a unit circle?

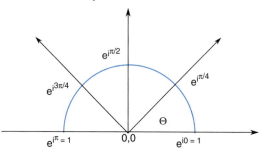

10. What is the total lift in steady level flight on an airfoil 21.0 m wingspan travelling at 150. m/s if the air density is 0.525 kg/m^3 and the circulation is 55.0 m^2/s?

11. In Example 6.8, the SFC for a Boeing 787 is 0.587 lbm/h/lbf leading to a fuel consumption rate of 12,680 lbm/h. If a standard aviation jet fuel (JP-4) costs \$3.50/US gallon, and has a density of 6.84 lbm/US gallon, how much does fuel cost for each hour of level flight?

12. An aircraft has a specific fuel consumption SFC of 0.450 lbm/h per lbf of thrust. Assume this is a constant during normal steady cruise conditions. The thrust over this period is a constant 5120 lbf. On a flight lasting 2.00 hours, how much fuel is consumed by this aircraft? How much is the cost of fuel over this 2-hour flight? Assume 1.00 lbm of fuel $= 0.150$ US gallons and that it costs \$3.50/US gallon. Ignore the weight changes due to fuel consumption en-route.

13. Most large passenger planes fly at a cruising speed of about $0.85 \times$ Mach 1.[20] The Mach number is defined as the ratio of actual speed V to the speed of sound, c, as $M = V/c$. The Mach number is an important variable in most speeding objects moving at or exceeding the local speed of sound. The speed of sound varies directly with the square root of the absolute temperature $(c \sim \sqrt{T})$. At 20 °C, the speed of sound is 343 m/s or 767 mph. At an altitude of 10,000 m, the atmosphere has cooled down to about -45 °C $= 228$ K. What is the speed of sound at 10,000 m in m/s and mph?

14. Refer to Example 6.8: How much *extra* fuel does it take if another passenger (including luggage) is added to the plane for a total mass gain of 175 lbm? Assume the flight time is 5.25 h and the SFC is constant during level cruising at 0.587 (lbm/h)/(lbf of thrust). The original thrust was 21,600 lbf.

15. A new development in airfoil design is the so-called "scimitar winglet" as seen below on a United Airlines airplane.

---

[19]You can see that the translation between polar coordinates and Cartesian coordinates is easy using standard trig functions, e.g., $x = R \cos\theta = \cos\theta$ and $y = R \sin\theta = \sin\theta$ if $R = 1$ etc.

[20]Ernst Mach was an Austrian physics known especially for his work on shock waves. You have heard a shock wave when lightning strikes nearby.

Actually this particular picture shows a *double* scimitar winglet, one facing upward and the other one facing downward. Boeing states that these winglets save several percent in drag (and thus fuel) by reducing the size of vortices shed from the wingtip in normal flight. On a Boeing 737, these winglets stand as much as 3.5 m over the wingtip surfaces.

What is the net improvement in lift-over-drag ratio given the unmodified plane had a lift-to-drag ratio of 17.0 while the drag has been reduced by 4.00%? Your solution must also compensate for the combined 220.0 lbm of two winglets. Assume the unmodified weight of the plane was 176,000 lbf.

16. Using the results of Exercise 15, what is the reduction in the fuel rate of consumption in steady level flight? Assume the fuel rate is proportional to the thrust when in steady level flight.

17. Nevil Shute was a British/Australian author and aeronautical engineer highly successful in both careers. He wrote an influential book entitled "No Highway" about a post-World War II British airliner he called the "Reindeer."[21]

The narrator is a hard driving government-employed aeronautical engineer, Dr. Scott, who takes over a laboratory staffed by over-the-hill specialists, most of whom he quickly retires. But one metallurgist, Mr. Honey, is a quiet and unassuming scientist (and iron willed). He is working on stress fracture[22] of the tail plane of the Reindeer. Mr. Honey says the tail unit will catastrophically fail in a certain number of flight hours and Dr. Scott wonders if Mr. Honey is just another over-the-hill specialist. In fact, a full-scale experiment of Mr. Honey's is already underway to test for stress fractures on the tail section. Mr. Honey's projection for a catastrophic failure is already overdue.

Then a Reindeer crashes killing all on board in remote Canada (while unfortunately carrying a Soviet ambassador and causing an international incident).

Could Mr. Honey have been right all along? Should Dr. Scott have grounded the Reindeer when he *first* heard of Mr. Honey's theory? Or was his impression of Mr. Honey as ineffectual correct?

---

[21]The world's first passenger jet was the Comet. (Reindeer and Comet seem to be related by a well-known Santa poem!) The Comet *did* suffer from stress fatigue cracking and several aircraft went down with 100% fatalities by the time the problem was fixed.

[22]Take a piece of a tin can and bend it back and fro. It will eventually crack. This is a stress fracture.

Is there an ethical dilemma here? If Dr. Scott thinks Mr. Honey is wrong, why has Mr. Honey been allowed to waste the government's money on a large and useless experiment? On the other hand, if Dr. Scott believes Mr. Honey, why didn't he ground the Reindeer immediately?

Use an Engineering Ethics Matrix with these top line choices for Dr. Scott (you can add more of your own).

Action for Dr. Scott Canons (add below)	Fire Mr. Honey	Encourage Mr. Honey's expt.	Ground all Reindeers until fracture possibility rules out	Effects on British aeronautical engineering	Report to a newspaper reporter

Chemical Engineering

Source: iStockphoto, 09-27-11 © Kali Nine LLC

7.1 INTRODUCTION

Chemical engineering is a branch of engineering that applies physical (and increasingly biological) sciences as well as mathematics in the design, development, and maintenance of large-scale chemical processes that convert raw materials into useful or valuable products. Chemical engineers aid in the manufacture of a wide variety of products such as fuels, fertilizers, insecticides, plastics, explosives, detergents, fragrances, flavors, electronic chip manufacture, biotechnology, and pharmaceuticals.

The three primary physical laws underlying chemical engineering design are (1) **the conservation of mass**, (2) **the conservation of momentum**, and (3) **the conservation of energy**. The movement of mass and energy around a chemical process is evaluated using mass and energy balances. In more advanced texts, some of these basic ideas are further applied in a powerful methodology called *unit operations*, but for the moment, we confine ourselves to simpler techniques: the movement of mass and energy around a chemical process is evaluated using mass balances and energy balances, as illustrated in the following section.

7.2 CHEMICAL ENERGY CONVERSION

How is the chemical energy of a fuel, such as gasoline or natural gas, converted into the kinetic energy of an automobile moving down the road or the electrical energy from a turbine-generator set used to power a television or a refrigerator? The conversion in both cases is part of a multistep process, and in this chapter, we consider the first step in that process, the conversion of chemical energy into thermal energy.

Combustion is the oxidation of a fuel to generate heat, perhaps also accompanied by the emission of light. For engineers, oxidation is basically the chemical reaction of a substance with oxygen.[1] Combustion may be either slow or rapid, depending on the circumstances.

Other processes can be analyzed in a similar way to combustion. Whereas combustion processes require that one keeps tabs on the fate of atoms of reactant species, some processes, such as those found in oil refineries, can be analyzed by keeping tabs on molecules. A prime example of the latter is distillation, a major process that is used in oil refineries to separate useful compounds from crude petroleum.

The key points to understanding combustion and the field of chemical engineering are the concepts of:

- **Atoms, molecules**, and **chemical reactions**.
- **kmol**, a name for a very large number of atoms or molecules.
- **Stoichiometry**, a "chemical algebra" method.
- **Air-to-fuel ratio**, a measure of the amount of air present when combustion occurs.
- **Heating value**, the amount of heat produced by the combustion of a fuel.

One useful principle that is very helpful is the discovery by the Italian scientist Avogadro that equal volumes of gases at the same conditions of temperature and pressure contain the same number of molecules.

7.3 ATOMS, MOLECULES, AND CHEMICAL REACTIONS

An atom (from the Greek meaning "cannot be sliced") is the smallest possible piece of a chemical element. Although now known to be sliceable, atoms are still used as the basic building blocks of matter in almost all engineering models. In combustion, as in all chemical reactions, the number and type of participating atoms must remain constant. A molecule is the smallest possible piece of a chemical compound. Molecules are made of atoms. In combustion, as in other chemical reactions, the number of molecules present typically does not remain constant, since molecules can be "sliced." This mathematics of molecules, where 2 plus 1 can equal 2, is presented symbolically in chemical equations, such as the combination of two hydrogen molecules with one oxygen molecule to form two molecules of water:

$$2H_2 + O_2 \rightarrow 2H_2O \tag{7.1}$$

This equation means that two molecules of hydrogen (symbol H_2, meaning a molecule contains two atoms of hydrogen, each symbol H) combine with one molecule of oxygen (symbol O_2), which is a molecule that contains two atoms of oxygen, each symbol O. The result of these molecules reacting is to form two molecules of water, a molecule with symbol H_2O; each molecule of which contains two atoms of hydrogen and one of oxygen. Note that there are four hydrogen and two oxygen atoms to the left and four hydrogen and two oxygen atoms to the right of the arrow.

7.4 THE MOL AND THE KMOL

Since molecules are extremely small entities, it takes enormous numbers of them to provide useful amounts of energy for powering automobiles or performing any macroscopic task. So rather than counting molecules by

[1]Chemists have generalized this definition to mean the loss of an electron from an atom or molecule, but for most engineering purposes, this definition given is sufficient.

ones or twos, they are counted in very large units called mols,[2] or even larger units called kmols[3] (which are thousands of mols).

> *The mol is defined to be the amount of substance containing as many "elementary entities" as there are atoms in exactly 0.012 kg of pure carbon-12 (the kmol is a factor 10^3 larger).*

Just as a dozen eggs is a way of referring to exactly 12 eggs, a mol is a way of referring to 6.0221367×10^{23} molecules, which is the number of elementary entities in exactly 0.012 kg of carbon. This number of elementary entities is very large, indeed, and is referred to as **Avogadro's Number** (N_{Av}). Obviously, in a kilomole, the number of elementary entities is 6.0221367×10^{26}.

Elementary entities may be such things as atoms, molecules, ions, electrons, or other well-defined particles or groups of such particles. The mole unit is therefore nothing but an alternate unit to counting individual elementary particles, and it is useful in the analysis of chemical reactions. Continuing our dozen-egg analogy, the elementary entities might consist of five individual chicken eggs and seven individual turkey eggs. If so, notice that not every egg has the same mass. The atomic masses[4] of some common elements correct to three significant figures are given in Table 7.1. They are measured *relative* to the mass of carbon-12 (written C^{12} or C-12), which is **exactly** 12 (an integer). In addition, the number of mols n of a substance with mass m and a molecular mass M is given by

$$n = \frac{m}{M} \tag{7.2}$$

Table 7.1 Atomic Masses of Some Common Elements to Three Significant Figures

Hydrogen, H	1.00	Nitrogen, N	14.0
Oxygen, O	16.0	Helium, He	4.00
Carbon, C	12.0	Argon, Ar	40.0
Sulfur, S	32.1	Chlorine, Cl	35.5

Many gases are divalent (i.e., chemically combined as a paired set), such as hydrogen, oxygen, and nitrogen molecules, written H_2, O_2, and N_2, respectively (and their molecular masses are 2.00, 32.0, and 28.0, respectively). Therefore, every kmol of water has a mass of 18.0 kg, since the atomic mass of every hydrogen atom is (approximately) 1.00 kg/kmol and the atomic mass of every oxygen atom is 16.0 kg/kmol.

---

[2]The mol is an abbreviation for *mole*, which in turn is an abbreviation for the word *molecule*. Without a prefix, mol/mole always means a *gram* mole that contains Avogadro's Number of elementary entities.

[3]Note that we can also define a *kilogram mole* (or kg mole, further abbreviated as kmol) as 1000 mol. Because of its virtual universal acceptance, the fundamental molar unit is the mol rather than the more logical SI-compatible kmol.

[4]The term *g atom* (gram atom) is sometimes used to refer to the mass of atoms in Avogadro's Number of the particular atom in question so that 1.00 g atom of oxygen atoms has a mass of 16.0 g. However, the definition of *mol* in terms of "elementary entities" is inclusive, and we need not use the "g atom" terminology.

Example 7.1

a. How many mols of water are in 10.0 kg of water?

b. How many kmols of water are in 10.0 kg of water?

Need: Number of mols and kmols in 10.0 kg of H_2O.

Know: Atomic masses of O and H are 16.0 and 1.00, respectively.

How: The number of moles n of a substance with a mass m that has a molecular mass M is given by $n = m/M$.

Solve: The molecular mass of water (H_2O) is $M = 2 \times (1.00) + 1 \times (16.0) = 18.0$ kg/kmol $= 18.0$ g/mol.[5] Then, for 10.0 kg of water,

> **a.** 10.0 kg $= 100. \times 10^2$ g, then $n = m/M = 100. \times 10^2$ [g]/[18.0 g/mol] $= \mathbf{556\ mol}$.
>
> **b.** $n = m/M = 10.0$ [kg]/[18.0 kg/kmol] $= \mathbf{0.556\ kmol}$.

Example 7.2

Determine the effective molecular mass of air, assuming it is composed of 79.0% nitrogen and 21.0% oxygen.

Need: Molar mass of air with 21.0% O_2 and 79.0% N_2. Therefore, air is a mixture. A kilomole of a mixture of "elementary entities" must still have Avogadro's Number of elementary particles, be they oxygen or nitrogen molecules. So, we need the combined mass of these two kinds of elementary entities in the correct ratio, each of which has a different mass.

Know: Molar mass of O_2 is 32.0 kg/kmol, and the molar mass of N_2 is 28.0 kg/kmol.

How: Proportion the masses of each constituent according to their concentration.

Solve: $M_{air} = \%N_2 \times M_{N_2} + \%O_2 \times M_{O_2} = 0.790 \times 28.0 + 0.210 \times 32.0 = \mathbf{28.8\ kg/kmol}$. Note: We have defined a kilomole of air (even though "air" molecules *per se* do not exist), but in so doing, we preserved the notion that every mole should have Avogadro's Number of entities.

7.5 STOICHIOMETRY

The goal of many important engineering models is determining the energy that can be provided by the combustion of a particular kind of fuel. For example, how does an automobile engine achieve by combustion the conversion of some 80. kJ per second of chemical energy into the same quantity of thermal and mechanical energy. The model of combustion presented in this chapter helps answer that question.

A first step in the model is writing the sort of symbolic chemical reaction described previously. A combustion reaction has the following general form:

$$\text{Fuel} + \text{oxygen} \rightarrow \text{reaction products} \tag{7.3}$$

For example, the equation for the formation of water $2H_2 + O_2 \rightarrow 2H_2O$ is a combustion reaction, with hydrogen as the fuel and water (H_2O) as the reaction product.

For most of the combustion reactions, we consider the fuel is a **hydrocarbon**, a mixture of compounds containing only two chemical elements: hydrogen and carbon. For example, "natural gas" is mostly methane (CH_4),

---

[5]And, equally, the molecular mass of water is 18.0 lbm/lbm mol $= 18.0$ ton/ton mol, etc.

which has four hydrogen atoms bonded to each carbon atom. Gasoline is a mixture of more than a hundred hydrocarbons, with chains of carbon and hydrogen atoms containing from 4 to 12 carbon atoms. For many purposes, it can be conveniently modeled by considering it to consist only of molecules of isooctane, C_8H_{18}, with a molar hydrogen-to-carbon ratio[6] of $18/8 = 2.25$.

The energy history of the modern world might be summed up as an increase in the hydrogen-to-carbon ratio of the predominant fuel. Coal, the dominant fuel in the nineteenth century, has about one atom of hydrogen per atom of carbon. The twentieth century saw increasing use of petroleum-based fuels (with about two hydrogen atoms per carbon atom). At the turn of the twenty-first century, wealthy societies rely increasingly on natural gas (with about four hydrogen atoms per carbon atom). A possible goal for the twenty-first century is the "hydrogen economy," with an infinite hydrogen-to-carbon ratio (that is, no carbon in the fuel at all). Advocates of this continuing reduction of the carbon content of fuels point out that the combustion of carbon produces carbon dioxide, a major "greenhouse gas" implicated in global warming. Nuclear, solar, wind, and hydroelectric energy are possible contenders for primary noncarbon energy sources (indeed, these are also the first two entries in the National Academy of Engineering list of Engineering Challenges for the twenty-first century).

In general, in an engineering analysis, we know the fuel we want to "burn" and (most of) the reaction products that result. But we do not initially know the number of kilogram mols of oxygen needed to combine with the fuel, and we do not know the number of kilogram mols of each kind of reaction product that will result.

We can express this situation symbolically by assuming that we have just one kilogram mole of fuel and putting undetermined coefficients in front of the other chemicals present to represent those unknown amounts. For the example we have been considering, this would look as follows:

$$H_2 + aO_2 \rightarrow bH_2O \tag{7.4}$$

Determining the numerical value of the coefficients a and b is done by a "chemical algebra" called **stoichiometry** (Greek for "component measuring"). It relies on a key fact mentioned earlier: the number of each kind of atom in a chemical reaction remains constant. So, we simply write an equation for each type of atom that expresses this equality. Then, we solve for the unknown coefficients just as we solve any other set of simple algebraic equations. In the reaction shown in Equation (7.4), for example, we have:

- **Hydrogen equation**: $2 = 2b$ (because there are 2 kmol of hydrogen atoms on the left and $2b$ kmol of hydrogen atoms in the b kmol of water molecules on the right).
- **Oxygen equation**: $2a = b$ (because there are $2a$ kmol of oxygen atoms on the left and b kmol of oxygen atoms in the b kmol of water molecules on the right).

Solving those two equations with two unknowns yields $b = 1$ and $a = \frac{1}{2}$. Substituting this back into our original reaction results in the **stoichiometric equation**:

$$H_2 + 0.5O_2 \rightarrow H_2O \tag{7.5}$$

You will note that this equation looks slightly different from the one that was presented earlier ($2H_2 + O_2 \rightarrow 2H_2O$), but you can reassure yourself that it is actually the same in meaning as the earlier equation by dividing each of the coefficients of that equation by 2. Note, too, that 0.5 kmol of O_2 makes perfectly good sense, whereas there is ambiguity[7] if we think we can equate $\frac{1}{2}$ molecule of oxygen (i.e., 0.5 O_2) with one atom of O.

---

[6]By mass, the hydrogen-to-carbon ratio is approximately 0.158.
[7]We learn later that the *energetics* of the reaction is dependent on the *state* of the reactants and products; O_2 is markedly different in this respect from 2O just as CO_2 is different from C and O_2.

The process of finding the stoichiometric coefficients can be more difficult as the number of atoms in the participating molecules increases and yet more complicated when there are repeat molecules on both sides of the chemical reaction. Consider some common fossil fuels: natural gas (methane, CH_4), coal[8] (which we assume is approximately $CHO_{0.1}S_{0.05}N_{0.01}+ash$), oil $\sim CH_2$, and many others. Most novices believe incorrectly that they can reliably find stoichiometric coefficients by inspection. Experienced engineers always use a systematic method. We use a systematic method using tabular entries.

Example 7.3

Common ether, better identified as diethyl ether, can be written as $H_3CH_2COCH_2CH_3$ but can be more simply written[9] as $C_4H_{10}O$. What are the stoichiometric coefficients to burn it completely?

$$C_4H_{10}O + aO_2 \rightarrow bCO_2 + cH_2O$$

Need: Stoichiometric coefficients, a, b, and c.
Know: You could certainly find a, b, and c by trial and error.
How: The calculation can be systematized into a simple tabular method of finding three equations in the three unknowns a, b, and c, as per Table 7.2.

Table 7.2 Tabular Method of Solving Stoichiometric Combustion Problems

Atoms	LHS	RHS	Solution
C	4	b	$b=4$
H	10	$2c$	$c=5$
O	$1+2a$	$2b+c$	$a=b+c/2-\frac{1}{2}=6$

Note: LHS = left-hand side of the equation and RHS = right-hand side.

Solve: $a=6$, $b=4$., and $c=5$. Therefore, $C_4H_{10}O+6O_2 \rightarrow 4CO_2+5H_2O$.

Notice that you have three unknowns (a, b, and c) and three equations: one for C, one for H, and one for O. Hence, you can solve it uniquely by equating the LHS with the RHS for each element.

In this book, we exclusively use this "tabular" method of determining stoichiometric coefficients. One secret is to make sure that simple balances, such as for C and H in this illustration, each involving only one unknown, are solved *before* those that involve more than one variable, such as for O in this instance.

Make a quick scan of your solution when complete to confirm that you have a "mass balance," that is, that you have neither created nor destroyed atoms. In this case, you can quickly spot that there are 4 C atoms (or kmols, etc.), 10 H atoms, and 13 O atoms on both sides of the equation. If other are elements present, you can use one additional row in the table to balance each element (one equation per unknown).

---

[8] Coal is very variable and has significantly different composition and properties, depending on its geologic age and its location.
[9] The chemical compositions of "organic" chemicals can be summarized as "Hill formulae," in which we first write the number of carbon atoms, then the number of hydrogen atoms, and the rest of the atoms in alphabetical order. What this loses in chemical structure, it makes up in simplicity for retrieval of data pertaining to a complex molecule.

7.5.1 The Air-to-Fuel Ratio

Although rockets in space chemically react fuel with an oxygen source that they have to haul at takeoff, most combustion reactions on earth react fuel with the oxygen in the air. In almost all their combustion models, engineers treat air as a mixture that has 3.76 kmol of nitrogen molecules (N_2) for every kmol of oxygen molecules (O_2), ignoring the other molecules, such as carbon dioxide, water vapor, and argon, present in much smaller quantities. This approximate composition of air works out to be 79.0% N_2 and 21.0% O_2 expressed equally as volume fraction or mole fractions. (They are the same if Avogadro's Law is obeyed.[10])

Many models also incorporate the fact that, in typical combustion systems, 100–200 parts per million (ppm) of the nitrogen in the air react with oxygen in the air to form nitrogen oxides that are an important cause of acid rain. However, in this introduction to engineering, we ignore this quantitatively small but important reaction and treat all the nitrogen in products as in its inert divalent form N_2. Therefore, the N_2 in air is considered totally unreactive in this simple treatment of combustion. Also, the N_2 in the reactant air remains unchanged and leaves N_2 as a product.

To design efficient combustion systems, it is important to know the amount of air needed to burn each gallon or kilogram of fuel. This amount of air is determined by the **air-to-fuel ratio**. To determine it, we begin by including those 3.76 kmol of nitrogen that accompany each kmol of oxygen in the stoichiometric equation. This is done simply by multiplying the stoichiometric coefficient for oxygen by 3.76 and using that as the stoichiometric coefficient for nitrogen. Since, in our model, the nitrogen does not take part in the combustion reaction, it usually[11] has the same coefficient on both sides of the reaction. Therefore,

$$H_2 + [\tfrac{1}{2}O_2 + \tfrac{1}{2} \times 3.76N_2] \rightarrow H_2O + \tfrac{1}{2} \times 3.76N_2$$
or
$$H_2 + \tfrac{1}{2}[O_2 + 3.76N_2] \rightarrow H_2O + 1.88N_2 \qquad (7.6)$$

The term $O_2 + 3.76\ N_2$ (in the square brackets) is *not* one mol of air even though only 1 mol of it, the oxygen, is actually participating in the reaction; in fact, it is 4.76 mol of an oxygen plus nitrogen mixture. It can be thought of as being 4.76 mol of air in the sense that 1.00 mol of air, as is any mol, always contains Avogadro's Number's worth of elementary entities irrespective of their identity as an oxygen or a nitrogen molecule; hence, 4.76 "mol of air" contains $4.76 \times N_{Av}$ elementary particles.

The previous stoichiometric equation makes it possible to determine the air-to-fuel ratio in two ways: as a ratio of numbers of molecules (called the molecular or molar air-to-fuel ratio) and as a ratio of their masses (called the mass air-to-fuel ratio). We use the nomenclature $(A/F)_{molar}$ and $(A/F)_{mass}$ to distinguish between these two dimensionless ratios. Since these ratios have no units, they are best written explicitly as $(A/F)_{molar} = $ [kmol of air/kmol of fuel] and $(A/F)_{mass} = $ [kg of air/kg of fuel], respectively, to emphasize what is being stated.

---

[10]After an extensive period of experimentation, the Italian chemist Count Amado Avogadro (1776–1856) proposed in 1811 that equal volumes of different gases at the same temperature and pressure contained equal numbers of molecules. It was not generally accepted by the scientific community until after Avogadro's death. The law is equivalent to claiming that each component in the gaseous mixture behaves as an *ideal gas*, a constraint that is best assumed at low pressures and high temperatures for most gases.

[11]The exception is the fuel-bound nitrogen case, e.g., burning ammonia gas according to $NH_3 + 1.5[O_2 + 3.76\ N_2] \rightarrow 3H_2O + \mathbf{6.64}\ N_2$ (and not $\mathbf{5.64}\ N_2$).

Example 7.4

Determine the molar and mass stoichiometric air-to-fuel ratios for the combustion of hydrogen.

Need: $(A/F)_{molar}$ = ____ mol of air per mol of hydrogen and $(A/F)_{mass}$ = ____ mass of air per mass of hydrogen.

Know: Stoichiometric equation: $H_2 + \frac{1}{2} [O_2 + 3.76 N_2] \rightarrow H_2O + 1.88 N_2$.

How: Use the stoichiometric equation to ratio mols; then, multiply by the relative masses of atoms from Table 7.1.

Solve: $A/F = \dfrac{\frac{1}{2}(1 + 3.76)}{1} = $ **2.38 kgmol of air/kgmol H$_2$**

To express the *mass* ratio, simply assign the molecular masses to each component; that is, $(A/F)_{mass} = \frac{1}{2} \times (1 \times 32.0 + 3.76 \times 28.0)/(1 \times 2.00)$ {[kmol O$_2$] [kg O$_2$/kmol O$_2$] + [kmol N$_2$][kg N$_2$/kmol N$_2$]}/[kmol H$_2$] [kg H$_2$/kmol H$_2$] = **34.3** **[kg air/kg H$_2$]** of mass in which we have identified the air term as that in braces, { }.

Since the unit method is clumsy in this example, we can use the concept of a molar air mass of M_{air} = **28.8 kg air/ kmol air** to simplify the calculation. Hence, $(A/F)_{mass}$ = 2.38 × 28.8/(1 × 2.00) [kmol air] [kg air/kmol air]/[kmol H$_2$] [kg H$_2$/kmol H$_2$] = **34.3 [kg air/kg H$_2$]** (as previously).

The fuel-to-air ratios are simply the inverses of the air-to-fuel ratios:

- $(F/A)_{molar}$ = $1/(A/F)_{molar}$ = 1/2.38 = 0.42 [kmol H$_2$/kmol air]
- $(F/A)_{mass}$ = $1/(A/F)_{mass}$ = 1/34.3 = 0.029 [kg H$_2$/kg air]

The molar and the mass air-to-fuel ratios have different uses in engineering analysis. For example, the molar air-to-fuel ratio is useful to a design engineer determining the volume of air involved[12] in a combustion process and therefore the basis for the dimensions of air intake and exhaust passages. The mass air-to-fuel ratio is useful to the engineer who wants to calculate how much fuel to provide, perhaps in estimating a vehicle's fuel economy. An environmental engineer may also calculate the masses of different types of pollutants that might result from a combustion process.

The stoichiometric fuel-to-air ratio is invariant for a given fuel composition, while actual fuel-to-air ratios can be varied by an engineer deciding just how much air or fuel to add. The term **equivalence ratio**, symbol φ (Greek phi), is defined as $\varphi = (F/A_{actual})/(F/A_{stoichiometric})$. It is useful, since it quickly tells you the overall conditions of the burn. Stoichiometric combustion means that the equivalence ratio is 1.0. In the jargon of the combustion engineer, excess fuel is described as *rich* and excess air as *lean*, for example, equivalence ratios of φ = 1.05 (rich) and φ = 0.95 (lean), respectively, the lean being preferred to ensure none of the fuel escapes in a semiburnt state.

7.6 THE HEATING VALUE OF HYDROCARBON FUELS

Beyond the stoichiometric fuel ratios, chemistry can also answer the question, "How much energy can be obtained by the combustion of 1 kg of fuel under stoichiometric conditions?" That, in turn, is a key part of answering a typical question, "How many miles per gallon of fuel is it possible for an automobile to travel?"

To determine the amount of energy that can be obtained by the combustion of 1 kg of fuel under stoichiometric conditions, the type of fuel being used must be specified. For simplicity, assume that fuel is composed of only one kind of molecule with x atoms of carbon and y atoms of hydrogen, and its combustion is described by the following stoichiometric equation:

$$C_xH_y + [x + (y/4)]O_2 \rightarrow xCO_2 + (y/2)H_2O \qquad (7.7)$$

---

[12]Recall Avogadro's law that moles and volumes are proportional to each other.

We simplify our study of the energy content of hydrocarbon fuels by *asserting* that our typical hydrocarbon fuel has a heating value of 45,500 kJ/kg, irrespective of precise composition (see Table 7.3). This says the energy to dissociate each molecular bond of these hydrocarbons is about equal, in particular for the C–C bonds. This is obviously not true of the smallest hydrocarbon member, CH_4, or indeed, if the fuels also contain significant oxygenates (such as alcohols), but it is sufficiently accurate for "linear" hydrocarbons.

Some fuels have different molecular structure than others; for example, "aromatic" fuels such as benzene (C_6H_6) form hexagonal molecules and are structurally unlike the more linear "aliphatic" fuels such as isooctane (C_8H_{18}) and cetane ($C_{16}H_{34}$). Gasoline is typically modeled as isooctane, where $x=8$ and $y=18$. Using Equation (7.7), you can show that 1 kg of gasoline will react with 3.5 kg of oxygen and produce 3.1 kg of carbon dioxide and about 1.4 kg of water (while releasing 45,560 kJ of energy—see Table 7.3).

Table 7.3 Heating Values of Hydrocarbon Fuels

Fuel	Heating Value (kJ/kg)
Methane, CH_4	55,650
Propane, C_3H_8	46,390
Isobutane, C_4H_{10}	45,660
Gasoline, "C_8H_{18}"	45,560
Diesel fuel, "$C_{16}H_{34}$"	43,980
Benzene, "C_6H_6"	42,350
Toluene, "$C_6H_5CH_3$"	42,960

Example 7.5

An engine burns a hydrocarbon with a composition C_7H_{17} at the rate of 25.0 lbm/h. How many horsepower (hp) of thermal energy are produced?

Need: hp of thermal energy by burning linear hydrocarbons of composition C_7H_{17}.

Know: Heat of combustion for linear hydrocarbons is 45,500 kJ/kg.

How: Power = fuel consumption rate × fuel energy

Fuel consumption rate $= 25.0 \times 0.454 \times 1/3600$ [lbm/h] [kg/lbm] [hour/s] $= 3.15 \times 10^{-3}$ kg/s

Solve: Power $= 3.15 \times 10^{-3} \times 45,500$ [kg/s] [kJ/kg] $= 143$ [kJ/s] $= 143$ kW $=$ **192 hp**.[13]

If this were an automotive application, all we would need is the conversion efficiency of the heat rate to mechanical power to be able to calculate the mpg achieved.

7.7 CHEMICAL ENGINEERING: HOW DO YOU MAKE CHEMICAL FUELS?

Why don't you pull your car up to an oil well and fill it up? Perhaps because you don't have an oil well in your backyard. But suppose that you did. Could the oil that comes directly out of the ground run your car? The short answer to the question is no. Your car would sputter to a stop if filled up from an oil well. In response to that negative answer, chemical engineers design oil refineries that convert crude oil from the well into gasoline at the

---

[13]Note: 1 hp $=0.746$ kW.

pump (and many other useful and related compounds). Gasoline is simply one component of crude oil that has been tailored to keep your car running smoothly and efficiently.

We saw that chemical energy is typically supplied to automobiles mostly in the form of hydrocarbons. These molecules, made of hydrogen and carbon atoms, combine with oxygen to yield carbon dioxide, water, and energy in the form of heat. The most convenient sources of hydrocarbons lie concentrated in deposits under the ground, the remains of animals and plant matter that lived hundreds of millions of years ago. The name for these hydrocarbon deposits, *crude oil*, encompasses a wide range of substances. At one extreme, hydrocarbon deposits contain just light gases containing only a few carbon atoms per molecule of hydrocarbon. Sometimes, just one carbon atom joins to four hydrogen atoms forming a hydrocarbon molecule called *methane* or *natural gas*. At the other extreme, crude oil contains a thick viscous sludge[14] with molecules containing dozens of carbon and hydrogen atoms. Some typical hydrocarbon molecules are shown schematically in Figure 7.1.

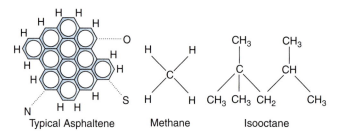

Typical Asphaltene   Methane   Isooctane

FIGURE 7.1 Some of the molecules present in crude oil.

Crude oil from an oil well would fill your tank with an impractical mixture of molecules. The heavy viscous ones would clog the fuel system, and they would not ignite in the combustion chamber. Crude oil also contains significant amounts of foul-smelling, smog-producing, sulfur and nitrogen compounds. But, the lightest hydrocarbons in crude oil evaporate easily and explode readily. A better choice is the Goldilocks' solution: fill the tank with molecules that are not too big, not too small. For automotive engines, this means molecules that have about 5–10 carbon atoms and 10–25 hydrogen atoms. This class of molecules includes the compounds collectively called *gasoline*. The ideal molecule for providing chemical energy to the Otto Cycle,[15] the cycle used in most automobile engines, is isooctane, whose nonstructural composition is C_8H_{18}.

The principal tool introduced earlier in this chapter is stoichiometry, that is, the conservation of matter measured in appropriate aggregates. It is used with control boundaries chosen to emphasize input and output streams of various processes.

Example 7.6

An oil deposit contains crude oil that is a mixture of, by moles (proportional to the number of molecules), 7.0% C_5H_{12}, 33.% C_8H_{18}, 52.% $C_{16}H_{34}$, and 8.0% $C_{38}H_{16}$. What is the percentage of each substance by mass?

Need: By mass, $C_5H_{12} = $____%, $C_8H_{18} = $____ %, $C_{16}H_{34} = $____%, and $=$____%.
Know-How: Imagine a total of 100 kmol. Divide them among the various species:

---

[14]Some crude oils are even solids at room temperature.
[15]The engine type in most cars.

7.0 kmol C_5H_{12}, 33. kmol C_8H_{18}, 52. kmol $C_{16}H_{346}$, and 8.0 kmol $C_{38}H_{16}$. Now determine the mass of each species by multiplying number of [kmol] by molecular mass in [kg/kmol]:

$$7.0 \times (60+12) = 500 \text{ kg } C_5H_{12},$$
$$33. \times (96+18) = 3800 \text{ kg } C_8H_{18},$$
$$52. \times (192+34) = 12,000 \text{ kg } C_{16}H_{34}, \text{ and}$$
$$8.0. \times (456+16) = 3800 \text{ kg } C_{38}H_{16}$$

For a total of $500+3800+12,000+3800 = 20,000$ kg

Solve: By mass,

$$C_5H_{12} = (500/20,000) \times 100 = \mathbf{2.5\%},$$
$$C_8H_{18} = (3,800/20,000) \times 100 = \mathbf{19.\%},$$
$$C_{16}H_{34} = (12,000/20,000) \times 100 = \mathbf{60\%},$$
$$C_{38}H_{16} = (3,800/20,000) \times 100 = \mathbf{19.\%}.$$

(It is recommended to use a tabular format to solve this problem, and it is especially helpful to use a spreadsheet in problems of this nature.)

7.7.1 Process Engineering

In the early decades of the twentieth century, a new idea emerged. Chemical engineers, such as Arthur D. Little, an independent consultant, and W.W. Lewis, a professor at MIT, looked at chemical manufacturing plants in a new way. They viewed the plants, however complicated, as arrangements of simple elements. These elements, which served as a sort of alphabet for chemical plant construction, came to be known as **process steps** and were analyzed using the concept of *unit operations*. It turned out that many process steps had seemed to be quite different but were closely related in their basics. Therefore, several chemical engineering processing steps, such as distillation, absorption, stripping, and liquid extraction, are all described by essentially the same mathematics; here, we use distillation as representative.

7.7.1.1 *Distillation*

In this section, we conduct a simplified analysis of a basic chemical engineering process, **distillation**. The first major process in most refineries is a set of distillation columns (see Figure 7.2). This is a purely *physical* separation process that relies on the different relative volatilities among the various components of crude oil. Generally, they are arranged in sequence with the most volatile components removed first (such as methane, pentane, and isooctane), and in the last distillation columns, the least volatile, such as asphaltenes, are removed. But, because distillation is a physical treatment, the various *molecules* retain their identity. This is *not* true of *chemical* treatments, in which the molecules are broken and reformed into other molecules, such as in a combustion process.

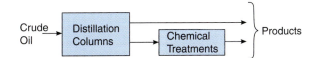

FIGURE 7.2 Process diagram for a refinery.

Distillation is conceptually a simple process operation. The key unit operation is the fact that condensed liquid (condensate) flows down the distillation column while vapor of a different composition flows up (Figure 7.3).[16]

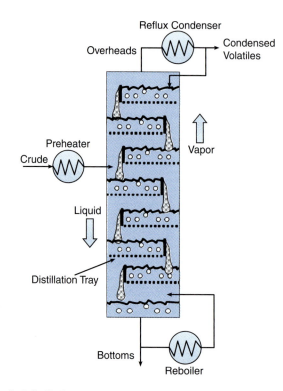

FIGURE 7.3 Schematic of a typical distillation column.

Two process diagram symbols for a distillation unit are shown in Figure 7.4. The more detailed one is the-more literal, as it includes the reflux condenser and the reboiler; however, often we do not care about the internal flows and just need the inlet and outlet flow streams. Again, you should imagine the distillation column symbol as surrounded by a control surface "dotted line," through which the arrows denoting the flows penetrate. That control surface reminds us that over enough time, the mass into a distillation unit must equal the mass out. For these two control surfaces, the internal flows are irrelevant and the two diagrams are completely equivalent.

---

[16]The driving force for distillation is the difference in composition between the liquid condensate and the surrounding vapor. A distillation column is made up of pipes, pumps, and heat exchangers, plus a large vertical column containing trays. The trays temporarily hold back the liquid condensate to expose it to up-flowing vapor of a different composition. The vapor contacts the liquid by flowing through an arrangement that does not allow passage of the liquid. For the feed, a heat exchanger adds enough heat to evaporate some of the crude oil into vapor. At the middle of the column, the hot crude oil and some of its vapor are pumped in. The vaporized components go up the column and are washed by condensate flowing down to extract any remaining amounts of the less-volatile components. The condensate is produced by condensing the vapor at the top of the column by a "reflux condenser" (another heat exchanger). The greater amount of the vapor that is condensed and recycled at the top of the column, the purer the "overheads" become. The bottom part of the column operates similarly. Condensate runs down the column and contacts up-flowing hot vapor with the result that additional volatile components are extracted from it. This vapor is produced by heating the product "bottoms" in a "reboiler" (just another heat exchanger). The greater amount of the bottoms that are heated and vaporized, the more volatiles are stripped from the bottoms. The overhead product at the top of the column is then richer in the volatile components and the bottoms product is richer in the heavier components than the original crude oil. While 100% separation can never be achieved, we can get close by having many trays and large reflux condensers and reboilers.

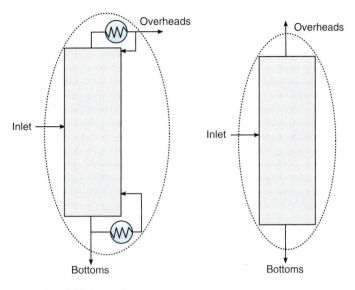

FIGURE 7.4 Process diagrams of a distillation column.

Example 7.7

A crude oil contains four components by mass: 10.% asphaltenes ($C_{38}H_{16}$), 5.0% light gases, 40.% isooctane (C_8H_{18}), and 45.% cetane ($C_{16}H_{34}$). An input stream of 1.00×10^4 kg/h of this crude oil is fed to a distillation column in an oil refinery. Assume the distillation process separates out a top and a bottom stream. If the bottom stream consists of 100% ashphaltenes, what are the flow rates and composition of all of the process lines?

Need: Composition and rate of flow of each component in each line.

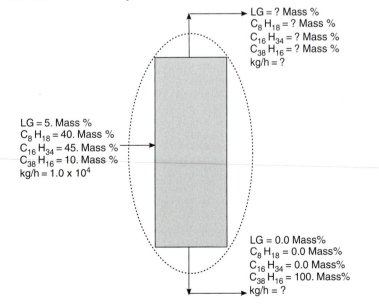

Know: Principle of conservation of mass across control volume. Also that the distillation process is a physical one, and therefore, the type and quantities of molecules are preserved.

How: Use a process diagram showing inputs and outputs of each stream and calculate the required unknowns.

Solve: Since we know the mass flow rate of the input stream is 1.00×10^4 kg/h of crude oil, we can easily calculate the amount of each component in the feed:

Inlet mass flow rate of light gases $=(5.0/100) \times 1.00 \times 10^4 = 5.0 \times 10^2$ kg/h
Inlet mass flow rate of isooctane $=(40./100) \times 1.00 \times 10^4 = 4.0 \times 10^3$ kg/h
Inlet mass flow rate of cetane $=(45./100) \times 1.00 \times 10^4 = 4.5 \times 10^3$ lb/h
Inlet mass flow rate of asphaltenes $=(10./100) \times 1.00 \times 10^4 = 1.0 \times 10^3$ kg/h

In this case, the bottoms are just the asphaltenes, so the total overheads are $1.0 \times 10^4 - 1.0 \times 10^3 = 9.0 \times 10^3$ kg/h and the percent composition can be deduced from the inlet mass percentages as

Mass% of light gases in the overheads $=(5.0 \times 10^2)/(9.0 \times 10^3) \times 100 = 5.6\%$
Mass% of isooctane in the overheads $=(4.0 \times 10^3)/(9.0 \times 10^3) \times 100 = 44.\%$
Mass% of cetane in the overheads $=(4.5 \times 10^3)/(9.0 \times 10^3) \times 100 = 50.\%$

Flow Stream	Light Gases	Isooctane	Cetane	Asphaltenes
	Percent by mass			
Inlet	5.0%	40%	45%	10%
Overheads	5.6%	44%	50%	0%
Bottoms	0%	0%	0%	100%
	Flow rate (kg/h)			
Inlet	5.0×10^2	4.0×10^3	4.5×10^3	1.0×10^3
Overheads	5.0×10^2	4.0×10^3	4.5×10^3	0
Bottoms	0	0	0	1.0×10^3

Notice the overhead steam is not pure; it consists of three components in a mass ratio of light gases to isooctane to cetane of 5.6 to 4.4 to 50. Since each of these components is potentially specialized (and valuable) product, all three are further refined in standard refinery practice. A series of distillation columns normally follows.

7.8 MODERN CHEMICAL ENGINEERING

We are limited in how many topics we can introduce between the covers of any book. Suffice it to say that chemical engineering today is much more than just process engineering. For example, chemical engineers are responsible for the availability of the high-quality materials essential for a modern world. Chemical engineers are now engaged in the development and production of a diverse range of products that include high-performance materials used in aerospace, automotive, biomedical, electronic, environmental, and space applications. Some examples include ultrastrong fibers; new fabrics, adhesives, and composites for vehicles; biocompatible materials for implants and prosthetics; gels for medical applications; and pharmaceuticals. Chemical engineers today also work on biological and bioengineering projects, such as understanding biopolymers (proteins) and developing artificial organs.

SUMMARY

Determining useful engineering parameters, such as the miles per gallon achievable by an automobile, requires calculating the amount of thermal energy (heat) available from the chemical energy contained in a fuel. This calculation requires the mastery of several concepts from the field of chemistry. Those concepts include **atoms**, **molecules**, and **chemical reactions**; a name for a very large number of atoms or molecules called the **kmol**; a chemical algebra method called **stoichiometry** for keeping track of the numbers of atoms that occur in those reactions; a measure of the amount of air needed for combustion called the **air-to-fuel ratio**; and finally, a way to understand the **heating value** of that fuel.

The term **equivalence ratio**, φ, is defined as $(F/A_{actual})/(F/A_{stoichiometric})$, is used to quickly assess combustion conditions: a "fuel-rich" process operates at $\varphi > 1$ and a "fuel-lean" one at $\varphi < 1$. The principles of stoichiometry and stoichiometric balances have further utility when applied to physical, as opposed to chemical, processes. Many oil refinery processes fall into this category, including distillation to separate useful products from crude oil.

EXERCISES

1. How many kmol are contained in 3.0 kg of ammonia NH_3? (**A: 0.18 kmol** to two significant figures.)

2. How many kmol are contained in 1.0 kg of nitroglycerine $C_3H_5(NO_3)_3$?

3. What is the mass of 5.0 kmol of carbon dioxide CO_2? (**A: 2.2×10^2 kg.**)

4. What is the mass of 1.00 kmol vitamin B1 disulfide $C_{24}H_{34}N_8O_4S_2$?

5. The effective molecular mass of air is defined as the mass of a kilomole of elementary particles of which 78.09% are nitrogen molecules, 20.95% are oxygen molecules, 0.933% are argon atoms, and 0.027% are carbon dioxide molecules. What is the effective molecular mass of air? (Watch your significant figures!) What other factor could affect the effective molecular mass of air?

6. A gallon of gasoline has a mass of about 3.0 kg. Further, a kilogram of gasoline has an energy content of about 45,500 kJ. If an experimental automobile requires just 10. kW of power to overcome air resistance at steady speed of 30. miles an hour, and *if* there are no other losses, what would the gas mileage of the car be in miles per gallon? (**A: 110 mpg** to only two significant figures, as problem is stated.)

7. Determine the value of the stoichiometric coefficients for the combustion of an oil (assumed molecular formula CH_2) in oxygen: $CH_2 + aO_2 = bCO_2 + cH_2O$. Confirm your answer is correct! (**A: $a = 1.5$, $b = 1$, $c = 1$.**)

8. Determine the value of the stoichiometric coefficients for the combustion of coal in oxygen given by the stoichiometric equation

$$CHN_{0.01}O_{0.1}S_{0.05} + aO_2 = bCO_2 + cH_2O + dN_2 + eSO_2$$

9. Determine the value of the stoichiometric coefficients for the combustion of natural gas in air:

$$CH_4 + a(O_2 + 3.76N_2) = bCO_2 + cH_2O + dN_2$$

10. Using the stoichiometric coefficients you found in Exercise 9, determine the molar air-to-fuel ratio $(A/F)_{molar}$ for the combustion of natural gas in air. (**A: $(A/F)_{molar} = 9.52$ kmol of air/kmol of fuel.**)

11. Using the results of Exercises 9 and 10, determine the *mass* air-to-fuel ratio $(A/F)_{mass}$ for the combustion of natural gas in air. (**A: 17.2 kg of air/kg of fuel.**)

12. Determine the mass air-to-fuel ratio $(A/F)_{mass}$ for the combustion of an oil (represented by CH_2) in air.

13. Determine the mass air-to-fuel ratio $(A/F)_{mass}$ for the combustion of coal in air represented by the equation

$$CHN_{0.01}O_{0.1}S_{0.05} + a(O_2 + 3.76N_2) = bCO_2 + cH_2O + dN_2 + eSO_2.$$

14. Determine the mass air-to-fuel ratio $(A/F)_{mass}$ for the combustion of isooctane C_8H_{18} in air. Its stoichiometric equation is

$$C_8H_{18} + 12.5(O_2 + 3.76N_2) = 8CO_2 + 9H_2O + 47.0N_2.$$

Exercises 15–20 use engineering considerations to give insight on the "global warming" issue.

15. Assume 1 kg of isooctane produces 45,500 kJ of energy. How much carbon dioxide is released in obtaining a kilojoule of energy from the combustion of isooctane in air? (**A: 6.79×10^{-5} kg CO_2/kJ.**)

16. Assume a kilogram of hydrogen produces 1.20×10^5 kJ of thermal energy. How much carbon dioxide is released in obtaining a kilojoule of energy from the combustion of hydrogen in air?

17. The amount of carbon dioxide levels in the atmosphere has been increasing for several decades. Many scientists believe that the increase in carbon dioxide levels could lead to "global warming" and might trigger abrupt climate changes. You are approached by a future U.S. President for suggestions on what to do about this problem. Based on your answers to Exercises 15 and 16, what might be the *simplistic* approach to reducing the rate of increase of the amount of carbon dioxide in the atmosphere due to combustion in automobiles and trucks?

18. Give three reasons why your suggested solution in Exercise 17 to global warming has *not* already been adopted.

19. A monomeric formula for wood cellulose is $C_6H_{12}O_6$ (it repeats a hexagonal structure based on this formula). The energy produced by burning wood is approximately 1.0×10^4 kJ/kg. Determine the amount of carbon dioxide released in obtaining a kilojoule of energy by the combustion of wood in air. (**A: 1.5×10^{-4} kg of CO_2 released/kJ of energy produced.**)

20. Repeat Exercise 19 for the combustion of coal. For simplicity, assume coal is just carbon, C, with a HV of 32,800 kJ/kg.

21. In 1800, the main fuel used in the United States was wood. (Assume wood is cellulose whose representative repeating formula is $C_6H_{12}O_6$.) In 1900, the main fuel used was coal (assume coal in this example can be approximated by pure carbon, C). In 2014, the main fuel used is oil (assume it can be represented by isooctane). Assume that, in 2100, the main fuel will be hydrogen (H_2). Use a spreadsheet to prepare a graph of carbon dioxide released per kilojoule of energy produced (y-axis) as a function of year (x-axis) from 1850 to 2050.

22. A pipe carries a mixture of, by kmols (which are proportional to the number of molecules), 15.% C_5H_{12}, 25.% C_8H_{18}, 50.% $C_{16}H_{34}$, and 10.% $C_{32}H_{40}$. What is the percentage of each substance by mass?

23. A crude oil contains just three components by weight: 15.% asphaltenes, 1.0% light gases (containing about 1–5 carbon atoms + hydrogen), and the remainder light "distillate," that is, pure isooctane. This crude is fed

to the distillation columns in an oil refinery at a rate of 1.0×10^4 kg/h. Assume the distillation process perfectly separates the stream of gas between a mixture of light gas and isooctane in the top stream. Draw a process diagram of this distillation column and indicate on the drawing the flow rates and composition by component of the output streams.

24. The glass pipe system shown in the following figure has a liquid flowing in it at an unknown rate. The liquid is very corrosive, and flow meters are correspondingly very expensive. Instead, you propose to your boss to add a fluorescent dye via a small pump and to deliver 1.00 g/s of this "tracer" and activate it by UV light. By monitoring the dye fluorescence downstream, you determine its concentration is 0.050 wt.%. What is the unknown flow rate in kg/h? (This is a useful method of indirectly measuring flows.)

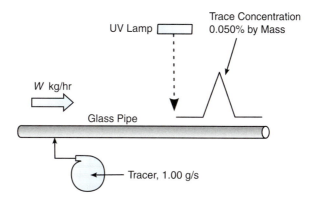

25. You have worked for a petroleum company producing automotive fuels for several years, and in your spare time at home, you developed a new fuel composition that has a higher heating value than ordinary gasoline. When you began work at this company, you signed a "confidentiality agreement" that gave them all of your intellectual property. What are your obligations to your employer? (Give your solution using the Engineering Ethics Matrix.)

   a. It is your work on your time, so you have no obligations to your current employer.
   b. Everything you used to develop your new fuel you learned on the job at your company, so your work really belongs to them.
   c. Your company's confidentiality agreement requires you to provide them with all your work.
   d. Give your idea to a friend and let him pursue it while keeping you as a silent partner.

26. As a production engineer for a large chemical company, you need to find a new supplier for a specific commodity. Since this contract is substantial, the salespeople you meet with are naturally trying to influence your purchasing decision. Which of the following items are ethical in your opinion?[17] (Give your solution using the Engineering Ethics Matrix.)

   a. Your meeting with a salesperson extends over lunch, and she pays for the lunch.
   b. In casual conversation at the sales meeting, you express an interest in baseball. After the meeting, a salesperson sends you free tickets to your favorite team's game.

---

[17]How do you distinguish between a token gift and a bribe? Where do you draw the line?

c. After your meeting, a salesperson sends a case of wine to your home with a note thanking you for the "useful" meeting.

d. As a result of the sales meeting, you are invited on an all-expenses-paid trip to China to visit the salesperson's manufacturing facility.

27. You are attending a national engineering conference as a representative of your company. A supplier to your company has a booth at the conference and is passing out small electronic calculators to everyone who comes to their booth. The calculators are valued at about $25, and you are offered one. What do you do? (Give your solution using the Engineering Ethics Matrix.)

a. Accept the gift; since it has a small value, there is no need to report it.

b. Accept the gift and report it to your supervisor.

c. Decline the gift, explaining that you work for one of their customers.

d. Ask someone else to get one of the calculators for you.

28. It is December 1928 and your name is Thomas Midgley, Jr.[18] You have just invented a new miracle refrigerant composed of chlorinated fluorocarbons (CFCs) that will make your company a lot of money. At midnight on Christmas Eve, you are visited by three spirits who show you your past, present, and future. In your future, you discover that the chlorine in your miracle refrigerant will eventually destroy the Earth's ozone layer and put the entire human race at risk. What do you do? (Give your solution using the Engineering Ethics Matrix.)

a. Claim the entire vision was due to a bit of undigested meat and forget it?

b. Run to the window and ask a passerby to go buy a large goose?

c. Destroy all records of your CFC work before anyone can use it?

d. Start work developing nonchlorinated hydrocarbon refrigerants?

29. It is 2021, and you are a process engineer at a large oil refining company. The world is rapidly moving toward a hydrogen energy economy, and your company has been trying to develop an efficient and cost-effective way to extract hydrogen from crude oil. You found a low-cost process, but it produces a considerable amount of undesirable chemical by-product pollutants. However, during your work on this project, you also discovered an effective and inexpensive way to extract hydrogen directly from seawater. You realize that revealing this process would effectively eliminate the world demand for petroleum and would probably cause serious financial damage to your company. What do you do? (Give your solution using the Engineering Ethics Matrix.)

a. Quit your job and start your own hydrogen producing company.

b. Talk to your supervisors and reveal your process to them to see if they wish to pursue implementing it as part of their company.

c. Contact a patent lawyer not associated with your current employer and try to patent this potentially lucrative new process.

d. Without your employer's permission, publish an article in a well-read chemical or energy magazine revealing your process and giving it to the world free of charge.

---

[18]See an abbreviated story of his life: http://www.uh.edu/engines/epi684.htm.

Civil Engineering

Source: www.londonhotels.com/london/attractions/must-see/tower-bridge.html.

8.1 INTRODUCTION

Civil engineering is one of the oldest and broadest engineering professions. It focuses on the infrastructure necessary to support a civilized society. The Roman aqueducts, the great European cathedrals, and the earliest metal bridges were built by highly skilled forerunners of the modern civil engineer. These craftsmen of old relied on their intuition, trade skills, and experience-based design rules, or heuristics, derived from years of trial and error experiments but rarely passed on to the next generation.

In contrast, today's civil engineers bring to bear on these problems a knowledge of the physical and natural sciences, mathematics, computational methods, economics, and project management. Civil engineers design and construct buildings, transportation systems (such as roads, tunnels, bridges, railroads, and airports), and facilities to manage and maintain the quality of water resources. Society relies on civil engineers to maintain and advance human health, safety, and our standard of living. Those projects that are vital to a community's survival are often publically funded to ensure that they get done, even where there is no clear or immediate profit motive.

8.2 WHAT DO CIVIL ENGINEERS DO?

Civil engineers may find themselves just as easily on the computer, optimizing a design, or in the field, monitoring progress and interacting with customers, contractors, and construction team members. They are involved throughout the life cycle of the project: exploration and measurement of on-site environmental conditions, design and computer modeling, management of construction, system management of operation (e.g., of traffic or water resources), demolition, and recycling of materials.

An important challenge confronted by all civil engineers is the need to deal with the uncertainty of highly variable environmental conditions. Informed design decisions by civil engineers based on a historical knowledge of environmental extremes are our first line of defense against hurricane winds, earthquakes, droughts, flooding, periods of extreme heat or cold, and nonuniform or time-dependent soil properties. When actual conditions exceed predictions of those extremes, disasters happen.

In this chapter, you will gain a flavor of the fundamentals underpinning three traditional areas of civil engineering:

- **Structural engineering** deals with the analysis and design of structures that support and resist loads. Structural engineers are most commonly involved in the design of buildings, bridges, tunnels, heavy machinery, and vehicles or any item where structural integrity affects the item's function and safety. For example, buildings must be designed to endure heavy wind loads as well as changing climate and natural disasters.
- **Geotechnical engineering** deals with the behavior of earth materials. It uses principles of soil mechanics and rock mechanics to investigate underground conditions and materials. Using this information, they design foundations and earthworks structures for buildings, roads, and many other types of projects.
- **Water resources engineering** deals with designing the infrastructure to manage and utilize the nation's water supply. Applications range from irrigation, navigation, and hydropower generation to water treatment plants and the network of pipes that convey clean, safe drinking water to our taps. It also involves the treatment of domestic and industrial wastewaters before discharging them back into the environment.

New civil engineering fields and applications are constantly emerging as new technologies replace old ones and priorities shift. For example, environmental engineering, earthquake engineering, and alternative energy engineering (the development of alternative energy systems including new energy extraction methods) have all become more prominent in recent years.

8.3 STRUCTURAL ENGINEERING

Civil engineers strive to create efficient structures by minimizing cost or weight subject to constraints on **stiffness** (resistance to deformation) and **strength** (resistance to breaking). Today, the stiffness and strength of a prospective structure can be rapidly evaluated using computational methods. Many of the fundamentals underlying these computational models, such as free-body diagrams (FBD) and the equations of static equilibrium, have been around for a long time. For example, the American civil engineer Squire Whipple was analyzing truss bridges in the 1830s. But, the limited computing power of those times placed severe restrictions on the range of bridge designs that could be safely built.[1] With the development of the computer and its ability to solve huge systems of equations quickly, the variety of structures that can be accurately modeled now seems unlimited.

---

[1]For a perspective on early American bridge failures see Henry Petroski, *To Engineer Is Human*, (New York: Vintage Books, 1992).

8.3.1 Truss Structures and the Method of Joints

A **truss** is a special type of structure renowned for its high strength-to-weight and stiffness-to-weight ratios. This structural form has been employed for centuries by designers in a myriad of applications ranging from bridges and race car frames to the International Space Station.

Trusses are easy to recognize—lots of straight slender struts joined end-to-end to form a lattice of triangles, such as the bridge in Figure 8.1. In large structures, the joints are often created by riveting the strut ends to a gusset[2] plate, as shown in Figure 8.2. A structure will behave like a truss only in those regions where the structure is fully triangulated; locations where the struts form other polygonal shapes (e.g., a rectangle) may be subject to a loss of stiffness and strength.

FIGURE 8.1 Truss bridge in Interlaken, Switzerland.

FIGURE 8.2 Joint formed by riveting a gusset plate to converging members.

---

[2]A gusset plate is a thick steel plate used for joining structural components. It may be triangular, square, or rectangular. A recent failure of one in Minneapolis caused a bridge to collapse with multiple deaths and injuries.

The special properties of a truss can be explained in terms of the loads being applied to the individual struts. Consider the three general types of end loadings shown in Figure 8.3: tension, compression, and bending. If you were holding the ends of a long thin steel rod in your hands and wanted to break it or at least visibly deform it, bending would be the way to go. Therefore, if we could eliminate bending of the struts as a potential failure mode, the overall strength and stiffness of the truss would be enhanced. This is precisely the effect of the truss geometry on the structure, as the stiff triangular lattice serves to keep any bending induced in the struts to a minimum.

If we accept that all the members of a truss are in either tension or compression, each joint can be idealized, without loss of accuracy, by a **single pin** connecting the incoming members. The pin implies that the members are free to rotate with respect to the pin, thereby avoiding introduction of any bending as the structure deforms. Both the **live load**[3] (due to traffic) and the **dead load** (due to the weight of the members) must be applied at the pins to preserve the "no bending" assumption in our mathematical model. Often, the weights of the members are assumed to be negligible compared to the live load.

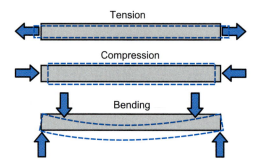

FIGURE 8.3 Different end-loading possibilities. The dashed line represents the deformed shape produced by the applied forces.

The forces acting at the ends of each strut in a truss can be found from static equilibrium using a specialized approach known as the **Method of Joints**. This method derives its name from the fact that the static analysis is based on the FBDs of the pins. You may recall from physics that a FBD is a drawing in which the component is shown isolated from its environment with the effects of adjoining members and other external forces (e.g., gravity) replaced by equivalent forces. The feasibility of basing the analysis on the FBDs of just the pins is demonstrated in Figure 8.4, which shows the FBDs of the pin and struts composing a typical model of a joint. All the struts are assumed to be in tension, in accordance with the recommended sign convention, while subscripts uniquely associate the forces with the members. To minimize the number of unknowns, forces exerted by adjoining parts are shown as equal in magnitude and opposite in direction as required by Newton's Third Law. Equilibrium of the struts yields the trivial result that the end forces are equal and so may be dropped from the analysis.

---

[3]In 1987, on the 50th anniversary of the opening of the Golden Gate Bridge, the bridge was closed to traffic and opened to pedestrians. So many people (hundreds of thousands) celebrated on the bridge that its live load was the weight of these people. This justifiably scared the bridge operators into rapid stress calculations as the bridge's arch had visibly flattened under the load!

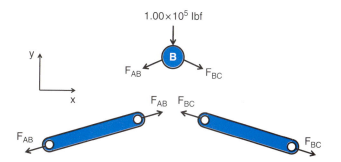

FIGURE 8.4 A mathematical model of a typical truss joint. The gusset plate has been modeled as a pin to indicate that no bending is present. The pin is held in equilibrium by forces exerted by the connecting members and the known live load. Since we already know the end forces of each strut are equal, their free-body diagrams need not be included in the final static analysis.

8.3.2 Example Using the Method of Joints

Our goal in applying the method of joints is to determine the forces on the members constituting the truss bridge of Figure 8.5. The magnitude of the applied force is based on a conservative estimate of the **live load** due to pedestrian traffic.

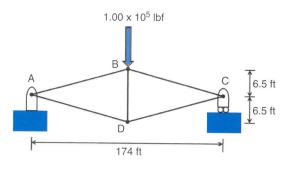

FIGURE 8.5 Truss bridge.

    The first step is to draw the **free-body diagrams**. The FBDs of the pins are shown in Figure 8.6. We assume all members are in tension and use subscripts to associate the forces with the members. Forces exerted by the ground supports on the pins are labeled as A_x, A_y, and C_y. There are two force components at pin A, since this support prevents the movement of the pin in two directions. The rollers under the support at C mean this support does not limit the horizontal movement in the x direction at C (i.e., $C_x=0$). The assumed directions of these support reactions are guesses; later, negative signs in the final calculated results tell us where those guesses were incorrect.

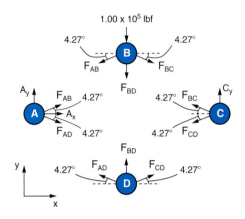

FIGURE 8.6 Pin FBDs.

The next step is to check that the total number of unknowns associated with the pins is equal to the total number of available equilibrium equations. Only when they match will it be possible to solve the equations. Such a problem is referred to as being **statically determinate**. For each pin, there are two equations governing *static* **equilibrium**, representing the summation of forces in two coordinate directions:

$$\sum_{i=1}^{N} F_{xi} = 0 \quad \sum_{i=1}^{N} F_{yi} = 0 \tag{8.1}$$

where N is the total number of forces acting on the joint, F_{xi} is the x component of the ith force, F_{yi} is the y component of the ith force, and $\sum$ indicates a *summation*[4] over the N force *components*. The x and y components of a force are found by projecting it onto orthogonal axes located at the tail of the force vector, as illustrated in Figure 8.7. Checking determinacy, we find that there are eight unknowns ($F_{AB}, F_{BC}, F_{CD}, F_{AD}, F_{BD}, A_x, A_y$, and C_y) and eight equations (four pin FBDs with two equilibrium equations per FBD). This tells us in advance that the equations will be solvable.

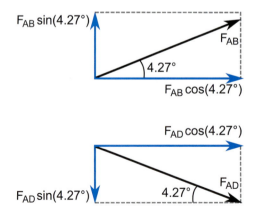

FIGURE 8.7 Components of the forces acting on joint A.

---

[4]Pronounced "sigma."

The determinacy check clears the way to write the equilibrium equations. The FBD of joint A in Figure 8.6 yields the following two equations describing equilibrium:

$$\text{Pin } A: \sum F_x = 0 \text{ and then we have } F_{AB}\cos\left(4.27°\right) + F_{AD}\cos\left(4.27°\right) + A_x = 0$$

$$\sum F_y = 0 \text{ and then we have } F_{AB}\sin\left(4.27°\right) - F_{AD}\sin\left(4.27°\right) + A_y = 0$$

where the signs are determined by the directions of the force components relative to the coordinate axes. Thus, x components that point to the right are positive and those that point to the left are negative, and y components that point up are positive and those that point down are negative. The equilibrium equations for the remaining pins are

$$\text{Pin } B: \sum F_x = 0, \text{ then } -F_{AB}\cos\left(4.27°\right) + F_{BC}\cos\left(4.27°\right) = 0$$

$$\sum F_y = 0, \text{ then } -F_{AB}\sin\left(4.27°\right) - F_{BC}\sin\left(4.27°\right) - F_{BD} - 1.00 \times 10^5 = 0$$

$$\text{Pin } C: \sum F_x = 0, \text{ then } -F_{BC}\cos\left(4.27°\right) - F_{CD}\cos\left(4.27°\right) = 0$$

$$\sum F_y = 0, \text{ then } F_{BC}\sin\left(4.27°\right) - F_{CD}\sin\left(4.27°\right) + C_y = 0$$

$$\text{Pin } D: \sum F_x = 0, \text{ then } -F_{AD}\cos\left(4.27°\right) + F_{CD}\cos\left(4.27°\right) = 0$$

$$\sum F_y = 0, \text{ then } F_{AD}\sin\left(4.27°\right) + F_{CD}\sin\left(4.27°\right) + F_{BD} = 0$$

8.3.3 Solution of the Equations Using Excel

The solution of these equations by hand is both cumbersome and subject to error. We, therefore, turn to a computational approach based on first translating the equations into a **matrix** form that can be read by a computer and then solving the equations with the matrix operations available in a spreadsheet.

Two intermediate steps are required to set up the matrix equations. First, any constants in the equations need to be shifted to the right-hand side of the equal ($=$) signs. Of course, this changes the sign on those constants. Second, the coefficients multiplying the unknown forces need to be identified. The resulting right-hand side constants (RHS) and the coefficients of the forces are listed as follows for each equation (EQ). The order of the forces across the top of the table is totally arbitrary (Table 8.1).

Table 8.1 Matrix Setup of Equations

EQ	F_{AB}	F_{BC}	F_{CD}	F_{AD}	F_{BD}	A_X	A_Y	C_Y	RHS
1	cos(4.27)	0	0	cos(4.27)	0	1	0	0	0
2	sin(4.27)	0	0	−sin(4.27)	0	0	1	0	0
3	−cos(4.27)	cos(4.27)	0	0	0	0	0	0	0
4	−sin(4.27)	−sin(4.27)	0	0	−1	0	0	0	1.00×10^5
5	0	−cos(4.27)	−cos(4.27)	0	0	0	0	0	0
6	0	sin(4.27)	−sin(4.27)	0	0	0	0	1	0
7	0	0	cos(4.27)	−cos(4.27)	0	0	0	0	0
8	0	0	sin(4.27)	sin(4.27)	1	0	0	0	0

The matrix form of the equilibrium equations, given below, was extracted directly from this table.

$$\begin{bmatrix} \cos(4.27) & 0 & 0 & \cos(4.27) & 0 & 1 & 0 & 0 \\ \sin(4.27) & 0 & 0 & -\sin(4.27) & 0 & 0 & 1 & 0 \\ -\cos(4.27) & \cos(4.27) & 0 & 0 & 0 & 0 & 0 & 0 \\ -\sin(4.27) & -\sin(4.27) & 0 & 0 & -1 & 0 & 0 & 0 \\ 0 & -\cos(4.27) & -\cos(4.27) & 0 & 0 & 0 & 0 & 0 \\ 0 & \sin(4.27) & -\sin(4.27) & 0 & 0 & 0 & 0 & 1 \\ 0 & 0 & \cos(4.27) & -\cos(4.27) & 0 & 0 & 0 & 0 \\ 0 & 0 & \sin(4.27) & \sin(4.27) & 1 & 0 & 0 & 0 \end{bmatrix} \begin{Bmatrix} F_{AB} \\ F_{BC} \\ F_{CD} \\ F_{AD} \\ F_{BD} \\ A_x \\ A_y \\ C_y \end{Bmatrix} = \begin{Bmatrix} 0 \\ 0 \\ 0 \\ 1.00 \times 10^5 \\ 0 \\ 0 \\ 0 \\ 0 \end{Bmatrix}$$

The square brackets [] denote a matrix with multiple rows and columns; the braces { } denote a single-column matrix commonly referred to as a **vector**. For ease of reference, we can express this equation using shorthand matrix notation as

$$[A]\{\mathbf{x}\} = \{\mathbf{b}\} \tag{8.2}$$

where the coefficient matrix $[A]$ and the right-hand side vector $\{\mathbf{b}\}$ come straight from the table. The order of the unknown force magnitudes in the vector $\{\mathbf{x}\}$ must correspond to the force labels across the top of the table so that the same multiplications are implied. Because the coefficients of $[A]$ premultiply the variables in $\{\mathbf{x}\}$, this operation is known as a **matrix multiplication**.

The unknowns can be isolated by premultiplying both sides of Equation (8.2) by the **matrix inverse** of $[A]$. Denoted by $[A]^{-1}$ and having the same dimension as $[A]$, the matrix inverse can be found using one of the special functions available in Excel. Premultiplying a matrix by its inverse is equivalent to premultiplying a number by its reciprocal. Therefore, Equation (8.2) simplifies to

$$[A]^{-1}[A]\{\mathbf{x}\} = [A]^{-1}\{\mathbf{b}\} \quad \text{or} \quad \{\mathbf{x}\} = [A]^{-1}\{\mathbf{b}\} \tag{8.3}$$

Thus, having found $[A]^{-1}$ using Excel, we can then apply another special function in Excel to perform this matrix multiplication and, in so doing, determine the values of the unknown force magnitudes.

The details of how to apply Excel to this example are summarized in the following procedure. Results are shown in the worksheet of Figure 8.8.

1. Open an Excel worksheet.
2. Enter the values of $[A]$ into a block of cells with 8 rows and 8 columns.
3. Enter the values of $\{\mathbf{b}\}$ into another nearby block of cells with 8 rows and 1 column.
4. Highlight an empty block of cells with 8 rows and 8 columns. Then, type

=MINVERSE(B4:I11),

and instead of pressing Enter, press the **Ctrl-Shift-Enter** keys simultaneously to compute the inverse of $[A]$ and has the results appear in the highlighted block of cells. Note that B4:I11 defines the range of cells belonging to $[A]$, and MINVERSE is a function that computes the inverse of the matrix defined by the range. This range could also have been entered by highlighting the cells of $[A]$.
5. Highlight another empty block of cells with 8 rows and 1 column. Then, type

=MMULT(B14:I21, L4:L11)

and press the following keys simultaneously.

Ctrl-Shift-Enter

This multiplies $[A]^{-1}$ and $\{b\}$ in accordance with Equation (8.3) and has the results appear in the highlighted column of cells. Note that the function MMULT requires us to define two ranges of cells, one for each of the matrices being multiplied.

	A	B	C	D	E	F	G	H	I	J	K	L	M
1	Solution of the Equations for the Truss Bridge												
2													
3													
4		0.997	0	0	0.997	0	1	0	0			0	
5		0.0745	0	0	−0.0745	0	0	1	0			0	
6		−0.997	0.997	0	0	0	0	0	0			0	
7	[A] =	−0.0745	−0.0745	0	0	−1	0	0	0		{b} =	1.00E+05	
8		0	−0.997	−0.997	0	0	0	0	0			0	
9		0	0.0745	−0.0745	0	0	0	0	1			0	
10		0	0	0.997	−0.997	0	0	0	0			0	
11		0	0	0.0745	0.0745	1	0	0	0			0	
12													
13													
14		0	0	−0.7523	−3.3557	−0.5015	0	−0.2508	−3.356			−3.36E+05	=F$_{AB}$
15		0	0	0.2508	−3.3557	−0.5015	0	−0.2508	−3.356			−3.36E+05	=F$_{BC}$
16		0	0	−0.2508	3.3557	−0.5015	0	0.25075	3.3557			3.36E+05	=F$_{CD}$
17	inv[A] =	0	0	−0.2508	3.3557	−0.5015	0	−0.7523	3.3557		{x} =	3.36E+05	=F$_{AD}$
18		0	0	0.0374	−0.5	0.0747	0	0.03736	0.5			−5.00E+04	=F$_{BD}$
19		1	0	1	0	1	0	1	0			0.00E+00	=A$_x$
20		0	1	0.0374	0.5	0	0	−0.0374	0.5			5.00E+04	=A$_y$
21		0	0	−0.0374	0.5	0	1	0.03736	0.5			5.00E+04	=C$_y$
22													

FIGURE 8.8 Excel worksheet for solving the equations.

The results from Excel are displayed along with the structure in Figure 8.9. Since the geometry and external force share the same plane of symmetry, the results are also symmetrical, with the upper members in compression (as indicated by the minus signs) and the lower members in tension.

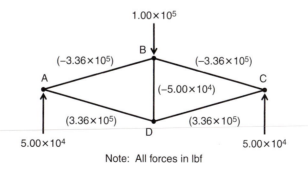

FIGURE 8.9 Results for the truss bridge. Member forces are indicated in parentheses.

These steps just carried out for the truss—(1) draw the FBD, (2) check determinacy, (3) write the equilibrium equations, and (4) solve the equations—define a general methodology for equilibrium analysis that is applicable to any statically determinate structure.

8.4 GEOTECHNICAL ENGINEERING

In the previous example, we assumed that the bridge was secured at each end to rigid supports. This is a big assumption if we consider that the ground beneath those supports may be subject to consolidation (i.e., compaction), the influences of a rising water table, or the possibility of the soil failing. The same concerns are present when designing the foundation for a building, a highway, or a dam.

Therefore, we may see that, to design and construct a proper foundation, the geotechnical engineer must have a working knowledge of both **soil mechanics** and **foundation engineering**. The aim of soil mechanics is to predict how the soil will respond (e.g., consolidate or fail) to water seepage and forces at the construction site. With knowledge of the forces exerted on the proposed foundation by the surrounding soil well in hand, a strong and stable foundation can be designed and constructed using the principles of foundation engineering.

8.4.1 Properties of Soils

Unlike human-made materials like steel or concrete, whose properties can be tailored to the application, the geotechnical engineer has limited control of the properties of the soil at the proposed site. Rather, it is more a matter of determining the soil properties through exploration and testing and reacting to those properties when designing the foundation.

The **Tower of Pisa** in Italy is a classic example of the role soil properties can play in the design and stability of a foundation. The tower was built over a period of 197 years, after starting construction in 1173. If the work had progressed more quickly, the tower would have fallen over during construction, due to settling of the soil below. Instead, the builders had time to compensate for the observed tilting and built the upper floors along a more vertical axis. Tilting of the tower continued through the last century, though at a much slower rate than during construction. The history of tilt can be understood in terms of progressive consolidation of the soil layers (shown in Figure 8.10) beneath the tower. Initial tilting has been traced to the highly compressible[5] upper clay

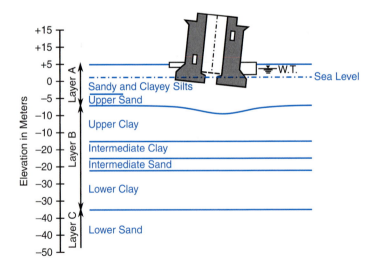

FIGURE 8.10 Soil layers below the Tower of Pisa.

---

[5]Highly compressible=deforms easily when squeezed.

layer. With time, consolidation of the upper clay stabilized, enabling construction to continue. More recent tilting movements have been attributed to Layer A, whose soil properties are believed to be sensitive to seasonal rises and falls of the water table.[6]

The compressibility of soils and their sensitivity to the presence of water are the results of a highly porous and mobile microstructure, as depicted in Figure 8.11. A sponge, in the dry and saturated states, provides a good but imperfect analogy. The grains of a soil are free to associate and disassociate as it plastically and permanently deforms under load. In contrast, the skeleton of a sponge, while highly flexible, will return to its original shape when the load is removed.

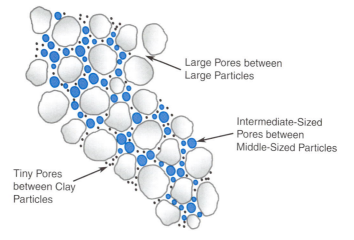

FIGURE 8.11 Soil structure consisting of large particles (sand) and small particles interspersed with air- or water-filled pores.

While the structure of a soil typically varies from point to point, average properties can be defined to characterize the porous microstructure. We begin by introducing the general definition of **weight density**[7]:

$$\gamma = \frac{W}{V} \tag{8.4}$$

which defines the weight (W) of a substance per unit volume (V). In the case of soil, the total volume (V) consists of two parts: the volume of the solid particles (V_s) and the volume of the pores (V_p), with corresponding weights W_s and W_p, respectively. Another property, **porosity** (n), relates pore volume to the total volume:

$$n = \frac{V_p}{V} \tag{8.5}$$

With these two definitions in hand, we can proceed to define the dry and saturated densities of soil. For **dry soil**, since the pores are air filled, $W_p = 0$ and all the weight is due to the solid particles:

$$\gamma_{dry} = \frac{W_s}{V} \tag{8.6}$$

By contrast, the pores in **saturated soil** are completely filled with water. The weight of the water in the pores is significant and so must be accounted for when defining the weight density of saturated soil:

$$\gamma_{sat} = \frac{W_s + W_p}{V} = \gamma_{dry} + n\gamma_w \tag{8.7}$$

where γ_w is the weight density of water. Use was made of Equation (8.6), the equality $W_p = \gamma_w V_p$, and the definition of porosity in arriving at Equation (8.7) (Example 8.1).

Example 8.1

The typical dry weight densities and porosities of three common soil types are given in Table 8.2. Calculate the weight density of each soil type when fully saturated with water.

Table 8.2 Dry Weight Densities of Different Soils

Soil Type	Dry Density (lbf/ft³)	Porosity
Sand	94.8	0.375
Clay	74.9	0.550
Silt	79.9	0.425
Water	62.4	—

Need: Saturated soil densities (γ_{sat}) of sand, clay, and silt.
Know: The dry weight densities and porosities of sand, clay, and silt (from Table 8.2) and the density of water.
How: Apply Equation (8.7).
Solve:

$$\text{Sand: } \gamma_{sat} = \gamma_{dry} + n\gamma_w = 94.8 + (0.375)(62.4) = 118. \text{ lbf/ft}^3$$
$$\text{Clay: } \gamma_{sat} = \gamma_{dry} + n\gamma_w = 74.9 + (0.550)(62.4) = 109. \text{ lbf/ft}^3$$
$$\text{Silt: } \gamma_{sat} = \gamma_{dry} + n\gamma_w = 79.9 + (0.425)(62.4) = 106. \text{ lbf/ft}^3$$

8.4.2 Effective Stress Principle

The goal of this section is to introduce the **effective stress principle**, considered by many to be the most important concept in soil mechanics. Here, we apply it to investigate how pressure naturally varies with depth in soil.

In general, **pressure** is caused by forces pressing against a surface, as in Figure 8.12, and is defined as follows:

$$p = \lim_{A \to 0} \frac{F_n}{A} \tag{8.8}$$

where F_n is the resultant force acting normal to a surface of area A and p is the associated pressure; the symbol "$\lim_{A \to 0}$" is read as "in the limit as A approaches zero"; this means that pressure can vary from point to point on the surface. The magnitude of pressure is indicative of the tendency of F_n to compact, or consolidate, the volume of material under A. When pressure acts on an internal surface, it is called **stress**, though it has the same definition and units as pressure.

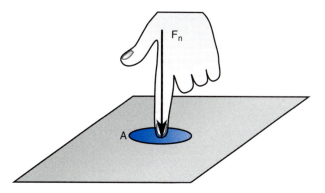

FIGURE 8.12 Pressure is what you feel when your fingers presses with force F_n on the contact patch of area A (the portion of the surface in contact with the finger, about the size of the finger tip).

Before proceeding, we need to make a few assumptions. For the sake of simplicity, we assume that the soil properties are uniform over the range of depths under consideration; in Example 8.2, we will generalize the equations to two layers, each with its own set of soil properties. We also assume that the water table has reached surface ground level so that all the soil is fully saturated. Finally, we assume that nothing human-made on top of the soil is bearing down on it, such as a building or road. In other words, the source of all pressures below ground is the soil itself.

To determine pressure, we must first find the force, which is best done using a FBD. Our system is the large chunk of soil shown in Figure 8.13, which extends from the surface down to a depth, z, and has the lateral dimensions L by T. The depth is kept general so that our final result is applicable to all depths; the exact lateral dimensions are of no consequence, as will soon become evident. Just two vertical forces are acting on our system: the downward acting weight of the soil, as given by weight density times volume, and a resisting force, F_n, exerted on the lower face by the soil just below. Static equilibrium in the vertical direction requires that these opposing forces be equal in magnitude, as expressed by

$$F_n = \gamma_{sat} z L T \tag{8.9}$$

Substituting this force into Equation (8.8) and noting that $A = LT$ leads to an expression for pressure on the bottom face that varies linearly with depth:

$$\sigma = \gamma_{sat} z \tag{8.10}$$

where the symbol lower case sigma σ for stress has been used to denote the pressure, since it acts on a hypothetical surface "inside" the soil. Equation (8.10) defines a "total stress" in the sense that it accounts for forces exerted by both soil grains and water-filled pores.

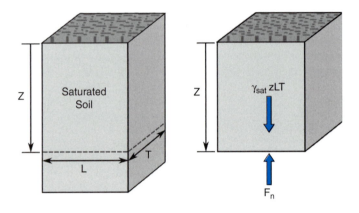

FIGURE 8.13 Soil section (left) being analyzed for stress on the horizontal plane (dashed line) at depth z. As the free-body diagram demonstrates, the weight of the saturated soil induces a reaction force, F_n, of equal magnitude on the bottom face of the block. This force is actually distributed uniformly over the bottom face creating a constant pressure (or stress).

Experiments have shown that the water pressure in the pores "does not" in fact contribute to the compaction of soil, such as occurred beneath the Tower of Pisa. We, therefore, remove its contribution to stress by first assuming water pressure in the pores varies with depth, as it would if the saturated soil were replaced by an equal volume of water:

$$p_p = \gamma_w z \tag{8.11}$$

In this context, water pressure is referred to as pore pressure (p_p). Subtracting the pore pressure from the total stress then yields the **effective stress** (σ'):

$$\sigma' = \sigma - p_p = (\gamma_{sat} - \gamma_w)z \tag{8.12}$$

This is the component of stress exerted by the soil skeleton. According to the **effective stress principle**, it is the effective stress, not the total stress, which determines if the soil consolidates or fails.

Even though we subtracted pore pressure to define effective stress, Equation (8.12) still models the buoyancy force exerted on the soil skeleton by the water.[8] In other words, the effective stress at a given depth is smaller for saturated soil than for dry soil because of the soil skeleton's tendency to float in water.

Example 8.2
The two top layers of a soil sample taken from a potential building site are shown in the first figure. If the thicknesses of the layers are $H_A = 2.50$ ft and $H_B = 2.00$ ft, determine (a) the effective stress at a depth of $Z_A = 2.00$ ft in Layer A and (b) the effective stress at a depth of $Z_B = 3.00$ ft in Layer B. Assume that both layers are fully saturated with weight densities of $\gamma_A = 78.6$ lbf/ft³ and $\gamma_B = 91.2$ lbf/ft³. The weight density of water is $\gamma_w = 62.4$ lbf/ft³.

---

[8]According to Archimedes' Principle, the upward-acting buoyancy force is equal to the weight of the water displaced by the soil skeleton.

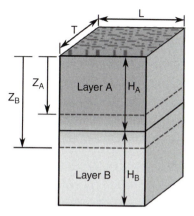

Need: Find the values of effective stress at depths of 2.00 ft and 3.00 ft.

Know: The depths of interest are $Z_A = 2.00$ ft and $Z_B = 3.00$ ft; the layer thicknesses are $H_A = 2.50$ ft and $H_B = 2.00$ ft; the weight densities of saturated soil and water are $\gamma_A = 78.6$ lbf/ft³, $\gamma_B = 91.2$ lbf/ft³, and $\gamma_w = 62.4$ lbf/ft³; both soil layers are fully saturated.

How: (a) Since all of the soil above $z = Z_A$ is of a uniform composition and fully saturated, Equation (8.12) applies.

(b) With two layers of soil above $z = Z_B$ to consider, Equation (8.12) cannot be used directly. Therefore, rederive Equation (8.12), this time including the weights of each layer above $z = Z_B$ in the FBD. Follow the same procedure that was used to develop Equation (8.12): (1) draw the FBD of the soil section, (2) find F_n from the summation of forces in the vertical direction, (3) substitute F_n into the definition of pressure to get total stress, and (4) subtract pore pressure from total stress to get effective stress.

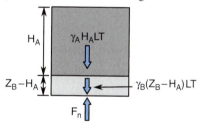

Solve: (a) From Equation (8.12),

$$\sigma' = (\gamma_A - \gamma_w)z_A = (78.6 - 62.4)(2.00)\ \left[\text{lbf/ft}^3\right][\text{ft}] = 32.4\ \text{lbf/ft}^2$$

(b) The FBD now includes two weight vectors counteracted by the normal force on the bottom face: Static equilibrium in the vertical direction yields the normal force:

$$F_n = \gamma_A H_A LT + \gamma_B (Z_B - H_A)LT$$

Divide by the area of the bottom face to get total stress:

$$\sigma = \gamma_A H_A + \gamma_B (Z_B - H_A)$$

Subtract the pore pressure to get the effective stress and substitute in these values:

$$\sigma' = \gamma_A H_A + \gamma_B (Z_B - H_A) - \gamma_w Z_B$$

$$\sigma' = (78.6)(2.50) + (91.2)(3.00 - 2.50) - (62.4)(3.00)\left[\text{lbf/ft}^2\right] = \textbf{54.9 lbf/ft}^2$$

Note that the effective stress in this example is much smaller than the pore pressure, yet it is the effective stress that determines if the soil skeleton fails or consolidates.

8.5 WATER RESOURCES ENGINEERING

The availability of clean water is taken for granted by most of us who live in the more temperate regions of the Northern Hemisphere. But, that common perception is changing rapidly as connections to the global economy strengthen and multiply. With modernization often associated with increased water usage, such as due to rising standards of cleanliness and the demands of industry, there are now concerns that water is becoming an increasingly scarce commodity in some regions of the world (see Figure 8.14). The goal of **sustainability** of the Earth's water supplies, that is, long-term balancing of need with availability, is one that can be achieved only through international cooperation and well-planned water resources engineering.

Two questions are central to every water resources project: (1) How much water is needed? and (2) How much water can be delivered? The answer to the first question is deeply rooted in the social and economic needs of the region. The answer to the second question is achieved through the application of the fundamentals of water resources engineering.

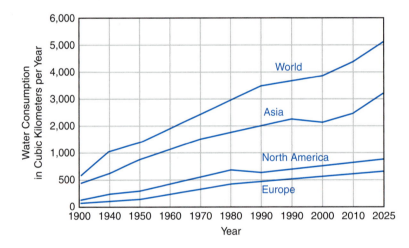

FIGURE 8.14 Forecast of water scarcity in 2025.

8.5.1 Reservoir Capacity

One method for achieving a sustainable water supply is to build a water reservoir. When properly designed, a reservoir can provide a steady supply of water that reliably meets the needs of regional consumers, even under adverse environmental conditions, such as drought.

To explore how this can be achieved, we consider the hypothetical reservoir site shown in Figure 8.15. A stream feeds into the reservoir site and serves as the main source of water. On the downstream end of the reservoir, there are plans to construct a triangular-shaped dam of height H and width W to contain the water within. The reservoir itself will be triangular in cross section, linearly increasing in depth along the length L from near zero at the stream inlet to H at the dam, consistent with the native topography.

The maximum volume of water that can be stored in the reservoir, and retrieved, is known as **capacity**, which is typically expressed in units of volume (e.g., acre-ft). The capacity of the reservoir in Figure 8.15 is given by the following formula:

$$V = \frac{1}{6} LWH \tag{8.13}$$

or more generally by

$$V = \frac{C_1 C_2}{6} H^3 \tag{8.14}$$

We express the proportions of the reservoir in terms of H as follows: $L = C_1 H$ and $W = C_2 H$, where C_1 and C_2 are constants (Example 8.3).

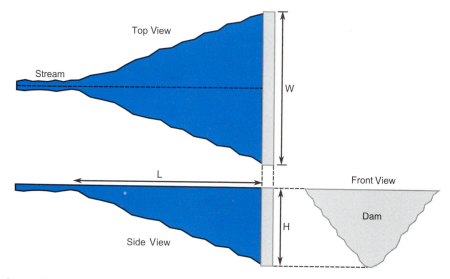

FIGURE 8.15 Reservoir geometry used to determine capacity.

Example 8.3

Develop an expression that relates the instantaneous height of the water level in the reservoir of Figure 8.15 to the volume of water stored within. Assume reservoir dimensions of $H = 100.$ ft, $W = 300.$ ft, and $L = 3000.$ ft.

Need: An expression for stored volume (V_s) in terms of instantaneous depth (h).

Know: The proportions of the reservoir ($H = 100.$ ft, $W = 300.$ ft, $L = 3000.$ ft) and a formula for volume, Equation (8.13).

How: Derive an equation for the volume of water in the partially filled reservoir, recognizing that the dimensions of this smaller volume have the same relative proportions as the completely filled reservoir.

Solve: First, let the dimensions l, w, and h of the partially filled volume correspond to L, W, and H of the filled reservoir, respectively. Since both volumes have the same relative dimensions, we know that

$$l/h = L/H = 3000./100. = 30.0, \text{ then } l = 30.0h$$
$$w/h = W/H = 300./100. = 3.00, \text{ and then } w = 3.00h$$

Substituting both results into the equation for volume leads to the desired expression in terms of h:

$$V_s = \frac{1}{6} lwh = \frac{1}{6}(30.0h)(3.00h)h = 15.0h^3$$

Similar relationships are used to monitor the amount of water stored in a reservoir in real time. In actual practice, the local terrain is usually too complex to be approximated by so simple a geometric shape. Under such circumstances, computational methods are used in conjunction with topographical surveys to generate curves of reservoir volume versus h.

8.5.2 Conservation of Mass

The water level in the reservoir continuously rises and falls as environmental conditions and consumer demand change over time. If this were not the case, there would be no need for a reservoir. Any changes in water level depend on the amount of water coming in (mass inflow, m_{in}) and the amount of water going out (mass outflow, m_{out}) during a given time interval (Δt), as related through **conservation of mass**:

$$\Delta m_{in} - \Delta m_{out} = \Delta m_{stored} \tag{8.15}$$

in which Δm_{stored} is the net mass of water added to the reservoir during time Δt; m_{in} is mainly attributable to the incoming stream water; m_{out} goes toward meeting various consumer demands (e.g., drinking water) and diverting water downstream; and Δt is typically defined to be a month. Thus, the total amount of water in the reservoir at a particular instant in time may be found by summing up the changes in stored mass starting from some initial time:

$$m_{stored}^n = \sum_{i=1}^{n} \Delta m_{stored}^i \tag{8.16}$$

where Δm_{stored}^i is the change in mass associated with the ith time interval, m_{stored}^n is the total (or cumulative) mass stored in the reservoir after n number of time intervals, and $\sum_{i=1}^{n}$ indicates that the mass increments i through n mass are to be summed (Example 8.4).

Example 8.4

Monthly data for inflow from a feeder stream at a potential reservoir site are available in the form of a table (which follows). Average monthly water use by regional consumers has been estimated and also appears in the table. Plot expected water storage over the time span of the historical record, allowing for negative storage.

Month	Inflow $(\times 10^9 \text{ ft}^3)$	Demand $(\times 10^9 \text{ ft}^3)$	Net Inflow $(\times 10^9 \text{ ft}^3)$	Cumulative Storage $(\times 10^9 \text{ ft}^3)$
1	2.36	3.11	−0.75	−0.75
2	2.49	2.85	−0.36	−1.11
3	1.66	3.37	−1.71	−2.82
4	2.62	3.63	−1.01	−3.83
5	13.58	3.89	9.69	5.86
6	3.45	5.18	−1.73	4.13
7	4.48	5.96	−1.48	2.65
8	4.20	6.48	−2.28	0.37
9	4.25	4.15	0.10	0.47
10	4.92	3.63	1.29	1.76
11	3.16	3.37	−0.21	1.55
12	2.70	3.11	−0.41	1.14

Need: Water storage plotted as a function of time

Know: Monthly values of inflow and consumer demand for a 12-month period

How: Calculate the net inflow for each month using Equation (8.15). Then, for each month, add storage increments, up to and including that month's contribution, to define cumulative storage values as per Equation (8.16). Finally, plot results using Excel.

Solve: Results, shown in the table, can be obtained using Excel without the use of special functions.

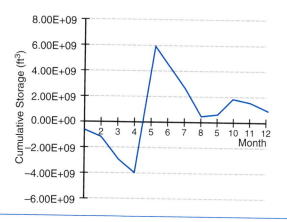

8.5.3 Estimation of Required Capacity and Yield

An estimate of required reservoir capacity may be found from monthly data for mass inflow and consumer demand using a graphical method known as **sequent-peak analysis**. The sequent peaks are the succession of next highest peaks, or maxima, that occur as you move from left to right (i.e., chronologically) along the cumulative storage versus time curve, as illustrated in Figure 8.16. Between each pair of sequent peaks, the required storage is calculated as the difference between the first sequent peak and the lowest valley in the interval. Required **capacity** is then the largest of these differences over the span of the historical record being examined. If the calculated capacity exceeds the economic or physical constraints of the project, expectations of demand may have to be adjusted down.

Another important design parameter, **yield**, defines how fast the water can be drawn out of the reservoir without compromising its ability to meet future demands. It therefore typically has units of volume per unit time (e.g., m^3/s). Yield is calculated as the average inflow over a span of time, which is often taken to be the period of lowest inflow from the stream on record:

$$Y = \frac{\frac{1}{M}\sum_{i=1}^{M}\Delta m_{in}^{i}}{\Delta t} \tag{8.17}$$

where Y is the yield, M is the number of time intervals spanning the period in question, and Δt is the time associated with each equal-sized time interval.

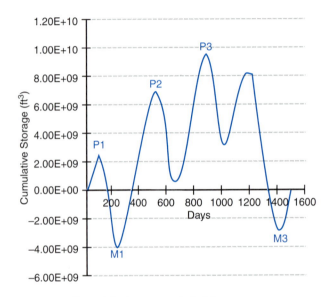

FIGURE 8.16 Cumulative storage versus time plotted over a period of 4 years.

Sequent peaks (P1, P2, P3) and their corresponding minima (M1, M2, M3) have been indicated. The sequent peaks are defined such that $P1 < P2 < P3$. This graph was adapted from historical data available online from the U.S. Department of the Interior (Examples 8.5 and 8.6).

Example 8.5

Assuming Figure 8.16 represents cumulative storage (i.e., inflow minus demand summed over time) at a proposed reservoir site, estimate the required reservoir capacity.

Need: Reservoir capacity based on Figure 8.16.
Know: We have 4 years' worth of historical data for cumulative storage.
How: First, identify sequent peaks and associated minima by reference to Figure 8.16. Compute the difference between each sequent peak and its minimum. The largest difference is the required capacity.
Solve: Values corresponding to sequent peaks and their minima are listed in the table that follows, along with computed differences. The largest difference defines the required capacity.

No.	Peak ($\times 10^9$ ft^3)	Min ($\times 10^9$ ft^3)	P − M ($\times 10^9$ ft^3)
1	2.51	−3.95	6.46
2	6.70	0.59	6.11
3	9.44	−2.27	11.7

$$C = \max_{i=1}^{3} (P_i - M_i) = 11.7 \times 10^9 \text{ ft}^3$$

Example 8.6

Estimate the yield based on the historical data provided in Example 8.4. Also, calculate how many individual consumers could be serviced by such a reservoir if daily indoor per capita water usage is 69.3 gallons.

Need: An estimate of the yield based on the flow data of Example 8.4 and an estimate of the number of individual customers whose needs could be met by a reservoir of this size.

Know: One year's worth of flow data; per capita water usage is 69.3 gallons/day.

How: Substitute monthly inflow values into Equation (8.17) to estimate yield. One year is not a lot of data to work with, so to be safe, ignore the peak monthly inflow (month 5). Divide this estimate of yield by per capita water usage to determine the potential number of customers.

Solve: Basing the yield on 11 months of inflow data, we obtain

$$Y = \frac{\dfrac{1}{11}(2.36 + 2.49 + 1.66 + 2.62 + 3.45 + 4.48 + 4.20 + 4.25 + 4.92 + 3.16 + 2.70) \times 10^9}{1} = 3.30 \times 10^9 \text{ ft}^3/\text{month}$$

To estimate the number of customers, first convert 69.3 gallons/day to units of ft^3/month:

$$U = 69.3 [(\text{gal}/\text{day})/\text{customer}] \times 0.134 [\text{ft}^3/\text{gal}] \times 30 [\text{day}/\text{month}]$$

$$U = 279. [(\text{ft}^3/\text{month})/\text{customer}]$$

With the units now compatible, divide the yield by the monthly per capita usage:

$$N = Y/U = 3.30 \times 10^9 / 279. = 11.8 \times 10^6 \text{ customers}$$

SUMMARY

In this chapter, we examined representative applications of three traditional areas of civil engineering. First, we learned a general methodology that a **structural engineer** uses to determine the forces in a truss. The same methodology can be applied to find the forces in a structure composed of hundreds of triangulated members, provided of course that a computer and an equation solver, such as the one in Excel, are available to solve the resulting system of equilibrium equations.

Second, we learned that the nonuniform composition and highly porous structure of soils lead to complex behaviors that can challenge **geotechnical engineers** tasked with the job of designing a stable foundation. From the **effective stress principle**, we know that it is the properties of the soil skeleton, and not the pressurized water in the pores, that determine if the soil compacts or fails. This observation greatly simplifies the design of experiments aimed at measuring the properties of soils under varying conditions.

Finally, we learned that one method used by **water resources engineers** to achieve a sustainable water supply in the face of variable demand and environmental conditions is to build a reservoir. One of the keys to accurate estimates of required storage (capacity) and delivery rates (yield) is an accurate historical record of the outflows (i.e., demand, seepage, and evaporation) and inflows (from incoming streams).

EXERCISES

Problems for Structural Engineering

1. When estimating the live load for a new bridge design, you want that estimate to be **conservative**. In other words, you want to err on the safe side by basing the estimate on the worst possible scenario, such as bumper-to-bumper traffic, the heaviest vehicles, or the worst environmental conditions. With this in mind, estimate the live load for the following proposed bridge designs.
   a. A foot bridge with a 174. ft span and two separate 4. ft wide lanes to allow for pedestrian traffic in both directions.
   b. A four-lane highway bridge with a 300. ft span.

2. For the given structure, do your best to identify all of its members as beams, compression members, or tension members.

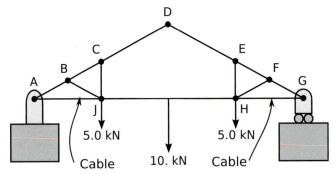

For each of the trusses shown in Exercises 3–7, determine the forces on all members by the Method of Joints. Use a spreadsheet to solve the equations.

3.

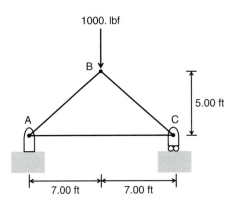

4.

5.

6.

7.

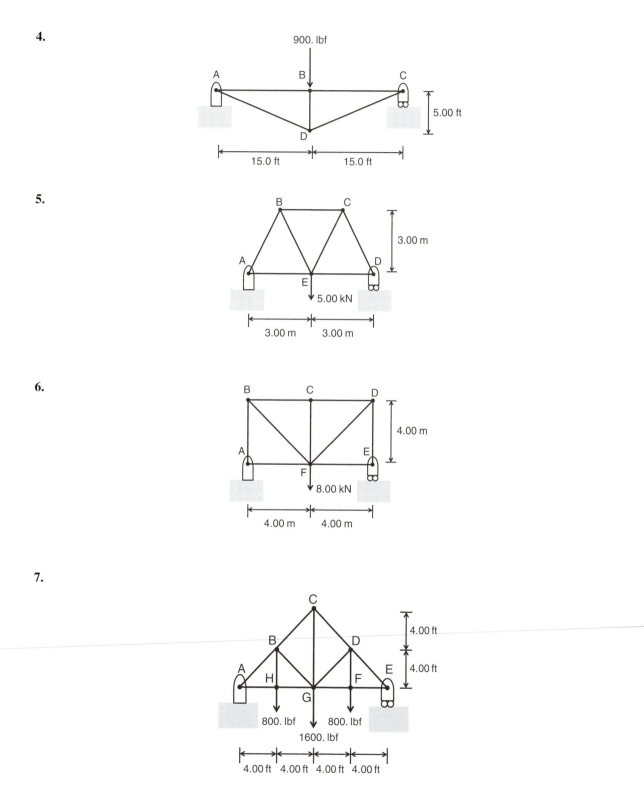

8. A truss that is pinned at one end and on rollers at the other is statically determinate if it satisfies the following equation:

$$0 = 2j - m - 3$$

where j is the number of joints and m is the number of members. This is a quick way to check for static determinacy without drawing the FBDs. Apply this equation to

a. verify that the trusses in Exercises 3–7 are statically determinate and

b. sketch three statically determinate truss bridges with the following number of members: (i) 11 members, (ii) 15 members, and (iii) 19 members.

9. Download the West Point Bridge Designer software from http://bridgecontest.usma.edu and use it to design a truss.

Problems for Geotechnical Engineering

Refer to this table for Exercises 10 and 11.

Dry Weight Densities of Different Soils		
Soil Type	**Dry Density (lbf/ft^3)**	**Porosity**
Sand	103.	0.423
Clay	87.2	0.615
Water	62.4	—

10. A recently extracted soil sample is 1.50 ft^3 in volume and composed of saturated sand. Calculate (a) the weight density of the saturated sand, (b) the weight of the solid particles in the soil sample, and (c) the weight of the water in the soil sample. (Refer to the preceding table of dry densities.)

11. The soil at a proposed construction site is composed of saturated clay. Calculate (a) the weight density of the saturated clay, (b) the total stress at a depth of 20.0 ft, (c) the effective stress at a depth of 20.0 ft, and (d) the buoyancy force on the soil skeleton. (Refer to the preceding table of dry densities.)

12. In arriving at Equation (8.12), we assumed that the water table was even with ground level. Now assume that the water table is located at a distance, d, below ground level and develop a new expression for effective stress valid for $z \geq d$. (Hint: Follow a procedure similar to the one used in Example 8.2b; that is, (1) draw the FBD of the soil section, this time including the weights of both the dry and saturated layers of soil, (2) find F_n from summation of forces in the vertical direction, (3) substitute F_n into the definition of pressure to get total stress, (4) redefine pore pressure to be $p_n = \gamma_w(z - d)$, and (5) subtract pore pressure from total stress to obtain effective stress in terms of γ_{sat}, γ_{dry}, γ_w, z, and d.

13. Referring to Example 8.2 in the section on geotechnical engineering, use Excel or another spreadsheet to plot the depth, z, versus the effective stress for values of z ranging from 0.00 to 4.50 ft. Feel free to use any results obtained in Example 8.2 to assist you with creating the graph.

14. The top three layers of soil at the proposed site of a new building are shown in the following figure, along with each layer's thickness and saturated weight densities. Assuming the water table is even with ground level, calculate the effective stress at the following depths: (a) $z = 4.00$ ft, (b) $z = 8.00$ ft, (c) $z = 12.0$ ft.

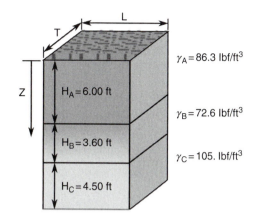

$\gamma_A = 86.3$ lbf/ft³

$H_A = 6.00$ ft

$\gamma_B = 72.6$ lbf/ft³

$H_B = 3.60$ ft

$\gamma_C = 105.$ lbf/ft³

$H_C = 4.50$ ft

Problems for Water Resources Engineering

15. A proposed reservoir has the geometry described by Figure 8.15 and Equation (8.13). The local topography constrains the dimensions H, W, and L of this geometry to have the relative proportions 1 to 5 to 20, respectively. Assuming the required capacity has been estimated at 5.00×10^8 ft³, use Excel to plot the water level height versus the stored volume for water level heights ranging from 0 to H.

16. The average flow rates into and out of a small reservoir are given in the following table for each hour of a 5 h time period. If, at the start of this time period, 90,000. ft³ of water is being stored in the reservoir, how much is left after 5.0 h?

Hour	Flow In (ft³/s)	Flow Out (ft³/s)
1	0.436	0.645
2	0.567	0.598
3	0.734	0.553
4	0.810	0.541
5	0.832	0.583

17. Use sequent-peak analysis, with the cumulative storage versus the time graph provided to estimate reservoir capacity.

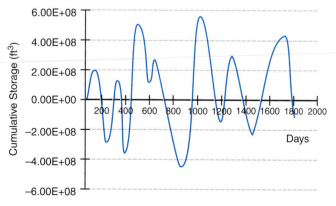

18. The monthly inflow of water to a proposed reservoir site is listed in the following table for a critical 3-year period associated with low rainfall. Average consumer demand for water over the same 3-year period was estimated to 2.86×10^8 ft^3/month. Based on these data, determine
    a. an estimate of yield in ft^3/month;
    b. an estimate for reservoir capacity in ft^3;
    c. the required dimensions of the reservoir if it has the geometry described by Figure 8.15 and Equation (8.13), and H, W, and L have the relative proportions 1 to 4 to 40, respectively.

Yr 1 Inflow ($\times 10^7$ ft^3)	Yr 2 Inflow ($\times 10^7$ ft^3)	Yr 3 Inflow ($\times 10^7$ ft^3)
1.26	1.42	1.72
0.985	1.27	1.68
1.01	3.06	2.01
1.78	6.85	4.87
4.86	4.71	7.93
3.50	4.98	3.97
5.91	6.21	4.97
4.33	4.79	3.38
3.01	4.03	8.34
2.37	4.21	3.76
1.89	2.23	2.89
1.35	1.90	2.26

Engineering Ethics Problems

19. In 1985, a judge found the structural engineers for the Hyatt Regency Hotel guilty of "gross negligence" in the July 17, 1981, collapse of two suspended walkways in the hotel lobby that killed 114 and injured 200 people. Many of those killed were dancing on the 32-ton walkways when an arrangement of rods and box beams suspending them from the ceiling failed.

    The judge found the project manager guilty of a conscious indifference to his professional duties as the Hyatt project engineer who was primarily responsible for the preparation of design drawings and review of shop drawings for that project. He also concluded that the chief engineer's failure to closely monitor the project manager's work betrayed "a conscious indifference to his professional duties as an engineer of record." Responsibility for the collapse, it was decided, lay in the engineering design for the suspended walkways. Expert testimony claimed that even the original beam design fell short of minimum safety standards. Substantially, less safe, however, was the design that actually was used.[9]

    Use the Engineering Ethics Matrix to analyze the ethical issues that occurred in this case.

20. Sara, a recent graduate, accepts a position at a small engineering design firm. Her new colleagues form a tightly knit, congenial group, and she often joins them for get-togethers after work.

    Several months after she joins the firm, the firm's president advises her that his wife has objected to her presence on the staff, feeling that it is "inappropriate" for a young, single female to work and socialize with

---

[9]Modified from http://wadsworth.com/philosophy_d/templates/student_resources/0534605796_harris/cases/Cases/case68.htm.

a group of male engineers, many of whom are married. The president's wife encouraged him to terminate Sara's employment, and although the president himself has no issues with Sara or her work, he suggests to her that she should look for another employer.

Everyone in the firm becomes aware of the wife's objections, and Sara begins to notice a difference in her work environment. Although her colleagues are openly supportive of her, she nevertheless feels that the wife's comments have altered their perception of her. She stopped receiving invitations to her company's parties and was excluded from after-hours gatherings. Even worse, while she had previously found her work both interesting and challenging, she no longer receives assignments from the firm's president and she begins to sense that her colleagues are treating her as someone who will not be a long-term member of the staff. Believing that she is no longer taken seriously as an engineer and that she will have little opportunity to advance within the firm, she begins searching for a new position. However, before she can do so, her supervisor announces a downsizing of the firm, and she is the first engineer to be laid off.[10]

Use the Engineering Ethics Matrix to examine potential ethics violations for an engineering employer to exclude and ultimately discharge an employee on the basis of sex, age, or marital status.

---

[10]Modified from http://pubs.asce.org/magazines/ascenews/2008/Issue_01-08/article6.htm.

Computer Engineering

9

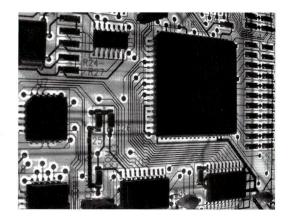

9.1 INTRODUCTION

Why do cars, trucks, planes, ships, and almost everything else today contain computers? After all until the 1960s we got along perfectly well without computers. As late as 1965, the question "Should a car or truck or plane or ship or anything else contain a computer?" would have seemed ridiculous. In those days, a computer was not only much more expensive than a car, but also bigger.

9.2 MOORE'S LAW

In the late 1960s, the integrated circuit was invented, and computers began to shrink in both size and cost. In the 1990s, it was also asserted that "if a Cadillac had shrunk in size and cost as fast as a computer did since 1960, you could now buy one with your lunch money and hold it in the palm of your hand."[1] The basic message, however, was and remains valid. From the 1960s until the present, integrated circuits, the building blocks of computers, have doubled their computing power every year or two. This explosive progress is the result of a technological trend called *Moore's law* that states that the number of transistors on an integrated circuit "chip" will double every year or two. The name honors electronics pioneer Gordon Moore, who proposed it in 1965. Moore's law is not a law of nature, but an empirical rule of thumb, and it has held true for four and a half decades.

Today's computers are so small that, in most applications size is no longer an issue. For example, a computer for controlling the air/fuel mixture in an automobile is typically housed along with its power supply and communication circuitry in an assembly about the size of a small book. As a result, people reliably control not just the hundreds of horsepower of an automobile, but the delivery of energy routinely available to them in modern industrial societies while minimizing pollution emitted in the course of energy conversion.

---

[1] Although no one has ever explained how you could drive around town in a Cadillac that you could hold in the palm of your hand.

The subjects of control and binary logic have taken on a life of their own in the form of computers and other information systems. Indeed, such systems have become so pervasive as to lead many people to assert that the industrialized world changed in the late twentieth century from an industrial society to an information society.

The computers that carry out these logical and arithmetical tasks are essentially collections of electrical/electronic circuits. In more descriptive terms, they are boxes of switches that are turning each other on and off. So the question for this chapter might be rephrased as follows: How can switches turning each other on and off respond to inputs and perform useful calculations? We will give a modest answer to this question by the end of this chapter.

Answering that question introduces the topics of digital logic and computation, which are central parts of computer engineering. The term digital can apply to any number system, such as the base ten system used in ordinary arithmetic. However, today's computers are based on a simpler number system: the base two or binary system. Therefore, this chapter focuses on binary logic and computation.

Computation was implemented first by mechanical devices, then by analog computers, and today by digital computers. A particularly convenient way to use computation to accomplish control is through the use of binary logic. A means of summarizing the results of binary logic operations is through truth tables that can be expressed as electrical logic circuits. This technique is useful not only in control, but also in other areas of computation, such as binary arithmetic and binary codes. Binary arithmetic and information are the basis of computer software. Binary logic also enables us to define the engineering variable *information*. In this chapter, you will learn how to perform the above operations. These topics will be mostly illustrated by reference to a familiar automotive application: deciding when a seat belt warning light should be turned on.

As an afterword, the process by which actual computers do these things, the hardware-software connection, will be qualitatively summarized.

The topics in this chapter extend far beyond single computers. Millions of *embedded* computers (i.e., computers placed within other systems and dedicated to serving those systems) now can be found in thousands of applications. Electronic and computer engineers have the jobs of reproducing this computer population, which brings forth a new generation every 2 or 3 years. The jobs these engineers do range widely. Some develop the electronic circuits that are the basis of electronic computation. Others devise new computer architectures for using those circuits. Yet others develop the "software," which are the instructions that make computers perform as intended. Many other engineers apply these tools to everything from appliances to space vehicles.

9.3 ANALOG COMPUTERS

For hundreds of years, engineers proved ingenious and resourceful at using mechanical devices to control everything from the rotation of water wheels and windmills to the speed of steam engines and the aiming of guns on battleships. However, such mechanical systems had major disadvantages. They were limited in their speed and responsiveness by the mechanical properties of the components and they had to be custom designed for every control challenge.

By the twentieth century, this had led engineers to seek more flexible, responsive, and general types of controls. As a first step, engineers put together standardized packages of springs, wheels, gears, and other mechanisms. These systems also proved capable of solving some important classes of mathematical problems defined by differential equations. Because these collections of standard mechanisms solved the problems by creating a mechanical analogy for the equations, they were called analog computers. Bulky and inflexible, they often filled an entire room and were "programmed" by a slow and complex process of manually reconnecting wires and components in order to do a new computation. Despite these drawbacks, in such applications as predicting the tides, determining the performance of electrical transmission lines, or designing automobile suspensions, analog computers marked a great advance over previous equation-solving methods.

9.4 FROM ANALOG TO DIGITAL COMPUTING

Meanwhile, a second effort, underway since about 1800, sought to calculate solutions numerically using arithmetic done by people. Until about 1950, the word *computer* referred to a person willing to calculate for wages. Typically, these were selected members of the workforce whose economic status forced them to settle for relatively low pay. It was not until about 1900 that these human computers were given access to mechanical calculators and not until the 1920s that practical attempts at fully automated mechanical calculations were begun.

Meanwhile, as early as the 1820s, beginning with the ideas of the British scientist Charles Babbage, attempts had been made to do arithmetic accurately using machinery. Because these proposed machines operated on digits as a human would (rather than forming analogies), they were called *digital* computers. Digital techniques were also used to control machinery. For example, the French inventor Joseph Marie Jacquard (1752–1834) used cards with holes punched in them as a digital method to control the intricate manipulations needed to weave large complex silk embroideries.

Humans learn digital computing using their ten fingers.[2] Human arithmetic adapted this 10-digit method into the decimal system. Digital computers can be built on a decimal basis, and some of the pioneers, such as Babbage in the 1840s and the team at the University of Pennsylvania who, in the 1940s built a very early electronic digital computer, the ENIAC,[3] adopted this decimal system.

9.5 BINARY LOGIC

However, computer engineers quickly found that a simpler system less intuitive to humans proved much easier to implement with electronics. This is the binary system, based on only two digits: 1 and 0. The Jacquard loom was such a binary system, with a hole in a card representing a 1 (one) and the lack of a hole representing 0 (zero). In other applications, turning on a switch might represent a 1, and turning it off might represent a 0. The logic and mathematics of this system were developed mainly by the British mathematician George Boole (1815–1864), whose contributions were so important that the concept is often referred to as Boolean algebra.

Binary logic begins with statements containing variables symbolized by letters such as X or Y. The statement can assert anything whatsoever, whether it is "The switch is open" or "There is intelligent life on a planet circling the star Procyon." The variable representing the statement can be assigned either of two values, 1 or 0, depending on whether the statement is true or false. (We will adopt the convention of using the value 1 for true statements and the value 0 for false ones.)

Binary logic permits three, and only three, operations to be performed, **AND, OR,** and **NOT**:

AND (sometimes called "intersection" and indicated by the symbol • or *) means that, given two statements X and Y, if both are true, then $X \bullet Y = 1$. If either one is false, then $X \bullet Y = 0$. For example, the statement "It is raining (X) and the sun is out (Y)" is true only if both it is raining and the sun is out.

---

[2] The word *digitus* is literally "finger" in Latin, indicating the discrete nature of such counting schemes.
[3] Whether ENIAC was the first electronic computer is open to debate. Two years before ENIAC Tommy Flowers, a British post office engineer, designed and built "Colossus" to break German military codes in WWII; many attribute this to be the first electronic computer.

OR (sometimes called "union" and indicated by the symbol +) means that, given two statements X and Y, if either one or both are true, then $X + Y = 1$, while only if both are false is $X + Y = 0$. For example, the statement "It is raining (X) *or* the sun is out (Y)" is true in all cases except when *both* are not true.[4]

NOT (sometimes called "negation" and indicated here by the postsymbol $'$ as in X', but sometimes indicated in other texts with an overbar as in $\bar{X}$) is an operation performed on a single statement. If X is the variable representing that single statement, then $X' = 0$ if $X = 1$, and $X' = 1$ if $X = 0$. Therefore, if X is the variable representing the statement "It is raining," then X' is the variable representing the statement "It is not raining."

These three operations provide a remarkably compact and powerful tool kit for expressing any logical conditions imaginable. A particularly important type of such a logical condition is an "if-then" relationship, which tells us that **if** a certain set of statements has some particular set of values, **then** another related statement, often called the **target statement**, has some particular value.

Consider, for example, the statement "If a cold front comes in from the south or the air pressure in the north remains constant, but not if the temperature is above 50 °F, then it will rain tomorrow." The target statement and each of the other statements can be represented by a variable. In our example, X can be the target statement "It will rain tomorrow," and the three other statements can be expressed by $A =$ "a cold front comes in from the south," $B =$ "the air pressure in the north is constant," and $C =$ "the temperature is above 50 °F." The connecting words can be expressed using their symbols. Thus, this long and complicated sentence can be expressed by the short and simple **assignment** statement:

$$X = A + B \bullet C'$$

This is not a direct *equivalence* in which information flows both ways across the equation. Specifically, in an assignment,[5] the information on the right-hand side is assigned to X, but not vice versa. This distinction is required because of the way that computers actually manipulate information.

However, as written above, the statement still presents a problem. In which order do you evaluate the operations? Does this make a difference? A simple example will show that it *does* make a difference! Consider the case $A = 1, B = 0$, and $C = 1$. Suppose the symbol $+$ is used first, the symbol $\bullet$ next, and the symbol $'$ last. Then in the preceding statement, $A + B = 1 + 0 = 1$; $(A + B) \bullet C = 1 \bullet C = 1 \bullet 1 = 1$; and $C' = 1' = 0$. Then $X = 0$.

However, if the symbols are applied in the reverse order ($'$ first, $\bullet$ next, and $+$ last), then $C' = 1' = 0$; $B \bullet C' = 0 \bullet 1 = 0$; and $A + B \bullet C' = 1 + 0 = 1$. Then $X = 1$.

So to get consistent results when evaluating logic statements, a proper order must be defined. This is similar to standard precedence rules used in arithmetic.

That order is defined as follows:

1) All NOT operators must be evaluated first, then

2) all AND operators (starting from the right if there is more than one), and finally

3) all OR operators (starting from the right if there is more than one) are evaluated.

---

[4]Technically, this operation is called an *inclusive or* to distinguish it from the operation *exclusive or*, which gives a value 0 when both X and Y are true, as well as when *both* are false. In our example, this is when it is raining and the sun is out as well as when it is not raining and the sun is not out.

[5]In some programming languages such logic statements are written with an *assignment* command, ": =" and not just an "=" command so that our Boolean statement could be written as $X := A + B \cdot C'$ to reinforce the fact that these are not reversible equalities.

If a different order of operation is desired, that order must be enforced with parentheses, with the operation within the innermost remaining parentheses being evaluated first, after which the parentheses is removed. In our example above, the proper answer with explicit parentheses would have been written $X=A+(B\bullet C')$ and evaluated as $1+(0\bullet 1')=1$. However, the value of the expression $X=(A+B)\bullet C'$ is $X=(1+0)\bullet 1'=1\bullet 1'=1\bullet 0=0$.

Example 9.1

Consider the following statement about a car: "The seat belt warning light is on." Define the logic variable needed to express that statement in binary logic.

Need: Logic variable (letter) = "…" where the material within the quotes expresses the condition under which the logic variable has the value 1 = true and 0 = false.

Know-how: Choose a letter to go on the left side of an equivalence. Express the statement in the form it would take if the content it referred to is true. Put the statement on the right side in quotes.

Solve: W = "the seat belt warning light is on."

This example is trivially simple, but more challenging examples can arise quite naturally. For example, if there are two or more connected constraints on a given action, then the methods of Boolean algebra are surefire ways of fully understanding the system in a compact way.

Example 9.2

Consider the statement involving an automobile cruise control set at a certain speed (called the "set speed"): "Open the throttle if the speed is below the set speed and the set speed is not above the speed limit." Express this as a logic formula and evaluate the logic formula to answer the question: "If the speed is below the set speed and the set speed is above the speed limit, then will the throttle be opened?" Answer using these variables: the car's speed is 50. miles per hour, the set speed is 60. miles per hour, and the speed limit is 45 miles per hour.

Need: A binary logic formula expressing the statement "Open the throttle if the speed is below the set speed and the set speed is not above the speed limit," and an evaluation of the formula for the situation when the speed is 50. miles per hour, the set speed is 60. miles per hour, and the speed limit is 45 miles per hour.

Know: Any statement capable of being true or false can be represented by a variable having values 1 = true and 0 = false, and these variables can be connected by AND ($\bullet$), OR (+), and NOT (').

How: Define variables corresponding to each of the statements, and use the three connectors to write a logic formula.

Solve: Let X = "throttle is open."

Let A = "speed is below set speed."

Let B = "set speed is above speed limit."

Then the general logic formula expressing the target statement is $X=A\bullet B'$.

If the speed is 50. miles per hour, and the set speed is 60. miles per hour, then $A=1$. If the set speed is 60. miles per hour, and the speed limit is 45 miles per hour, then $B=1$. Substituting these values, the general logic formula gives $X=1\bullet 1'$.

Evaluating this in the proper order gives $X=1\bullet 0=0$.

In plain English, if the speed is below the set speed, and the set speed is above the control speed limit, the throttle will not open.

Example 9.3

Suppose we want to find a Boolean expression for the truth of the statement "W=the seat belt warning light should be on in my car," using all of the following Boolean variables:

Case 1

    D is true if the driver seat belt is fastened.

    Pb is true if the passenger seat belt is fastened.

    Ps is true if there is a passenger in the passenger seat.

Case 2

For actuation of the warning light, include the additional Boolean variable:M is true if the motor is running.

    Need: $W=?$

    Know-how: Put the W variable on the left side of an assignment sign $W=$ and then array the variables on the other side of the assignment sign.

Case 1

    a. Put D, Ps, and Pb on the right side of the assignment sign. Thus, the temporary (for now, incorrect) assignment statement is $W=D{\cdot}Ps{\cdot}Pb$.

    b. Connect the variables on the right side with the three logic symbols $\bullet$, $+$, and $'$ so that the relationship among the variables on the right side correctly represents the given statement.

A good way to do this is to simply put the symbols the way they should appear in the if statements. For example, since part of the if statement Pb contains the words "the passenger's seat belt is fastened," it should also contain a "not." The corresponding logic variable will appear as Pb'. Also, you only care about the passenger's seat belt if the passenger is sitting in the seat, that is, the intersection between these two variables.

    Solve: A solution in English is: "if the driver's seat belt is not fastened" or "if there is a passenger in the passenger seat" and "the passenger's seat belt in not fastened," then "the seat belt warning light should be on in my car," or $W=D'+Ps{\cdot}Pb'$.

Once written, a logic equation can now be solved for any particular combination of variables. This is done by first plugging in the variable and then carrying out the indicated operation.

Case 2

For activation of the warning light, the light will go on if the motor is running, and if either the driver's seat belt is not fastened or if there is a passenger in the passenger seat and that seat belt is not fastened. This is written as:

$$W = M \bullet (D' + Ps \bullet Pb')$$

9.6 TRUTH TABLES

It is often convenient to summarize the results of a logic analysis for all possible combinations of the values of the input variables of an "if ... then" statement. This can be done with a truth table. It is simply a table with columns representing variables and rows representing combinations of variable values. The variables for the "if" conditions start from the left, and their rows can be filled in systematically to include *all possible combinations* of inputs. The column at the far right represents "then." Its value can be computed for each possible input combination.

Example 9.4

Consider the condition in Example 9.2: open the throttle if the speed is below the set speed and the set speed is not above the speed limit. The "if" conditions are $A=$"speed is below the speed limit" and $B=$"set speed is above the speed limit." The "then" condition is $X=$"open the throttle." The truth table is set up as shown here.

A	B	B'	X = A • B'

Need: All 16 entries to the truth table.
Know: Negation operator, $'$ and the AND operator •.
How: Fill in all possible binary combinations of statements A and of B.
Solve: One convenient way of making sure you insert all the possible input values is to "count" in binary from all zeroes at the top to all ones at the bottom. (If you are not already able to count in binary, this is explained on the next page or so.) In this example, the top line on the input side represents the binary number 00 (equal to decimal 0), the second line is 01 (decimal 1), the third is 10 (decimal 2), and the fourth is 11 (decimal 3).[6]

A	B	B'	X = A • B'
0	0		
0	1		
1	0		
1	1		

Next the value of the "then" (open the throttle) is computed for each row of inputs (this is done here in two steps, first computing B', then computing $X=A\bullet B'$).

A	B	B'	X = A • B'
0	0	1	0
0	1	0	0
1	0	1	1
1	1	0	0

In English, this truth table is telling us that the only condition under which the control will open the throttle ($X=1$) is when both the speed is below the set speed and the set speed is below the speed limit. This is the way we *should* want our cruise control to operate!

Truth tables can also be conveniently expressed as electric circuits. Indeed, this capability is the essence of computing. This capability is further explored in the exercises.

---

[6]If the input had three columns, A, B, and C, you would count from 0 to 7 in binary (000 to 111). If four columns, the 16 entries would count from 0 to 15 in binary (000 to 1111), and so on.

9.7 DECIMAL AND BINARY NUMBERS

In the decimal[7] or base-10 number system, digits are written to the left or right of a dot called the decimal point to indicate values greater than one or less than one. Each digit is a placeholder for the next power of 10. The digits to the left of the decimal point are whole numbers, and as you move to the left every number placeholder increases by a factor of 10. On the right of the decimal point the first digit is tenths (1/10), and as you move further right every number placeholder is 10 times smaller (see Figure 9.1).

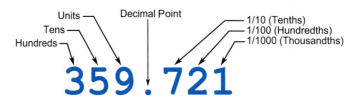

FIGURE 9.1 The structure of a decimal number.

The number **6,357** has four digits to the left of the decimal point, with "7" in the **units** place, "5" in the **tens** place, "3" in the **hundreds** place, and 6 in the **thousands** place. You can also express a decimal number as a whole number plus tenths, hundredths, thousandths, and so forth. For example, in the number 3.76, the 3 to the left of the decimal point is the "whole" number, the 7 on the right side of the decimal point is in the "tenths" position, meaning "7 tenths," or 7/10, and the 6 is in the hundredths position. So, 3.76 can be read as: "3 and 7 tenths and 6 hundredths."

There is nothing that requires us to have 10 different digits in a number system. The **base-10** number system probably developed because we have 10 fingers, but if we happened to have eight fingers instead, we would probably have a base-8 number system. In fact, you can have a **base-anything** number system. There are often reasons to use different number bases in different situations.

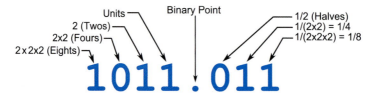

FIGURE 9.2 The structure of a binary number.

The binary[8] (base 2) number system is similar to the decimal system in that digits are placed to the left or right of a "point" to indicate values greater than 1 or less than 1 (see Figure 9.2). For binary numbers, the first digit to the left of the "binary point" is called the **units**. As you move further to the left of the binary point, every placeholder increases by a factor of **2**. To the right of the binary point the first digit is **half** (1/2), and as you move further to the right every placeholder number becomes **half again smaller**. Table 9.1 below shows a few equivalent decimal and binary whole numbers.

---

[7]The word "decimal" means "based on 10" (from the Latin *decima* meaning a tenth part).
[8]The word binary comes from "bi-" meaning two. We use it in words such as "bicycle" (two wheels) and "binocular" (two eyes).

Table 9.1 Decimal and binary numbers

Decimal	0	1	2	3	4	5	6	7	8	9	10	11	12	13	14	15
Binary	0	1	10	11	100	101	110	111	1000	1001	1010	1011	1100	1101	1110	1111

So, how do you convert binary numbers to decimal numbers? Several examples are given below to show the steps in converting several binary numbers to decimal numbers.

1. What is 1111 in decimal?
   - The "1" in the leftmost position is in the "$2 \times 2 \times 2$" (2^3) position, so that means $1 \times 2 \times 2 \times 2 = 8$
   - The next "1" is in the "2×2" (2^2) position, so that means $1 \times 2 \times 2 = 4$
   - The next "1" is in the "2" (2^1) position, so that means $1 \times 2 = 2$
   - The last "1" is in the units ($2^0 = 1$) position, so that means $1 \times 1 = 1$
   - Answer: $1111 = 8 + 4 + 2 + 1 = 15$ in decimal.
2. What is 1001 in decimal?
   - The "1" in the leftmost position is in the "$2 \times 2 \times 2$" (2^3) position, so that means $1 \times 2 \times 2 \times 2 = 8$
   - The next "0" is in the "2×2" (2^2) position, so that means $0 \times 2 \times 2 = 0$
   - The next "0" is in the "2" (2^1) position, so that means $0 \times 2 = 0$
   - The last "1" is in the units (2^0) position, so that means $1 \times 1 = 1$
   - Answer: $1001 = 8 + 0 + 0 + 1 = 9$ in decimal
3. What is 1.1 in decimal?
   - The "1" on the left side of the binary point is in the units (2^0) position, so that means $1 \times 1 = 1$
   - The 1 on the right side is in the "halves" (2^{-1}) position, so that means $1 \times (1/2) = 0.5$
   - So, 1.1 is "1 and 1 half" $= 1.5$ in decimal
4. What is 10.11 in decimal?
   - The "1" in the leftmost position is in the "2" (2^1) position, so that means $1 \times 2 = 2$
   - The "0" is in the units (2^0) position, so that means $0 \times 1 = 0$
   - The first "1" on the right of the point is in the "halves" (2^{-1}) position, so that means $1 \times (1/2) = 0.50$
   - The last "1" on the right side is in the "quarters" (2^{-2}) position, so that means $1 \times (1/4) = 0.25$
   - So, 10.11 is $2 + 0 + 1/2 + 1/4 = 2.75$ in decimal

A single binary digit (0 or 1) is called a "bit." For example, the binary number 11010 has five bits. The word bit is made from the words "binary digit." Bits are usually combined into 8-bit collections called byte. With an 8-bit byte, you can represent 256 values ranging from 0 to 255, as shown below:

$$0 = 00000000$$
$$1 = 00000001$$
$$2 = 00000010$$
....
$$254 = 11111110$$
$$255 = 11111111$$

Bytes often come with prefixes like kilo, mega, and giga, as in kilobyte, megabyte, and gigabyte. The Table 9.2 lists the actual sizes of these binary numbers.

Table 9.2 Binary number byte prefixes

Name	Size
Kilo (K)	$2^{10}=1,024$
Mega(M)	$2^{20}=1,048,576$
Giga (G)	$2^{30}=1,073,741,824$
Tera (T)	$2^{40}=1,099,511,627,776$
Peta (P)	$2^{50}=1,125,899,906,842,624$
Exa (E)	$2^{60}=1,152,921,504,606,846,976$
Zetta (Z)	$2^{70}=1,180,591,620,717,411,303,424$
Yotta(Y)	$2^{80}=1,208,925,819,614,629,174,706,176$

A terabyte hard drive actually stores 10^{12} bytes.[9] How could you possibly need a terabyte of disk space? When you consider all the digital media[10] available today (music, games, and video), it is not difficult to fill a terabyte of storage space. Terabyte storage devices are fairly common, and indeed there are some petabyte storage devices.

9.8 BINARY ARITHMETIC

The value of a **bit** depends on its position relative to the "binary point." For example, the binary number 11010.101 has a decimal value computed from the second and third rows of Table 9.3 as $16 \times 1 + 8 \times 1 + 4 \times 0 + 2 \times 1 + 1 \times 0 + 0.500 \times 1 + 0.250 \times 0 + 0.125 \times 1 = 26.625$.

Table 9.3 Converting the binary number 11010.101 to a decimal number

Placeholder	2^4	2^3	2^2	2^1	2^0	.	2^{-1}	2^{-2}	2^{-3}
Bit	1	1	0	1	0	.	1	0	1
Decimal value	16	8	4	2	1	.	1/2=0.500	1/4=0.250	1/8=0.125

Let us begin with the rules of binary addition.

$$0+0=0$$
$$0+1=1$$
$$1+0=1$$
$$1+1=10, \text{ so carry the 1 to the next bit and save the 0}$$

---

[9]The capacities of computer storage devices are typically advertised using their SI standard values, but the capacities reported by software operating systems uses the binary values. The standard SI terabyte (TB) contains 1,000,000,000,000 bytes $=1000^4$ or 10^{12} bytes. However, in binary arithmetic, a terabyte contains 1,099,511,627,776 bytes $=1024^4$ or 2^{40} bytes.

[10]Just 5 min of digital video requires about 1 gigabyte of storage.

For example, adding 010 (digital 2) to 111 (digital 7) gives:

$$
\begin{array}{r}
010 \\
+111 \\
\hline
\end{array}
$$

1001 (digital 9)

Binary addition is conceptually identical to decimal addition. However, instead of carrying powers of 10, one carries powers of 2. Here are the formalized steps to follow for the carry digits:

1. Starting at the right, $0+1=1$ for the first digit (no carry needed).
2. The second digit is $1+1=10$ for the second digit, so save the 0 and carry the 1 to the next column.
3. For the third digit, $0+1+1=10$, so save the 0 and carry the 1.
4. The last digit is $0+0+1=1$.
5. So the answer is 1001 (digital 9—you can see it is correct since decimal $2+7=9$).

Binary subtraction is conceptually identical to decimal subtraction. However, instead of borrowing powers of 10, one borrows powers of 2.

Rules of Binary Subtraction

$0–0=0$

$0–1=1$, and borrow 1 from the next more significant bit

$1–0=1$

$1–1=0$

The following examples illustrate "borrowing" in binary subtraction.

$$
\begin{array}{ccc}
10 & 100 & 1010 \\
-1 & -10 & -110 \\
\hline
1 & 10 & 100
\end{array}
$$

Can you complete Table 9.4? You can self-check against their decimal equivalents.

Table 9.4 Simple binary arithmetic examples

Example 1		Example 2		Practice 1		Practice 2	
Binary	**Decimal**	**Binary**	**Decimal**	**Binary**	**Decimal**	**Binary**	**Decimal**
1001	9	1001	9	1011		1011	
+101	+5	−101	−5	+110		−110	
1110	14	100	4				

The process of binary subtraction may be viewed as the addition of a negative number. For example, 3−2 may be viewed as $3+(-2)$. To do this you must determine the negative representation of a binary number. One way of doing this is with the one's complement.

The one's complement of binary number is found by changing all the ones to zeroes and all the zeroes to ones as shown below:

Number	One's Complement
10011	01100
101010	010101

To subtract a smaller number from a larger number using the one's complement method you:

1. Determine the one's complement of the smaller number,
2. Add the one's complement to the larger number,
3. Remove the final carry and add it to the result (this step is called the 'end-around carry').

Example 9.5

Do the following subtraction: 11001 (decimal 25) − 10011 (decimal 19).

Need: $11001 - 10011 = $ _____? (a binary number)

Know-how: Step1: The one's complement of 10011 is 01100.

Step2: Adding the one's complement to the larger number gives $01100 + 11001 = 100101$.

Step3: Removing the final carry and adding it to the result gives $00101 + 1 = 00110$.

Solve: $\mathbf{11001 - 10011 = 110}$. To verify that this is correct, convert each base-2 number to decimal and repeat the subtraction, or $\mathbf{25 - 19 = 6}$.

To subtract a larger number from a smaller number, the one's complement method is as follows:

1. Determine the one's complement of the larger number,
2. Add the one's complement to the smaller number (the result is the one's complement of the answer),
3. Take the one's complement of the result to get the final answer. Do not forget to add the minus sign since the result is negative.

Example 9.6

Do the following subtraction: 1001 (decimal 9) − 1101 (decimal 13)

Need: $1001 - 1101 = $? (a binary number)

Know-how: Step1: The one's complement of the larger number 1101 is 0010.

Step2: Adding the one's complement to 1001 gives $0010 + 1001 = 1011$.

Step3: Add a minus sign to the one's complement of 1011 to get $1001 - 1101 = -0100$.

Solve: $\mathbf{1001 - 1101 = -100}$. To verify that this is correct, convert each base-2 number to decimal and repeat the subtraction, or $\mathbf{9 - 13 = -4}$.

The rest of the familiar arithmetic functions can also be carried out in binary. Fractions can be expressed in binary by means of digits to the right of a binary point. Once again, powers of 2 take the role that powers of 10 play in digital arithmetic. Thus, the decimal fraction 0.5 (i.e., ½) is the binary fraction 0.1 and the decimal fraction ¼ is the binary fraction 0.01, and so on. A fraction that is not an even power of 1/2 can be expressed as a sum of binary numbers. Thus, the decimal ⅜ = 0.0011 in binary, which is the sum of decimal ¼ + ⅛.

Multiplication and division can be carried out using the same procedures as in decimal multiplication and "long division." The only complication is, again, systematically "carrying" and "borrowing" in powers of 2, rather than powers of 10.

Rules of binary multiplication

$$0 \times 0 = 0$$
$$0 \times 1 = 0$$
$$1 \times 0 = 0$$
$$1 \times 1 = 1, \text{ and no carry or borrow bits}$$

Example 9.7

If a powerful race car has an air to fuel ratio (A/F) of 15 [kg air]/[kg fuel] and the air intake draws in 1.5 kg of air per second, how much fuel must be injected every second? Solve in binary to five significant binary digits.

Need: Fuel rate = _____ kg fuel/s in binary?

Know-How: Fuel flow rate $= (F/A) \times 1.5$ kg air/s $= (1/15)$ [kg fuel]/[kg air] $\times 1.5$ [kg air/s] $= 0.10$ [kg fuel/s].

To illustrate binary arithmetic, let us break down this problem into two separate ones. While unnecessary to do this, one will illustrate binary division and binary multiplication.

The division problem will be 1/15 (decimal) $= 1/1111$ (binary), and the multiplication problem will take the solution of that problem and then multiply it by 1.5 (decimal) $= 1.1$ (binary).

Solve: Start with 1/1111, then binary multiply that answer by 1.1.

```
              .00010001            0.00010001
                                        ×1.1
      1111 | 1.00000000           0.00010001
             1111                 0.00001000
             10000                0.00011001
             1111
                 1
```

Therefore, the **fuel flow rate = 0.00011001 kg/s (in binary)**.
Checking in decimal: $(1/15) \times 1.5 = $**0.10** and $0.00011001 = 0/2 + 0/4 + 0/8 + 1/16 + 1/32 + 0/64 + 0/128 + 1/256 = $**0.098** (to get the closer answer of 0.10 we would need to use more significant (binary) figures.

Many times each second, a computer under the hood of an automobile receives a signal from an air flow sensor, carries out a binary computation such as the one shown in this example, and sends a signal to an actuator that causes the right amount of fuel to be injected into the air stream in order to maintain the desired air-fuel ratio. The result is much more precise and reliable control of fuel injection than was possible before computers were applied to automobiles.

9.9 BINARY CODES

We can now see how 0 and 1 can be used to represent "false" and "true" as logic values, while also of course 0 and 1 are numeric values. It is also possible to use *groups* of 0s and 1s as "codes." The now outdated Morse code is an example of a binary code, while the genetic code is based on just the pairings of four, rather than two, chemical entities known as bases.

Suppose we want to develop unique codes for the following nine basic colors: red, blue, yellow, green, black, brown, white, orange, and purple. Can we do this with a three-bit code [three 0s or 1s (bits)]? No, since there are only eight combinations of three bits (note: $2^3 = 8$): namely 000, 001, 010, 011, 100, 101, 110, and 111.

Table 9.5 contains a four-bit code that would work, but of course it is only one of several, since there are $2^4 = 16$ possible combinations of a four-bit code. If we have N bits, we can code 2^N different things.

Table 9.5 Assignment of binary numbers to colors

Color	Binary equivalent	Color	Binary equivalent
Red	0000	Brown	0101
Blue	0001	White	0110
Yellow	0010	Orange	0111
Green	0011	Purple	1000
Black	0100		

9.10 HOW DOES A COMPUTER WORK?

How can these abstract ideas of Boolean algebra, binary logic, and binary numbers be used to perform computations using electrical circuits, particularly switches (which we will study in a later chapter)? The modern computer is a complex device, and any answer we give you here is necessarily oversimplified. But the principles are sufficient to give you some insight. For this discussion we will need to know what a central processing unit, or CPU, does and what a computer memory is (which can take such forms as read-only memory or ROM and random access memory or RAM).

The "smart" part of the computer is the CPU. It is just a series of registers, which is nothing but a string of switches. These switches can change their voltage states from "off" (nominally a no voltage state) to "on" or +5 volts above ground. The early personal computers (or PCs) used only 8-bit registers and modern ones use 64 or 128, but we can think in terms of the 8-bit registers. (The notation x, y in Table 9.6 means either state x or state y.)

Table 9.6 Eight bit register

Bit #	7	6	5	4	3	2	1	0
Voltage	0 or 5 (0,5)	0 or 5 (0,5)	0 or 5 (0,5)	0 or 5 (0,5)	0 or 5 (0,5)	0 or 5 (0,5)	0 or 5 (0,5)	0 or 5 (0,5)
Bits	0 or 1 (0,1)	0 or 1 (0,1)	0 or 1 (0,1)	0 or 1 (0,1)	0 or 1 (0,1)	0 or 1 (0,1)	0 or 1 (0,1)	0 or 1 (0,1)

This register can contain 2^8, or 256, discrete numbers or addresses. What the CPU addresses is the memory in the computer. You can think of memory as a pigeonhole bookcase with the addresses of each pigeonhole preassigned.[11]

---

[11]M. Sargent III and R. L. Shoemaker, *The IBM PC from the Inside Out* (Reading, MA: Addison-Wesley Publishing Co. Inc., revised edition, 1986), p. 21.

If we have just 256 of these pigeonholes in our memory, our CPU can address each of them. If we want to calculate something, we write a computer code in a suitable language that basically says something like: add the number in pigeonhole 37 to that in pigeonhole 64, and then put the contents in pigeonhole 134. The binary bits can represent numbers, logic statements, and so on. All we then need to do is to be able to send out the results of our binary calculations in a form we can read. For example, we can decide that the contents of pigeonhole 134 is an equivalent decimal number.

If, in the above example, we identify the program input as what is in pigeonholes 37 and 64, and the program output is pigeonhole 134, the computer is constructed essentially as in Figure 9.3 to accomplish its mission.

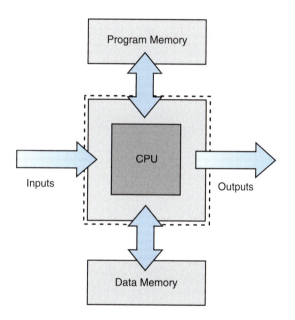

FIGURE 9.3 The CPU.

A computer is made up of hardware (electrical circuits containing such components as transistors). How the computer is instructed to execute its functions is managed by software (the instructions that tell the components what to do).

Part of the hardware of the computer that we have so far ignored is an internal clock. One might view the software as a list of instructions telling the computer what to do every time the clock ticks. This list of instructions includes the normal housekeeping functions that the computer carries out regularly (such as checking whether a user has entered in any keystrokes on the keyboard in the very short-time interval since the last time this was checked). But it can also include downloading into a portion of the computer's memory called program memory, a special list of instructions called a stored program, and then executing that stored program. Examples of stored programs are a word processor and a spreadsheet program.

Once this stored program is downloaded into the memory, the CPU carries out this stored program step by step. Some of those steps involve carrying out computations, which are executed in binary arithmetic by the CPU using the methods described in this chapter. In some cases, the result of the computation determines which of the stored program's instructions is the next one that should be carried out. This flexibility regarding the order in which instructions are carried out is the basis of the computer's versatility as an information processing system.

Software in this binary form is called machine language. It is the only language that the computer understands. It is, however, a difficult language for a human to write or read. So computer engineers have developed programs that can be stored in the computer that translate software from a language that humans understand into machine language.

The types of language that humans understand are called a higher level languages. These language consist of a list of statements somewhat resembling ordinary English. For example, a higher level language might contain a statement such as "if $x < 0$, then $y = 36$." Examples of higher level languages are C++, BASIC, and Java. Most computer programs are originally written in a high-level language.

The computer then translates this high-level language into machine language. This translation is typically carried out in two steps. First, a computer program called a compiler translates the statements of the high-level language into statements in a language called assembly language that is closer to the language that the computer understands. Then another computer program called an assembler translates the assembly language program into a machine language program. It is the machine language program that is actually executed by the computer.

The personal computers that we all use, typically by entering information through such devices as a keyboard or mouse, and producing outputs on a screen or via a printer, are called general-purpose computers. As the name suggests, they can be used for a wide range of purposes, from game playing to accounting to word processing to monitoring scientific apparatus.

Not all computers need this wide range of versatility. So there is another important class of computers directed at narrower ranges of tasks. These computers are called embedded computers (Figure 9.4).

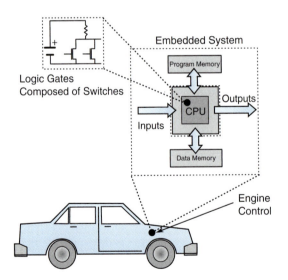

FIGURE 9.4 Schematic for an Embedded Automotive Computer.

As their name suggests, these computers are embedded within a larger system. They are not accessible by keyboard or mouse, but rather receive their inputs from sensors within that larger system. In an automobile, there may be dozens of embedded computers. There might be, for example, an embedded computer for controlling a car's stereo system, another recording data for automated service diagnostics, another for the operation of the brakes, another for the steering conditions, and yet another for fuel control.

SUMMARY

Although analog computers have been of some historical importance, digital computers do almost all important control jobs in today's advanced technologies. So an engineer must understand the principles of digital computation that rest on the immensely powerful concepts of **Boolean algebra, binary logic, truth tables, binary arithmetic, and binary codes**. These concepts enable an engineer to make a first effort at defining the concept of information. Binary arithmetic and information are the basis of computer software. These concepts make it possible to move on to the challenge of implementing digital controls using computation.

EXERCISES

1. A popular ditty of the late nineteenth and early twentieth-century railroad era was the following (sung to the tune of *Humoresque* by the nineteenth-century Czech composer Antonin Dvorak[12]):

    Passengers will please refrain—From flushing toilets—While the train—Is standing in the station—I love you.

    For the preceding ditty, define a variable S expressing whether or not the train is in the station, a variable M expressing whether or not the train is moving ("standing" meaning "not moving"), and a variable F expressing the fact that the toilet may be flushed. [A: $S=$ "the train is the station," $M=$ "the train is moving" (or you could use its negation, M' meaning the train is stationary), and $F=$ "the toilet may be flushed"]

2. For the ditty in Exercise 1, (a) express as a logic formula the conditions under which one may flush the toilet, (b) evaluate the formula you wrote in (a) for $M=1$ and $S=1$, and (c) express in words the meaning of your answer to (b). (Assume that any behavior not explicitly forbidden is allowed.)

3. Rework Example 9.2 to express and evaluate the logic formula to answer the question "If the speed is below the set speed and the set speed is above the speed limit, then will the throttle be opened?"

4. In Example 9.3, include the additional Boolean variable that for actuation of the warning light the driver's door must be closed (D_{door} is true if the driver's door is closed).

5. Consider the following logic variables for a car:
    $Db=$ "the driver's seat belt is fastened"
    $Pb=$ "the passenger seat belt is fastened"
    $W=$ "the seat belt warning should be on in my car"
    Write a sentence in English that expresses the logic equation $W=Db'+Pb'$. (A: $W=$ "If either the driver's seat belt is not fastened or the passenger's seat belt is not fastened, the seat belt warning light should be on.")

6. Consider the following logic variables for a car:
    $W=$ "the seat belt warning light is on"
    $D=$ "a door of the car is open"
    $Ps=$ "there is a passenger in the passenger seat"
    $K=$ "the key is in the ignition"
    $M=$ "the motor is running"
    $Db=$ "the driver's seat belt is fastened"
    $Pb=$ "the passenger seat belt is fastened"

[12]You can find the music at: http://www.youtube.com/watch?v=WmAZoexenx8.

Write a logic equation for W that expresses the following sentence: "If all the doors of the car are closed, and the key is in the ignition, and either the driver's seat belt is not fastened or there is a passenger in the passenger seat and the passenger's seat belt is not fastened, then the seat belt warning light should be on."

7. Consider the following logic variables for a car:

   M = "the motor is running"

   Db = "the driver's seat belt is fastened"

   In the early 1970s, the government ordered all seat belt warnings to be tied to the motor in a manner expressed by the following sentence: "If the driver's seat belt is not fastened, then the motor cannot be running."

   a. Write a logic equation for M in terms of Db that expresses this sentence.

   b. Write a truth table for that logic equation.

   (In practice, this seat belt light logic caused problems. Think of trying to open a manual garage door or pick up the mail from a driveway mailbox. The government soon retreated from an aroused public.) A: a. $M=Db$; b (see table below).

Db	M
0	0
1	1

8. Consider the following logic variables:

   A = "a customer at a restaurant orders an alcoholic beverage"

   I = "the customer shows proper identification"

   M = "a customer at a restaurant is over 21"

   Consider the sentence "If a customer at a restaurant is over 21 and shows proper identification, then she can order an alcoholic beverage." (a) Express this sentence as a logic equation, and (b) write a truth table for the logic equation.

9. Consider the following variables expressing a football team's strategy.

   T = "it is third down"

   L = "we must gain more than 8 yards to get a first down"

   P = "we will throw a pass"

   The team's strategy is expressed by this truth table.

T	L	P
0	0	0
0	1	1
1	0	1
1	1	1

Write a logic equation for P in terms of T and L (A: $P=T+L$).

Exercises 10 and 11 use electrical circuits to effect logical statements. If you are uncertain about electrical circuits, you should review ahead to the chapter on Electrical Engineering. In particular, an electrical switch (shown as an inclined line when open) means there is no current flowing from the battery. The circuit is off and is then said to be in a "0" or a "false" state; contrarily, when the switch is closed, current flows from the battery and a voltage appears across the lamp L. The circuit is now in an "on" position and is said to be "1" or a "true" state.

10. Consider the following electric circuit and the variables L="the light is on," A="switch A is closed," and B="switch B is closed." Express the relationship depicted by the electric circuit as a logic equation for L in terms of A and B. (A: $L=A+B$.)

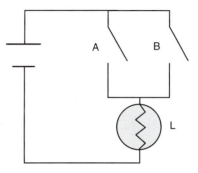

11. Consider the following circuit diagram and the variables L="the light is on," A="switch A is closed," B="switch B is closed," and C="switch C is closed." Write a logic equation for L in terms of A, B, and C.

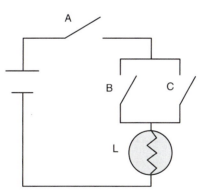

12. Consider the following logic variables for a car:
    1. W="the seat belt warning light is on"
    2. Ps="there is a passenger in the passenger seat"
    3. Db="the driver's seat belt is fastened"
    4. Pb="the passenger seat belt is fastened"
    Draw a circuit diagram for the logic equation $W=Db'+(Ps \cdot Pb')$.

13. Explain the following sentences.
    (a) "There are 10 kinds of people in the world, those who understand binary numbers, and those who don't."
    (b) "Binary is as easy as 1, 10, 11."

14. Convert the following numbers from binary to decimal: (a) 110, (b) 1110, and (c) 101011. Partial A: (see table below)

—	2^5	2^4	2^3	2^2	2^1	2^0	Decimal equivalent
—	32	16	8	4	2	1	—
(a) 110	0×32	0×16	0×8	1×4	1×2	0×1	6
(b) 1110							
(c)101011							

15. Convert the following numbers from decimal to binary: (a) 53, (b) 446, and (c) 1492. Partial A: (see table below)

Decimal	1024	512	256	128	64	32	16	8	4	2	1
Binary place	2^{10}	2^9	2^8	2^7	2^6	2^5	2^4	2^3	2^2	2^1	2^0
(a) 53	0	0	0	0	0	1	1	0	1	0	1
(b) 446											
(c) 1492											

16. Do the binary additions in the table below. Check your answer by converting each binary number into decimal. Partial A: the first addition

Binary	Decimal	Binary	Decimal	Binary	Decimal
1010	10	11101		10111	
+110	6	+10011		+10	
10000	16				

17. Do the binary subtractions in the table below. Check your answer by converting each binary number into decimal. Partial A: the first subtraction

Binary	Decimal	Binary	Decimal	Binary	Decimal
1010	10	11101		10000	
−110	−6	−10011		−1	
0100	4				

18. If a powerful race car has an $(A/F)_{\text{Mass}}$ of 12 (kg air)/(kg fuel), and the air intake draws in 1000 kg of air per second, how much fuel must be injected every second? Solve in binary to three significant binary digits. (A: 0.000101 kg)

19. Suppose we want to devise a binary code to represent the fuel levels in a car:
    a. If we need only to describe the possible levels (empty, 1/4 full, 1/2 full, 3/4 full, and full), how many bits are needed?
    b. Give one possible binary code that describes the levels in (a).
    c. If we need to describe the levels (empty, 1/8 full, 1/4 full, 3/8 full, 1/2 full, 5/8 full, 3/4 full, 7/8 full, and full), how many bits would be needed?
    d. If we used an 8-bit code, how many levels could we represent?

20. Construct a spreadsheet[13] that converts binary numbers from 0 to 111 to decimal numbers, print as formulae using the "control tilde" command. Check your spreadsheet against Exercise 14. (A: e.g., binary $110 \equiv 1 \times 2^2 + 1 \times 2^1 + 0 \times 2^0 =$ decimal 6.)

21. Construct a spreadsheet that converts decimal number 53 to binary. Print as formulae using the "control tilde" command. Check your spreadsheet against Exercise 15. **Hint:** 5 (decimal) can be divided by 2^2 to yield an integer "1" and remainder 1; 1 cannot be divided by 2^1 [therefore, integer "0"]; and "1" can be divided by 2^0 for the last integer "1." Check for a spreadsheet function that will divide two numbers and display their result with no remainder.

22. Construct a spreadsheet that does binary subtraction with one's complement. Test it on Exercise 17.

23. In the game "rock, scissors, and paper" we have the following rules:

> Rock breaks scissors
> Scissors cuts paper
> Paper covers rock

If we were to create a code to represent the three entities, rock, scissors, and paper, we would need two bits. Suppose we have the following code, where we call the first bit X and the second bit Y:

	X	**Y**
Rock	0	0
Scissors	0	1
Paper	1	0

Now if we have two players who can each choose one of these codes, we can play the game.
Examples

Player 1	Rock (0,0)	Player 2	Scissors (0,1)	Player 1 wins
Player 1	Scissors (0,1)	Player 2	Scissors (0,1)	Tie
Player 1	Rock (0,0)	Player 2	Paper (1,0)	Player 2 wins

---

[13]Spreadsheets have some built-in functions for decimal conversions to and from binary. It is recommended that you try first to use the actual mathematical functions described in this chapter and then *check* your answers using these functions to confirm those answers.

We see that there are three possible outcomes of the game: Player 1 wins, Tie, Player 2 wins. Complete the table below that describes all the possible outcomes of the game.

Player 1 X,Y	Player 2 X,Y	Player 1 wins	Tie	Player 2 wins
0,0	0,0	0	1	0
0,0	0,1	1	0	0
0,0	1,0			
0,1	0,0			
0,1	0,1			
0,1	1,0			
1,0	0,0			
1,0	0,1			
1,0	1,0			
	Totals			

24. A company purchased a computer program for your part-time job with them. The license agreement states that you can make a backup copy, but you can only use the program on one computer at a time. Since you have permission to make a backup copy, why not make copies for friends? What do you do? (Use the Engineering Ethics Matrix.)
    a. Go ahead, since your friends only use one computer at a time and these are backup copies.
    b. Make the backup copy, but sharing it with anyone clearly violates the license agreement.
    c. Ask your supervisor if you can use the backup copy at home, and then make as many copies as you wish.
    d. Use the program discretely, since software license agreements cannot be enforced anyway.

25. You are a software engineer at a small company. You have written a software program that will be used by a major manufacturer in a popular product line. Your supervisor asks you to install a "back door" into the program that no one will know about so that he can monitor its use by the public. What do you do? (Use the Engineering Ethics Matrix.)
    a. Install the back door, since it sounds like a fun experiment.
    b. Tell your supervisor that you cannot do it without authorization from the end user.
    c. Install the back door, but then deactivate it before the software is implemented.
    d. Stall your supervisor while you look for another job.

Electrical Engineering

10

10.1 INTRODUCTION

Electrical engineering is a field of engineering that deals with the study and application of electricity, electronics, and electromagnetism. One of the most convenient ways to make energy useful is by converting it to electromagnetic energy, usually called *electricity*. In a previous chapter, electromagnetic power was briefly introduced as the result of multiplying the instantaneous current by the voltage drop that produced it. This chapter expands that introduction by giving a more detailed, but still highly simplified, discussion of electromagnetic energy and some of its uses.

Though electricity is everywhere around us at all times, the easiest way of getting electricity to go where it is needed is by means of an arrangement called an **electrical circuit**. In this chapter, we learn what electrical circuits are and apply a very simple model for analyzing their operation. The model has as its basic variables **charge**, **current**, **voltage**, and **resistance**. This simplified treatment relies on a model consisting of **Ohm's Law**, the **Power Law**, and **Kirchhoff's Voltage** and **Current Laws**. Using these laws, we can analyze the operation of two classes of direct current circuits called series circuits and parallel circuits.

The field of electrical engineering is devoted to the design and analysis of electrical circuits to meet a wide range of purposes. Electrical engineers work in such varied areas as melting metals, keeping food frozen, projecting visual images, powering cell phones, taking elevators to the top of skyscrapers, and moving submarines through the ocean depths. Rather than attempt to describe the entire range of electrical engineering applications, we focus on one case study: the development of ever-faster **switches**. Different kinds of switches can be implemented in many ways, using the properties of electricity, such as the ability of a current to produce **magnetism** and the use of electric charge to control the flow of current.

10.2 ELECTRICAL CIRCUITS

An electric circuit is a *closed* loop of wire connecting various electrical components, such as batteries, light bulbs, switches, and motors. The word *circuit* should bring to mind the idea of a circle. We first recognize a natural phenomenon of electric charge, measured in a SI unit called **coulombs**, that flows through the electrically conductive wires just as water flows through pipes. This is the essence of the simple model we use for understanding electrical circuits. And, just as water is actually made up of very small particles that we can call water molecules, each molecule of which has the same mass, the charge in a wire is actually carried by very small particles called **electrons**, each of which has the same tiny electrical charge of -1.60×10^{-19} C (notice the minus sign). Because it is inconvenient to measure amount of water in numbers of molecules, instead we use gallons, liters, or whatever, which contain a huge number (about 10^{27}) of molecules. It is similarly inconvenient to measure charge in number of electrons, so we measure charge in coulombs. The coulomb charge is a huge number (about 10^{20}) of electrons. A car battery might hold 1 million or more coulombs of charge, while the mid-sized C cell used in a flashlight might hold 10,000 C.

Water would not be useful to us if it just sat in a reservoir, and electric charge is not useful to us if it just sits in a battery. Think of a Roman aqueduct system, in which water from a high reservoir is allowed to flow down to the city; it progressively loses its original potential energy as it flows. Electrons similarly flow from a high potential to a lower one. We call this potential **voltage**. Therefore, electric charges are pumped by a battery or generator through the wires for use in electrical components that we value such as lights and air conditioning, and then they return to the battery by means of wires for their reuse.

Just as the flow of water from a hose can be separated into two or more exiting streams totaling the amount of entering water, if trillions of electrons flow every second into an electrically conductive branch junction, you can be sure that the total number of electrons leaving the junction is exactly equal to the number of arriving electrons, a principle called **conservation of charge**. The flow of charge, a variable called **current**, is measured in the SI unit of **amperes**, where 1 A is equal to 1 C per second (and correspondingly equal to the flow of many trillions of electrons every second).

The hollow interior of a water pipe permits the water flow. Wires are not hollow, but they do allow the flow of electricity in the same way that a hollow pipe allows the flow of water when subjected to upstream pressure. We call materials used in wires **conductors**, which include metals. Other materials do not allow the significant flow of electricity through them. These materials are called **insulators**. They include such nonmetals as ceramics and plastics. Other materials, such as the **semiconductors**, have properties somewhere between those of conductors and insulators. We distinguish among conductors, semiconductors, and insulators by a property known as **resistance**. Electric current, in the form of electrons, flows because of the voltage applied to the electrons. The electrons in metal conductors flow from the negative to the positive direction in potential. However, traditionally (and unfortunately), the current is considered to flow from the positive pole of the battery to the negative one.[1] It was discovered well after this convention had been adopted that the current carriers in metals were electrons and flow in the opposite direction (see Figure 10.1).

---

[1]This convention was suggested by Benjamin Franklin in the mid-eighteenth century and is still used.

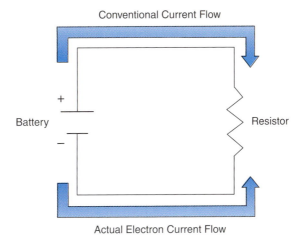

Conventional Current Flow

+

Battery

Resistor

−

Actual Electron Current Flow

FIGURE 10.1 Blame Ben Franklin!

This figure also introduces you to the visual symbol for a battery, with the longer line indicating the "plus" side of the battery and the shorter line representing the "minus" side of the battery.

The motion of water through a pipe may be caused by a pump, which exerts a pressure on the water forcing it to move through the pipe. In the same way, a battery, in addition to holding electrical charge, serves as a sort of electrical pump to force electrons to move through a wire. This electrical analog of "pressure" is the variable we call *voltage*, and it is measured in a SI unit called **volts**. Just as a hand pump might be used to apply a pressure of 12 psi to move water through a pipe, a car battery might be used to apply 12 V to move current through a wire.

Resistance is just the ratio of the voltage drop to the current. High resistance means that a large voltage drop is required to achieve a given current. Low resistance means that only a small voltage drop is required to achieve a given current. Resistive devices are called **resistors** (and often colloquially just *resistances*) and may be made from coils of wire alloy, pieces of carbon, or other materials that conduct electricity. In fact, all materials except electrical **superconductors**[2] exhibit resistance to current flow. Some metals, such as pure silver and pure copper, have a very low resistance to electron flow. Other metals, such as aluminum and gold, have somewhat higher resistance but are still used as conductors in some applications. We usually use copper for connections, since we do not want much resistance there. Metal alloys and semiconductors offer appreciable resistance to electron flow, and we use those where we wish to exploit that property.

To summarize, in our model of an electric circuit, the voltage (i.e., the electrical "pressure") pushes a moving current of charge though a conductor. This charge (i.e., electrons) never disappears but simply circles around the circuit again and again, doing its various functions each time around the circuit. As the electrons travel around the circuit, they are pumped to high potential energy by the voltage source and then lose this energy as they perform their functions.

Symbolically, the simplest possible electrical circuit consists of a battery, a wire, and an electrical component located somewhere along the wire, as shown in Figure 10.2.

---

[2]These materials allow the *unrestricted* flow of electrons.

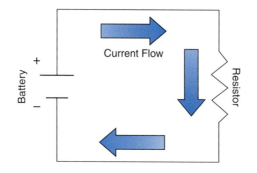

FIGURE 10.2 Current flow in a circuit.

The zigzag line represents a resistor, and the straight lines are the wires connecting it to the battery. As previously stated, the current is considered to flow from the positive pole of the battery to the negative one. Since a circuit goes in a circle and ends up where it started, the voltage must also return to its initial value after a trip around the circuit. Since the voltage is raised across the battery, it must be lowered back somewhere else in the circuit. In our simplest model, we assume that none of this voltage "drop" occurs in the wires, but rather all of it occurs in the component or components in the circuit (although we soon relax that restriction).

10.3 RESISTANCE, OHM'S LAW, AND THE "POWER LAW"

As water flows through a pipe, it experiences energy dissipation by coming into contact with the walls of the pipe, and this tends to slow down the flow. Pressure is needed to keep water moving through the pipe, and the magnitude of the change of the pressure is determined by the energy dissipated in the water.

In the same way, as charge flows through a wire, it experiences interactions that tend to reduce the voltage "pushing" the charge. We initially assume that these interactions do not occur in the connecting wires but are confined to the components hooked together by the wires. The interactions that reduce the voltage are described by resistance. It is measured in SI units called **ohms** and abbreviated with the uppercase Greek letter omega, Ω. The voltage drop (that is, an electrical "pressure drop") needed to keep the charges moving through the wire is determined by the resistance. This relation between current, voltage, and resistance constitutes the first basic law of our model of electrical circuits, **Ohm's Law**. It states that the resistance, which is defined as the ratio of voltage drop to current, remains *constant* for all applied voltage drops. Mathematically stated,

$$R = V/I \tag{10.1}$$

If V is in volts and I is in amperes, R is defined in the units of ohms $[\Omega] = [V]/[A]$. For example, consider a 2.00 Ω resistor made of a material that obeys Ohm's Law. If a voltage drop of 1.000 V is applied across the resistor, its current is exactly 0.500 A. If a voltage drop of 1000. V is applied, the current is exactly 500. A.

Few if any real materials obey Ohm's Law in this exact manner. The law is, however, a very useful approximation for most practical electric circuits. We now have a way of using numbers in our model of a circuit.

Example 10.1

Find the current flowing in amperes through the wire in a circuit consisting of a 12. V battery, a wire, and a 10. Ω resistor.

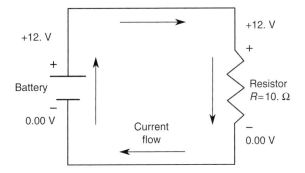

Need: Current flowing through the wire in amperes.

Know: Voltage provided by battery is 12. V, and the resistance is 10. Ω.

How: Sketch the circuit and apply Ohm's Law. Note we are using Franklin's convention for current direction but you can easily use the alternative that follows the actual electron flow.

Solve: The voltage drop across the resistor therefore is 12.0 V and, as given by Ohm's Law, $V = I \times R$ or $I = V/R = 12./10. = $ **1.2 A** (two significant figures).

Ohm's Law implies a second important property about electric circuits. Recall that, for the circuit as a whole, the total drop in voltage in the components other than the battery must be equal in magnitude but opposite in sign to the increase in voltage caused by the battery. Ohm's Law then enables us to find this voltage drop, including, if necessary, its implied sign.

In continuing our water analogy, hydraulic resistance to water flow may be thought of as flow restrictions in the water pipe—think of a water valve quarter open. However, opening the valve say from one quarter open to one half open reduces its hydraulic resistance. Since water is forced through pipes under pressure, it provides energy that might be used, for example, to power a water wheel to run your washing machine. To attempt this, it would be useful to know the relation between water pressure, amount of water flow, and power output from the water wheel. In the same way, electric charge can provide energy to run a wide variety of electrical appliances, from light bulbs to computers. To attempt this, it is useful to know the relation between voltage, current, and power.

That relation is the second leg of our model—the "**Power Law**." We have seen that electrons with some charge—say, Q coulombs—"fall" through an electric potential—say, ΔV—in a resistor. This is analogous to a mass m falling through a gravitational potential Δh to produce a gravitational energy change $mg\Delta h$. The equivalent electric work done is $Q\Delta V$. The power produced is therefore $Q/t \times \Delta V$, in which Q/t is the rate of charge flow, which we call *current*. The "Power Law" simply states that electric power is given by: **Power = Current × Voltage**, or

$$P = I \times V \qquad (10.2)$$

Note that power is not a new variable, but it is the same old one that we used earlier in talking about energy flows. We inserted quotes around "Power Law" to emphasize that it is not an independent law but perfectly derivable from the definition of power as the rate of working. Power is still measured in watts, where a watt is a joule per second. This fact allows us to relate our model of electrical circuits to our earlier energy models. The analogy between water flow and electricity is thus reasonably complete (see Table 10.1).

Table 10.1 Analogy Between Water Flow in a Pipe and Electricity Flow in a Wire

Water Flow in a Pipe	Electricity Flow in a Wire
Pressure	Voltage
Flow rate	Current
Hydraulic resistance	Electrical resistance
Power in water stream = pressure drop × flow rate	Power in electrical circuit = voltage drop × electric current

Evidence of the basic "Power Law" relation is provided by a second important property of a resistor in addition to its ability to limit the flow of electricity. That property becomes evident if we force enough charge through the resistance at a high enough voltage and observe that the resistor may then "glow" (unless it melts first!). This is the basis of a very useful invention (now being phased out as too inefficient). The light bulb (technically the incandescent lamp) was invented not, as popularly believed, by Thomas Edison in 1879 but by Humphrey Davy in England and August De La Rive in France in about 1810.[3]

10.4 SERIES AND PARALLEL CIRCUITS

There are two basic types of electric circuits. In a **series circuit**, shown in Figure 10.3a, the current takes a single path. In a **parallel circuit**, shown in Figure 10.3b, the wire branches out into two or more paths and these paths subsequently join up again. In a parallel circuit, we should envision the current as being divided at the point where the wires branch out then joining together again where the wires come together.

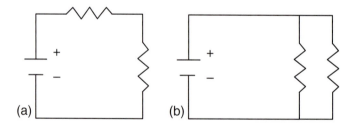

FIGURE 10.3 Series circuit (a) and parallel circuit (b).

As previously stated, by the conservation of charge principle, the current in each of the two branches has to add up to the current that went in; and conversely, the total current that goes out of the two branches has to equal the total current that went in.

The mathematics for determining the voltages and currents in series and parallel circuits rests on two very simple principles: (1) two resistors in a series have the same current passing through each of them, and (2) two resistors in a parallel have the same voltage drop across each of them.

---

[3]However, neither scientist took his observations to a practical conclusion, as did Edison.

Example 10.2

Consider the series circuit in Figure 10.3a, and assume that the battery voltage equals 12. V and that each resistor has a resistance of 100. Ω. What is the current in the circuit and the voltage drop across each resistor?

Need: The current in the circuit and the voltage drop across each resistor.

Know: Battery voltage $= 12$. V; resistance of each resistor $= 100$. Ω.

How: Call the unknown current I and the resistance in each resistor R. By the first principle just stated, the current I must be the same in each series resistor. So, by Ohm's Law, the voltage drop across each resistor must be $I \times R$. Since the total voltage drop around the circuit must be zero, $V(\text{battery}) + V(\text{resistor}_1) + V(\text{resistor}_2) = 0$, or V $(\text{battery}) - I \times R_1 - I \times R_2 = 0$. Solve for the unknown I.

Solve: 12. V $- I \times 100$. $\Omega - I \times 100$. $\Omega = 12$ V $- I \times 200$. $\Omega = 0$, or $I = \mathbf{0.060}$ **A**. Then the **voltage drop** across each resistor is $V = I \times R = 0.060 \times 100. = \mathbf{6.0}$ **V**.

Notice that in this example we could have simply added the two 100. Ω resistances together and used their sum of 200. Ω directly to get the current by Ohm's Law in one step: $I = 12./200. = 0.060$ A. In a series circuit, the equivalent series resistance R is always the sum of the individual resistances: $\mathbf{R_{eq} = R_1 + R_2 + R_3 + ...}$

Example 10.3

Consider the parallel circuit in Figure 10.3b with the battery voltage $= 12$. V and each resistor having a resistance of 100. Ω. What is the current drawn from the battery?

Need: The current drawn from the battery.

Know: Battery voltage $= 12$. V; resistance of each resistor $= 100$. Ω.

How: Call that current the unknown, I. By the second principle stated, the voltage drop must be the same across each parallel resistor. Since the total voltage drop around the circuit must be zero, $V(\text{battery}) = V(\text{resistor}_1) = V$ (resistor_2).

Now assume that the current divides between the two resistors with I_1 going through one of the resistors and I_2 going through the second. By Ohm's law, $V(\text{battery}) = V(\text{resistor}_1) = I_1 \times R_1 = V(\text{resistor}_2) = I_2 \times R_2$. Solve for I_1 and I_2. Since the current coming out of the branches is the sum of the current in each branch, $I = I_1 + I_2$.

Solve: 12. V $= I_1 \times 100$. $\Omega = I_2 \times 100$. Ω. Therefore, $I_1 = I_2 = 0.12$ A (since here both resistors have the same value). Then, $I = I_1 + I_2 = \mathbf{0.24}$ **A**.

Note that, in this case, the parallel circuit draws more current than the series circuit. This is true for parallel circuits. It's the same thing as adding a second hose to a garden faucet; provided the supply pressure remains constant, we spray more water (analogous to an electric current) with two hoses than with one.

The value of a single resistor equivalent to a group of resistors connected in parallel is easily computed by observing that, since the total current I is equal to the sum of the currents in each of the parallel resistors, $I = I_1 + I_2 + I_3 + ...$ Since each resistor is exposed to the same voltage, using Ohm's Law as $I = V/R$ gives $I = V/R_{eq} = V/R_1 + V/R_2 + V/R_3 + ...,$ or $\mathbf{1/R_{eq} = 1/R_1 + 1/R_2 + 1/R_3 + ...},$ where R_{eq} is the equivalent resistance of all the resistors in parallel. Therefore, the equivalent resistance in Example 10.3 is $1/R_{eq} = 1/100. + 1/100. = 2/100. = 1/50.0,$ so $R_{eq} = 50.0$ Ω. Then, $I = V/R_{eq} = 12./50.0 = 0.24$ A as before.

One particular form of series circuit enables us to add more realism to our simple model. Earlier, wires were modeled as conductors with zero resistance. However, a real wire has a small resistance, which is typically proportional to the length of the wire and inversely proportional to its cross-sectional area. In more realistic circuit models, this resistance of the wire is modeled by a resistance element (our now familiar wiggly-line symbol) inserted at an arbitrary location into the wire. This resistance henceforth is called the *wire* resistance, and the resistance of a component such as a lightbulb is called the *load* resistance.

Example 10.4

For the circuit shown, assume (1) the wire resistance (actually distributed along the whole length of the wire) totals 1.0 Ω, (2) the load resistance is 10. Ω, and (3) the battery voltage is 6.0 V. Compute the efficiency of the circuit, where the efficiency is defined as the power dissipated in the in the load divided by the power produced by the battery.

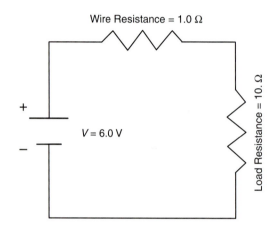

Need: Efficiency of circuit=(power dissipated in load)/(power produced by battery).

Know: Battery voltage=6.0 V; wire resistance=1.0 Ω; load resistance= 10. Ω.

How: Computing efficiency requires two computations of power. Computation of power, using the "Power Law" $P=I \times V$ requires knowing the voltage across and the current through each circuit element. So, first, find the voltage across and current through each element using Ohm's Law. Then, compute the power. Then, compute the efficiency.

Solve: In a series circuit, the total resistance is the sum of the individual resistors.

So, $R=R(\text{wire})+R(\text{load})=10.+1.0=11.$ Ω. The battery voltage is 6.0 V, so, by Ohm's Law, $I=V/R=6.0/11.=0.55$ A.

Now, applying the "Power Law," $P(\text{battery})=V(\text{battery}) \times I=6.0 \times 0.55$ $[V][A]=3.3$ W. This is the *total* power drained from the battery.

Applying Ohm's Law to the wire's resistance, its overall voltage drop is $V(\text{wire})=1.0 \times 0.55$ $[\Omega][A]=0.55$ V.

Applying the "Power Law" to wire losses, $P(\text{wire})=0.55 \times 0.55$ $[V][A]=0.30$ W.

Therefore, the "load" power=$3.3-0.30=3.0$ W.

So the **efficiency** is $3.0/3.3=$**0.91** (note that efficiency, being a ratio, has no units).

One task of an engineer might be to maximize the efficiency of application of power, since higher efficiency generally results in lower cost, and lower cost is generally desired by customers. The previous exercise gave one version of that challenge. The next example gives another.

Edison, in 1879, applied the "Power Law" and as a result made the first *useful*[4] lightbulb. He did so by recognizing that a lightbulb is simply a resistor that converts electricity first into heat then renders part of that heat (typically only about 5%) into visible light. In the example that follows, treat the resistor (indicated by the zigzag line in the figure) inside the bulb (indicated by a circle) as an ordinary, if very high-temperature, resistor that obeys our two electrical laws. The lightbulb's resistor is called its *filament*.

Example 10.5

In the diagram shown, assume the battery delivers a constant 10.0 A of electric current whatever the voltage across the filament may be. If one wishes to design the lightbulb to have a power of 100. W, what should the resistance of the lightbulb be? What is the voltage drop across the filament?

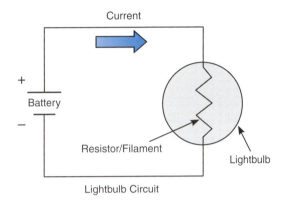

Lightbulb Circuit

Need: $R =$ _____ Ω and $V =$ _____ V.
Know: $P = 100.$ W and $I = 10.0$ A.
How: $P = I \times V$ and $V = R\,I$.
Solve: $P = I \times V$ or $V = P/I = 100./10.0\,[W]/[A] = \mathbf{10.0\ V}$ to three significant figures. Therefore, $\mathbf{R} = V/I = 10.0/10.0$
$[V/A] = \mathbf{1.00\ \Omega}$ (to three significant figures).

If the conductor wire in this example also had a 1.00 Ω resistance, as much power would be dissipated in it as in the bulb's filament. It would also have "consumed" another 10.0 V of voltage, so the battery would have had to supply 20.0 V.

Having resistance in the wire comparable to that of the filament complicates the problem. The key to Edison's problem of maximizing the bulb's efficiency was in maximizing the efficiency of the *entire* circuit, rather than concentrating on the bulb alone. Higher efficiency meant lower costs, eventually low enough for electricity and the lightbulb to become mass consumer products.

In all the circuits we discussed so far, the current always flows in the same direction (shown, according to convention, flowing from positive to negative around a circuit). These are called **direct current** circuits.

---

[4]Edison's major achievement came from considering the entire electrical production, distribution, and lighting problem as a single system design problem. The choice of a filament resistance was only a part of this greater problem. His rivals were unsuccessful mostly because they were concerned only with producing lamps. (Although Edison's claims actually followed those of an English engineer Joseph Swan, who had already invented a near identical bulb and an electrical distribution system about a year before Edison.)

However, another class of electrical circuits allows the current to reverse its direction many times a second. These **alternating current** circuits are the basis for the vast number of electrical technologies beyond the light-bulb. They range from the 1000 MW generators that supply electricity, to the motors that put electricity to work, to the radio wave generators that enable us to send out radio signals or to thaw frozen food in minutes in microwave ovens.

10.5 KIRCHHOFF'S LAWS

Gustav Robert Kirchhoff (1824–1887) was a German physicist who contributed to the fundamental understanding of electrical circuits, spectroscopy, and the emission of thermal radiation by heated objects. Kirchhoff formulated his circuit laws through experimentation in 1845, while still a university student.

10.5.1 Kirchhoff's Voltage Law

Kirchhoff's Voltage Law is a form of the Conservation of Energy Law. Kirchhoff's Voltage Law can be stated as

The algebraic sum of the voltage drops in a closed electrical circuit is equal to the algebraic sum of the voltage sources (i.e., increases) in the circuit.

Mathematically, Kirchhoff's Voltage Law is written as

$$\sum V_{(closed\ loop)} = \sum IR_{(closed\ loop)} = 0 \qquad (10.3)$$

where $\sum$ is the summation symbol.

A *closed loop* can be defined as any path in which the originating point in the loop is also the ending point for the loop. No matter how the loop is defined or drawn, the sum of the voltages in the loop must be zero. Both loop 1 and loop 2 in Figure 10.4 are closed loops within the circuit. The sum of all voltage drops and rises around loop 1 equals zero, and the sum of all voltage drops and rises in loop 2 must also equal zero. There is also a third closed loop in this circuit, the one that goes around the outside from A to B to C to D and back to A again. But this loop is not independent of the other two, so it provided no additional information.

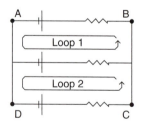

FIGURE 10.4 Closed loops in a simple electrical circuit.

10.5.1.1 Voltage dividers

If two resistors are in series, there is a voltage drop across each resistor, but the current through both resistors must be the same. A simple circuit with two resistors in series with a source (voltage supply) is called a **voltage divider**.

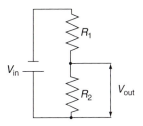

FIGURE 10.5 Voltage divider circuit.

In Figure 10.5, the voltage V_{in} is progressively dropped across resistors R_1 and R_2. If a current I flows through the two series resistors, by Ohm's Law, $I = V_{in}/(R_1 + R_2)$. Then, $V_{out} = IR_2$ or

$$V_{out} = V_{in} \times [R_2/(R_1 + R_2)] \tag{10.4}$$

Hence, V_{in} is "divided," or reduced, by the ratio $R_2/(R_1 + R_2)$.

Example 10.6

Suppose you want to build a voltage divider that reduces the voltage by a factor of 15. You visit your local RadioShack store and find that it has 5.0, 10., 15., and 20. Ω resistors in stock. Which ones do you choose to make your voltage divider in the simplest and most economical way?

Need: An inexpensive voltage divider such that $V_{out}/V_{in} = 1/15. = 0.067$.

Know: The voltage divider Equation (10.4), and the supply of resistors.

How: Since we know that $V_{out}/V_{in} = R_2/(R_1 + R_2) = 0.067$, we could just try the values of the resistors in stock to see if we can get close to the required ratio. This is a trial and error method and, in the general case, well suited to a spreadsheet analysis.

Solve: By inspection, we see that R_1 must be much bigger than R_2 to make V_{out}/V_{in} a small number. So we choose the smallest available resistor, 5.0 Ω, for R_2 and solve the voltage divider equation for R_1. This gives $R_1 = 70 \ \Omega$. Using the available resistors, we can come close to this value by connecting seven 10 Ω resistors in series with a 5.0 Ω resistor. This makes $R_1 = 70. \ \Omega$ and $V_{out}/V_{in} = 0.067$. Of course, if each of the 10. Ω resistors has a *tolerance* of $\pm 1. \ \Omega$, the range on $V_{out}/V_{in} = 0.06-0.08$.

10.5.2 Kirchhoff's Current Law

Kirchhoff's Current Law is a form of the Conservation of Charge Law. Kirchhoff's Current Law can be stated as

The algebraic sum of all the currents at a node must be zero,

where a "node" is any electrical junction in a circuit. Kirchhoff's Current Law can be written mathematically as

$$\sum I_{(at\ a\ node)} = 0 \tag{10.5a}$$

where $\sum$ is the summation symbol. Stated differently,

The sum of all the currents entering a node is equal to the sum of all the currents leaving the node,

or

$$\sum I_{(entering\ a\ node)} = \sum I_{(leaving\ the\ node)} \tag{10.5b}$$

Current flows through wires much like the analog of water flows through pipes. The amount of water that enters a branching junction in a pipe system must be the same amount of water that leaves the junction. The number of pipes in the branching junction does not change the net amount of water (or current in our case) flowing (see Figure 10.6).

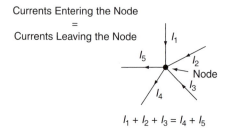

FIGURE 10.6 Currents entering and leaving a circuit node.

10.5.2.1 *Current dividers*

If two resistors are in parallel, the voltage across them must be the same, but the current divides according to the values of the resistances. A simple circuit with two resistors in parallel with a voltage source is called a **current divider**.

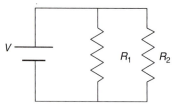

FIGURE 10.7 Parallel resistor current divider.

The equivalent resistance of two resistors R_1 and R_2 in parallel in Figure 10.7 is

$$1/R_{eq} = 1/R_1 + 1/R_2$$

and can be algebraically rearranged as

$$R_{eq} = (R_1 \times R_2)/(R_1 + R_2) \tag{10.6}$$

Now, if the voltage across these resistors is V, then the current I flowing in the circuit, before the division, is, according to Ohm's Law,

$$I = V/R_{eq} = V \times (R_1 + R_2)/(R_1 \times R_2)$$

The current through R_1 according to Ohm's Law is

$$I_1 = V/R_1$$

Dividing this equation by the previous one and solving for I_1 gives

$$I_1 = I \times R_2/(R_1 + R_2) \tag{10.7}$$

We can also show that

$$I_2 = I \times R_1/(R_1 + R_2) \tag{10.8}$$

Thus, the current I has been "divided," or reduced, by the resistance ratio.

Example 10.7

Find the current through and voltage across the 40. Ω resistor, R_3, in the circuit that follows.

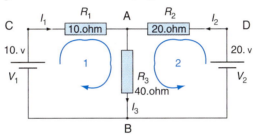

Need: $I_3 = $_____ A and $V_3 = $_____ V.

Know: Kirchhoff's voltage and current laws, Equations (10.3), (10.5a), and (10.5b).

How: This circuit has two nodes (A and B) and two independent loops. Write the Current Law equations for the two nodes and the Voltage Law equations for the two loops. Then, solve the resulting set of algebraic equations for the currents.

Using Kirchhoff's Current Law, at node A the equation is: $I_1 + I_2 = I_3$

Using Kirchhoff's Voltage Law around loops 1 and 2, the equations are

Loop 1: $10. = R_1 \times I_1 + R_3 \times I_3 = 10. \times I_1 + 40. \times I_3$
Loop 2: $20. = R_2 \times I_2 + R_3 \times I_3 = 20. \times I_2 + 40. \times I_3$

Since I_3 is the sum of $I_1 + I_2$, we can substitute this into the loop equations and rewrite them as

Loop 1: $10. = 10. \times I_1 + 40. \times (I_1 + I_2) = 50. \times I_1 + 40. \times I_2$
Loop 2: $20. = 20. \times I_1 + 40. \times (I_1 + I_2) = 40. \times I_1 + 60. \times I_2$

Solve: Now, we have two simultaneous equations that can be solved algebraically to give the values of I_1 and I_2. Algebraically solving the loop 1 equation for I_2 and substituting this expression into the loop 2 equation allows us to solve for I_1 as: $I_1 = -0.14$ A. Now, substituting this value of I_1 into either loop equation allows us to solve for I_2 as $I_2 = +0.43$ A. Since $I_3 = I_1 + I_2$, the current flowing in resistor R_3 is $I_3 = -0.14 + 0.43 = 0.29$ A, and from Ohm's Law, the voltage drop across the resistor R_3 is $0.29 \times 40. = 11$ V.

The negative sign we got for I_1 in this example simply means that we initially assumed the wrong direction for the current flow, but the numerical value is still correct. In fact, the 20. V battery is actually "charging" the 10. V battery (i.e., putting electrical energy back into the battery by forcing an electric current through it).

10.6 SWITCHES

Here, we concentrate on a humble circuit element whose importance was appreciated by Edison: the switch. In our water flow model, a switch is equivalent to a faucet. It turns the flow on or off. Surprisingly, this simplest of components is the basis of the most sophisticated of our electrical technologies, the computer. In its essence, a computer is nothing more than a box filled with billions of switches, all turning each other on and off. The faster the switches turn each other on and off, the faster the computer can do computation. The smaller the switches, the smaller the computer can be. The more reliable the switches, the more reliable is the computer. The more efficient the switches, the less power the computer uses. So the criteria engineers have used in developing better switches are higher speed, smaller size, greater reliability, and increased efficiency.

In the late nineteenth century, a few farsighted individuals, such as the British economist William Stanley Jevons and the American mathematician Allen Marquand, proposed that the humble switch could be used to do something remarkable: carry out calculations in the field of philosophy and mathematics, called *logic*. (The mathematics of logic was discussed in Chapter 9.) Here, we stick to the underlying electrical technology: the switch.

Unlike the elements we discussed so far, the switch has two **states**. As indicated in Figure 10.8a, a switch can be open, in which case it presents an infinite resistance to the flow of electric current (that is, no current flows at all), or a switch can be closed, in which case ideally it presents zero resistance to an electric current.

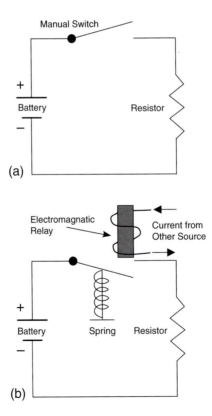

FIGURE 10.8 (a and b) Simple switches.

Jevons and Marquand envisioned using switches to automate logic, but the technology of their time was not up to the task. One of the pioneers who turned their idea into twentieth-century reality was George R. Stibitz. Along with other inventors in the 1930s, he accomplished automation of computation using a type of switch component called a *relay*.

A **relay** is a switch operated by an electromagnet, Figure 10.8b. The relay simply closes the switch against the spring when the external power source is turned on. When the external power source is turned off, the spring returns the switch to its open position. It got its name from its original use, relaying telegraph messages over longer distances than could be accomplished by a single circuit. The relay consists of two parts. The first is an electromagnet powered by one circuit, which we call the **driving circuit**. The second part is another circuit with a metal switch in it, called the **driven circuit**. The driven circuit is placed so that, when the driving circuit is on, its electromagnet attracts the moving metal part of the driven circuit's switch. This results in closing the gap in the driven circuit and turning it on.

Relays continue to be used in control systems because they are highly reliable, but they are high in power demands, slow, and bulky. In the twentieth century, the basis of a faster switch emerged. It was based on an invention used in radios, a circuit element called the *vacuum tube*. Vacuum tubes were used as switches because they were fast. However, they were inefficient, bulky, and unreliable. By the 1940s, an even more revolutionary idea was already in the wings. It emerged in initial form in 1946, when three physicists at Bell Laboratories— John Bardeen, Walter Brittain, and William Shockley—invented the transistor, a solid-state switch developed to replace the vacuum tube in communications. How the transistor works is a topic for an advanced course in electronic engineering or physics. For our purposes, we can carry on the analogy with the relay and the vacuum tube. Figure 10.9 shows one important type of transistor, the metal oxide semiconductor field effect transistor (MOSFET), a type that came along several years after the original one invented by Bardeen, Brittain, and Shockley. In the MOSFET, a driving circuit, called a *gate*, plays the role the magnet plays in the relay, and the source plays the role of a heated metal grid in the vacuum tube. Together they open and close (like a switch) a driven circuit that determines the state of the output.

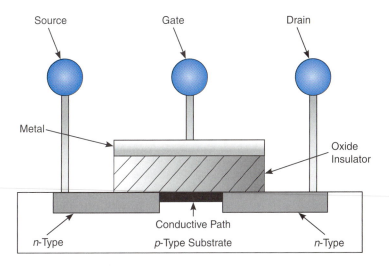

FIGURE 10.9 Principles of MOSFET construction.

A transistor is made of semiconductors—materials that have properties midway between those of a conductor and those of an insulator. Most of today's transistors use semiconductors made out of the element silicon. By "doping" silicon—that is, by adding trace amounts of other elements—its electrical conductivity can be controlled. As a simple model of a particularly useful type of transistor called the *field effect transistor*, we can imagine two such pieces of doped silicon, a "source" that serves as a source of electrical charge and a "drain" that serves as a sink into which the charge flows. The source and drain are connected by a semiconductor channel, which is separated by an insulator from a metal electrode called a *gate*. Some of the semiconductor is doped with material that sustains an excess of electrons and is called *n-type* (*n* for "negative"). Other semiconductors are doped with materials that sustain a deficiency of electrons called *p-type* (*p* for "positive"). When the gate is given a positive voltage in the *npn* MOSFET shown in Figure 10.9, this promotes the flow of charge from source to drain, effectively making a thin layer of continuous *n*-type conductive channel between the source and the drain. When the gate is given a negative voltage, current cannot flow from source to drain and the transistor acts as an open switch.

Transistors (switches) are smaller, faster, more efficient, and more reliable than vacuum tubes. They made possible practical hearing aids, pocket-sized portable radios, and computers that were the size of refrigerators instead of filling entire rooms. But, to get to the modern computer, one more step was needed. This was the **integrated circuit**, independently coinvented by Nobel Prize-winning engineer Jack Kilby and Intel cofounder Robert Noyce in the early 1960s.

A modern integrated circuit contains millions (or billions) of transistors connected together to form a logic circuit on a piece of silicon the size of a fingernail. This is done by a process called **photolithography**, which literally means "using light to write on a stone." The circuit designer first makes a large drawing of the circuit. Optical lenses are then used to project a very small image of the drawing onto a fingernail-sized chip of silicon, which is coated with special chemicals.

Modern integrated circuits might contain the level of detail that would be seen on a street map of the entire state of California. Packing all this complexity into a small space launched one of the most dramatic technological revolutions in history. Since 1960, the number of electronic computations possible per second and the number of transistors that can be packed onto a single chip have doubled every one or two years (and since 1990, rather than slowing, the pace has actually speeded up).

Integrated circuits are the hardware that made personal computers possible. But integrated circuits by themselves do not constitute a computer. What enables a collection of switches to carry out computations? That is the job for a sequence of instructions, called **software**. A crucial distinction needed to understand computers is **hardware** versus software. Hardware is the collection of mechanical, electrical, or electronic devices that make up a computer. Among the hardware in a modern computer, two elements are crucial: (1) a system called a **central processing unit** or **CPU**, which carries out binary logic and arithmetic and (2) memory, often called **RAM** (random access memory) or **ROM** (read only memory), which store the ones and zeroes corresponding to inputs and outputs of the CPU. Both CPU and memory are integrated circuits composed of many millions of switches (see Chapter 9).

SUMMARY

The flow of electricity though wires can be modeled in the same way one might treat the flow of water through pipes. The model has as its basic variables **charge**, **current**, **voltage**, and **resistance**. The simplest model consists of **Ohm's Law**, the **Power Law**, and **Kirchhoff's Voltage** and **Current** Laws. Using this model, one can analyze the operation of a particular class of electric circuits called **direct current** circuits and, in particular, two classes of direct current circuits called **series circuits** and **parallel circuits**.

An important type of electric circuit component is the **switch**. On the hardware side, the development of the computer can be viewed as the search for ever-faster switches, based first on **relays**, then on **vacuum tubes**, then on **transistors**, and finally, on the form of electronics called **solid-state integrated circuits**. This form of electronics provides the hardware, which in conjunction with a sequence of instructions, called **software**, makes modern computation and control possible.

EXERCISES

If no circuit sketch is already given, it is highly recommended you *first draw each circuit* to answer these problems. It is generally useful to notate your diagram with what you think you know, such as voltages and currents at each point.

Note the equivalence among these electrical units:

$$[\Omega] = [V]/[A], \ [W] = [V][A] = [\Omega][A]^2 = [V]^2/[\Omega]$$

The circuits for Exercises 1–8 are given in the figure below.

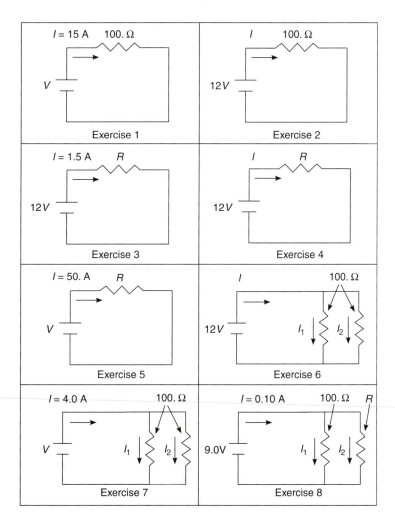

1. For the circuit shown, find the voltage V. (**A: 15×10^2 V.**)

2. For the circuit shown, find the current I.

3. For the circuit shown, find the resistance R.

4. For the circuit shown, a power of 100. W is dissipated in the resistor. Find the current I. (**A: 8.3 A.**)

5. For the circuit shown, a power of 100. W is dissipated in the resistor. Find the resistance, R.

6. For the circuit shown, find the current I. (**A: 0.24 A.**)

7. For the circuit shown, find the voltage V.

8. For the circuit shown, find the resistance R.

9. For the circuit that follows, a power of 100. W is dissipated in each resistor. Find the current I and the voltage V.

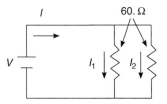

For Exercises 10–19, draw the circuit and solve for the unknown quantity requested. It is suggested that you include a small solid circle to identify each point where you can specify the local voltage.

10. A circuit consists of a 3.0 V battery and two resistors connected in series with it. The first resistor has a resistance of 10. Ω. The second has a resistance of 15. Ω. Find the current in the circuit. (**A: 0.12 A.**)

11. A circuit consists of a 12. V battery and a resistor connected in series with it. The current is 105. A. Find the resistance.

12. A circuit consists of a 9.0 V battery and two parallel branches, one containing a 1500 Ω resistor and the other containing a 1.0×10^3 Ω resistor. Find the current drawn from the battery. (**A: 0.015 A.**)

13. A circuit consists of a 12. volt battery attached to a 1.0×10^2 Ω resistor, which is in turn connected to two parallel branches, each containing a 1.0×10^3 Ω resistor. Find the current drawn from the battery.

14. An automobile's 12. V battery is used to drive a starter motor, which for several seconds draws a power of 3.0 kW from the battery. If the motor can be modeled by a single resistor, what is the current in the motor circuit while the motor is operating? (**A: 250 A.**)

15. An automobile's 12. V battery is used to light the automobile's two headlights. Each headlight can be modeled as a 1.00 Ω resistor. If the two headlights are hooked up to the battery in series to form a circuit, what is the power produced in each headlight? Why would you *not* wire car lights in series?

16. An automobile's 12. V battery is used to light the automobile's two headlights. Each headlight can be modeled as a 1.00 Ω resistor. If the two headlights are hooked up to the battery in parallel to form a circuit, what is the power produced in each headlight? Why is a parallel circuit preferred for this application?

17. An automobile's 12. V battery is used to light the automobile's two headlights. Each headlight can be modeled as a single resistor. If the two headlights are hooked up to the battery in parallel to form a circuit and each headlight is to produce a power of 100 W, what should the resistance of each headlight be? (**A: 1.4 Ω per bulb**.)

18. Suppose one of the headlights in Exercise 17 suddenly burns out. Will be the power produced by the other headlight increase or decrease?

19. Suppose a car has headlights operating in parallel but with different resistances—one of 2.0 Ω and the other of 3.0 Ω. Suppose the headlight parallel circuit is connected in series with a circuit for a car stereo that can be modeled by a 1.0 resistor. What is the current being drawn from a 12. V battery when both the lights and the stereo are on? (**A: $I = 5.5$ A**.)

20. Consider the circuit that follows. What are the currents I_0 through I_6? (**Hint:** What are the voltages V_a and V_b? Note the symmetry between $R_1/R_2 = 5.0/5.0$ and $R_3/R_4 = 10.0/10.0$)

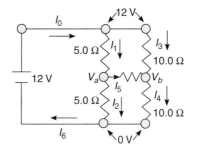

21. Determine the output voltage, V_{out}, in Figure 10.5 if $V_{in} = 90$ V, $R_1 = 10.$ Ω, and $R_2 = 50.$ Ω.

22. Use the same RadioShack stock resistors in Example 10.6 to design the construction of a voltage divider that reduces the voltage by a factor of 30.

23. Use Kirchhoff's Voltage and Current Laws to determine the voltage drop and current in each of the three resistors that follow if $R_1 = 10.$ Ω, $R_2 = 20.$ Ω, and $R_3 = 30.$ Ω.

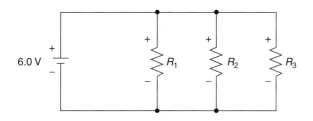

24. If the resistors in Exercise 23 are all equal to 10. Ω ($R_1 = R_2 = R_3 = 10.$ Ω), what is the current supplied by the 6.0 V battery?

25. Determine the currents I_1 and I_2 in the current divider illustrated in Figure 10.7 if $V = 50.3$ V, $R_1 = 125.$ Ω, and $R_2 = 375.$ Ω.

26. Repeat Example 10.7 with the resistor R_1 changed to 100. Ω, and all other values remaining unchanged.

27. Repeat Example 10.7 with V_1 increased to 100. V and V_2 reduced to 5.0 V.

28. Determine the currents I_1, I_2, I_3, and I_4 in the figure that follows.

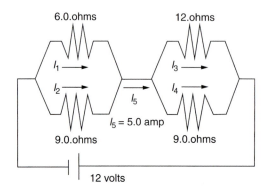

29. It is the last semester of your senior year, and you are anxious to get an exciting electrical engineering position in a major company. You accept a position from company A early in the recruiting process, but you continue to interview, hoping for a better offer. Then, your dream job offer comes along from company B. More salary, better company, more options for advancement. It is just what you have been looking for. What do you do?

(a) Accept the offer from company B without telling company A (just don't show up for work).

(b) Accept the offer from company B and advise company A that you have changed your mind.

(c) Write company A and ask them to release you from your agreement.

(d) Write company B thanking them for their offer and explain that you have already accepted an offer.
(Summarize using the Engineering Ethics Matrix.)

30. A female student in your class mentions to you that she is being sexually harassed by another student. What do you do?

(a) Do nothing; it is none of your business.

(b) Ask her to report the harassment to the course instructor.

(c) Confront the student accused of harassment and get his or her side of the issue.

(d) Talk to the course instructor or the college human resource director privately.
(Summarize using the Engineering Ethics Matrix.)

Industrial Engineering

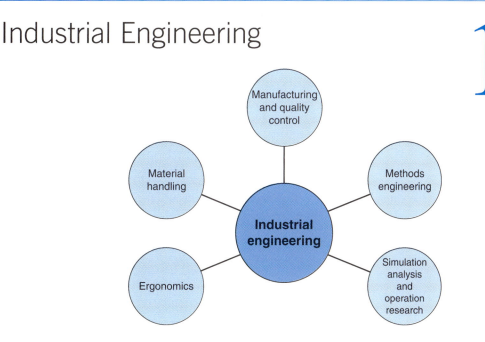

11.1 INTRODUCTION

Industrial engineering[1] is one of the most diverse fields of engineering because the word "Industry" has a very broad meaning. It refers to the production of any economic goods within an economy. Because there are so many different kinds of "industries," they are divided into the following three different categories.

- **Primary industries.** These industries deal with the *extraction of resources* directly from the Earth such as farming, mining, petroleum, and logging.
- **Secondary industries.** These industries are involved in *manufacturing products* from the resources provided by the primary industries. They include the manufacturing of steel, automobiles, furniture, food, electronics, and so forth.
- **Tertiary industries.** These industries compose the *service industry*. They include education, health care, package delivery, software development, financial institutions, government organizations, and so forth.

In general, industrial engineers develop integrated systems consisting of people,[2] knowledge, equipment, energy, and materials. Examples of activities where industrial engineering is used include designing assembly

---

[1]Originally, the name "industrial engineering" applied only to manufacturing. Today, it has grown to encompass any procedure that includes how any process, system, or organization operates. Some universities have changed the name "industrial engineering" to a broader term such as "production engineering" or "systems engineering."

[2]When industrial engineers work in one of the secondary industries, they are often called manufacturing engineers. When they work in one of the primary or tertiary industries, they are called industrial engineers.

lines and workstations; process efficiency analysis, developing a new financial algorithm for a bank; streamlining emergency room use in a hospital; planning complex distribution schemes for materials or products; and shortening lines at a bank, a hospital, or a theme park.

Modern industrial engineering began in the twentieth century. The first half of the century was characterized by mass production with an emphasis on increasing efficiency and reducing costs. In 1909, Frederick Taylor published his now famous book on the *Theory of Scientific Management*, which included an analysis of human labor, a systematic definition of methods, and tools and training for employees. Taylor's time and motion studies set standard times for workers to complete a task that subsequently increased productivity, reduced labor costs, and increased the wages of the employees. In 1913, the moving assembly line developed by Henry Ford produced major increases in manufacturing productivity by reducing the assembly time of a car from 12.5 to 1.5 hours. In addition, he was a pioneer in providing financial incentives for his employees to increase productivity.

In the second half of the twentieth century, the Japanese management theories highlighted issues of product quality, rapid delivery time, and manufacturing flexibility to improve customer satisfaction.[3] In the 1990s, the industrial globalization process emphasized supply chain management and customer-oriented business process design.

Today, industrial engineering can be broken down into the following categories:

(1) Manufacturing and quality control
(2) Methods engineering
(3) Simulation analysis and operations research
(4) Ergonomics
(5) Material handling

11.2 MANUFACTURING AND QUALITY CONTROL

Once a product is designed, industrial engineers guide it through the manufacturing process. Since this textbook contains a separate chapter on manufacturing, students are encouraged to review this chapter before proceeding. Manufacturing information topics will not be repeated here.

Quality control can be broken down into the areas of (a) statistical analysis, (b) probability theory, (c) reliability analysis, and (d) design of experiments (DOE).

11.2.1 Statistical Analysis

Statistics is a branch of mathematics in which groups of measurements or observations are studied. Some of the more common descriptive terms used in statistics are defined below:

Population—a group of items that is being studied.
Sample—a small group of items selected from a population.
Random samples—occur when members have an equal chance of being selected.
Data[4]—numbers or measurements that are collected in a population.
Variables—characteristics that allow us to distinguish one individual from another.
Constants—items whose value never changes.

---

[3]The Japanese also came up with the term *Kaizen*, which means "good change" to represent continuous improvement.
[4]The word "data" is plural (therefore, we should say "data are" "not data is"). Data represent a set of measurements. A single element of that set is called a "datum" as in "there is only one datum point."

Example 11.1
Which of the following is a variable?

(a) Scores obtained on a math exam.

(b) The number of days in the year.

(c) The time it takes you to get to this class.

Solution

Need: To decide which of these three items is a variable.

Know: A "variable" has characteristics that allow us to distinguish one individual from another.

How: Apply the definition of "variable" to these three items.

Solve: Items (a) and (c) are variables because they are not constants. Item (b) is a constant.

Some of the common mathematical terms used in statistics are defined in Table 11.1.

Table 11.1 Some of the Common Mathematical Terms Used in Statistics

Term	Definition	Formula	
Mean	The **mean** $\bar{x}$, or **average**, is 1 divided by the number of samples N multiplied by the sum of all the data points	$\bar{x} = \dfrac{1}{N}\sum_{i=0}^{i=N} x_i$	(11.1)
Median	The **median** M is the middle value of a set of data containing an odd number of values, or the average of the two middle values of a set of data with an even number of values	Median of a set $x_1, x_2, \ldots, x_M, \ldots, x_N$ $M =$ Middle value x_M if N is an odd number $M =$ the average of the two middle values if N is and even number	(11.2)
Variance	The **variance** σ^2 of a set of data is 1 divided by the number of samples (data) minus 1, multiplied by the sum of the data points subtracted from the mean and then squared	$\sigma^2 = \dfrac{1}{N-1}\sum_{i=0}^{i=N}(x_i - \bar{x})^2$	(11.3)
Standard deviation	The **standard deviation** σ is the square root of the variance	$\sigma = \left[\dfrac{1}{N-1}\sum_{i=0}^{i=N}(x_i - \bar{x})^2\right]^{1/2}$	(11.4)
Standard error	The **standard error** of the mean is the standard deviation divided by the square root of the number of samples	$\mathrm{SE} = \dfrac{\sigma}{\sqrt{N}}$	(11.5)
Z-score	The **Z-score** is the number of standard deviations an observation or datum is *above* the mean	$Z = \dfrac{x_i - \bar{x}}{\sigma}$	(11.6)

Various computer programs such as Microsoft Excel contain routines that perform these calculations.

Example 11.2

Ten employees of a department store earn the following weekly wages: $400, $650, $660, $625, $660, $650, $680, $630, $670, $650.

    (a) Find the median weekly wage.

    (b) What is the mean weekly wage?

    (c) Find the standard deviation of the set of wages.

Solution

    Need: The median, mean, and standard deviation of the given set of wages.

    Know: The set of wages in question.

    How: Use Equations (11.1), (11.2), and (11.4).

    Solve: **(a)** The wages can be listed in order as $400, $625, $630, $650, $650, $660, $660, $665, $670, $680. Since this set has an even number of items, in this case Equation (11.2) gives the **median** as the average of the two middle values, or ($650+$660)/2 = **$655**.

    (b) The **mean** (or **average**) weekly wage is computed from Equation (11.1) as $\bar{x} = (1/N)\sum_{i=0}^{i=N} x_i$

$$\text{Mean weekly wage} = \frac{1}{10}\sum_{i=0}^{i=10}(\$400 + \$625 + \cdots + \$680) = \$589$$

    (c) **The standard deviation** of the set of wages is computed from Equation (11.4) as:

$$\sigma = \left[\frac{1}{N-1}\sum_{i=0}^{i=N}(x_i - \bar{x})^2\right]^{1/2} = \frac{1}{(10-1)}\sum_{i=0}^{i=10}\left[(\$400 - \$589)^2 + (\$625 - \$589)^2 + \cdots + (\$680 - \$589)^2\right]^{1/2} = \$30.8$$

11.2.2 Probability Theory

Probability is a number that represents the likelihood that a specific event will occur. It is the ratio of the number of actual occurrences to the number of possible occurrences.

$$P(A) = \text{Probability of event } A \text{ happening} = \frac{\text{Number of ways } A \text{ can happen}}{\text{Total number of possible outcomes}} \qquad (11.7)$$

When $P(A) = 0$, then event A cannot occur (e.g., the probability that you will grow younger rather than older is zero), and when $P(A) = 1$, then event A is certain to occur (e.g., the probability that you will grow older and not younger). There are a number of mathematical rules that govern probability theory. The rules of probability are easily understood by using dice and card games.

11.2.2.1 Probability addition

The probability addition rule is used when we have independent "or" events. Suppose that you throw a single die.[5] There are six possible outcomes: **1, 2, 3, 4, 5, and 6**. The probability of getting a 6 is

$$P(6) = \frac{\text{The number of ways of getting a 6}}{\text{The total number of possible outcomes from the die}} = \frac{1}{6}$$

and the probability of getting any other value is $P(5) = P(4) = P(3) = P(2) = P(1) = 1/6$.

Since the values in the face of a die are all independent, the probability of getting either a 5 *or* a 6 is

$$P(5 \text{ or } 6) = P(5) + P(6) = 1/6 + 1/6 = 2/6 = 1/3$$

If you choose a single card from a pack of 52 cards, the probability of getting an ace of spades is P(ace of spades) $= 1/52$. The probability of getting any of the four aces is P(any ace) $= 4/52$. The probability of getting any of the 12 picture cards (jack, queen, or king) is P(any picture card) $= 12/52$.

Since all the cards in the deck are unique, the probability of getting any of the four aces *or* any of the 12 picture cards is

$$P(\text{ace or picture}) = 4/52 + 12/52 = 16/52$$

Notice that the "or" event probabilities can be computed by **adding** their individual probabilities if the events are "mutually exclusive," meaning that none of the "or" events can occur at the same time (i.e., you cannot have an ace that is also a picture card). So, before adding "or" probabilities, you need to make sure that the events cannot simultaneously occur and are mutually exclusive (i.e., they are independent and each excludes the other). The complete addition rule when A, B, C, … are mutually exclusive is

$$P(A \text{ or } B \text{ or } C \text{ or}...) = P(A) + P(B) + P(C) + \cdots \tag{11.8}$$

11.2.2.2 Probability multiplication

The multiplication rule is used when we have independent "and" events. For example, the probability of getting a 6 *and* a spade is

$$P(6 \text{ and spade}) = P(6) \times P(\text{spade}) = 1/6 \times 1/4 = 1/24$$

Notice that the "and" event probabilities can be computed by *multiplying* their individual probabilities if the events are "mutually exclusive," meaning that none of the "and" events can occur at the same time. For example, the probabilities $P(6)$ and $P(\text{spade})$ are independent of each other and therefore are mutually exclusive. Knowing whether or not one has occurred has no effect on the probability of the other happening. In this example, knowing that the thrown die produced a 6 has no effect on what happens when we draw a card from the deck, these two events are statistically independent. The complete probability multiplication rule when A, B, C are mutually exclusive is

$$P(A \text{ and } B \text{ and } C \text{ and}...) = P(A) \times P(B) \times P(C) \times \cdots \tag{11.9}$$

---

[5]Dice is plural, and die is singular.

Example 11.3

There are five marbles in a bag of which four are blue and one is red. What is the probability that you can pick a blue marble from the bag?

Solution

 Need: Probability of picking a blue marble.
 Know: The number of blue and red marbles in the bag
 How: Use Equation (11.7) to compute the probability.
 Solve: The number of ways that a blue marble will be chosen is 4, and the total number of possible outcomes is 5 (there are five marbles in the bag). Then, Equation (11.7) gives

$$\text{Probability of getting a blue marble} = \frac{4}{5} = 0.8$$

11.2.3 Reliability Analysis

The reliability of a product (or a system) is defined as the probability that a product will perform its required function under specified conditions for a certain period of time. If we have a large number of a certain product that we can test over time, then the reliability of that product at time t is given by

$$R(t) = \frac{\text{Number of products that have NOT failed by time } t}{\text{Total number of products being tested at time } t = 0} \qquad (11.10)$$

At time $t = 0$, the number of products that have not failed is equal to the number of products being tested. Therefore, the reliability at $t = 0$ is $R(0) = 1 = 100\%$. Note that the reliability, $R(t)$, will continuously decline as t increases and products fail.

The **failure rate** of a product at time t is defined as the probability of a product failing at time t. It is represented by the symbol $F(t)$. Let

N_F is the number of products that failed during the time interval $t + \Delta t$ and N_S is the number of products that have NOT failed by time t.

If Δt is small and the failure rate does not change during this time interval, then the failure rate at time t can be calculated as:

$$F(t) = \frac{\text{Number of product failures between } t \text{ and } t + \Delta t}{(\text{Number of products that have NOT failed by time } t) \times \Delta t} = \frac{N_F}{N_S \times \Delta t} \qquad (11.11)$$

For example, if 250 components have not failed by $t = 8$ hours and 3 components fail during the next 1 hour, then the failure at 8 hours is

$$F(8) = \frac{3}{250 \times 1} = 0.012 = 1.2\% \text{ per hour}$$

While it is relatively easy to define failure rates for individual product components such as springs or gears, it becomes more difficult for complex products like automobiles. For example, a failure in a complex product can be complete breakdown (a "catastrophic" failure) to relatively minor easily repaired failure. While failures in complex products may be repairable, they may not be repairable for individual product components.

The bath tub curve shown in Figure 11.1 represents the failure pattern seen in many products. The vertical axis is the failure rate at each point in time and the horizontal axis is the test time (or the time since the product was released for sale to the public). The bath tub curve is divided into three regions: early failure, useful life, and wear out.

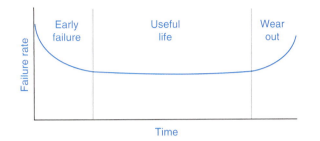

FIGURE 11.1 Failure rate bath tub curve.

(1) **Early failure.** This is the early failure (or break-in) stage. During this period, failures typically occur because products were not designed properly or manufacturing flaws occurred. Failure rates at the beginning of the early failure stage are high, but then they decrease with time as the early failures are removed.

(2) **Useful life.** This is the center stage of the bath tub curve and is characterized by a constant failure rate. This stage is the most significant stage for reliability prediction and evaluation activities.

(3) **Wear out stage.** This is the final stage where the failure rate increases as the products begin to wear out and break down. When the failure rate becomes high after a long period of use, repair or replacement of parts should be performed.

Example 11.4

Would you expect the bath tub curve to apply to:

(a) new automobiles
(b) human beings?

Solution

Need: Interpretation of the bath tub curve for new automobiles and human beings.

Know: How the bath tub curve is defined.

How: Intuition.

Solve: (a) No, because new automobiles are not likely to have a catastrophic early failure. The failure rate of a new car's components may be high as faults from the design or manufacturing process are found and fixed via recall. However, if the manufacturer tested for early faults, then the early failure stage will not occur.

  (b) The bath tub pattern does apply to human beings because failure rates (i.e., death rates) are higher in both infancy and old age.

Many organizations keep failure and repair records for the products they produce. This information can then be used to determine failure rates for those products. Existing failure rate data are also useful when designing similar new products.

Under certain assumptions, the failure rate for a complex system is simply the sum of the individual failure rates of its components. This permits testing of individual components or subsystems, whose failure rates are then added to obtain the total system failure rate.

Example 11.5

Thousand transistors are placed on life test, and the number of failures in each time interval is recorded in the table below.

Time interval (hours)	0 -100	100-200	200-300	300-400	400-500	500-600	600-700	700-800	800-900	900-1000
Number of failures	160	86	78	70	64	58	52	43	42	36

(a) Find the reliability and the failure rate at 0, 100, 200, ..., 1000 hours.
(b) Plot the failure rate $F(t)$ as the transistors get older. Does this show the bath tub pattern of failure?
(c) Plot the reliability $R(t)$ vs time.

Solution

Need: The reliability and failure rate of the transistors plus a plot of their values vs time.

Know: The test results given in the table.

How: Use Equations (11.10) and (11.11).

Solve: (a) At time $t=0$, $N_F=160$ failures, $N_S=1000$ nonfailures, and $\Delta t=100$ hours. Using Equation (11.10), we get

$$R(t=0) = \frac{\text{Number of products that have NOT failed by time } t=0}{\text{Total number of products being tested at time } t=0} = \frac{1000}{1000} = 100\%$$

and Equation (11.11) gives the failure rate as:

$$F(t=0) = \frac{160}{(1000)(100)} = 0.0016 = 0.16\% \text{ per hour}$$

The table below shows the remaining calculation results.

Time t	Nonfailures N_S	Failures in Next 100 Hours N_F	Reliability $R(t)$ (%)	Failure Rate $F(t)$ (Failures/ Hour)
0	1000	160	100	0.16
100	840	86	84.0	0.10
200	754	78	75.4	0.10
300	676	70	67.6	0.10
400	606	64	60.6	0.11
500	542	58	54.2	0.11
600	484	52	48.4	0.11
700	432	43	43.2	0.10
800	389	42	38.9	0.11
900	347	36	34.7	0.10
1000	311		31.1	

(b) The failure rate information from part (a) is shown in the plot below.

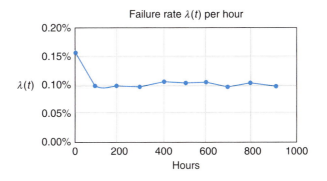

(c) The reliability information from part (a) is shown in the plot below.

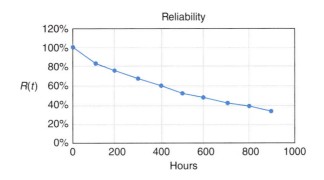

This result shows that the initial failure rate is 0.16% per hour and then quickly drops to around 0.10% per hour. This is the early failure phase. From then on, the failure rate is fairly constant and represents the useful life phase. There is no wear out phase data, but there are still 311 components left after the 1000 hour test. The wear out phase could become apparent if the tests were to be prolonged.

When the failure rate is constant, the reliability graph is a straight horizontal line and represents the middle phase of a bath tub curve. When there is little break-in failure (early failure), a constant failure rate can be effectively used to predict the reliability of a product to a particular time. If the failure rate F is a constant (i.e., independent of t), then the reliability $R(t)$ at time t is given by

$$R(t) = e^{-Ft} \tag{11.12}$$

where F is the constant failure rate.

For example, if there are initially 1000 components and their failure rate is constant at $F = 10\%$ per hour, then using Equation (11.12), the reliability after 3 hours, with $F = 0.1$/hour, is

$$R(t) = e^{-3F} = e^{-0.3} = 0.741 = 74.1\%$$

The **Mean Time to Failure** (MTTF) is the average time an item may be expected to function before failure and applies to nonrepairable items. The MTTF is simply the average of all the times to failure.

$$\text{Mean time to failure} = \text{MTTF} = \frac{\text{Total time of operation, } T}{\text{Total number of failures during time } T} \tag{11.13}$$

For example, if four items are tested and have lasted 3900, 4100, 4300, and 3700 hours, respectively, then the MTTF is

$$\text{MTTF} = \frac{3900 + 4100 + 4300 + 3700 \text{ hours}}{4 \text{ failures}} = \frac{16{,}000 \text{ hours}}{4 \text{ failures}} = 4000 \text{ hours per failure}$$

The **Mean Time Between Failures** (MTBF) applies to repairable items. Since this refers to the mean time "between" failures, it is defined as:

$$\text{Mean Time Between Failures} = \text{MTBF} = \frac{\text{Total time of operation, } T}{\text{Total number of failures during time } T} \tag{11.14}$$

Remember, MTBF is for repairable items, and MTTF is for nonrepairable items.

For example, if an item has failed four times over a period of 8000 hours. Then its MTBF is 8000/4 = 2000 hours. Notice that the MTTF is the same as the MTBF. Thus, for a constant failure rate,

$$F(t) = \frac{1}{\text{MTBF}} = \frac{1}{\text{MTTF}} \tag{11.15}$$

For example, the item above fails, on average, once every 2000 hours, so the probability of failure for each hour is obviously 1/2000. This result depends on the failure rate being constant. Equation (11.15) can also be written the other way round as MTBF (or MTTF) $= 1/F(t)$. For example, if the failure rate is 0.0002 per hour, then MTBF (or MTTF) $= 1/0.0002 = 5000$ hours.

Example 11.6

The equipment in a furniture manufacturing plant has a MTBF of 1000 hours. What is the probability that the equipment will operate for a period of 500 hours without failure?

Solution

Need: The probability that the equipment will operate for a period of 500 hours without failure.

Know: The mean time between failures (MTBF) is 1000 hours.

How: Use Equation (11.12).

Solve: Assuming the constant failure rate (useful life) model, the failure rate

$F(t) = 1/\text{MTBF} = 1/1000 = 0.001$. Then, from Equation (11.12),

$R(500 \text{ hours}) = e^{-500 \times 0.001} = e^{-0.5} = 0.61$, and the probability of the equipment operating for 500 hours without failure is 61%.

11.2.4 Calculating System MTBFs

The MTTF of any system with many independent components can be calculated if you have the MTTF of each component in the system. For example, you can calculate the MTTF of a computer if you have the MTTF of the CPU board, disk drives, memory chips, etc. However, computer MTTF analysis typically focuses on disk drive

MTTFs because components with moving parts (such as disk drive actuators and motors) typically have significantly lower MTTFs than nonmoving components (such as CPU boards and memory chips). Large server configurations containing many disk drives obviously have lower MTTFs.

A server's MTTF is calculated from the MTTFs of the independent components that make up the server as follows:

$$\text{MTTF} = \frac{1}{\dfrac{1}{N_1} + \dfrac{1}{N_2} + \dfrac{1}{N_3} + \dfrac{1}{N_4} + \cdots + \dfrac{1}{N_x}} \tag{11.16}$$

where N_1 is the MTTF of component #1, N_2 is the MTTF of component #2, and so forth, and x is the total number of components in the system.

If one component, such as a disk drive, has a significantly lower MTTF than the other components in the system, its MTTF dominates the overall server MTTF. For example, if a computer has a CPU MTTF of 50,000 hours (5.8 years) and a single drive with an MTTF of 10,000 hours, then, using Equation (11.16) and neglecting all of the other components, the computer has a MTTF of:

$$\text{Computer MTTF} = \frac{1}{\dfrac{1}{\text{CPU MTTF}} + \dfrac{1}{\text{Disk MTTF}}} = \frac{1}{\dfrac{1}{50,000} + \dfrac{1}{10,000}} = 8300 \text{ hours}$$

which is close to the disk drive's MTTF.

Even if all the system components have high MTTFs, the system's overall MTTF is reduced in direct proportion to the number of components in the system. For example, the MTTF of a computer subsystem consisting of two disk drives with identical 10,000 hour MTTFs is

$$\text{Disk Drive Subsystem MTTF} = \frac{1}{\dfrac{1}{10,000} + \dfrac{1}{10,000}} = 5000 \text{ hours}$$

which is exactly half the MTTF of each disk drive. Similarly, a 10-drive configuration MTTF is one-tenth the MTTF of a single drive, or 10,000 hours, and a 100-drive configuration is reduced to just 100 hours. This is why component redundancy in a complex system is important.[6]

11.2.5 Design of Experiments

DOE sounds like an odd topic for an industrial engineer, but it is a very important concept. Manufacturing processes are not carried out in a laboratory under controlled conditions. They are carried out under chaotic factory conditions where many uncontrolled variables can affect the quality of the manufactured part. So, it is imperative that an industrial engineer understands the relationships between all the input process variables and all of the required design factors in the finished product. For example, in a metal cutting process, the cutting speed, feed rate, depth of cut, and coolant flow are all input variables. The surface finish and dimensional tolerances of the finished part specified by the design engineer are the required output factors.

How do industrial engineers test for the influence of input process variables on the results of a manufacturing process? How do they carry out meaningful experiments in a complex manufacturing environment to determine

---

[6]A system with "redundancy" has extra components that provide backup in case of a component failure. A common example of computer redundancy is extra (redundant) independent disk drives.

which of the many input variables has the most influence on the desired output or product quality? Manufacturing equipment is expensive and cannot be taken out of service for long periods of time for extensive experimentation. A common approach employed by many industrial engineers today is the "one-variable-at-a-time" (OVAT) approach, where just one input variable is changed at a time keeping all other input variables fixed. This approach depends upon guesswork and luck to be successful. "OVAT" experiments often are unreliable, inefficient, time consuming (expensive), and may yield false operating conditions for the process.

DOE deals with designing, conducting, and statistically analyzing data to determine how the input variables influence the desired outcome. It allows changing more than one input variable at a time to speed up the experimentation, and it will indicate when variables interact with each other.

A very early DOE occurred in 1747. While serving as surgeon on HM Bark Salisbury, James Lind carried out a controlled experiment to develop a cure for scurvy.[7] He selected 12 men from the ship, all suffering from scurvy, and divided them into six pairs, giving each group different additions to their basic diet for a period of 2 weeks. The treatments were all remedies that had been proposed at one time or another. They were:

(1) A quart of cider every day
(2) Twenty five drops of sulfuric acid three times a day upon an empty stomach
(3) One half-pint of seawater every day
(4) A mixture of garlic, mustard, and horseradish in a lump the size of a nutmeg
(5) Two spoonfuls of vinegar three times a day
(6) Two oranges and one lemon every day

The men who had been given citrus fruits[8] (item 6 above) recovered dramatically within a week. The others experienced some improvement, but nothing comparable to the citrus fruits, which proved to be substantially superior to the other treatments.

When analyzing a manufacturing process, experiments (or "tests") are often used to evaluate which process input variables have a significant effect on the process output. A well-designed experiment (or "test") can provide answers to:

- What are the key input variables that influence the process output?
- What are the input variable settings that would produce the best output results?
- What are the main variable interactions?
- What variable settings would produce the most stable process output?

The four steps of a DOE are

(1) Design the experiment
(2) Collection of data
(3) Statistical analysis of the data
(4) Conclusions and recommendations made as a result of the experiment

DOE is a powerful technique for understanding a process, studying the impact of variables or factors affecting the process and for providing continuous quality improvement possibilities. DOE has proved effective in improving the process yield, process capability, process performance, and reducing process variability.

---

[7]Patients with scurvy develop anemia, exhaustion, swelling in some parts of the body, and sometimes ulceration of the gums and loss of teeth. The name scurvy comes from the Latin *scorbutus*. This disease existed in ancient Greek and Egypt.
[8]Hence, British sailors were "limeys," a term that has expanded to mean ordinary Brits.

Example 11.7

An engineer is interested in increasing the yield of a plastic injection molding process. The process seems to be governed by two key process variables, injection pressure and temperature. The engineer decides to perform an OVAT experiment to study the effects of these two variables on the process yield. The first experiment was to keep the temperature T_1 constant and vary the pressure from p_1 to p_2. The results of these experiments are shown in Table A. The next experiment was to keep the pressure p_1 constant and vary the temperature from T_1 to T_2. The results of these experiments are shown in Table B.

Table A. The Effect of Varying Pressure on Process Yield

Test	Temperature	Pressure	% Yield
1	$T_1 = 500$ °F	$p_1 = 2000$ psi	76
2	$T_1 = 500$ °F	$p_2 = 2500$ psi	88

Table B. The Effect of Varying Temperature on Process Yield

Test	Temperature	Pressure	% Yield
3	$T_1 = 500$ °F	$p_1 = 2000$ psi	76
4	$T_2 = 800$ °F	$p_1 = 2000$ psi	86

The engineer concludes from these experiments that the maximum yield of the process will be 88% when $T_1 = 500$ °F and $p_2 = 2500$ psi. Is this a reasonable conclusion?

Solution

Need: An assessment of the engineer's conclusion.

Know: The data in Tables A and B.

How: Using the principles of DOE.

Solve: The problem with this analysis is that the engineer doesn't know what the yield will be when $T_2 = 800$ °F and $p_2 = 2500$ psi. This condition did not occur in the tests, and it could have had a higher yield than that at T_1 and p_2. The engineer was unable to study this combination or determine whether or not any interaction occurred between temperature and pressure.

Interaction between input variables exists when the effect of one variable on the output is different at different values of another variable. The difference in yields between the tests 1 and 2 provides an estimate of the effect of pressure. The effect of pressure on the yield was 12% (i.e., 76–88%) when temperature was kept constant at $T_1 = 500$ °F. But there is no guarantee that the effect of pressure will be the same when the temperature changes.

Similarly, the difference in yields between tests 3 and 4 provides an estimate of the effect of temperature. The effect of temperature was 10% (i.e., 76–86%) on the yield when pressure was kept constant at $p_1 = 2000$ psi. It is reasonable to expect that we would not get the same effect of temperature when the pressure changes. The OVAT approach to experimentation can be misleading and may lead to unsatisfactory conclusions in real-life situations.

In order to obtain a reliable and predictable estimate of the effects of input variables, it is important that input variables are simultaneously varied over their levels. In the previous example, the engineer should have varied the levels of temperature and pressure simultaneously to obtain reliable estimates of the effects of temperature and pressure. To carry out a complete DOE analysis of the plastic molding process in the previous example, the engineer would still perform four experiments as shown in Table 11.2.

Table 11.2 Complete DOE Analysis for Example 11.7

Test	Temperature (°F)	Pressure (psi)	Yield (%)
1	500	2000	76
2	800	2000	86
3	500	2500	88
4	800	2500	??

In general, the number of experiments (tests) required for a DOE with N variables tested at M levels of each variable is

$$\text{Number of DOE Experiments} = N^M \tag{11.17}$$

So, if you have 2 variables and want to test them over 2 levels of each variable, you will need $2^2 = 4$ experiments. Similarly, if you have 3 variables to be tested over 2 levels, you need $3^2 = 9$ experiments.

11.3 METHODS ENGINEERING

Methods engineering is concerned with the selection, development, and documentation of the methods by which work is to be done. It includes the analysis of input and output conditions, choosing the processes to be used, operations and work-flow analyses, workplace design, tool and equipment selection, human motion analysis and standardization, and the establishment of work time standards.

Methods engineering is the design of the productive process in which a person is involved. The task of the methods engineer is to decide where humans will be utilized in the process of converting raw materials to finished products, and how workers can most effectively perform their tasks. The terms "operation analysis," "work design and simplification," and methods engineering are frequently used interchangeably.

A methods engineering procedure utilizes five separate stages to ensure that an existing process is fully analyzed before a new process is introduced. The five stages are:

(1) project selection
(2) data acquisition
(3) analysis
(4) development
(5) implementation

Stage 1 (project selection) is the point at which a methods engineer will be tasked with improving a process. In this stage, a project is identified that either requires an efficiency enhancement or is a new production process where reliability, accuracy, and efficiency are of particular importance.

Stages 2 and 3 (data acquisition and analysis) are concerned with collecting and analyzing data from an existing activity or a production line. These data can include output records, detailed design drawings, and marketplace demand or performance records. Data that are collected during the second stage is analyzed in stage 3 to establish the optimum man-to-machine ratio and production line outputs. This third stage can also suggest improvements in working conditions, production floor layout, and materials handling.

Developing the key improvements identified during the data analysis exercise is carried out in **stage 4**. It is in this stage that man-to-machine ratios and the number of operators allocated to a machine or the number of machines to a single operator are established and developed into a workable process. Once final improvements have been established, the economic benefit to the company must be detailed before being presented to management for **stage 5** implementation of the method.

11.4 SIMULATION ANALYSIS AND OPERATION RESEARCH

Simulation analysis is an analytical tool that can be used to evaluate the performance of an existing or proposed system under different operating conditions of interest. Simulation analysis can be used before an existing system is modified or a new system is constructed to reduce the chances of failure, eliminate unforeseen problems, and optimize performance. It can be used to answer questions like: What is the best design for a new communications network? What are the material resource requirements? How will a highway perform when the traffic increases by 50%? What will be the impact of a failure?

The United States Department of Homeland Security manages the National Infrastructure Simulation and Analysis Center[9] (NISAC). NISAC conducts computer modeling, simulation, and analysis of the nation's 18 critical infrastructure areas and focuses on the challenges posed by interdependencies and the consequences of disruption.

Simulation analysis is the imitation of the operation of a real-world process or system over time. The act of simulating a system requires that a model of the system be developed. This model represents the key characteristics or behaviors/functions of the selected physical or abstract system or process. The model represents the system itself, whereas simulation analysis represents the operation of the system over time.

Simulation is used in many contexts, such as simulation of technology for performance optimization, safety engineering, testing, training, education, and video games. Often, computer experiments are used to study simulation models. Simulation is also used with scientific modeling of natural systems or human systems to gain insight into their functioning. Simulation can be used to show the eventual real effects of alternative conditions and courses of action. Simulation is also used when the real system cannot be engaged because it may not be accessible, or it may be dangerous or unacceptable to engage, or it is being designed but not yet built, or it may simply not exist.

Physical simulation refers to simulation in which physical objects are used to model the real system. These physical objects are often chosen because they are smaller or cheaper than the objects in the actual system. These models are instrumented and tested under variety of conditions typical of those to which the actual system may be exposed.

Interactive simulation is a simulation in which a physical simulation includes human operators, such as in a flight or driving simulator and in computer games.

Computer simulation is an attempt to model a real or hypothetical situation on a computer. By changing variables in the simulation, the computer program forecasts results about the behavior of the system. Computer simulation has become a common engineering modeling tool. It is now part of most engineering computer-aided drafting and design (CAD) programs (e.g., ProEngineer, SolidWorks, ANSYS, etc.) and produces stress and deformation information for various loading conditions.

As a simple simulation analysis illustration, the cable clamp shown in Figure 11.2 is first drawn in a CAD program, a finite element mesh is developed by the program, and finally, the stress and deformation of the clamp

---

[9]NISAC was incorporated by the USA Patriot Act of 2001 into the Department of Homeland Security upon its inception in March 2003. For more information, see http://www.dhs.gov/about-national-infrastructure-simulation-and-analysis-center.

is computed for various loads. The deformation of the clamp can also be animated to show how the stresses develop for various loading conditions.

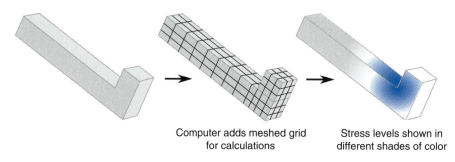

Computer adds meshed grid
for calculations

Stress levels shown in
different shades of color

FIGURE 11.2 Simulation analysis of a cable clamp.

Operations Research (OR) is a strange title for an industrial engineering topic that does not involve any actual research. The name comes from the military during the Second World War. It was used in England to describe the process they used to research (military) operations. After the war, the process moved into mainstream operational planning and the name stuck. Industrial engineers today use OR methods to produce optimal solutions to complex human-technology interaction problems. These problems often require dealing with a large number of variables and constraints.

OR can be applied to areas such as:

- scheduling planes and crews, pricing tickets, passenger reservations
- automobile traffic routing and planning
- financial credit scoring, marketing, banking operations
- deployment of emergency medical services
- material transportation networks
- energy and management

OR also deals with decisions involved in optimizing the allocation of resources such as material, skilled workers, machines, money, and time. Rather than using trial and error methods on the system itself, a mathematical model of the system is developed and then manipulated to find optimum operating conditions. Many mathematical methods and tools have evolved to model and solve complex problems in this field, including linear programming, game theory, queuing theory, network analysis, decision analysis, multifunction analysis, and so forth.

Operational research mathematical models can be either *deterministic* or *stochastic*. A deterministic model contains no random variables. The output is "determined" once the set of input quantities and relationships in the model have been specified. Stochastic models, on the other hand, have one or more random input variables. For this kind of model, probability and statistics are important to measure the prediction uncertainty.

Linear programming is a *deterministic* method to achieve the optimum outcome (such as maximum profit or lowest cost) in a given mathematical model for a set of constraints. It is one of the most widely used tools in OR today. It has been used successfully as a decision-making aid in almost all industries, including financial and service organizations. The "linear" part of the name refers to the fact that the optimization output (i.e., maximization or minimization) is assumed to be a linear function of the input variables. That is, it must involve only the first powers of the input variables with no cross products.

If an OR system has only two input variables, a simple graphical method can be used to solve its linear programming model. If there are more than two input variables, the problem needs to be solved using more advanced analytical methods. The graphical method is illustrated in the following example.

Example 11.8

You are going to manufacture a new product. The product is a Widget, and to make it, you need to use a lathe, a milling machine, and a drill press (see Chapter 12 for examples of these machines). The Widget contains only two parts, A and B. Since your machine shop is already busy, you survey the current work load and determine that the machine tools you need are available only for the times shown below.

Machine Tool	Availability (minutes per day)
Lathe	440
Milling machine	336
Drill press	280

You also determine the time needed on each machine tool to produce parts A and B. This is shown below.

Part	Lathe	Milling	Drilling
A (minutes)	25	8	10
B (minutes)	8	12	8

Finally, the profit that you can make on a Widget is based on the estimated unit profit of each of its parts. The profit you can make on each part A is $25 per unit, and the profit on each part B is $10 per unit, so the total profit for your Widget is $25 \times A + $10 \times B$. What is the optimum daily production for parts A and B that will maximize your profit on the Widget?

Solution

Need: The optimum daily production of parts A and B that will maximize profit.

Know: The machine tool availability and machining times required for each part.

How: Using linear programming methods.

Solve: We can write the following set of operational equations for making the Widget.

$$25A_{\text{lathe}} + 8B_{\text{lathe}} = 440 \text{ minutes of turning on the lathe}$$

$$8A_{\text{milling}} + 12B_{\text{milling}} = 336 \text{ minutes of milling}$$

$$10A_{\text{drilling}} + 8B_{\text{drilling}} = 280 \text{ minutes of drilling}$$

Assuming a linear relation between machine tool operations, we can graph these equations with the number of A parts made per day on the horizontal axis and the number of B parts made per day on the vertical axis. The A and B axis intercepts can be computed as follows. When $B=0$, then

$$A_{\text{lathe}} = 440/25 = 17.6$$

$$A_{\text{milling}} = 336/8 = 42$$

$$A_{\text{drilling}} = 280/10 = 28$$

and when $A=0$,

$$B_{\text{lathe}} = 440/8 = 55$$

$$B_{\text{milling}} = 336/12 = 28$$

$$B_{\text{drilling}} = 280/8 = 35$$

These intercepts are used to plot the straight lines on the A-B graph shown below using an Excel spreadsheet.

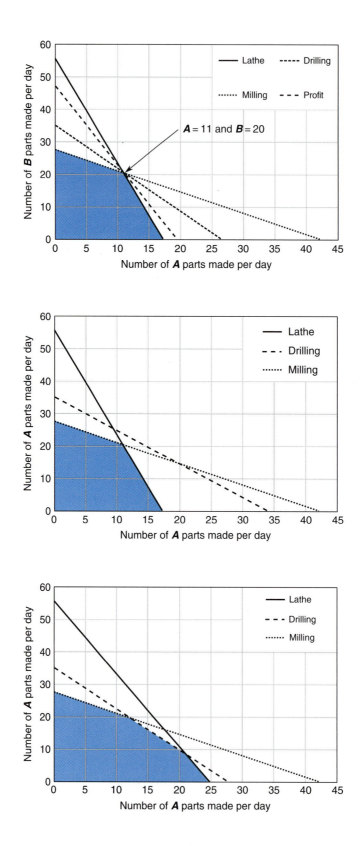

The shaded area in this figure is called the "feasible region" (or solution space) of the problem. Any point inside this region is called a feasible solution and provides values of A and B that satisfy all the requirements. In this case, the "maximum" feasible solution occurs at the intersection of the machining operation lines on the graph, or when 11 units of A and 20 units of B are produced per day. Then, since the total profit for your Widget is $\$25 \times A + \$10 \times B$, the maximum profit is

$$\text{Maximum Profit} = \$25 \times 11 + \$10 \times 20 = \$475$$

11.5 ERGONOMICS

Industrial engineers also design the work area in which humans interact with machines. This is known as ergonomic engineering. There are two general divisions of ergonomics:

- **Occupational ergonomics** (human factors) is concerned with the strength capabilities of the human body in performing manual work (such as lifting, turning, and stretching), and with the environmental effects of temperature, humidity, vibration, and so forth on the human worker. This is done to reduce occupational injuries and increase work productivity.
- **Cognitive ergonomics** (task analysis) is concerned with understanding the behavior of humans as they interact with machines. This information is used to design machine display interfaces and controls to support operator needs, to limit their workload, and to promote awareness of the operation.

11.5.1 Occupational Ergonomics

Occupational ergonomics uses information about human ability in the design of tools, machines, jobs, and work environment for productive, safe, and effective human use. Occupational ergonomics provides methods for optimizing tasks in the workplace. The design of workplaces requires the understanding of the ergonomic principles of posture and movement that produce a safe, healthy, and comfortable work environment. Worker posture and movement is dictated by the task and the body's muscles, ligaments, and joints needed in carrying out the task. Poor posture and movement can produce stress that results in damage to the neck, back, shoulder, wrist, and other work-related injuries.

One of the most common work-related injuries is musculoskeletal disorder. This disorder results in persistent pain, loss of functional capacity, and work disability. For example, a poorly designed tool can adversely impact overall worker performance, create injuries, and produce human task error. Industrial engineers evaluate these tools to determine potential sources of injury and attempt to improve them to fit the needs and workflow of workers.

For example, a worker should be able to operate the electric drill shown in Figure 11.3 with one hand. To do this safely and comfortably, the weight of the tool should not exceed about two pounds (1 kg) and the center of gravity should be near the center of the gripping hand. The tools should be easy to hold in any position needed. Drills that are front-heavy can produce stress in the wrist and forearm.

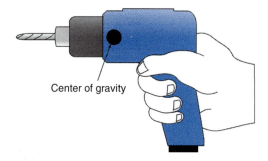

FIGURE 11.3 The ergonomics of a hand drill.

Example 11.9

For each of the consumer items below, identify three ways in which they could be ergonomically improved?

(a) Cell phone

(b) Alarm clock

(c) The electric drill in Figure 11.3

Solution (Note: these are not the only improvements you can think of)

Need: Representative ergonomic improvements for a cell phone, alarm clock, and electric drill.

Know: The basics of ergonomics.

How: Apply the basics or ergonomics.

Solve: (a) Simplify the programming to make it easier to use all the phone's functions.

(b) Prevent the user from turning off the alarm when he or she tries to activate the snooze option.

(c) Move the heavy motor back over the center of the hand.

11.5.2 Cognitive Ergonomics

A time and motion study for repetitive tasks is a method for establishing worker productivity standards in which:

- a complex task is broken into small, simple steps
- the sequence of movements taken by the worker in performing those steps is carefully observed to eliminate wasteful motions
- the precise time taken for each movement is measured

A time and motion study is a method to determine the best way to complete a certain task. The simplest way to conduct it is to break the task into a series of individual steps and record the motion and time required for each step. This should be done repeatedly in order to refine the process of completing a task. You can then determine the "normal" or average time for a task by recording how much time is devoted to each part of the task.

This can be done as follows:

(1) List the steps needed to perform the task

(2) Discuss them with the worker

(3) Measure each step with a stopwatch as the worker performs the task

(4) Repeat the complete process at least 10 times
(5) Compute the mean and standard deviation of each step and of the complete task
(6) Be aware of worker disruptions and learning curves

From these measurements, production and delivery times and prices can be computed, and worker incentive schemes can be devised.

11.6 MATERIAL HANDLING

Material handling is the field concerned with solving the problems of movement, storage, control, and protection of materials throughout the processes of cleaning, preparation, manufacturing, distribution, consumption and disposal of all related goods, and their packaging. The focus of material handling is on the methods, mechanical equipment, systems, and controls used to achieve these functions.

There are many ways by which material handling has been defined, but the simplest definition is that material handling means providing the *right* amount of the *right* material at the *right* place at the *right* time for the *right* cost. The **primary** objective of a material handling system is to reduce the unit cost of production. Other objectives are to:

(1) reduce manufacturing cycle time
(2) reduce delays and damage
(3) promote safety and improve working conditions
(4) maintain or improve product quality
(5) promote productivity by having material move short distances, in a straight line, and use the force of gravity whenever possible
(6) encourage increased use of facilities by purchasing adaptable equipment and developing a preventive maintenance program
(7) control inventory

A material handling system often involves warehouses, conveyors, part sorters, feeders, and manipulators. Material handling systems range from simple pallet rack and shelving projects to complex conveyors that use computer-controlled storage and retrieval systems with automatic guided vehicles.

A material handling system is a fundamental part of a flexible manufacturing system because it connects the different manufacturing processes used to create a finished product. Due to the automated nature of the whole production process, a material handling system must respond in concert with all the requirements of the manufacturing processes. It is something that goes on in every manufacturing plant all the time. It applies to the movement of raw materials, parts in process, finished goods, packing materials, and disposal of scraps. In general, hundreds and thousands tons of materials are handled daily requiring a large amount power.

The cost of material handling contributes significantly to the total cost of manufacturing. The importance of reducing material handling is greater in those industries where the ratio of material handling cost to processing cost is large. Today, material handling is rightly considered as one of the most profitable areas for reducing manufacturing costs. A properly designed and integrated material handling system provides tremendous cost saving and customer service improvement.

A properly designed material handling system will improve customer service; reduce inventory; shorten delivery time; and lower overall handling costs in manufacturing, distribution, and transportation. Some items to consider in designing a material handling system are

- **Performance objectives**—Define the performance objectives and functional requirements of the system at the beginning of the design process.
- **Standardization**—All material handling methods, equipment, controls, and software should be standardized and able to perform a range of tasks in a variety of operating conditions.
- **Ergonomics**—The environment should support the abilities of a human worker, reduce repetitive and strenuous manual labor, and promote worker safety.
- **Space**—Develop the efficient use of space within a facility by keeping work areas organized and maximizing storage density.
- **Automation**—Automated material handling technologies should be used when possible to improve material handling efficiency.

The total cost of material handling per unit is the sum of the following:

(1) Cost of material handling equipment—both fixed cost and operating cost calculated as the cost of equipment divided by the number of units of material handled over the working life of the equipment.
(2) Cost of labor—both direct and indirect.
(3) Cost of maintenance of equipment, damages, lost orders, and expediting expenses.

Example 11.10

Aluminum castings at a small engine manufacturing plant are processed through several different machining operations. Initially, these operations took 3 hours per casting. An engineer subsequently installed a unified machining cell that carried out the required drilling, milling, boring, and polishing operations on a casting in just minutes. The plant manager was pleased that the cycle time was reduced from 3 hours to a few minutes.

Then, the plant manager noticed that the engineer specified the batch size for the new machining cell to be 10,000 castings. Since the castings were heavy, the material handling system would need to deliver 10,000 casting batches to the cell using a fork lift. The cell would then automatically machine multiple castings at the rate of 250 per hour. Once the batch of 10,000 was completed, a fork lift would be called back to deliver a new batch of 10,000 castings and transfer the finished batch of 10,000 machined castings to the assembly line. What problems did this new machining cell create?

Solution

Need: Problems produced by the new machining center.
Know: Cell batch size = 10,000 castings with 250 castings machined per hour.
How: Material handling basics.
Solve: Because the process batch size of 10,000 was incorrectly assumed to be equal to the overall material handling batch size, this created a large backup of finished castings at the assembly line. This required extra space at the assembly line and produced material handling problems at that point. The parts backup at the assembly line would be reduced if the machining cell material handling batch size was reduced from 10,000 castings to a value compatible with the assembly line capacity.

Materials delivered in the wrong sequence will directly result in increased backup (called "work in progress"). Product design and process planning should be done before the layout of the factory and its material handling system. In assembly lines, the sequence in which the parts arrive at an assembly station should match the sequence in which they can be assembled. This is very important when pallets are loaded with a particular part and moved from one station to another with a fork lift.

A **conveyor** is a common material handling system that moves materials from one location to another. Conveyors are able to safely transport materials from one processing station to another, and they can be installed almost anywhere. They are also much safer than using a forklift or other machine to move materials, and they can move loads of all shapes, sizes, and weights.

The layout of a conveyor system is determined by the relative positions of processing stations and existing floor space. The width of the conveyor is determined by the size of the part carriers. Usually, parts are put in bins or boxes (called "part carriers") that travel on the conveyor. These carriers are usually rectangular, and their size is determined by the size of each part and the number of parts put in each carrier. Given the size of the carrier, it is easy to determine the width of the conveyor. The speed of a conveyer is dictated by its loading and unloading speeds as illustrated in the following example.

Example 11.11

A conveyor belt has boxes attached to it to carry parts to processing stations. The conveyor moves at speed V. The required loading speed of the conveyer is P parts per minute, and the conveyor length used by each part box carrier is L. There are N unloading stations that have an average unloading time of M parts per minute per station.

(a) How should the conveyer speed be set relative to the part arrival rate?
(b) What is the relation between the rate at which parts are loaded on the conveyer and the rate at which they are unloaded?

Solution

Need: The conveyer speed and the relation between parts loaded and parts unloaded.
Know: Basic mechanics of motion (physics).
How: Use the concepts of material handling.
Solve: (a) Parts should arrive at the loading station somewhat slower than they can be loaded (to prevent material backup), so the conveyor speed should be set so that the carrier arrival rate at the loading station (V/L) is greater than the part arrival rate (P).

(b) In order for the unloading stations to properly unload the conveyor, the number of parts unloaded per minute should be somewhat greater than the parts loaded per minute on the conveyer, or $N \times M > P$.

Note that if every third carrier is empty ($PL/V = 2/3$), then an unloading station with unload times between L/V and $1.5\,L/V$ will have sufficient capacity. However, it will not be able to unload every second consecutive part carrier that arrives, so there will be filled carriers continuously circulating on the conveyor.

SUMMARY

Industrial engineering is the field of engineering concerned with the design, analysis, and operation of systems that range from a single piece of equipment to large businesses. Industrial engineers serve in areas that range from the production of raw materials to manufacturing to the service industry. Industrial engineers determine the most effective way to utilize people, machines, materials, information, and energy to make a product or provide a service.

An important area of industrial engineering is manufacturing. This topic is covered in a separate chapter in this textbook. In this chapter, we covered five other important areas of industrial engineering: **quality control, methods engineering, simulation analysis** and **operation research, ergonomics,** and **material handling**.

EXERCISES

1. The median household income in 2012 for the following six nations is shown in the table below.

Nation	Luxembourg	United States	Singapore	Bahrain	Venezuela	Liberia
Median income	$52,493	$43,585	$32,360	$24,633	$11,239	$781

    (a) Find the median household income of these six nations.
    (b) Find the mean household income of these six nations.
    (c) Find the standard deviation of the household income of these six nations.

2. An engineer collects data that consists of the following four observations: 1, 3, 5, and 7. What is the (a) variance, (b) standard deviation, and (c) standard error of this data?

3. Every year, 1000 engineers compete in a sports car road race. The mean (average) finishing time is 55 minutes, with a standard deviation of 10 minutes. Justin and Cindy completed the race in 61 and 51 minutes, respectively. Barry and Lisa had finishing times with Z-scores of −0.3 and 0.7, respectively. Since the Z-score is the number of standard deviations *above* the mean, then the lowest Z-score corresponds to the fastest car. List the drivers, fastest to the slowest, based on their Z-scores.

4. A final examination in an engineering class has a mean (average) score of 75 and a standard deviation of 10 points. If Laura's Z-score is 1.20, what was her score on the test?

5. We want to know the chances of choosing a particular student type based on four demographic characteristics. The demographics of the students are illustrated in the table below.

	Caucasian (*c*)	Minority (*m*)	Total
Male	55	40	95
Female	65	40	105
Total	120	80	200

    (a) What is the probability of choosing a Caucasian student?
    (b) What is the probability of choosing a minority student?
    (c) What is the probability of choosing a Caucasian male?
    (d) What is the probability of choosing a Caucasian female?

6. Suppose the number of parts that can be machined on a lathe in 1 day ranges from 0 (the lathe is not operating) to 250 (the best lathe operator). By collecting data, the following probabilities for machining parts in the first hour were determined.

Number of parts machined in 1 day	Probability
0	0.10
1-50	0.05
51-100	0.10
101-150	0.30
151-200	0.43
201-250	0.02

This table tells us that the probability of machining between 1 and 50 parts in 1 day is 5.0%, and the probability of machining between 50 and 100 parts in 1 day is 10%. What is the probability of machining "at least" 101 parts in 1 day?

(a) What is the probability of machining at least 101 parts in 1 day?

(b) What is the probability of not machining at least 101 parts in 1 day?

7. A manufacturer wishes to order material from a wholesale supplier. The supplier has three phone lines with three different numbers that operate *independently*. The probability of phone 1 being busy is 0.80, the probability of phone 2 being busy is 0.80, and the probability of phone 3 being busy is 0.80. The manufacturer calls to order a large supply of material. What is the probability that the phone lines will be busy?

8. Suppose 10 devices are tested for 500 hours. During this test, 2 devices fail.
   (a) What is the MTBF?
   (b) What is the MTTF?
   (c) What is the failure rate?
   (d) What is the probability that any one device will be operational when $t = $ MTBF?

9. A product has a constant failure rate of 0.2% per 1000 hours of operation.
   (a) What is its MTTF?
   (b) What is the probability of it successfully operating for 10,000 hours?

10. Engineers carry out a reliability test to develop a warranty policy. The test finds that the failure rate is approximately constant with $F(t) = 1/8750 = 1.14 \times 10^{-4}$ failures per hour.
   (a) Determine the reliability function $R(t)$
   (b) Determine the MTTF.

11. A new automotive valve spring design is tested in continuous use and is found to have a constant failure rate of $F = 0.002$ failures per hour.
   (a) Determine the probability of failure $(1 - R(t))$ within the first hour of use.
   (b) Determine the probability of failure within the first 100 hours.
   (c) Determine the probability of failure within the first 1000 hours.

12. An electronic circuit board has the failure rate described by: $F(t) = 0.05/(t^{1/4})$ per year (with "t" in years).
   (a) Determine the reliability function $R(t)$ and the failure probability distribution function.
   (b) Determine the reliability for 1 year of use.
   (c) Determine the fraction of failed circuits in 3 years.

13. A cell phone manufacturer tested 10,000 phones in a reliability evaluation program. Each phone was turned on-and-off 160 times each day to mimic extreme phone usage for a week. Based on a failure-to-perform criterion, the following failure data were obtained for the first 10 days of test:

Day	1	2	3	4	5	6	7	8	9	10
Failures	180	120	100	70	60	50	40	30	0	10

For the first day, the failure rate is: $F(1) = 180/10,000 = 0.018$ per day.
For the second day, the failure rate is: $F(2) = 120/(10,000 - 180) = 120/9820 = 0.0122$ per day.
For the third day, the failure rate is: $F(3) = 100/(10,000 - 180 - 120) = 100/9700 = 0.010$ per day.
And so on.

(a) Plot $F(t)$ from $t=0$ to $t=10$ days.

(b) Fit these data with a curve of the form $F(t)=at^{-b}$ and determine the values of a and b.

(c) Determine the reliability function $R(t)$

(d) Determine the mean-time-to-failure, MTTF.

14. If you have a MTBF of 160 days, it doesn't mean you can expect an individual device to operate for 160 days before failing. MTBF is a statistical measure and can't predict anything for a single unit. However, if you have 1000 devices with this same MTBF operating continuously in a factory, (a) how often would you expect one to fail and (b) how long would it take for 20 failures to occur?

15. Using a DOE process, how many experiments do you need to perform if you have:

(a) 2 variables and want to test them over 2 levels of each variable

(b) 2 variables and want to test them over 3 levels of each variable

(c) 3 variables and want to test them over 3 levels of each variable

(d) 5 variables and want to test them over 4 levels of each variable

16. List the five stages in methods engineering.

17. Determine the maximum feasible profit in Example 11.8 if the drilling time for part B was 10 minutes instead of 8 minutes. Assume all the remaining conditions are unchanged.

18. Determine the maximum feasible profit in Example 11.8 if the lathe time for part A was reduced from 25 to 10 minutes. Assume all the remaining conditions are unchanged.

19. Freddy works on an assembly line and uses a handheld pneumatic impact wrench. The assembly line makes up to 1400 products a day and it takes about 3 seconds for Freddy to tighten each of seven different components. Freddy often has to use poor posture to attach some of the parts. After a few weeks, Freddy found that he was leaving work with shoulder and neck pain. Using ergonomic principles assesses Freddy's working conditions and suggests at least four workplace modifications that will reduce Freddy's problems.

20. Janet is a sales engineer at your company. Much of her work involves using a telephone and a computer to make appointments, call customers, and respond to e-mails. Janet would often hold the telephone between her shoulder and ear while talking on the phone and typing on the computer. Also, Janet's computer screen was difficult to read because of glare and reflections from light through the window in her office. After working for 8 months, Janet found that she was leaving work with an aching shoulder, sore eyes, and a headache. Using ergonomic principles assesses Janet's working conditions and suggests at least three workplace modifications that will reduce Janet's problems.

21. A worker's hand was seriously injured while trying to clear a reoccurring blockage in a computer-controlled machining station. Access to the machining station is through a door in its enclosure, and all machining operations are stopped when this door is opened. The worker had obtained an override key, so he could open the door and enter the enclosure without stopping production. Using ergonomic principles identified at least three steps that can be taken to help prevent a recurrence.

22. A company uses reusable wooden boxes 36″ square and 48″ long to ship their product. Recently, the cost of purchasing a new shipping box has increased from $32.50 to $81.00, and the maintenance cost for existing boxes has increased from $15 to $21 per year. Also, extra truck runs and outside trucking services are required to return the wooden boxes. As a material handling engineer, how would you improve this system by suggesting new methods for shipping the product?

23. The storage area at your company is presently filled to capacity with 20,000 cases of products. Your company recently increased its manufacturing capability by 100%, and the finished goods inventory storage area is expected to increase by the same amount. The present storeroom is 250 feet by 375 feet with a 30 foot ceiling height. The product is packaged in cases 15 inches long, 24 inches high, and 32 inches wide, and you cannot stack these cases more than 6 feet high without damage. As a material handling engineer, suggest methods for increasing the storage area to accommodate the contemplated increase in finished goods inventory.

24. One of your first assignments when you graduate with an engineering degree and accept a well-paying position at a prominent company is working in the product testing laboratory. Your manager wants you to do a OVAT experiment on the products reliability. You know that a DOE testing method would save the company time and money and produce more accurate results. Your manager has never heard of DOE and insists you proceed with a long and expensive OVAT experiment. What do you do? (Use the Engineering Ethics Matrix.)
    (a) Go over your manager's head and talk to his boss.
    (b) Ignore your manager's orders and carry out a DOE without his knowledge.
    (c) Ask a colleague for help in convincing your manager to your point of view.
    (d) Confront your manager in writing (with a cc to his/her boss) saying that he/she doesn't understand modern engineering testing methods and should listen to you.

25. You now have an entry level position as an engineer at a hand tool manufacturing company. You are put on a team of engineers that is about to release the design of a new battery-powered screw driver to manufacturing. By handling the prototype, you notice that the screw driver has a heavy battery and that the direction that the shaft turns is activated by the user twisting his/her wrist at a severe angle. You point this out at a team meeting and suggest that the design may have serious ergonomic problems that could result in user wrist injury during long-term use. The team quietly listens to your comment and replies that profit, not user ergonomics, is their goal. What do you do? (Use the Engineering Ethics Matrix.)
    (a) Reassert your position by suggesting that customer satisfaction will also lead to profits.
    (b) Contact the company lawyer as a whistle blower about a possible dangerous product.
    (c) Talk to the team leader in private and suggest an ergonomics expert be added to the team.
    (d) Send an anonymous e-mail to the company president warning him/her about possible law suits.

Manufacturing Engineering

12.1 INTRODUCTION

Virtually, everything we use at home, at work, and at play was manufactured. Manufactured goods are everywhere: aircraft, bicycles, electronics, coat hangers, automobiles, refrigerators, toys, clothing, cans, bottles, cell phones, and so on. Even the humble pencil and the paper clip[1] are triumphs of manufacturing processes.

12.2 WHAT IS MANUFACTURING?

You probably have a general idea about manufacturing processes, but let's look at the word itself. The word *manufacture* derives from two Latin words: *manu* (meaning "by hand") and *factum* (meaning "made"). We generally think of manufacturing taking place in a factory (an abbreviation of the eighteenth-century word *manufactory*).

> *Manufacturing is the process of converting (either by manual labor or by machines) raw materials into finished products, especially in large quantities.*

Manufacturing covers a wide variety of processes and products. It involves the production of many types of goods that range from food to microcircuits to airplanes to health-care equipment. The number and complexity of the processes involved in the production of these items varies. Some products are basic, such as flour, cheese, leather, and iron. Some are merely altered, such as metal ingots, Portland cement, and chemicals like gasoline.

---

[1]Henry Petroski, *The Evolution of Useful Things: How Everyday Artifacts—From Forks and Pins to Paper Clips and Zippers—Came to Be as They Are* and *The Pencil: A History of Design and Circumstance*, (New York: Alfred A. Knopf, 1992 and 1989, respectively).

Others are moderately changed, such as wire rods, metal pipes, glass bottles, soap, cloth, and paper; while things like vehicles, computers, medicines, and cell phones are elaborately transformed.

Manufacturing began more than 10,000 years ago when nomadic people settled in one geographic area and began to plant and grow food and domesticate animals. Skilled artisans did the manufacturing, and later, the guild system (the predecessors of modern trade unions) evolved to protect their privileges and trade secrets.

Today's manufacturing engineers apply scientific principles to the production of goods. Manufacturing engineers design the processes and systems to make products with the required functionality, high quality, at the lowest price, and in ways that are environmentally friendly. Figure 12.1 illustrates a typical manufacturing process, from artisans to their market.

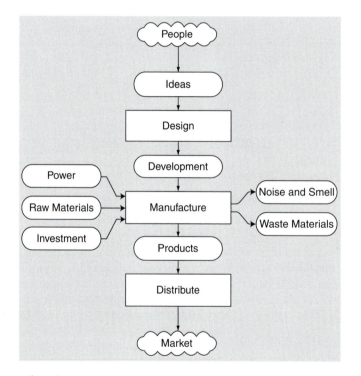

FIGURE 12.1 A typical manufacturing process.

In this chapter, we discuss several machining processes, including **drilling**, **lathe work**, **milling**, **welding**, **extrusion**, **pultrusion**, **blow molding**, and **thermoforming**. Modern manufacturing methods include **Just-In-Time** (JIT) inventory control and **flexible**, **lean**, and **life-cycle** manufacturing. A key concept in manufacturing is the recognition of **variability** and **tolerance**, which we describe by introducing statistical methods that include a relatively recent paradigm called **Six Sigma**. An important variable in all of this is the **standard deviation** which is central to **statistical** methods.

12.3 EARLY MANUFACTURING

Before the Industrial Revolution, the production of goods for sale occurred in homes on farms and in villages to provide additional income. This "cottage" industry provided extra income to a farming family that would take in weaving, sewing, dyeing, pins, and so forth that was then sold to a retailer.

In 1776, the Scotsman Adam Smith[2] (1723–1790) published his influential book *An Inquiry into the Nature and Causes of the Wealth of Nations*. This was one of the first studies on the production techniques in the colonial era. In his book, he discussed how the division of labor could significantly increase production. One example he used was the making of pins. One worker could probably make only 20 pins per day. But, if 10 people divided up the 18 steps required to make a pin, they could make 48,000 pins per day (240 times as many). By the early 1900s, the assembly line method of mass production developed by Henry Ford represented the perfect application of Smith's ideas.[3]

Eli Whitney (1765–1825) is most famous for his invention of the cotton gin.[4] However, the gin was a minor accomplishment compared to his perfection of the concept of interchangeable parts. At the time, the United States had limited human resources and scarce manufacturing labor. Whitney developed the concept of interchangeable parts about 1799 when he took a contract from the U.S. Army for the manufacture of 10,000 muskets at the unbelievably low price of $13.40 each. He is credited for inventing what is called the *American system of manufacturing*—the combination of power machinery, interchangeable parts, and the division of labor that would become one of the foundations of the Industrial Revolution.

12.4 INDUSTRIAL REVOLUTION

The Industrial Revolution[5] began in Great Britain during the second half of the eighteenth century and spread through Europe and the United States in the nineteenth century. In the twentieth century, industrialization extended to Asia and to the Pacific Rim. While modern manufacturing processes and economic growth continue to spread throughout the world, many people have yet to experience the benefits of improved economic and social conditions typical of the Industrial Revolution.

The key technology that brought about the Industrial Revolution was the invention and subsequent improvement of the steam engine. First invented by the English blacksmith Thomas Newcomen in the late seventeenth century, it was initially used to pump water out of flooded tin mines and then applied to coal mines. Increased coal production subsequently led to the smelting of iron, which in turn produced a vast array of new metals and associated technologies.

The Industrial Revolution produced sweeping social changes,[6] including the movement of people to cities, the availability of a greater variety of material goods, and new ways of doing business. Goods that had been made in the home began to be manufactured at lower cost in a factory in a city. The resulting economic development combined with superior military technology made the nations of Europe and the United States the most powerful in the eighteenth- and nineteenth-century world.

---

[2]Adam Smith is also known for his explanation of how rational self-interest and competition can lead to economic well-being and prosperity without government intervention. The French expression *laissez-faire* (roughly translatable as "to leave alone") became closely associated with his name. His work also helped to create the modern academic discipline of economics and provided one of the best-known rationales for free trade and capitalism.

[3]On June 25, 2006, Bill Gates presented a copy of Adam Smith's *Wealth of Nations* to fellow billionaire Warren Buffett after Buffett announced that he would donate much of his wealth to The Bill and Melinda Gates Foundation.

[4]This term *gin* is a contraction of the word "engine," since Whitney's device was a creative machine (engine) for removing seeds from hand-picked cotton.

[5]The Industrial Revolution is called a *revolution* because it significantly and rapidly changed society. Over the course of human history, there has been only one other group of changes as significant as the Industrial Revolution. This is what anthropologists called the *Neolithic Revolution*, which took place in the later part of the Stone Age. In the Neolithic Revolution, people moved from social systems based on hunting and gathering to much more complex communities that depended on agriculture and the domestication of animals. This led to the rise of permanent settlements and, eventually, urban civilizations. The Industrial Revolution brought a shift from the agricultural societies created during the Neolithic Revolution to modern industrial societies.

[6]The excesses of the early Industrial Revolution were documented in Charles Dickens's novels.

Between 1850 and 1900, machine tool technology advanced due, in part, to the development of engineering documentation through accurate detailed drawings. However, as products moved from one process to the next in factories, a number of unresolved questions arose: What should happen between processes? How should sequential processes be arranged? How should processes function as a system? What was the best way for workers to carry out their tasks?

In the 1890s, Frederick W. Taylor (1856–1915) began to study the process of how work was carried out. Taylor felt that the industrial management was poor and that it could be formulated as an academic discipline. He thought that the best results would come from a partnership between management and workers. He felt that, by working together, there would be no need for trade unions. Taylor's "scientific" management process consisted of four principles:

1. Replace rule-of-thumb[7] methods with techniques based on a scientific study of the tasks.
2. Train workers rather than have them train themselves.
3. Provide detailed instruction and supervision for each worker.
4. Managers apply scientific management principles to planning the work and the workers perform the tasks.

Taylor's ideas became known as **scientific management**. The concept of applying science to management was useful; however, Taylor unfortunately ignored the science of behavior. It seems that he had a peculiar attitude toward workers. He felt that they were not very bright and generally incapable of understanding what they were doing (this was true even for rather simple tasks[8]).

In the early 1900s, Frank Gilbreth (1868-1924) developed the concept of a **time and motion study** and invented **process charting** to focus attention on the entire work process, including non-value-added steps that occur between production processes. He became aware of these problems when he found ways to make brick-laying (his first trade) faster and easier. He later collaborated with his future wife, Lillian Moller Gilbreth[9] (1878–1972), to study the work habits of factory and clerical workers in several of industries to find ways to increase output and make their jobs easier. She brought psychology into the mix by studying the motivations of workers and how their attitudes affect their work. She and her husband Frank were genuine pioneers in the field of industrial engineering.

Around 1910, Henry Ford developed the first all-inclusive manufacturing plan. As a farm boy in Michigan, he saw how animals were "disassembled" by a line of workers in slaughterhouses. By simply reversing this process, he created an "assembly" line process for manufacturing products. He took all the elements in a manufacturing system—people, machines, tooling, and supplies—and arranged them in a long continuous line to efficiently manufacture his early (Model T) automobiles. By the 1920s, you could buy a new Ford car for only $300 (about $7300 in 2014 inflation-adjusted dollars). Ford was so successful that he quickly became one of the world's richest men. He is considered by many to be the first practitioner of "just-in-time" and lean manufacturing.

After the 1920s, the Ford system began to break down and an aging Henry Ford refused to change the system. Prosperity, the growth of labor unions, and competition from General Motors and Chrysler began to weaken the Ford Motor Company. Annual model changes, multiple colors, and options were not embraced in Ford factories. General Motors developed new business and manufacturing strategies for managing and producing a variety of

---

[7]The term probably originated from wood workers who used the length of their thumb as a "ruler" for measurement.
[8]"I can say, without the slightest hesitation," Taylor told a congressional committee, "that the science of handling pig-iron is so great that the man who is … physically able to handle pig-iron and is sufficiently phlegmatic and stupid to choose this for his occupation is rarely able to comprehend the science of handling pig-iron."
[9]She was the first woman elected into the National Academy of Engineering. She served as an advisor to Presidents Hoover, Roosevelt, Eisenhower, Kennedy, and Johnson on matters of civil defense, war production, and rehabilitation of the physically handicapped.

cars (Chevrolet, Pontiac, Buick, and Cadillac). By the mid-1930s, GM had passed Ford in dominating the automobile market.

Edward Deming, an American quality control expert, was a consultant to the occupying forces in Japan after World War II. He introduced statistical quality control methods that allowed Japanese industries to manufacture very high quality products. Deming's methods at first were not understood or appreciated by the rest of the world, until Japan eventually became a world leader in exporting superior goods and technology. Motorola expanded Deming's statistical quality approach to reduce its defective products to an amazing 1 defect in 3.4 million products. They call their process **Six Sigma**, which has now been adopted by much of European and American manufacturing industry.

12.5 MANUFACTURING PROCESSES

Modern manufacturing can be divided into the following four main categories: subtractive, additive, continuous, and net shape.

12.5.1 Subtractive Processes

Subtractive processes involve material removal via machining (e.g., turning, milling, boring, grinding, cutting, and etching). The three principal machining processes are classified as **turning**, **drilling**, and **milling**.

12.5.1.1 Turning

Turning is a process that uses a lathe (see Figure 12.2) to produce "solids of revolution." The lathe can be operated by a person or a computer. A lathe that is controlled by a computer is known as a CNC (**computer numerical control**) machine. CNC is commonly used with many other types of machine tools in addition to the lathe.

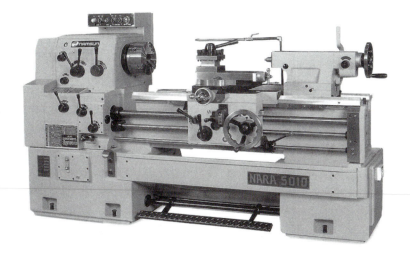

FIGURE 12.2 A typical lathe.

In a turning operation, the part is rotated (i.e., turned) while it is being machined. Turning produces straight, conical, curved, or grooved workpieces such as shafts, spindles, and pins. The majority of turning operations involve the use of a single-point cutting tool, such as that shown in Figure 12.3.

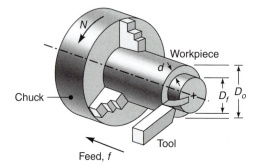

FIGURE 12.3 The basic elements of a turning operation.

The *material removal rate* (MRR) is defined as the amount of material removed per unit time, and it typically has units of mm^3/min or in^3/min. In turning, each revolution of the workpiece removes a nearly ring-shaped layer of material. The material removed has a cross-sectional area equal to the axial distance the tool travels in one revolution (called the **feed**, f) multiplied by the depth of the cut, d. Then the volume of material removed per revolution is approximately equal to the product of the circumference and the cross-sectional area of the workpiece, or

$$\text{Material removed per revolution of the workpiece} = (\pi D_{avg})(fd),$$

where D_{avg} is the average diameter of the workpiece during the cutting operation, or (see Figure 12.3).

$$D_{avg} = \frac{(D_o + D_f)}{2} \tag{12.1}$$

If the workpiece is rotating at N RPM, then the material removal rate in turning is

$$MRR_{turning} = \pi D_{avg} fdN \tag{12.2}$$

The **cutting speed**, $V_{cutting}$, is the speed at which material is removed, and in turning, it is

$$V_{cutting} = \pi D_{avg} N \tag{12.3}$$

Then, we can write the material removal rate as

$$MRR_{turning} = fdV_{cutting} \tag{12.4}$$

In the design of manufacturing operations, the time required to machine a machined part directly affects the manufacturing cost of the part. In turning, the cutting tool has a **feed rate** (FR) in mm/min or in/min, determined by the tool's feed f times the rotational speed N of the workpiece, or

$$FR = fN \tag{12.5}$$

If the axial cutting distance along the workpiece is L, then the time required to make the cut is

$$t_{machining} = \frac{L}{FR} \tag{12.6}$$

Example 12.1

A shaft 6.0 in. long with an initial diameter of 0.50 in. is to be turned to a diameter of 0.48 in. on a lathe. The shaft rotates at 500 RPM on the lathe, and the cutting tool feed rate is 4.0 in./min. Determine

1. The cutting speed.
2. The material removal rate.
3. The time required to machine the shaft.

Need: $V_{cutting}$, $MRR_{turning}$, and $t_{machining}$.

Know: $L=6.0$ in., $D_o=0.50$ in., $D_f=0.48$ in., $N=500$ RPM, and the feed rate, FR$=4.0$ in./min

How: The cutting speed, material removal rate, and cutting time are given by Equations (12.3), (12.4), and (12.6).

Solve: 1. Equation (12.3) gives the cutting speed as $V_{cutting}=\pi D_{avg}N$, where $D_{avg}=(D_o+D_f)/2=(0.50+0.48)/2=0.49$ in. Then, $V_{cutting}=\pi(0.49)(500.)=$ **770 in/min**

2. The depth of the cut is $d=(D_o-D_f)/2=(0.50-0.48)/2=0.010$ in, and the feed is $f=$FR$/N=4.0/500$. [in./min]/[rev/min]$=0.008$ in./rev. Then, Equation (12.4) gives the MRR in turning as

$$MRR_{turning}=\pi(0.49)(0.008)(0.010)(500.)[in][in][in/rev][rev/min]=0.062\ in^3/min$$

3. The time required to machine the shaft is given by Equation (12.6) as

$$t_{machining}=L/FR=6.0/4.0[in]/[in/min]=1.5\ min$$

The power required by a machining operation, $P_{machining}$, is equal to the material's average machining energy per unit volume removed (see Table 12.1) multiplied by the material removal rate, or

$$P_{machining}=(\text{Average Machining Energy per Unit Volume})\times MRR \tag{12.7}$$

Table 12.1 Average machining energy per unit volume of material removed

Material	Average machining energy per unit volume	
	$W\cdot s/mm^3$	$HP\cdot min/in^3$
Cast iron	3.3	1.2
Steels	5.5	2.1
Stainless steel	3.5	1.4
Aluminum alloys	0.70	0.28
Magnesium alloys	0.45	0.15
Nickel alloys	5.8	2.2
Titanium alloys	3.5	1.4

Source: Condensed from Table 21.2 in Kalpakjian and Schmid, Manufacturing Engineering and Technology, 5th ed. (Upper Saddle River, NJ: Pearson Prentice Hall, 2006).

and since the machining power is the product of the cutting torque (in $N \cdot m$ or $ft \cdot lbf$) and the rotational speed of the workpiece (in radians/min[10]), then the cutting or machining torque is given by

$$T_{cutting} \text{ or } T_{machining} = P_{machining}/(2\pi N) \tag{12.8}$$

Example 12.2
Determine the machining power and cutting torque required to machine the shaft in Example 12.1 if it is made from stainless steel.

Need: $T_{cutting}$ and $P_{cutting}$ for the shaft in Example 12.1.
Know: The material is stainless steel, and from Example 12.1, we know that $MRR_{turning} = 0.062$ in³/min, and $N = 500$. RPM.
How: Using Table 12.1 and Equations (12.7) and (12.8).
Solve: From Table 12.1, the average machining energy per unit volume of material removed is 1.4 HP·min/in³. Then, Equation (12.7) gives the machining power as

$$P_{machining} = 1.4(0.062)[HP \cdot min/in^3][in^3/min] = 0.087 HP.$$

Now, Equation (12.8) gives the cutting torque as

$$T_{cutting} = P_{machining}/(2\pi N) = 0.087/(2\pi \times 500)[HP]/[rad/min] = 2.8 \times 10^{-5} HP \cdot min.$$

Since 1 HP = 33,000 ft·lbf/min, the cutting torque is

$$T_{cutting} = 2.8 \times 10^{-5}(33000)[HP \cdot min][ft \cdot lbf/(HP \cdot min)] = 0.92 ft \cdot lbf.$$

12.5.1.2 Drilling
Drilling is the process of using a cutting tool called a *drill bit* to produce round holes in materials such as wood or metal. The common twist drill shown in Figure 12.4 (the one sold in hardware stores) has a point angle of 118°. This angle works well for a wide array of drilling operations. A steeper point angle (say, 90°) works well for plastics and other soft materials. A shallower point angle (say, 150°) is suited for drilling steels and other tough materials. Drills with no point angle are used in situations where a blind, or flat-bottomed hole, is required. Figure 12.5 shows a typical drilling machine, called a *drill press*.

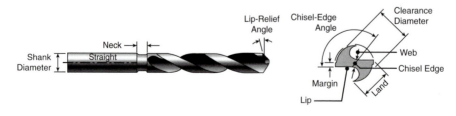

FIGURE 12.4 The characteristics of a common twist drill.

---

[10]Since there are 2π radians in one revolution (360°), you can convert N in revolutions per minute (RPM) to radians per minute simply by multiplying it by 2π. Also, note that radians are dimensionless and do not appear in the final dimensions of the answer.

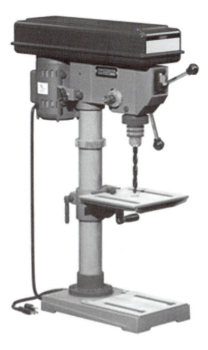

FIGURE 12.5 A typical drill press.

Creating holes in a workpiece is a relatively common requirement. Drills are particularly good for making deep holes. The **feed**, f, of a drill is the distance the drill penetrates the workpiece in one revolution of the drill. The speed of the drill is the product of its feed and rotational speed, N, or

$$V_{drill} = fN \qquad (12.9)$$

The cross-sectional area of a drill with a diameter D is $\pi D^2/4$, and the material removal rate of a drill is the product of its cross-sectional area and its speed, or

$$MMR_{drill} = (\pi D^2/4)fN \qquad (12.10)$$

Equations (12.7) and (12.8) can also be used to calculate the drilling power and torque.

Example 12.3
We need to drill a 10. mm hole in a piece of titanium alloy. The drill feed is 0.20 mm/revolution and its rotational speed is 600. RPM. Determine

(1) The material removal rate.
(2) The power required.
(3) The torque on the drill.

Need: MMR_{drill}, $P_{machining}$, and $T_{machining}$.
Know: $D = 10.$ mm, $f = 0.20$ mm/rev, and $N = 600.$ RPM.
How: The material removal rate is given by Equation (12.10). The power and torque are given by Equations (12.7) and (12.8). The machining energy per unit volume can be found in Table 12.1.

Solve: 1. Equation (12.10) gives

$$\text{MMR}_{\text{drill}} = (\pi D^2/4)fN$$
$$= (\pi 10.^2/4)(0.2)(600.)[\text{mm}^2][\text{mm}/\text{rev}][\text{rev}/\text{min}]$$
$$= 9400\text{mm}^3/\text{min} = 160\text{mm}^3/\text{s}$$

2. From Table 12.1 for titanium alloys, the machining energy per unit volume of material removed is 3.5 W · s/mm³. Then, Equation (12.7) gives

$$P_{\text{machining}} = 3.5(160)[\text{W} \cdot \text{s}/\text{mm}^3][\text{mm}^3/\text{s}] = 560\text{W}.$$

3. From Equation (12.8), the torque on the drill is

$$T_{\text{machining}} = P_{\text{machining}}/(2\pi N) = 560/(2\pi \times 600.)[\text{W}]/[\text{radians}/\text{min}]$$
$$= 0.15\text{W} \cdot \text{min} = 0.15(60)[\text{W} \cdot \text{min}][\text{s}/\text{min}] = 9.0\text{W} \cdot \text{s} = 9.0\,(\text{J}/\text{s})\text{s}$$
$$= 9.0\text{J} = 9.0\text{N} \cdot \text{m}$$

Equations (12.5) and (12.6) can also be used to determine the time it takes to drill the hole to a specific depth. For example, if we were required to drill the hole in Example 12.3 to a depth of 50. mm, then the drill feed rate would be FR = fN = 0.20(600.) [mm/rev][rev/min] = 120 mm/min and the time required to drill the hole 50. mm deep would be $t_{\text{machining}}$ = L/FR = 50./120 [mm]/[mm/min] = 0.42 min or 25 s.

12.5.1.3 Milling

Milling is carried out on a machine tool called a *milling machine* used for the shaping of metal and other solid materials. It has a cutting tool that rotates about the spindle axis similar to a drill. However, a drill moves along only one axis, but on a milling machine, the cutter and workpiece move relative to each other, generating a tool path along which material is removed along three axes. Often, the movement is achieved by moving the workpiece while the cutter rotates in one place, as shown in Figure 12.6. Milling machines may be manually operated, mechanically automated, or digitally automated via CNC (computer numerical control).

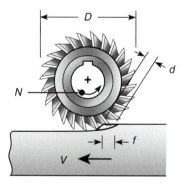

FIGURE 12.6 The characteristics of a common slab milling cutter.

Milling machines (Figure 12.7) can perform a vast number of operations, some of them with quite complex tool paths, such as slot cutting, planning, drilling, and routing.

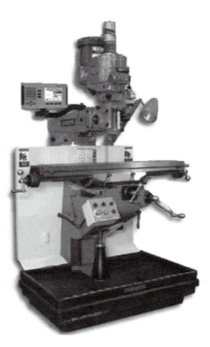

FIGURE 12.7 A typical milling machine.

Figure 12.6 illustrates the basic shape and operation of a milling cutter. The cutting speed, $V_{milling}$, of a cutter of diameter D rotating at N RPM is

$$V_{milling} = \pi D N \qquad (12.11)$$

The feed rate of the cutter is the rate at which its axis moves along the workpiece and can be calculated from the feed per tooth f, the rotational speed of the cutter N, and the number of teeth on the cutter n as

$$FR = f \times N \times n \qquad (12.12)$$

The material removal rate for a workpiece of width w, a depth of cut d, and a cutter feed rate, FR is

$$MRR_{milling} = w \times d \times FR \qquad (12.13)$$

The time required to mill a length L of material at a feed rate, FR, is

$$t_{milling} = L/FR \qquad (12.14)$$

Example 12.4

A flat piece of cast iron 4.0 inches wide and 10. inches long is to be milled with a feed rate of 20. inches/min. with a depth of cut of 0.10 in. The milling cutter rotates at 100. RPM and is wider than the workpiece. Determine

(**1**) The material removal rate.
(**2**) The machining power required.
(**3**) The machining torque required.
(**4**) The time required to machine the workpiece.

Need: $MMR_{milling}$, $P_{machining}$, $T_{machining}$, and $t_{milling}$.

Know: The material is cast iron with a length of $L=10.$ inches and a width of $w=4.0$ inches. The feed rate (FR)$=20.$ inches/min, $N=100.$ RPM, and the depth of the cut $(d)=0.10$ in.

How: Equation (12.13) provides the material removal rate. Equations (12.7) and (12.8) can be used to find the cutting power and torque, and Equation (12.14) gives the machining time.

Solve: **1.** From Equation (12.13), we get

$$MRR_{milling} = wd\mathrm{FR} = (4.0)(0.10)(20.)[\text{inches}][\text{inches}][\text{inches/min}] = 8.0\,\text{in}^3/\text{min}.$$

2. From Table 12.1 for cast iron, we find that the machining energy per unit volume of material removed is 1.2 HP·min/in^3. Equation (12.7) gives the machining power as

$$P_{machining} = 1.2(8.0)\left[\text{HP}\cdot\text{min}/\text{in}^3\right]\left[\text{in}^3/\text{min}\right] = 9.6\,\text{HP}.$$

3. From Equation (12.8), the torque on the milling cutter is

$$T_{machining} = P_{machining}/(2\pi N) = 9.6/(2\pi \times 100.)[\text{HP}]/[\text{radians/min}] = 0.015\text{HP}\cdot\text{min}$$
$$= 0.015(33000)[\text{HP}\cdot\text{min}][\text{ft}\cdot\text{lbf/HP}\cdot\text{min}] = 500\text{ft}\cdot\text{lbf}.$$

Equation (12.6) gives the machining time as

$$t_{machining} = L/\mathrm{FR} = 10./(20.)[\text{in}]/[\text{in/min}] = 0.75\,\text{min} = 45s.$$

12.5.2 Additive Processes

Additive processes involve material being added, such as in joining (e.g., welding, as in Figure 12.8, soldering, and gluing), rapid prototyping, stereolithography, 3D printing, and the application of composite layers of resin and fiber. Additive machining processes add material to a base object to create complex shapes. This is far less expensive than cutting an intricate product from a solid block of material by the subtractive process.

FIGURE 12.8 Welding with an oxyacetylene torch.

12.5.3 Continuous Processes

Continuous processes involve a product that is continuously produced, such as in the extrusion of metals and plastics and the pultrusion of composites. **Extrusion** (Figure 12.9) is a continuous process of manufacture used to create objects of a fixed cross-sectional profile by pushing or drawing material through a die of the desired cross section. In extrusion, a bar or metal or other material is forced from an enclosed cavity through a die orifice by a force applied by a ram. The extruded product has the desired, reduced cross-sectional area and a good surface finish so that further machining is not needed. Extrusion products include rods and tubes with varying degrees of complexity in cross section.

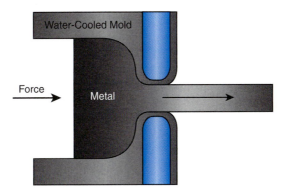

FIGURE 12.9 An extrusion process.

    Pultrusion (Figure 12.10) is a similar continuous process of manufacturing materials with constant cross section, except the material is **pulled** through a process or a die to its final shape. Pultrusion is the only continuous manufacturing process available for obtaining a high-quality composite profile, with high mechanical properties. Pultruded products are normally composed of high-performance glass or carbon fibers embedded in a polymer matrix.

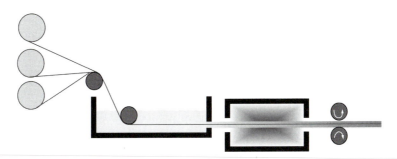

FIGURE 12.10 A typical pultrusion process.

12.5.4 Net Shape Processes

A **net shape** process occurs when the output is at or near its final shape, such as stamping, forging, casting, injection molding, blow molding, and thermoforming.

Stamping is a metalworking process in which sheet metal is formed into a desired shape by pressing or punching it on a machine press. Stamping can be a single stage operation, where only one stroke of the press produces the desired form, or could occur through a series of these stages. The most common stamping operations are piercing, bending, deep drawing, embossing, and extrusion.

Casting is a manufacturing process by which a liquid material is poured into a mold containing a cavity of the desired shape and allowed to solidify. The mold is then opened to complete the process. The casting process is subdivided into two distinct subgroups: expendable and nonexpendable mold casting. Casting is most often used for making complex shapes that would be otherwise difficult or uneconomical to make by other machining methods.

Injection molding is a manufacturing process used to make parts by injecting molten plastic at high pressure into a mold. Injection molding is widely used for manufacturing a variety of parts, from the smallest gears to entire automotive body panels. Injection molding is the most common method of production today and is capable of tight tolerances. The most commonly used thermoplastic materials are polystyrene, polypropylene, polyethylene, polycarbonate, and polyvinyl chloride (PVC, commonly used in extrusions such as pipes, window frames, and the insulation on wiring). Injection molding can also be used to manufacture parts from aluminum, zinc, or brass in a process called *die casting*. The melting points of these metals are, of course, much higher than those of plastics, but it is often the least expensive method for mass producing small metal parts.

Blow molding is a manufacturing process in which hollow plastic parts are formed. The blow molding process begins with melting the plastic and forming it into a tube-like piece with a hole in one end, into which air can be injected. It is then clamped into a mold, and air pressure pushes the plastic into the mold. Once the plastic has cooled and hardened, the mold opens up and the part is ejected.

Thermoforming (Figure 12.11) is a manufacturing process used with plastic sheets. Plastic sheets or film is converted into a finished part by heating it in an oven to its forming temperature and then stretching it on a mold and cooling it. A thermoform machine typically utilizes vacuum to draw the plastic onto the mold in the forming process.

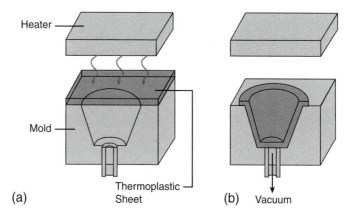

FIGURE 12.11 Thermoforming a drinking cup. A flat plastic sheet is placed over the mold and heated in illustration (a). In illustration (b), the sheet is drawn into the mold with a vacuum.

Example 12.5

The energy needed to manufacture a reusable cup is significantly greater than that required to manufacture a disposable cup, as shown in the table. For a reusable cup to be an improvement over a disposable cup on an energy basis, we have to use it multiple times. Determine the number of uses required for the energy per use of a reusable cup to become less than for a disposable cup (this is called the *breakeven point*) if it takes about 180 kJ of energy to wash a single cup in a commercial dishwasher. (Data from Professor Martin B. Hocking at the University of Victoria, Canada.)

Energy Required to Manufacture a Cup	
Cup Type	**kJ/cup**
Ceramic	14,000
Plastic	6,300
Glass	5,500
Paper	550
Foam	190

Need: The breakeven point for a reusable cup.

Know: Manufacturing energy from the given table and that it takes about 180 kJ of energy to wash a single cup in a commercial dishwasher.

How: If a cup is used 10 times, then each use costs 1/10 of the manufacturing energy. If it is used 100 times, then each use costs just 1/100 of the manufacturing energy. However, to reuse a cup, it has to be cleaned, and the energy required for each cleaning needs to be added to its manufacturing energy. The energy required per cup use can be computed from:

$$\text{Energy per use} = \text{Wash energy} + (\text{Manufacturing energy})/(\text{Number of uses})$$

Solve: Since it takes about 180 kJ of energy to wash a single cup in a commercial dishwasher, the total amount of energy per cup use is given by the preceding equation. So, how many times would you have to use a ceramic cup to "breakeven" with the manufacturing energy used in producing a foam cup?

Energy per use = 190 kJ/cup (Foam)
Wash energy = 180 kJ
Manufacturing energy = 14,000 kJ/cup (Ceramic)

Then, $190 = 180 + 14{,}000/N$, where N = number of uses (and washes) of the ceramic cup. Solving for N gives $N = 1400$. Therefore, you would have to use and wash the ceramic cup 1400 times to equal the manufacturing energy of a single foam cup.

The table that follows shows how the energy per use of the three reusable cups declines the more you use them.

Number of Times a Reusable Cup Must Be Used to Break Even with a Disposable Cup		
Reusable Cup	**Paper**	**Foam**
Ceramic	38	1,400
Plastic	17	630
Glass	15	550

These results are extremely sensitive to the amount of energy the dishwasher uses for cleaning each cup. The energy calculation for the dishwasher, requiring 180 kJ/cup per wash, is barely less than the manufacturing energy of the foam cup, 190 kJ/cup. If even a slightly less energy efficient dishwasher is used, then the reusable cups would never break even with the foam cup.

In situations where cups are likely to be lost or broken and thus have a short average lifetime, disposable cups are the preferred option *if* energy usage is the main criterion.

12.6 MODERN MANUFACTURING

What makes a good manufacturing process? Today, the overall performance of a company is often dictated by the design of its manufacturing process and facility. It turns out that four key elements are important in designing a good manufacturing process.

The first key element is the *movement of materials*. A well-designed facility results in efficient material handling, small transportation times, and short queues. This, in turn, leads to low work-in-process levels, effective production management, decreased cycle times and manufacturing inventory costs, improved on-time delivery performance, and higher product quality. The second key item is *time*. The setup time plus process time from order to shipping is extremely important in meeting customer demands and keeping a constant flow of product. The third key element is *cost*. Material, labor, tooling, and equipment costs must all be monitored and controlled. The fourth key element is *quality*. Customers today demand high-quality products, so deviations from design specifications must be kept to a minimum.

Modern manufacturing philosophies such as JIT manufacturing, flexible manufacturing, lean manufacturing, and life-cycle manufacturing contain all these elements. They are discussed in detail in the following material.

12.6.1 Just-in-Time Manufacturing

JIT manufacturing is easy to understand. Things are planned to occur *just in time*. For example, consider your activities today. You left your room *just in time* to walk to your first class. That class ended *just in time* for you to go to your next class, which ended *just in time* for you to have lunch. JIT processes may seem easy, but achieving them in a manufacturing environment is often very difficult.

In a manufacturing process, parts should arrive at the factory just in time to be distributed to the work stations, and they should arrive at a work station just in time to be installed by a worker. Finally, the factory should produce finished products just in time to be handed to a waiting customer. This eliminates any inventory of parts and products, and in theory, JIT does not need *any* inventory of raw materials, parts, or finished products.

JIT originated in Japan with the Toyota Motor Company, and JIT was initially known as the *Toyota Production System*. After World War II, the president of Toyota wanted to compete aggressively with American automobile production. What he found was that one American car worker could produce about nine times as much as a Japanese car worker. By further studying the American automobile industry, he found that American manufacturers made large quantities of each item (parts, engines, car model, etc.) before changing to a new item. American car manufacturers also stocked all the parts needed to assemble their cars.

Toyota felt that this would not work in Japan because the small Japanese market wanted small quantities of many different car models. Consequently, the company developed a production system that provided parts to the workers only when they were needed (i.e., just in time). It analyzed the concept of "waste" in general terms that included wasted time and resources as well as wasted materials. It subsequently identified the following seven wasteful activities that could be minimized or eliminated:

1. **Transportation** (moving products that are not required to perform the processing).
2. **Inventory** (all components, work-in-progress, and finished product not being processed).
3. **Motion** (workers or equipment moving more than is required to perform the processing).
4. **Waiting** (workers waiting for the next production step).
5. **Overproduction** (production ahead of demand).
6. **Overprocessing** (due to poor product design).
7. **Defects** (the wasted effort involved in inspecting for and fixing defects).

A number of Japanese jargon terms are associated with JIT that are in common use today. For example, **Andon** refers to trouble lights that immediately signal to the production line that there is a problem, and the production line is stopped until the problem is resolved. In the Toyota system, the Andon light indicating a stopped production line is hung from the factory ceiling so that it can be clearly seen by everyone. This raises the profile of the problem and encourages rapid attention to a solution. However, when General Motors instituted an Andon light, U.S. workers were reluctant to take the responsibility for stopping a production line, and defective products were produced. General Motors later solved this problem by allowing workers to signal that they had a problem without stopping the production line.

The JIT philosophy involves the elimination of waste in its many forms, a belief that ordering and stocking costs can be reduced, and always striving to progress through continuous improvement. The elements of a JIT process typically include the following:

- Meet regularly with the workers (daily/weekly) to discuss work practices and solve problems.
- Emphasize consultation and cooperation with the workers rather than confrontation.
- Modify machinery to reduce setup time.
- Reduce buffer stock.
- Expose problems, rather than have them covered up.

It is not necessary to apply JIT to all stages of a process. For example, we could keep large stocks of raw material but operate the production process internally in a JIT fashion (hence eliminating work-in-progress stocks).

12.6.2 Flexible Manufacturing

A flexible manufacturing system (FMS) is a manufacturing system that contains enough flexibility to allow the system to rapidly react to production changes. This flexibility is generally considered to fall into two categories. The first is machine flexibility. This allows the system to be changed to produce new product types and to change the order of operations executed on a part. The second is routing flexibility. This consists of the ability to use multiple machines to perform the same operation on a part as well as the system's ability to absorb large-scale changes, such as in volume, capacity, or capability.

Most FMS systems consist of three main systems: a **material handling system** to optimize the flow of parts, a **central control computer** that controls material movement, and the **working machines** (often automated CNC machines or robots).

The use of robots in manufacturing industries has many benefits. Each robotic cell is connected with a material handling system, which makes it easy to move parts from one robotic cell to another. At the end of processing, the finished parts are routed to an automatic inspection cell and removed from the flexible manufacturing system. This process is illustrated in Figure 12.12.

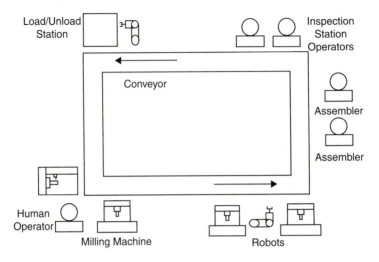

FIGURE 12.12 A flexible manufacturing system.

The main advantage of an FMS is its high flexibility in managing manufacturing resources like time and effort to manufacture a new product. The best application of an FMS is found in the mass production of small sets of products.

12.6.3 Lean Manufacturing

Lean manufacturing is the optimal way of producing goods through the removal of waste and implementing flow, as opposed to batch processing. Lean manufacturing is a generic process management philosophy derived mostly from Toyota and focuses mainly on reduction of the seven wastes originally identified by Toyota (see Section 12.6.1, on JIT).

Lean manufacturing is focused on getting the right things to the right place at the right time in the right quantity to achieve perfect work flow while minimizing waste and being flexible and able to change. All these concepts have to be understood, appreciated, and embraced by the workers who build the products and own the processes that deliver the value. The cultural and managerial aspects are just as important as, and possibly more important than the actual tools or methodologies of production itself. Lean manufacturing tries to make the work simple enough to understand, to do, and to manage.

The main principles of lean manufacturing are zero waiting time, zero inventory, internal customer pull instead of push, reduced batch sizes, and reduced process times.

12.6.4 Life-Cycle Manufacturing

Today, the term *green* is synonymous with environmental sustainability. Sustainability is the development and application of processes or the use of natural resources that can be maintained indefinitely. As public environmental awareness increased, manufacturing facilities began examining how their activities were affecting the environment.

To understand the environmental impact of a manufacturing process, you need to examine all the inputs and outputs associated with the entire existence of the product. This "cradle to grave" analysis is called **life-cycle analysis**, or simply LCA. LCA normally ignores second-generation impacts, such as the energy required to fire the bricks used to build the kilns used to manufacture the raw material. The concept of conducting a detailed examination of the life cycle of a product or a process in response to increased environmental awareness on the part of the public, industry, and government is relatively recent.

Just as living things are born, age, and die, all manufactured products have an analogous life cycle. Each stage of a product's development affects our environment, from the way we use products to the way we dispose of them when we are finished with them (see Figure 12.13). Looking at a product's life cycle helps engineers understand the connections between the Earth's natural resources, energy use, climate change, and waste disposal. Product life-cycle analysis focuses on the processes involved in the entire production system, including

- The design and functionality of the product.
- The extraction and processing of raw materials.
- The processes used in manufacturing.
- Packaging and distribution.
- How the product is used by the customers.
- Recycling, reuse, and disposal of the product.

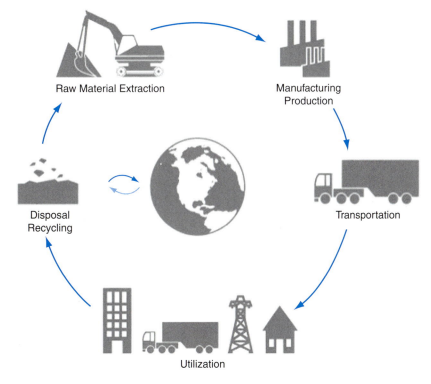

FIGURE 12.13 Life-cycle manufacturing.

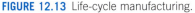

Engineers use life-cycle analysis during product design and development, and manufacturers use it during the production stage. However, LCA has its greatest potential to reduce environmental impacts during the design stage when 70% of the total product cost is determined (see Figure 12.14).

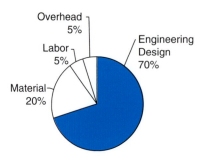

FIGURE 12.14 The influence of engineering design on the cost of manufacturing.

Life-cycle analysis uses detailed environmental and energy estimates for the manufacture of a product, from the mining of the raw materials to production and distribution, end use and possible recycling, and disposal. LCA enables a manufacturer to measure how much energy and raw materials are used; and how much solid, liquid, and gas waste is generated at each stage of a product's life.

12.7 VARIABILITY AND SIX SIGMA

Unintentional variability is a fact of life. As a simple example, it's instructive to have the students in the class each measure the same thing with the same instrument. We find that there will be slightly different results. For example, take a brand new unsharpened pencil and a metal ruler inscribed in cm. and mm. Each member of the class should measure the length of the pencil and record the number. Then, when all data have been collected, write all the measurements on the blackboard. Table 12.2 shows some typical numbers.

Table 12.2 Raw measurements of pencil length

Length (mm)	Length (mm)	Length (mm)
21.2	22.7	21.0
21.0	20.1	21.2
20.7	20.9	20.8
21.7	20.9	21.2
21.7	21.5	21.2

Note that the measured lengths are not precisely the same. Indeed, one person reported a number much greater than all the other measurements (22.7 mm) and another measured 20.1 mm, which is also quite a bit smaller than most of the other measurements. **Outliers** are quite common due to inattention or misreading of instruments.

Which reading is correct? Any, none, or some combination? Most of us would agree that the average or mean defined as the total of all the measurements divided by the number of measurements has less bias than does any individual measurement.

Let's analyze what we have measured. The mean or average measured length is

$$\bar{x} = \left(\sum_{i=1}^{N} \frac{x_i}{N} \right) = 21.2 \text{mm}$$

So, you say the pencil is 21.2 mm long. But is it? Suppose we drop the outlier of 22.7 mm. The mean of the remaining data is then 21.1 mm. Is that correct? Or, should we drop the other outlier of 20.1 mm, giving a new mean of 21.3 mm. Again, is that correct?

Let's try something else. Subtract each measurement from the mean of 21.2 mm, as in Table 12.3.

Table 12.3 Mean value of pencils minus measured length

Mean—Measurement (mm)	Mean—Measurement (mm)	Mean—Measurement (mm)
0.0	−1.5	0.2
0.2	1.1	0.0
0.5	0.3	0.4
−0.5	0.3	0.0
−0.5	−0.3	0.0

If you add all these "deviations," they will total 0.0 (with ±round-off error in the average), since after all, this is what we mean by an average. So, let's try a different approach—plot these deviations as a frequency histogram with the frequency of each measurement (say, in discrete 0.2 mm "bins") versus the measured values, as in Figure 12.15.

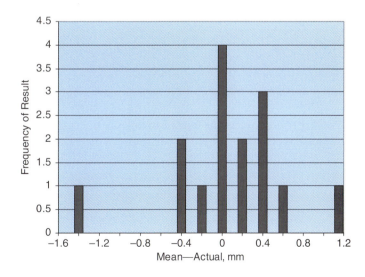

FIGURE 12.15 Histogram of deviations from mean.

The histogram shows that the most frequent measurements are centered about zero so that many, but not all, of the people in the class did think the actual length of the pencil was about 21.2 mm.

But, is ignoring all those who did not measure exactly 21.2 mm the best we can do? Suppose we square these deviations—some of which were positive and some which were negative—so that each squared term is then positive. An individual squared term is still a poor measure of all the data points. We need a measure that eliminates the bias of a single measurement just as the mean reduced the bias of a single number. So, add all the squares, take the square root of them, and then divide by the number of points.[11]

$$\sigma = \sqrt{\frac{\sum_{i=1}^{N}(\bar{x}-x_i)^2}{N}} \qquad (12.15)$$

The Greek letter sigma σ defined in Equation (12.15) is called the **standard deviation** of the data. In our case, $\sigma = 0.6$ mm, but what does that signify?

A typical statistical[12] analysis of variables starts with a large sample generalization of the histogram in Figure 12.15 known as the **normal error curve**, the **bell curve**, the **normal distribution curve**, or the **Gaussian**[13] **distribution** (Figure 12.16).

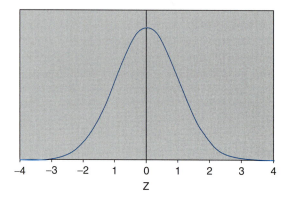

FIGURE 12.16 Normal error curve or normal distribution curve.

The ordinate is the probability of occurrence (the fraction of times or frequency) that the measurements lie in some small range. The abscissa Z is defined by the difference between the mean divided by the standard deviation, or

$$Z = \frac{(x_i - \bar{x})}{\sigma} \qquad (12.16)$$

---

[11]For mathematical reasons, if we use all N points of data to determine the mean of the data, we should divide by $N-1$ rather than N; however, for proper use of statistical methods, the number of samples should be "large" and whether we divide by N or $N-1$ should made no difference. If it does make a noticeable difference, you should reevaluate whether you should be using statistical methods at all! $N \geq 30$ is usually sufficient.

[12]Mathematical statistics was one of the methods introduced to Japanese manufacturers by an American statistician, W. Edwards Deming, just after World War II to improve the quality of their products; it has been so successful that many large international companies emulated them.

[13]Carl Friedrich Gauss (1777-1855) was one of the greatest mathematicians of all time.

The theoretical equation describing the ordinate is:

$$\text{Frequency} = \frac{\exp(-0.5Z^2)}{\sigma\sqrt{2\pi}} \tag{12.17}$$

The quantity defined by Equation (12.17) is usually referred to as a **normal distribution**.

Why go to these lengths to plot this curve in normal coordinates? The reason is that, with this coordinate system, the area under the curve (taken out to $Z = \pm\infty$) is 1. This means that there is 100% probability that the measurement is somewhere under the curve. In this standardized form, the abscissa can be thought of as the number of standard deviations corresponding to each point. The utility of this normalized coordinate is that

- For $Z = \pm 1$ standard deviations, the area under the normal error curve contains **68%** of the data.
- For $Z = \pm 2$, the area contains **95%** of the data.
- for $Z = \pm 3$, the area contains **99.7%** of the data.

This curve has the same general characteristics as the experimental points in Figure 12.15 although there are too few data points for an exact match.

Example 12.6

Convert the numerical data in Table 12.3 to Z coordinates. Then, plot its frequency versus Z. (This should be compared to the curve in Figure 12.16.)

Need: The frequency of occurrence for a normal distribution of pencil length measurements.
Know-How: It's easiest using a spreadsheet. (Patience! See Example 12.7.)
Solve: Use a spreadsheet to calculate and plot the results.

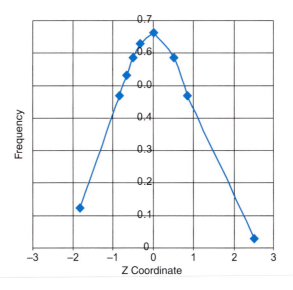

The areas under the tails of the normal error curve at abscissa values greater than $\pm 1\sigma, \pm 2\sigma, \pm 3\sigma$, and so forth contain the fraction of measurements that fail because they are too far from the mean. "Everyday" statistics, such as medical tests protocols and electorate polling, are usually accepted when 95% (2σ) of the data are included under the curve.

A "Six Sigma"[14] process means you are rejecting parts, processes, and the like for which $Z \geq \pm 6$. This means that very few defective parts are out of tolerance, since at these large abscissa values, the area under the tails of the normal error curve is very small (about 3.4 per million[15]). This can have a profound effect on costs.

Example 12.7

We manufacture 10,000 metal rods that we hope are 21.2 ± 0.2 mm long at a cost to us of $1.00 for each blank. How much each do the *acceptable* rods cost if their final machined measurements are normally distributed with a standard deviation of 0.6, 0.3, and 0.1 mm, respectively? In other words, what fraction of the rods' measurement is expected to be too far off the desired mean? Note: Your spreadsheet will have several functions that calculate areas under the normal error curve given Z specifications.

Need: Cost of the acceptable metal rods that are within the desired lengths between 21.0 and 21.4 mm and a mean of 21.2 mm.

Know: Lengths are normally distributed with $\sigma = 0.6$, 0.3, and 0.1 mm, respectively.

How: Use fraction of acceptable rods between $Z = \pm 0.2/\sigma$ or 0.333, 0.667, and 2.0 corresponding to a mean length of 21.2 mm and standard deviations of 0.6, 0.3, and 0.1 mm, respectively.

Solve: In Excel, the descriptions of the normal error curve are particularly murky. For a clearer explanation of NORMDIST and the NORMSDIST Excel functions, see http://www.exceluser.com/explore/statsnormal.htm, by Charley Kyd.

	A	B	C	D	E	F	G	H	I
1	Mean.mm	spread.mm	Std dev.mm	Z	Normsdist(Z)	Fraction passing	Number passing	Cost/rod	
2	21.2	0.2	0.6	0.333	0.631	0.261	2,611	$3.83	
3	21.2	0.2	0.3	0.667	0.748	0.495	4,950	$2.02	
4	21.2	0.2	0.1	2.00	0.977	0.954	9,545	$1.05	
5	21.2	-0.2	0.6	-0.33	0.369	0.261	2,611	$3.83	
6	21.2	-0.2	0.3	-0.67	0.252	0.495	4,950	$2.02	
7	21.2	-0.2	0.1	-2.00	0.023	0.954	9,545	$1.05	
8									
9		See:	http://www.exceluser.com/explore/statsnormal.htm						
10									
11	NORMSDIST(z)								
12									
13	NORMSDIST translates the number of standard deviations(z) into cumulative probabilities								
14									
15					.50				
16					.40				
17					.30				
18					.20				
19					.10				
20					.00				
21	To illustrate:			-4 -3 -2 -1 0 1 2 3 4					
22									
23	=NORMSDIST (-1) = 15.87%								
24									
25									
26									
27	Therefore, the probability of a value being within one standard deviation of the mean is the difference between these values of 68.27%								
28	This range is represented by the shaded area of the chart.								

	A	B	C	D	E	F	G	H
1	Mean, mm	Spread, mm	Std dev, mm	Z	Normsdist(Z)	Fraction passing	Number passing	Cost/rod
2	21.2	0.2	0.6	=B2/C2	=NORMSDIST(D2)	=IF(D2<=0, 1-2*E2, 1-(1-E2)*2)	=10000*F2	=10000/G2
3	21.2	0.2	0.3	=B3/C3	=NORMSDIST(D3)	=IF(D3<=0, 1-2*E3, 1-(1-E3)*2)	=10000*F3	=10000/G3
4	21.2	0.2	0.1	=B4/C4	=NORMSDIST(D4)	=IF(D4<=0, 1-2*E4, 1-(1-E4)*2)	=10000*F4	=10000/G4
5	21.2	-0.2	0.6	=B5/C5	=NORMSDIST(D5)	=IF(D5<=0, 1-2*E5, 1-(1-E5)*2)	=10000*F5	=10000/G5
6	21.2	-0.2	0.3	=B6/C6	=NORMSDIST(D6)	=IF(D6<=0, 1-2*E6, 1-(1-E6)*2)	=10000*F6	=10000/G6
7	21.2	-0.2	0.1	=B7/C7	=NORMSDIST(D7)	=IF(D7<=0, 1-2*E7, 1-(1-E7)*2)	=10000*F7	=10000/G7

---

[14]Six Sigma is a registered service mark and trademark of Motorola, Inc. Motorola reported over $17 billion in savings from Six Sigma as of 2006.

[15]There is some mathematical "chicanery" going on here because $Z \geq \pm 6\sigma$ actually means just two out of specification per *billion* tested. But one failure in 3.4 *million* is nothing to sneeze at!

Several points need to be made: rows 2, 3, and 4 actually answer the problem. Rows 5, 6, and 7 are included to show that the normal error curve is symmetrically distributed about the origin. It's also slightly easier just to double the tail area (see column F) of the normal error curve for a given negative Z to get the total (two-tailed) number of excluded metal rods.

Of special interest in this problem is the cost of acceptable rods. The blanks cost $1.00 each. If the machining is sloppy and we have to discard those with $Z=\pm0.333$, the net cost of rods in specifications is $3.89 each, since only 26.1% pass specifications. For $Z=\pm0.667$, just 49.5% pass and the cost is $2.02. However, if we can machine them to $Z=\pm2.0$ (meaning the standard deviation is smaller), the fraction passing is 95.4%, and the cost falls to $1.05 per rod.

The message of Example 12.7 is very clear: quality is not only good for the consumer, but it is economical for the manufacturer as well. In this respect, it is twice blessed.

Example 12.8

Your boss has a self-proclaimed improvement in your company's most profitable line, and to prove it, you are given the opportunity to test the product before and after your boss's alleged improvements. You test more than 50 of both the imputably improved widgets and the existing widgets. The data are given graphically in the figure marked Experiment Set A. The circles are the imputably improved widgets and the crosses are the original ones. Has your boss really improved the widget? (The heavy line is the mean of the data, and the dashed lines are $\pm2\sigma$ bounds, meaning that 95% of the data are within those bounds).

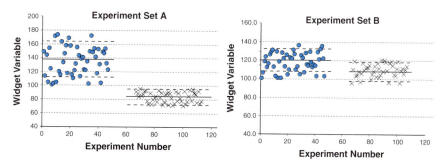

In a second trial, marked Experiment Set B, different widgets are subjected to comparative measurements. Again, the data with the circles are your boss's claimed improvements and the crosses are the original widget. Again, do the data show a real improvement?

Need: Are the data different in experiment A? (yes/no) Are the data different in experiment B? (yes/no)
Know: By eye, the means are different in between the experiments (139 vs. 82 in set A and 120 vs. 110 in set B).
How: 95% of the data are bounded by two sigma lines.
Solve: In set A, the dashed 2σ bounds do not overlap, so the difference between the means is probably real (i.e., statistically significant). So, in set A, the boss is correct—his widgets are better and the difference in the means (139 vs. 82) is real. (But, you can and should argue that the scatter in his data could have expensive downstream implications.).

In set B, the means (120 and 110) are apparently different, but their respective scatters overlap; this suggests that the data are probably not statistically different.

Answer: In set A, there is real improvement; in set B, there is no improvement. In set A, there is a potential price to pay in a larger standard deviation.

These simple graphical illustrations are related to some rigorous mathematical tests that answer whether "noisy" data are statistically the same or different. In most cases, the graphical method shown here is either sufficient in itself to distinguish the real from the noise or at least a way to visualize the situation before applying book formulas that accomplish the same thing.

SUMMARY

All products have some impact on the environment. Since some products use more resources, cause more pollution, or generate more waste than others, the aim is to identify those that are most harmful.

Even for those products whose environmental burdens are relatively low, a **life-cycle analysis** should help to identify those stages in production processes and in use that cause or have the potential to cause pollution and those that have a heavy material or energy demand.

Breaking down the manufacturing process into such fine detail can also be an aid to identifying the use of scarce resources, showing where a more sustainable product could be substituted.

Statistical methods are important for the control of processes. Manufacturing processes are not exactly on specification but have a scatter about their **mean**. The variability of such processes follows Gauss's **normal distribution**. The most important variable that describes the scatter in the consistency of manufacturing processes is the **standard deviation**. It relates to the number of rejects in a normal distribution of manufactured objects. The key variable is the normalized measurement, given by **Z**, which is defined as the measured deviation/standard deviation. For a normalized variable in which Z is between $\pm 1\sigma$, **68%** of the data is within specifications. For data in which Z is between $\pm 2\sigma$, **95%** is within specifications; and for data in which Z is between $\pm 3\sigma$, **99%** is within specifications.

EXERCISES

1. Discuss the building materials used by the three little pigs (straw, sticks, and bricks). Why were they chosen? Why did they fail? What was the environmental impact?

2. A casting process involves pouring molten metal into a mold, letting the metal cool and solidify, and removing the part from the mold. The solidification time is a function of the casting volume and its surface area, known as *Chvorinov's Rule*:

$$\text{Solidification Time} = K \times (\text{Volume}/\text{Surface Area})^2$$

where K is a constant that depends on the metal. Three parts are to be cast that have the same total volume of $0.015\ \mathrm{m}^3$, but different shapes. The first is a sphere of radius R_{sph}, the second is a cube with a side length L_{cube}, and the third is a circular cylinder with its height equal to its diameter ($H_{\text{cyl}} = D_{\text{cyl}} = 2R_{\text{cyl}}$). All the castings are to be made from the same metal, so K has the same value for all three parts. Which piece will solidify the fastest? The table that follows gives the equations for the volume and surface area of these parts.

Object	Surface Area	Volume
Sphere	$4\pi R_{\text{sph}}^2$	$(4/3)\pi R_{\text{sph}}^3$
Cube	$6L_{\text{cube}}^2$	L_{cube}^3
Cylinder	$2\pi R_{\text{cyl}}^2 + 2\pi R_{\text{cyl}} H_{\text{cyl}} = 6\pi R_{\text{cyl}}^2$	$\pi R_{\text{cyl}}^2 H_{\text{cyl}} = 2\pi R_{\text{cyl}}^3$

(**A.:** The cube will solidify the fastest, and the sphere will solidify the slowest.)

3. For a particular casting process, the constant K in Chvorinov's equation in Exercise 2 is 3.0 s/mm^2 for a cylindrical casting 150 mm high and 100 mm in diameter. Determine the solidification time for this casting.

4. Tool wear is a major consideration in machining operations. The following relation was developed by F. W. Taylor for the machining of steels:

$$VT^n = C$$

where V is the cutting speed; T is the tool life; the exponent n depends on the tool, the material being cut, and the cutting conditions; and C is a constant. If $n = 0.5$ and $C = 400$ mm/min$^{0.5}$ in this equation, determine the percentage increase in tool life when the cutting speed is reduced by 50%. (A: The tool life increases by 300%.)

5. For a particular machining operation, $n = 0.6$ and $C = 350$ mm/min$^{0.5}$ in the Taylor equation given in Exercise 4. What is the percentage increase in tool life when the cutting speed V is reduced by (a) 25% and (b) 74%?

6. Using Taylor's equation in Exercise 4, show that tool wear and cutting speed are related by

$$T_2 = T_1(V_1/V_2)^{1/n}$$

7. A shaft 150 mm long with an initial diameter of 15. mm is to be turned to a diameter of 13.0 mm in a lathe. The shaft rotates at 750. RPM in the lathe, and the cutting tool feed rate is 100. mm/min. Determine
   (a) The cutting speed.
   (b) The material removal rate.
   (c) The time required to machine the shaft.

8. Repeat the calculations in Example 12.2 for a magnesium alloy being machined at 700 RPM.

9. Determine the machining power and cutting torque required to machine a magnesium shaft if the material removal rate, MMR, is 0.1 in.^{3}/min and $N = 900$. RPM.

10. Determine the machining time required to turn a 0.2 m long shaft rotating at 300. RPM at a tool feed of 0.2 mm/rev.

11. A lathe is powered by a 5 HP electric motor and is running at 500. RPM. It is turning a 1.0 inch cast iron shaft with a depth of cut of 0.035 in. What maximum feed rate can be used before the lathe stalls?

12. A 2.0 inch diameter carbon steel shaft is to be turned on a lathe at 500. RPM with a 0.20 inches depth of cut and a feed of 0.030 in/rev. What minimum horsepower and torque must the lathe have to complete this operation?

13. Suppose the material used in Example 12.3 is an aluminum alloy and the drill rotational speed is 500. RPM. Recalculate the material removal rate, the power required, and the torque on the drill.

14. A drill press with a 0.375 inch diameter drill bit is running at 300. RPM with a feed of 0.010 in./rev. What is the material removal rate?

15. You need to drill a 20.0 mm hole in a piece of stainless steel. The drill feed is 0.10 mm/revolution and its rotational speed is 400. RPM. Determine
   (a) The material removal rate.
   (b) The power required.
   (c) The torque on the drill.

16. Repeat the calculations in Example 12.4 for stainless steel instead of cast iron. Use the same material dimensions and operating conditions and compare the machining times for the two materials.

17. A flat piece of cast iron 100. mm wide and 150. mm long is to be milled with a feed rate of 200. mm/min with a depth of cut of 1.0 mm. The milling cutter rotates at 100. RPM and is wider than the workpiece. Determine
    (a) The material removal rate.
    (b) The machining power required.
    (c) The machining torque required.
    (d) The time required to machine the workpiece.

18. A milling operation is carried out on a 10. inches long, 3.0 inches wide slab of aluminum alloy. The cutter feed is 0.01 inches/tooth, and the depth of cut is 0.125 in. The cutter is wider than the slab and has a diameter of 2.0 in. It has 25 teeth and rotates at 150. RPM. Calculate (a) the material removal rate, (b) the power required, and (c) the torque at the cutter.

19. A part 275. mm long and 75. mm wide is to be milled with a 10-toothed cutter 75 mm in diameter using a feed of 0.1 mm/tooth at a cutting speed of $40. \times 10^3$ mm/min. The depth of cut is 5.00 mm. Determine the material removal rate and the time required to machine the part.

20. In the face-milling operation shown in the figure that follows, the cutter is 1.5 in. in diameter and the cast iron workpiece is 7.0 in. long and 3.75 in. wide. The cutter has eight teeth and rotates at 350. RPM. The feed is 0.005 in./tooth and the depth of cut is 0.125 in. Assuming that only 70% of the cutter diameter is engaged in the cutting, determine the material removal rate and the cutting power required.

End Mill

21. In Example 12.5, we did not take into account recycling or disposal costs or benefits. Do you think his conclusions would change if these were included? Plot the net energy per use versus the number of uses for the reusable and disposable cups.

22. The equation for a normal error curve is

$$\frac{\exp\left(\left(-\frac{1}{2}\right)\left(\frac{(x-\mu)}{\sigma}\right)^2\right)}{\sigma\sqrt{2\pi}}$$

or in spreadsheet script: exp(−0.5*((x−μ)/σ)²)/(σ*sqrt(2*pi)).
Plot a normal distribution curve for the set of 100 numbers from 0 to 99. (Hint: Use the Chart-Column graphical representation. Observe its shape and calculate its mean and standard distribution. See Exercise 23.)

23. Students often are confused by the difference between a normal distribution curve, as in Exercise 22, and a *random* distribution. Excel has a variable, rand(), that generates random numbers between 0 and 1. (See the

Help menu for how to use it.) Plot a curve for the set of random numbers from 0 to 99. (**Hint:** Use the Chart-Column representation. Observe its shape and calculate its mean and standard distribution.)

24. Three-hundred widgets are manufactured with the following normal error distribution statistics: mean = 123 units and standard deviation = (a) 23 units, (b) 32.5 units, and (c) 41 units. How many units measure less than 140. units?

25. We wish to fit steel rods (mean, $\mu = 21.2$ mm; $\sigma = 0.20$ mm) into the jig that follows to make sure we can meet specifications. The process requires that 97.5% of all rods are to be within specifications. (Presume that no rod is so short that it cannot pass specifications.) What is the appropriate size L mm of the jaws of the jig? (Hint: 95% of (two-sided) normally distributed data are found within $\pm 2\sigma$ from the mean and therefore 2.5% of the rods are too large.)

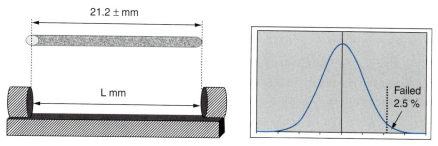

26. The jig in the Exercise 25 is OK, except now too many undersized rods are now passing specifications. The manufacturing section of your company wants to eliminate all rods that are 0.2 mm or more undersized. What fraction now passes specifications?

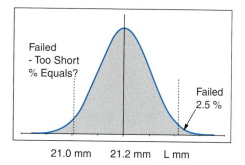

27. You perform 100 tests on one of two prototypes of the X15-24 line of products and a colleague performs another 100 tests on the other prototype.

Test Set Statistic	Your Test	Colleague's Test
Mean, μ	125.7	123.5
Standard deviation, σ	1.50	13.3

Your data are circles and your colleagues are crosses in the figure that follows.

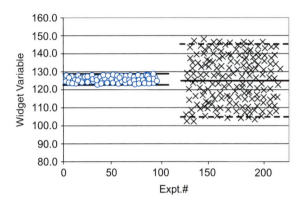

Your colleague notes that, because the means of the measurements are virtually identical, there is no difference between the two prototypes and therefore hers should be used because it will be cheaper to manufacture (which you concede). What counterarguments can you muster?

28. Quality control is more personal when the products are hand grenades, and it's your turn to learn how to throw one. You understand that, once you pull the pin, if you hold on too long, you might be severely injured or killed. However, if the fuse burns too long after you have thrown it, the enemy might have time to pick it up and throw it back at you, with results similar to holding it too long. Suppose the fuse burn time is designed to be 4.00 s with a standard deviation of 0.2 s and the grenade has been made to Six Sigma standards. What is the variability you can expect on the time for it to detonate after pulling its safety pin?

29. You are a new quality control engineer at a company that manufactures bottled drinking water. All the bottles and the filling water are checked hourly to make sure that there are no contaminants. Monday morning you notice that, for some reason, the water that was used to wash the bottles before filling was not tested over the weekend. Now you have several carloads of product ready to be shipped. What do you do? (Use the Engineering Ethics Matrix.)
(a) Have the shipment destroyed and start filling the order over again.
(b) Test random samples for the shipment, and if they pass, send the shipment.
(c) Tell your supervisor and let him or her decide what to do.
(d) Since the wash water has never been contaminated in the past, do nothing and release the shipment.

30. As a young engineer, you have been told not to trust the machinists who work for you. They make up excuses for bad parts that aren't true because they are lazy or incompetent. One day a machinist tells you his milling machine is "out of calibration" and that he can't make parts to specification unless the machine is repaired. What do you do? (Use the Engineering Ethics Matrix.)
(a) Replace him with a more competent machinist.
(b) Ask the machine shop supervisor to check the milling machine to see if there is a problem.
(c) Ask the machinist to show you exactly what he means by demonstrating the problem.
(d) Tell your supervisor you need a raise if you have to work with idiots like this.

Materials Engineering

Source: iStockphoto.com/Alwyn Cooper.

13.1 INTRODUCTION

Engineers select materials. Should a refrigerator case be made of steel or plastic? Should armor plate be a single sheet of steel or a lighter layered composite alloy? Should a transistor be made out of germanium or silicon? Should a telephone cable be made of copper or fiber optic glass? Should an artificial hip joint be made out of metal and, if metal, should it be titanium or stainless steel, or would a polymer composite be better? These materials selection problems are further examples of constrained optimization. An engineer finds the solution that best meets given criteria while also satisfying a set of constraints.

13.2 CHOOSING THE RIGHT MATERIAL

The criterion to be optimized when choosing the right material for an engineering application might be cost, weight, or performance or some index such as minimizing weight × cost with a requirement of a particular strength. The constraints (also called design requirements) typically involve such words as **elastic modulus** (which is closely related to **stiffness**), **elastic limit**, **yield strength** (sometimes abbreviated as plain "strength"), and **toughness**. While we all have loose ideas of what is meant by these terms, it is necessary to precisely express them as engineering variables. Those variables must, in turn, contain the appropriate numbers and correct units.

To help develop appropriate variables, numbers, and units, this chapter will introduce a new tool: the **stress-strain diagram**. It is a tool for defining elasticity, strength, and toughness as engineering variables, as well as a tool to extract the numerical values of those variables in materials selection.

Working through the examples and problems in this chapter will enable you to:

(1) Define **material requirements**;
(2) Relate two important classes of materials, **metals and polymers;**
(3) Understand the **internal microstructure of materials** that can be **crystalline and/or amorphous;**

(4) Use a stress-strain diagram to express materials properties in terms of the **five engineering variables:** stress, strain, elastic limit, yield strength, and toughness; and

(5) Use the results determined for those properties to carry out materials selection.

Humankind's reliance on the properties of materials for various applications, from weapons to shelter, has been around since the dawn of the ascendancy of our species.[1]

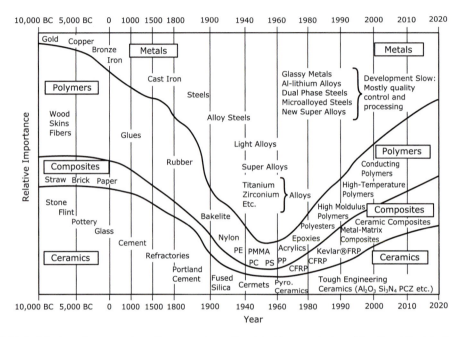

FIGURE 13.1 Materials since the accession of homo sapiens. Reprinted by permission of the Royal Society (London) and of Professor Ashby.

Figure 13.1 illustrates how various materials have evolved over time. The earliest were naturally occurring elements[2] and compounds, followed by what today is called the "Edisonian"[3] method (i.e., the method of trial and error), and today's modern materials designed by systematic investigation. Our interest here is to discover the principles behind new materials. Notice how the modern materials are classified by type: metals, polymers, composites, and ceramics. For the sake of brevity, we will confine ourselves to just the first two of these classes: metals and polymers.

Presumably you already have a picture of what is meant by a *metal*. Generally, metals are strong and dense. Metals also reflect light and conduct heat and electricity. On the other hand, polymers, popularly known as "plastics," are relatively weak, sometimes opaque and sometimes transparent, and generally do not conduct heat or electricity. We will shun the term *plastic* in favor of *polymer* in describing these materials, since as we shall see the word *plastic* is reserved to describe a particular behavior in materials.

---

[1]Ashby, M. F., Technology of the 1990s: Advanced Materials and Predictive Design. Phil. Trans. R. Soc. Lond., A322, 393, (1987).
[2]Most metals were discovered in ascending order of their melting points.
[3]Another homage to that inventive giant of the nineteenth century, Thomas Edison.

13.3 STRENGTH

What are the reasons for a material's strength? Material properties are directly related to their molecular properties and hence to the properties of their constituent atoms. The only fundamental example we will attempt to calculate is the breaking strength of a material, such as a piece of pure iron, as determined by its molecular structure. This calculation thus represents the upper limit to its strength. All failures at lower strength values are due to some defect in the material or its structure.

We will make a quick review of material structure such as whether a material is **amorphous** (meaning no structure observable at the microscopic level, which is 10^{-6} meters and smaller) or whether it is **crystalline** (which means it consists of definite microstructures that can be seen under a low power microscope or even by the naked eye).

Suppose you clamped the ends of a piece of pure iron and pulled it as shown in Figure 13.2 until it eventually breaks.

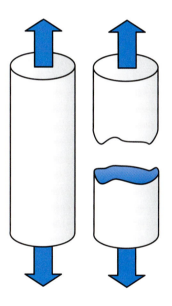

FIGURE 13.2 Failure under tension.

A straight pull is called **tension**. If the nominal cross-sectional area of the break is known, and the number of atoms in that plane is also known, one can, in principle, calculate the force to separate the atoms at the break zone and also the work required (i.e., force × distance) to break it.

In a crystal of iron, the geometry is particularly simple. A **crystal** is a representative repeat pattern of the atoms that make up the structure. Figure 13.3 shows the arrangement of the atoms in a perfect crystal of iron. If the crystals are large enough (i.e., there are trillions and trillions of atoms all arranged in the same pattern), these crystals can often be seen with low-power optical microscopes, and their atomic structure can be deduced using X-rays.

The crystal structure will arbitrarily extend in every direction in this idealized model (as in a single crystal). At the fracture plane shown in Figure 13.3, this works out to contain the planar projection of 1.8×10^{19} atoms per m^2.

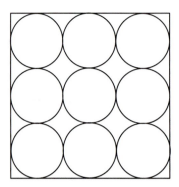

FIGURE 13.3 Structure of iron crystal. The centers of the atoms are 0.234×10^{-9} meters apart.

All we need to know now is the strength of each bond, and we will have a model of the theoretical strength of this material. We can estimate this quite easily by doing a simple thought experiment: Imagine you started with a lump of iron and then you heated it until it evaporated (just like boiling water but much hotter). The total energy absorbed represents the energy to break all the bonds in the solid iron and make individual atoms detach from the solid, and which we will use as a crude measure of the bond strength of all of those atoms in the original piece of iron. This works out to be about 6.6×10^{-19} J/atom of iron. Therefore, the *work* per unit area required to fracture a piece of iron is just about 12 J/m^2. This sounds quite modest.

Calculating the force required to break the material is another matter. Recall that work is force $\times$ distance. So, we need to consider how far do the atomic planes in iron need to be pulled apart to consider them fully separated. Atomic forces fall off very rapidly with distance and a reasonable estimate of the minimum distance required to produce fracture is 0.1-0.2 nm *beyond* the equilibrium center-to-center separation of the individual iron atoms (0.234 nm), as shown in Figure 13.4. At that additional separation, the individual iron atoms will no longer interact, and fracture has occurred.

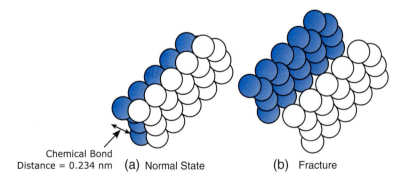

Chemical Bond
Distance = 0.234 nm   (a) Normal State   (b) Fracture

FIGURE 13.4 Before and after view of fracture along an atomic plane in iron.

The origin of the attractive force in metals is found in their atomic structure. Metals have free (negatively charged) electrons inhabiting the spaces between (positively charged) metal ions (see Figure 13.5). This electronic structure provides strong forces between the iron atoms and also explains why it is easy to get electrons to flow in metal wires.

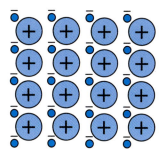

FIGURE 13.5 Metallic bonds.

We will equate the work done in separating the atoms beyond their equilibrium positions to a force × an *assumed* distance of 0.15 nm of separation. Using the 12 J/m^2 fracture work per unit area calculated above,

$$\text{Work/Area} \,(12\,\text{J/m}^2) = \text{Force/Area} \,(\text{N/m}^2) \times \text{Distance} \,(0.15 \times 10^{-9}\,\text{m})$$
$$\text{Therefore,} \quad \text{Force/Area} = 8.0 \times 10^{10} \approx 100 \times 10^{9}\,\text{N/m}^2$$

The units N/m^2 have the name **Pascal** [4] (abbreviated as **Pa**), but we are generally interested in large numbers and therefore use the *giga* prefix of 10^9. Our answer is therefore about 100 GPa (G being the prefix for *giga*). This force per unit area is equivalent to piling 100 billion apples on top of a 1 m × 1 m tray[5] here on Earth, and it is a much larger force/area than observed in practice.

Why was our estimate not seen experimentally? It's because real crystals are finite in extent and are usually randomly oriented. Figure 13.6 shows a polished piece of copper as viewed under a microscope. Individual random crystals of copper are quite obvious.

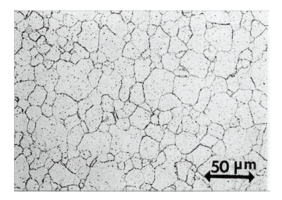

FIGURE 13.6 Polished copper showing individual crystals. (Courtesy of the Copper Development Association, Inc).

Fracture may occur between crystals at their relatively weak **grain boundaries** rather than within the crystals, resulting in lower fracture forces than the previous mathematical model.

In addition, not all materials are crystalline. Some have a completely random **amorphous** structure. Many materials may also consist of mixed phases such as crystals in a matrix of amorphous material.

---

[4] A French mathematician, scientist, inventor, and philosopher.
[5] The pressure of one atmosphere is 100,000 Pa, so we have to apply a million atmospheres in tension to fracture the metallic bond.

The other class of materials we will deal with is **polymers**. Polymers are repeating chains of small molecular assemblages, and the word *polymer* means "many molecular pieces." Generally, polymers are made of organic chemical links.[6] For example, the common polymer polyethylene consists of long chains of ethylene[7] molecules written as —[CH_2—CH_2]— strung together, each dash representing the net attraction of electrons between two adjacent carbon atomic nuclei (see Figure 13.7). Each carbon atom also has two off-axis hydrogen atoms and one on-axis C—C bond. Roughly speaking, the strength of a C—C bond is about 10% that of an iron-to-iron bond, so you might think that our model apparently implies a force/area of about 10 GPa as the upper limit to break a typical polymer bond. However, this is not true.[8]

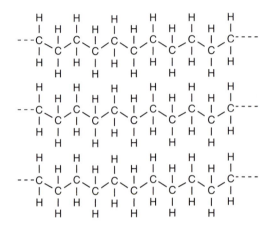

FIGURE 13.7 Regular array of polyethylene molecules.

What you need in this case is not the force between individual carbon atoms but that between adjacent polymer chains. While there are many kinds of possible molecular interactions, for the case of polyethylene, the kinds of forces are lumped under the name "van der Waals" forces. They are quite weak, about 0.1% of the strength of the metal bonds, or about 0.1 GPa. The upper limit for the strength of a polymeric material might therefore be of this magnitude.

13.4 DEFINING MATERIALS REQUIREMENTS

Consider the choice of material for a car's bumper. Defining material requirements for a bumper begins by understanding what a bumper is and what it does. A bumper is a structure often integrated into the main car's chassis; typically, it is 1–2 m wide, 0.1–0.2 m high, and 0.02–0.04 m thick and is attached to the front or rear of the car about half a meter above the ground.

---

[6]"Organic" meaning based on carbon, hydrogen, and a few other atoms such as nitrogen, oxygen, and sulfur.

[7]The chemical structure of ethylene gas is CH_2=CH_2, with the "=" standing for a "double bond" between the carbon atoms. If this double bond is broken and rejoined to an adjacent carbon, it makes the repeating polyethylene structure found in that solid material. (Milk bottles are commonly made of this material.)

[8]Regular arrays of a polymer are also called crystals. They can coexist with a matrix of an amorphous polymer, in good analogy to the structure of metals. Crystals in polymers are usually the right size to scatter light efficiently, so you know that a white, opaque polymer such as Teflon™ is crystalline, and a clear one such as polyethylene terephthalate or "PET" (used in soft drink bottles) is amorphous.

Figure 13.8 shows a "conceptual bumper" for which a material will be selected. A bumper does two main jobs:

(1) It survives undamaged at a very low-speed (1–2 m/s) collision.[9]
(2) It affords modest protection for the car, though sustaining major damage to itself, in a moderate-speed (3–5 m/s) collision.

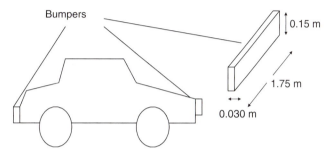

FIGURE 13.8 Conceptual bumpers.

What does an engineer mean by terms such as *collision, survive undamaged,* and *protection for the car*? Translation of such language into engineering variables, units, and numbers begins with the now familiar concept of energy conversion.

A collision is a rapid process of energy conversion. A moving car is carrying translational kinetic energy. A collision rapidly converts that TKE into **"elastic"** and **"plastic"** energy[10] of the materials of construction. Elastic energy is a form of stored potential energy, while plastic energy is converted, via internal collisions of the atoms in the material, into heat. In a properly designed bumper, these two processes of energy conversion will occur in the bumper—not in the car itself and certainly not in the passengers.

In a really low-speed collision, the energy conversion should leave the bumper looking as it did before the collision. In a moderate-speed collision, the bumper may be damaged or destroyed, but it should absorb and release the energy in a way that it leaves the car and the passengers undamaged. In a high-speed collision,[11] other methods of protecting the passengers are needed, as discussed in another chapter.

A wide range of bumper designs can achieve these energy conversion objectives. To select a design from these possibilities, an engineer first picks dimensions for a bumper. For simplicity, consider a bumper to be a rectangular solid object $1.75 \times 0.15 \times 0.030 = 0.0079$ m^3 (width, height, and thickness, respectively). For a given level of protection, the engineer will then seek the lightest material that allows a bumper of these dimensions to meet the design requirements.

Why does the engineer seek the lightest material? In the days of metal bumpers, a bumper provided a significant part (about 5%) of the mass of the car. The higher the mass of the car, the lower the gas mileage. Conversely, anything that reduces mass will increase gas mileage. So all other things being equal, a car with lighter

[9]Federal regulations call for protection against damage to the vehicle in a 2½ mph (1.18 m/s) collision; while this might seem to be a modest requirement, remember that cars are heavy!

[10]We'll properly define these terms later, but for the moment realize that plastic *in this context* is not a word meaning "polymers" and that *elastic* means to spring back like an elastic band.

[11]The Insurance Institute for Highway Safety tests and rates safety for passenger cars up to 35.0 mph (15.6 m/s); see http://www.iihs.org/default.htm.

bumpers will have a higher gas mileage than a car with heavier bumpers. Our optimization problem will be choosing the material that will give the lightest bumper of given dimensions that will do a bumper's job.

Solving that optimization problem begins by using the principle of energy conservation to translate the design requirements from words into variables, units, and numbers.

Example 13.1

Estimate the total energy that a 1.00×10^3 kg car has to absorb in a 2½ mph (1.12 m/s) collision with a perfectly rigid wall. If all of this is absorbed in the bumper, what is the "specific" energy absorbed (i.e., energy per m³ of bumper volume) given its dimensions of $1.75 \text{ m} \times 0.15 \text{ m} \times 0.030 \text{ m}$?

Need: Energy absorbed ____ in J and the specific energy absorbed ____ in J/ m³.

Know: Speed of car, dimensions of bumper.

How: The conservation of energy provides the energy absorbed, and the energy absorbed divided by the volume of the bumper gives the specific energy absorbed.

Solve: The bumper must absorb the total TKE of the vehicle from a 1.12 m/s collision. The vehicle's TKE is ½ $mv^2 = ½ \times 1.00 \times 10^3 \times 1.12^2$ [kg][m/s]$^2 = \mathbf{6.27 \times 10^2}$ **J**, and the bumper's volume $= 1.75 \times 0.15 \times 0.030 = 0.0079$ m³. Therefore, the **specific energy** absorbed by bumper = energy absorbed/volume of bumper = $6.27 \times 10^2 \times 10^{-6}/0.0079$ [J][1/m³][MJ/J] = **0.079 MJ/m³**.

Notice if we wanted 5.0 mph protection, the total energy absorbed goes to ~2500 J and the specific energy absorbed goes to 0.32 MJ/m³, and at 35 mph (15.6 m/s), the corresponding numbers are 0.12 MJ and 1.6×10^2 MJ/m³.

Is there a candidate bumper material that can absorb these amounts of energy elastically and rebound to its original dimensions? Or will there be sustained damage to the bumper or, even worse? For us, the choice of a bumper material boils down to a choice between a polymer and a metal. A polymer offers light weight at some sacrifice of strength. A metal offers high strength but with a penalty of higher weight. Which should an engineer choose? Answering these questions begins by considering the types of materials that nature and human ingenuity offer. Making that choice requires applying a new tool: the **stress-strain diagram**.

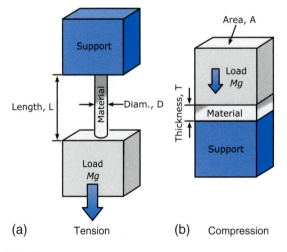

FIGURE 13.9 (a,b) Basic material tests.

The stress-strain diagram is a tool for using measurements of two new variables, **stress** and **strain**, to quantify the terms **modulus of elasticity**, **elastic limit**, **plasticity**, **yield strength**, and **toughness**. Two basic experiments[12] are shown in Figure 13.9. Materials are either stretched under tension, as in Figure 13.9a, or put into **compression**, as in Figure 13.9b. Add more weight, Mg, in either of the situations, and the wire stretches or the plate compresses accordingly. In this way, one can plot a diagram of the stretch in L or compression in T versus the applied load. The response is considered **elastic** if the material returns to its initial dimensions when the load is removed.

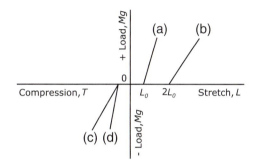

FIGURE 13.10 Elastic response of a material to tension and compression: (a) Tension, initial length L_0, diameter D; (b) Tension, initial length $2L_0$, diameter D; (c) Compression, thickness T, load area A; (d) Compression, thickness T, load area $2A$.

Figure 13.10 is a composite plot of four elastic experiments. In tension, the long rod (or wire, etc.) shown in curve (B) stretched in proportion to its load twice as much as did the shorter one shown in curve (A). The compression shown in curve (D) is half that shown in curve (C) and is also proportional to its load. This is an expression of **Hooke's Law.**[13]

$$(L - L_0) \propto L_0 \frac{Mg}{A} \tag{13.1}$$

Notice that we have used a convention that tension is a positive load so that stretching is a positive response to it, and the opposite is true for compression and the contraction response to it.

But why do we have to plot these data as clumsily as in Figure 13.10? Equation (13.1) is our clue that we can be more inclusive as well as more general. We write it as:

$$\frac{Mg}{A} \propto \frac{(L - L_0)}{L_0} \quad \text{or} \quad \frac{Mg}{A} = E \frac{(L - L_0)}{L_0} \quad \text{(thus defining } E\text{)} \tag{13.2}$$

The fractional (or for the percentage, $\times$ 100) stretching or compression—that is, the change in length divided by its original length, $(L - L_0)/L_0$—is called the **strain**. If the material stretches in tension, the strain is positive, and if the material is compressed, the strain is negative. We use the Greek letter epsilon (ε) to denote strain. Note that strain is a dimensionless number, since it is the quotient of two quantities with the length dimensions that in SI units are meters/meter but just as well could be inches/inch.

---

[12]A third mechanical test is to twist the material and measure its torsional material properties.
[13]Robert Hooke was a talented rival of Isaac Newton.

When the force causing the strain is divided by the applicable cross-sectional area, we get the **stress** in the material and denoted it by the Greek letter sigma (σ). Again, a tensile stress is positive, and a compressing stress is negative. Stress has SI units of N/m^2 (or Pa) and English units of lbf/in^2 (or psi). The constant E in Equation (13.2) is called **Young's modulus**[14] or the **Elastic modulus**.

Hooke's Law can now be written as:

$$\sigma = E\varepsilon \qquad (13.3)$$

Equation (13.3) is the way that engineers use Hooke's law, and it should be committed to memory in this form once it is understood.

Example 13.2

When a mass of 1000. kg of steel is carefully balanced on another piece of steel of depth 1.00 m, height 0.15 m, and thickness 0.030 m and sits on a rigid base, the length of the middle piece contracts by 0.000010 m.

   a. Is the middle piece of steel in tension or compression?
   b. What is the stress in the middle piece of steel?
   c. What is the strain in the vertical direction of the middle piece?

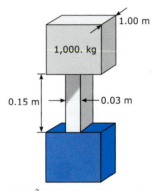

Need: (a) Type of Stress, (b) stress ____ N/m^2, and (c) strain ____ fraction.

Know: Mass of the object compressing the piece = 1000. kg. Initial thickness of sample is 0.15 m and it suffers a contraction of 1.0×10^{-5} m when the load is applied.

How: By definition, stress = force/area, and force = weight = Mg; the supporting cross-sectional area is $1.00 \times 0.030\ m^2 = 0.030\ m^2$. By definition, strain = (change in length)/(initial/length).

Solve: (a) The middle piece is in compression.

   (b) Stress = σ = weight/area = $-(1000. \times 9.81)/0.030\ [kg][m/s^2][1/m^2]$
                    $= -3.27 \times 10^5\ N/m^2 = -3.27 \times 10^5\ Pa$.

The negative sign indicates compression.

   (c) By definition, **strain** is the change in length divided by the initial/length or $\varepsilon = -1.0 \times 10^{-5}/0.15\ [m]/[m] =$
     -6.7×10^{-5} (ε is dimensionless, and the negative sign indicates compression).

---

[14]For historical reasons, this slope is also called Young's modulus after the early nineteenth century scientist and scholar Thomas Young who first suggested this approach to understanding materials.

Suppose one made a large number of measurements like those described in Example 13.2 but with differing loads. Suppose one then plotted the results on a graph, with strain as the horizontal axis and stress as the vertical axis. For each axis, the negative direction would be compression, and the positive direction would be tension. The resulting graph is called a **stress-strain diagram**. It captures in one diagram three of the most important properties of a material: elasticity, strength, and toughness.

The stress-strain diagram of an idealized steel[15] looks like Figure 13.11. It also defines some new terms: **yield stress** (or **yield strength**), **yield strain**, and **plastic deformation**.

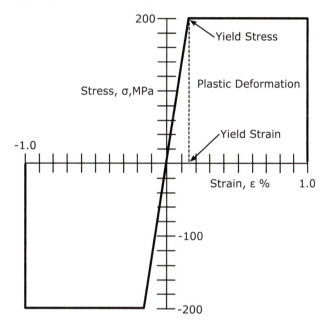

FIGURE 13.11 Simplified stress-strain curves for an idealized steel.

The **yield stress** is the maximum stress at the limit of elastic behavior in which the specimen returns to its initial length when the load is removed. In the case shown in Figure 13.11, it is 200. MPa. The **yield strain** is the strain corresponding to the elastic yield stress. But plastic deformation causes a *permanent* change in the properties of the steel when it passes beyond its elastic limits. Stretching here is called **plastic deformation**. For this idealized steel, compressive properties mirror the tensile ones.

Compare the yield stress in Figure 13.11 of 200 MPa to the theoretical failure value for iron calculated earlier at 100 GPa. In other words, our sample of steel is failing at a stress about 200 MPa or a factor of about 500 lower than its theoretical maximum. Presumably this failure is initiated at inclusions or at crystal boundaries rather than by pulling apart the rows of atoms of iron in the steel.

The elastic modulus E defined in Equation (13.3) is just the slope of the stress-strain curve. In Figure 13.11, it is $E = 200$ MPa/$0.0015 = 130$ GPa.

Another variable of interest is **toughness**. In ordinary speech, people sometimes confuse it with (yield) stress or strength. To an engineer, however, strength and toughness have very different meanings. A material can be strong without being especially tough. One example is a diamond. It can be subjected to great force and yet

---

[15]Basically iron + some additives + some processing.

return to its original shape, so it is therefore very strong. Yet, a diamond can be easily shattered and therefore is not very tough. A material can also be tough without being strong. An example is polycarbonate, a kind of polymer used in football helmets. It is relatively easily dented and therefore not very strong. Yet, it absorbs a great deal of energy per unit mass without shattering and is therefore very tough.

In this introductory textbook, the stress-strain curve for strains greater than the yield strain will be a horizontal line parallel to the strain axis. In actual stress-strain diagrams, the shape of that portion of the curve is more complicated. This region is called the **plastic** region, and when it is described with a horizontal line, it is called a **perfectly plastic** region. This overall model is called **perfectly elastic, perfectly plastic**. It's a useful simplification of how real materials behave.

The horizontal line in the perfectly plastic region extends out to a **maximum strain**. At that strain, the material "fails." The total strain at failure is as the sum of the elastic strain and the plastic strain. The important point is that a wire stretched beyond its maximum strain breaks. A plate compressed to its maximum strain shatters or spreads. The **toughness** of a material is defined as the area under the portion of the stress-strain curve that extends from the origin to the point of maximum strain.[16]

Because this area (shaded in Figure 13.12) is the result of stress, measured in N/m^2, and of strain, measured in m/m, toughness has the units of $N \cdot m/m^3$, or J/m^3 (that is, energy or work per unit volume). It is usually measured in a dynamic experiment involving a strike by a heavy hammer. Toughness is therefore the preferential criterion to be used to predict material failure if a sudden load is applied.

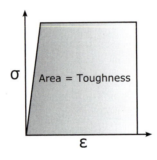

FIGURE 13.12 Toughness.

Toughness, defined as the work done until failure, represents the ability of the material to absorb energy. It is a prime variable for a large subset of materials research.

Example 13.3

Given the stress-strain diagram for steel in Figure 13.11, determine (a) its modulus under compression, (b) its yield strength under tension, (c) its toughness under tension to yield, and (d) its toughness under tension to 1.0% strain. Use appropriate SI units.

Need: E=____GPa under compression, σ=____ MPa at yield, and toughness at yield and 1.0% strain=____ MJ/m^3.
Know: Stress/strain curve in Figure 13.11; toughness is area beneath the stress/strain curve.
How: Hooke's law $\sigma = E\varepsilon$, + stress/strain curves.

---

[16] Actually, there are several different definitions of *toughness* used by different authors, in particular the ability of a material to withstand sudden impacts. Under certain circumstances, this may, in fact, measure the same thing as our definition.

Solve: **(a)** Need slope in compression region for σ/ε curve. $\mathbf{E} = -2.0 \times 10^2 \,[\text{MPa}]/-0.0015\,[\text{m/m}] = \mathbf{130\,GPa}$ (to graph-reading accuracy).

(b) Under tension, the **yield stress** is $\boldsymbol{\sigma = 2.0 \times 10^2}$ **MPa** (to graph-reading accuracy).

(c) Toughness under tension to the yield stress is equal to the triangular area,
$1/2 \times 0.0015\,[\text{m/m}] \times 2.0 \times 10^2\,[\text{MN/m}^2] = \mathbf{0.15\,MJ/m^3}$.

(d) **At 1% strain**, add rectangular area: $(0.01 - 0.0015) \times 200. = 1.7\,\text{MJ/m}^3$. Therefore, the **toughness** $= 0.15 + 1.7 = \mathbf{1.9\,MJ/m^3}$.

The no-damage elastic portion of the curve is only a small fraction of the total protection afforded the car's structure. You can also see that plastic deformation, which will manifest itself in permanent deformation or crushing,[17] affords a much better sink to dissipate energy than does the purely elastic behavior. Of course, whether a particular bumper design can distribute this energy deposition uniformly and avoid exaggerated *local* deformation is up to the skill of the engineer.

We're now in a position to see if a steel bumper satisfies our first constraint. What does the stress-strain diagram tell us about its ability to absorb the energy of a collision yet return to its original shape?

Example 13.4

Can the car's bumper from Example 13.1 absorb the TKE of the 2.5 mph collision assumed there? Or will it plastically deform? What if it is traveling at 5.0 mph? Also, what is the weight of the bumper if its density is 7850 kg/m³? Assume the mechanical behavior is described by the stress-strain diagram of Figure 13.11.

Need: Energy to be absorbed in bumper during collision = ____ MJ/m³ compared to energy of collision. Also, weight of bumper = ____ N.

Know: From Example 13.1: TKE $= 6.27 \times 10^2$ J at 2.5 mph; specific energy absorbed $= 0.079$ MJ/m³; steel bumper volume $= 0.0079$ m³; density of steel, $= 7850$ kg/m³.

At 5.0 mph, TKE $= 2500$ J and the specific energy absorbed is 0.32 MJ/m³. Also from Example 13.3, the elastic toughness $= 0.15$ MJ/m³ and plastic toughness is 1.8 MJ/m³ at 1.0% strain.

How: Compare specific energy to be absorbed to the capability of the bumper.

Solve: Assume the entire surface of the bumper contacts the wall simultaneously (i.e., the force on the bumper is uniform over its entire area).

At 2.5 mph, the specific energy to be absorbed is 0.079 MJ/m³ of steel. Steel can absorb up to 0.15 M J/m³, so it will elastically absorb this amount of energy—again if uniformly applied. If so, the bumper will bounce back to its original shape.

At 5.0 mph, the specific energy to be absorbed goes to 0.32 MJ/m³ of steel. The bumper will plastically deform since 0.32 MJ/m³ > 0.15 MJ/m³ but will not fail since 1.9 MJ/m³ > 0.32 MJ/m³.

The weight of this bumper is Mg = density × volume × g $= 7850 \times 0.0079 \times 9.81$ [kg/m³] [m³][m/s²] $= 610$ N, or the weight of a physically fit student!

---

[17]No doubt with an accompanying repair bill that will disappoint the vehicle's owner!

13.5 MATERIALS SELECTION

The previous section showed how the stress-strain diagram provides the information needed to determine if a material satisfies the design requirements. The question now becomes: Is there a material of less weight than steel that also satisfies the design requirements? To be specific, let us consider a polymer, a material that is much less dense than steel. Can it compete with steel in the bumper application? Again, our tool is the stress-strain diagram. Figure 13.13 and Table 13.1 are for a polycarbonate.

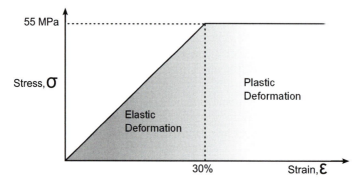

FIGURE 13.13 Stress and strain for a polycarbonate.

Table 13.1 Properties of a Polycarbonate

Elastic Yield, MPa	Strain Yield %	Density, kg/m^3
55.	30.	1300.

Note: Actually there are many kinds of polycarbonates; the properties shown in this table are that of a very common polycarbonate made from a monomer (BPA). It's ubiquitous, being used in crash helmets, CDs, DVDs, and bottles, as well as car bumpers. Another version of polycarbonate is used as a lens material for spectacles.

Notice that the polycarbonate is not as "strong" as steel because its yield strength is considerably less, but it can be stretched much further while elastically returning to its original strength. Let's repeat Example 13.4 using this material for the bumper.

Example 13.5

If the car's bumper in Example 13.1 is made of the polycarbonate described by Figure 13.13, can it absorb the TKE of the low-speed 2.5 mph collision, or will it plastically deform? What if it is traveling at 5.0 mph. What is the weight of this bumper if its density of polycarbonate is 1300 kg/m^3? Assume the compressive stress-strain diagram mirrors the tensile stress-strain diagram.

Need: Energy to be absorbed in bumper after collision = _____ MJ/m^3 compared to energy of collision and the weight of a polycarbonate bumper.

Know: From Example 13.4: TKE $= 6.27 \times 10^2$ J; specific energy absorbed $= 0.089$ MJ/m^3; bumper volume $= 0.0079$ m^3. At 5.0 mph, TKE $= 2500.$ J and the specific energy absorbed is 0.32 MJ/m^3.

How: Compare specific energy to be absorbed to capability of material to absorb it. Need to calculate the latter from the σ, ε diagram.

Solve: Calculate the elastic toughness by looking at Figure 13.13. It is given at yield by the triangular area,

$$1/2 \times 0.30 \, [\text{m/m}] \times 55. \, [\text{MN/m}^2] = 8.3 \, \text{MJ/m}^3.$$

Even at 5.0 mph, the specific energy to be absorbed is 0.32 MJ/m³, which is less than 8.3 MJ/m³. Thus, this bumper can survive elastically.

The weight of this bumper is $Mg=$ density $\times$ volume $\times g = 1,300 \times 0.0079 \times 9.81$ [kg/m³] [m³][m/s²] = 98. N, or much less than the weight of any student!

So, although polymer is not as "strong" as steel, its ability to compress further without breaking gives it adequate toughness to do the job. In addition, polymers are substantially less dense than steel. So, the polymer is the winner. In the last two decades, polymers have almost entirely replaced steel as the material from which automobile bumpers are made.

Of course, there are still plenty of options for the materials engineers to design the preferred configuration of the bumper. For example, the metal bumper may be manufactured with springs to absorb the impact. There is surely enough material in our steel bumper that we could use some of it in the form of coil springs, as in Figure 13.14.

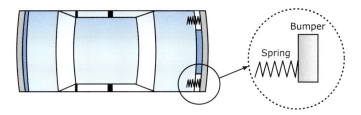

FIGURE 13.14 Mechanical design also influences the choice of materials.

A steel spring can be stretched or compressed much further than a flat piece of the same material. It's as if we traded the material for one with a lower modulus (less steep curve) and a much larger strain to yield. Furthermore, there is now plenty of room to design for the randomness of crashes—that is, how the vehicles will interact with an immoveable object such as a pole as indicated in Figure 13.15.

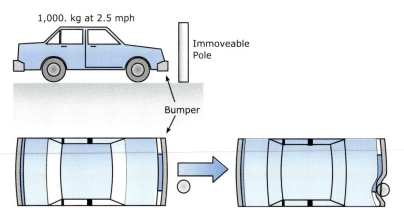

FIGURE 13.15 Not all crashes are head-on.

The material may yield locally in low-speed impacts.

This chapter contains just as a glimpse into the most basic considerations that would go into a material design. Of course, materials engineers have become very clever at rearranging the macroscopic properties of materials as they need them. In the car bumper case, the polymer may be backed with a second one that is in the form of a foam.

13.6 PROPERTIES OF MODERN MATERIALS

This discussion of selecting a material for a bumper gives a feel for the engineer's problem of material selection. However, elasticity, strength, and toughness, though very important, are not the whole story. In the decision whether to use silicon or germanium to make transistors, the electronic and thermal properties of materials play a key role. In the selection of copper versus fiber optic glass, information-carrying capacity becomes crucial. In the selection of steel versus polymers for refrigerators, considerations of appearance, manufacturability, and corrosion resistance become important. In the selection of a material for a hip joint transplant, compatibility with the human body becomes an essential materials requirement.

In addressing these varied requirements, twenty-first-century materials go far beyond the traditional categories of metals, polymers, and the other classes of materials shown in Figure 13.1. A wide range of "composite" materials now combine the properties of those original categories. Some are created by embedding fibers of one material within a matrix of a second material. This makes it possible to combine, for example, the high strength of a thin carbon fiber with the high toughness of a polymer matrix.

One of the newest areas of materials engineering are materials whose structure has been engineered at the nanometer scale (1 nm $= 10^{-9}$ m).[18] These materials are called **nanomaterials** and are important because the unique properties of materials at the nanoscale enable engineers to create materials and devices with enhanced or completely new characteristics and properties.

Many engineers consider nanotechnology to be the next industrial revolution because it is predicted to have enormous social and economic impact. Companies have already introduced nanotechnology in several consumer products. Examples of nanomaterials already on the market include nanoscale titanium dioxide used in some cosmetics and sunscreens, nanoscale silica being used as dental fillers, and nanowhiskers used in stain-resistant fabrics. Nanoclays and coatings are being used in a range of products from tennis balls to bikes to cars to improve bounce, strengthen high-impact parts, or render material scratchproof. Some nanomaterials even are based upon biological models, Figure 13.16.

FIGURE 13.16 Carbon nanotubes. Source: Dr. Michael De Volder Institute for Manufacturing, University of Cambridge, UK.

[18]As small as a nanometer is, it's still relatively large compared to atoms. Atoms, which are the basic building blocks of materials, are on the order of 0.1 to 0.5 nm in size. An atom's nucleus is much smaller again—about 0.00001 nm.

Nanomaterials are literally tailored molecule by molecule. This makes it possible, for example, to provide materials with precise combinations of electronic and optical properties. In one sense, however, the emergence of these new composites is actually a matter of going "back to the future." The original materials used by humanity some 10,000 years ago, such as wood, stone, and animal skins, are complicated natural composites. Figure 13.1 summed up this long-term history of humanity's materials use. Even with the new challenges of the twenty-first century, however, the traditional properties of elasticity, strength, and toughness will continue to play a central role in materials selection. As they do, the stress-strain diagram, the best representation of these properties, remains a crucial tool for the twenty-first century engineer.

SUMMARY

Materials engineers develop and specify materials. They do so using constrained optimization. Thus one property, such as performance or weight or cost, is to be optimized (maximized or minimized), subject to meeting a set of constraints (design requirements).

Materials selection begins by quantifying design requirements. It proceeds by searching the major classes of candidate materials, metals, polymers, composites, and so on for candidates capable of meeting design requirements. The candidates are then subjected to further screening, based on such characteristics as the **strain** they exhibit in response to **stress**. This particular characteristic is plotted on a graph called the **stress-strain diagram**. This stress-strain, or σ-ε diagram can then be used to determine key properties of the material: **modulus of elasticity**, **yield strength**, **plastic deformation**, **and toughness**. These properties, expressed in the appropriate units, provide the basis (the "know") for solving particular constrained optimization problems of materials selection.

EXERCISES

The figures below depict the situations described in Exercises 1–6.

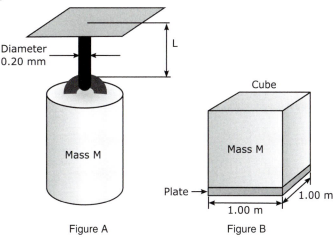

Figure A   Figure B

1. A mass of $M = 1.0$ kg is hung from a circular wire of diameter 0.20 mm as shown in Figure A. What is the stress in the wire? (**A: 0.31 GP.**)

2. If the wire in Figure A is stretched from 1.00 m to 1.01 m in length, what is the strain of the wire?

3. In Figure B, when a block of metal 1.00 m on a side is placed on a metal plate 1.00 m on a side, the stress on the plate is 1.00×10^3 N/m^2. What is the mass of the metal cube? (**A: 102 kg.**)

4. Suppose the plate described in Exercise 3 and Figure B was 0.011 m thick before the cube was placed on it. Suppose that placing the cube on it causes a compressive strain of 0.015. How thick will the plate be after the cube is placed on it? (**A: 0.011 or unchanged to three significant figures.**)

5. Suppose the wire in Figure A is perfectly elastic. When subjected to a stress of 1.00×10^4 Pa, it shows a strain of 1.00×10^{-5}. What is the elastic modulus (i.e., Young's modulus) of the wire?

6. A plate of elastic modulus 1.00 GPa is subjected to a compressive stress of -1.00×10^3 Pa as in Figure B. What is the strain on the plate? (**A: -1.00×10^6.**)

For Exercises 7–9, assume that a silicone rubber[19] has this stress-strain diagram.

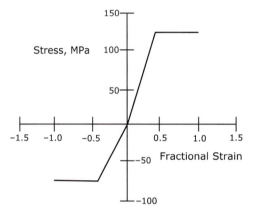

7. What is the yield strength under compression of the silicone?

8. A flat saucer made of silicon rubber has an initial thickness of 0.0050 m. A ceramic coffee cup of diameter 0.10 m and mass 0.15 kg is placed on a plate made of the same silicon rubber. What is the final thickness of the plate beneath the cup, assuming that the force of the cup acts directly downward and is not spread horizontally by the saucer?

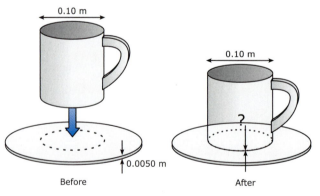

[19]Silicone rubbers are very flexible; their structure consists of polysiloxanes, >Si-O-, in which the Si atom also has two CH_3 chemical groups per atom (but that are not shown here); they are located off the main chain. Incidentally, you must distinguish *silicon* (a brittle element used in electronics) from *silicone*, the soft rubber described here.

9. What is the maximum number of such coffee cups that can be stacked vertically on the saucer and not cause a permanent dent in the plate? (**A: 4.3×10^5 cups**—a tough balancing act.)

10. A coat hanger is made from polyvinyl chloride.[20] The "neck" of the coat hanger is 0.010 m in diameter and initially is 0.10 m long. A coat hung on the coat hanger causes the length of the neck to increase by 1.0×10^{-5} m. What is the mass of the coat? The stress-strain diagram is provided. (**A: 1.3 kg.**)

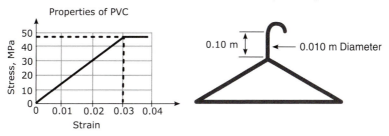

Properties of PVC

11. How many coats having the same mass as those in exercise 10 must be hung on the hanger to cause the neck to remain permanently stretched (i.e., plastically deformed) after the coats are removed?

For Exercises 12–15, imagine the idealized situation as shown in Figure A for an artillery shell striking an armor plate made of an (imaginary) metal called "armory." Figure B is the stress-strain diagram of armory. Assume the material diagram shows armory strained to failure.

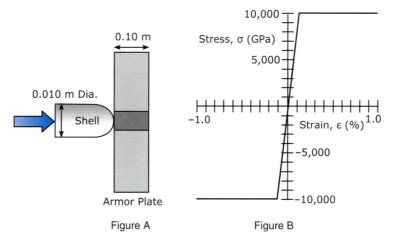

Figure A   Figure B

12. Suppose the shell has mass 1.0 kg and is traveling at 3.0×10^2 m/s. How much TKE does it carry?

13. Assume that the energy transferred in the collision between the shell and the armor plate in Exercise 12 affects only the shaded area of the armor plate beneath contact with the flat tip of the shell (a circle of diameter 0.0010 m) and does not spread out to affect the rest of the plate. What is the energy density delivered within that shaded volume by the shell? (**A: 5.7×10^9 J/m^3.**)

14. Which of the following will happen as a result of the collision in the Exercise 13? Support your answer with numbers.

---

[20]PVC is an inexpensive polymer of the monomer $CH_2 = CHCl$ with repeating unit $-CH_2(CHCl)-$.

a. The shell will bounce off without denting the armor plate.

b. The armor plate will be damaged but will protect the region beyond it from the collision.

c. The shell will destroy the armor plate and retain kinetic energy with which to harm the region beyond the plate. (**A: a**—but it is uncomfortably close to yielding and to a permanent set to the armor; an armored tank crew will at least get quite a headache.)

15. In exercise 14, what is the highest speed the artillery shell can have and still bounce off the armor plate without penetrating it?

Exercises 16–19 involve the situation depicted below. Consider a "micrometeorite" to be a piece of mineral that is approximately a sphere of diameter 1.0×10^{-6} m and density 2.0×10^3 kg/m^3. It travels through outer space at a speed of 5.0×10^3 m/s relative to a spacecraft. Your job as an engineer is to provide a micrometeorite shield for the spacecraft. Assume that if the micrometeorite strikes the shield, it affects only a volume of the shield 1.0×10^{-6} m in diameter and extending through the entire thickness of the shield.

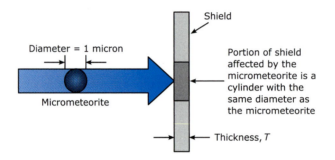

16. Using the stress-strain diagram for steel in Figure 13.11, determine the minimum thickness a steel micrometeorite shield would have to be to protect the spacecraft from destruction (even though the shield itself might be dented, cracked, or even destroyed in the process).

17. Using the stress-strain properties for a polymer (Figure 13.13 and Table 13.1), determine whether a sheet of this polymer 0.10 m thick could serve as a micrometeorite shield if this time the shield must survive a micrometeorite strike without being permanently dented or damaged. Assume the properties of the polymer are symmetric in tension and in compression. (**A: Yes**, it will survive unscathed.)

18. Suppose you were required to use a micrometeorite shield no more than 0.01 meters thick. What would be the required toughness of the material from which that shield was made if the shield must survive a micrometeorite strike without being permanently dented or damaged?

19. Suppose a shield exactly 0.01 m thick of the material in exercise 18 exactly met the requirement of surviving without denting or damage at a strain of -0.10, yet any thinner layer would not survive. What are the yield stress and elastic modulus of the material? (**A: $\sigma_{Yield} = -33$ MPa, $E = 0.33$ GPa.**)

20. Your company wants to enter a new market by reverse-engineering a popular folding kitchen step stool whose patent has recently expired. Your analysis shows that by simplifying the design and making all the components from injection-molded PVC plastic you can produce a similar product at a substantially reduced cost. However, when you make a plastic prototype and test its performance, you find that it is not as strong and does not work as smoothly as your competitor's original stool. Your boss is anxious to get your design into production because he/she has promised the company president a new high-profit item by the end of this quarter. What do you do? (Use the Engineering Ethics Matrix.)

a. Release the design. Nearly everything is made from plastic today, and people don't expect plastic items to work well. You get what you pay for in the commercial market.
b. Release the design, but put a warning label on it, limiting its use to people weighing less than 150 pounds.
c. Quickly try to find a different, stronger plastic that can still be injection molded and adjust the production cost estimate upward.
d. Tell your boss that your tests show the final product to be substandard and ask if he/she wants to put the company's reputation at risk. If he/she presses you to release your design, make an appointment to meet with the company president.

21. As quality control engineer for your company, you must approve all material shipments from your suppliers. Part of this job involves testing random samples from each delivery and making sure they meet your company's specifications. Your tests of a new shipment of carbon steel rods produced yield strengths 10% below specification. When you contact the supplier, they claim their tests show the yield strength for this shipment is within specifications. What do you do? (Use the Engineering Ethics Matrix.)
a. Reject the shipment and get on with your other work.
b. Retest samples of this shipment to see if new data will meet the specifications.
c. Accept the shipment, since the supplier probably has better test equipment and has been reliable in the past.
d. Ask your boss for advice.

Mechanical Engineering

14

Source: iStockphoto.com/Alex Mit

14.1 INTRODUCTION

Mechanical engineering is one of the most versatile disciplines in the broad field of engineering. Virtually, every aspect of life is touched by mechanical engineering. If something moves or uses energy, a mechanical engineer probably was involved in its design, testing, and production. Mechanical engineers also use economic, social, environmental, and ethical principles when they create new systems and products.

14.2 MECHANICAL ENGINEERING

Mechanical engineering typically uses the fields of **thermal design** and **machine design** (see Figure 14.1). Mechanical engineers use the principles in these fields in the design and analysis of mechanical systems such as cars, trucks, aircraft, ships, spacecraft, turbines, industrial equipment, robots, heating and cooling systems, medical equipment, and much more. They also design the automated machines that mass-produce these products.

In this chapter, we will explore the basic elements of the six subtopics shown in Figure 14.1. We begin by defining the meaning of each of the subtopics.

14.2.1 Thermal Design

Thermal design is the combined study of thermodynamics, fluid mechanics, heat transfer, and combustion.

- **Heat transfer** has applications in heat exchanger design, engine cooling, air conditioning, and cooling of electronics.

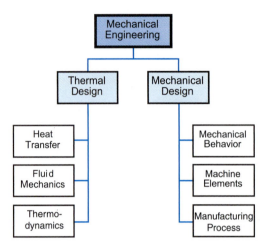

FIGURE 14.1 The structure of mechanical engineering.

- **Fluid mechanics** include fluid statics and fluid flow as applied to areas such as piping networks, hydraulic systems, bearings, turbo-machineries, and airfoils.
- **Thermodynamics** include studies of properties of pure and mixed substances, the ideal gas law, the conservation of energy, engine and refrigeration cycles, power plants, and combustion.

14.2.2 Machine Design

Machine design focuses on the basic principles of the following three areas:

- **Mechanical behavior** includes statics, dynamics, strength of materials, vibrations, reliability, and fatigue.
- **Machine elements** are basic mechanical parts of machines. They include gears, bearings, fasteners, springs, seals, couplings, and so forth.
- **Manufacturing processes** include areas such as computerized machine control, engineering statistics, quality control, ergonomics, and life-cycle analysis.

In this chapter, we will explore the basic elements of these six areas of mechanical engineering.

14.3 THE ELEMENTS OF THERMAL DESIGN

The concept of *heat* is somewhat mysterious. We talk of heat as though it is a substance. Even the phrase "*heat transfer*" implies that some "thing" is being moved from one place to another.

Through the eighteenth century, heat was thought to be a colorless, odorless, weightless fluid. It is from this era that we get the concept that heat is some sort of fluid that *flows* from one object to another. It was later discovered that heat is simply a form of energy (called *thermal energy*) and is not a fluid. However, the terminology of *heat flow* was already in common use and could not be changed.[1]

---

[1] A modern definition of *heat* is energy that is spontaneously transferred from a high-temperature system to a lower-temperature system.

14.3.1 Heat Transfer

When two objects at different temperatures come into thermal contact, they exchange thermal energy (heat). The hotter object spontaneously gives thermal energy to the colder body until their temperatures are equal, at which point they are said to be in a state of "thermal equilibrium."

Heat transfer is such a large and important mechanical engineering topic that most curricula have at least one required course in it. Heat transfer equations are always written as heat transfer *rate* (i.e., per unit time) equations, so to determine the amount of thermal energy transferred as a system undergoes a process from one equilibrium state to another you multiply it by Δt, the time required for the process to occur.

In engineering, **heat transfer** refers to the thermal energy that passes between objects by **thermal conduction** in which energy is transferred by molecular vibration, by **thermal convection** in which a fluid moves between regions of different temperature, or by **thermal radiation** in which energy is transmitted by electromagnetic radiation. These three heat transfer modes are shown in Figure 14.2 and briefly described below.

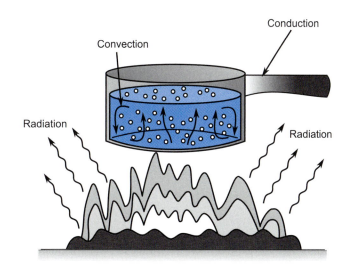

FIGURE 14.2 The three modes of heat transfer.

14.3.1.1 Thermal conduction

The basic equation for the rate of steady conduction heat transfer is called **Fourier's Law of Conduction**. For steady conduction heat transfer through a flat plate or a plane wall, Fourier's law is:

$$\left(\dot{Q}_{\text{cond}} \right)_{\text{plane}} = k_t A \left(\frac{T_{\text{hot}} - T_{\text{cold}}}{\Delta x} \right) \tag{14.1}$$

where $\dot{Q}_{\text{cond}}$ is the conduction heat transfer rate, k_t is the thermal conductivity of the material, A is the cross-sectional area perpendicular to the heat transfer direction, and Δx is the thickness of the plate or wall. The algebraic sign of this equation is such that a positive $\dot{Q}_{\text{cond}}$ always corresponds to heat transfer from the hot to the cold side of the plane or wall.

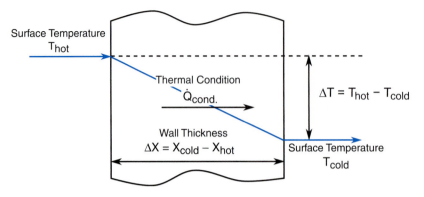

FIGURE 14.3 Thermal conduction through a wall.

Table 14.1 Thermal conductivity of various materials near 20 °C		
	Thermal conductivity k_t	
Material	**Btu^a/(h · ft · R)**	**W/(m · K)**
Air at atmospheric pressure	0.015	0.026
Hydrogen at atmospheric pressure	0.105	0.182
Liquid water	0.343	0.594
Engine oil	0.084	0.145
Mercury	5.02	8.69
Window glass	0.45	0.78
Glass wool	0.022	0.038
Aluminum	118.0	204.0
Copper	223.0	386.0
Carbon steel (1% carbon)	25.0	43.0

^{a}A Btu is an Engineering English quantity called British thermal units. It is the amount of heat to raise the temperature of 1 lbm of water by 1 °F. Also, a Btu is about 1 kJ of thermal energy, or more exactly, 1 Btu= 1055 J.

Example 14.1

The glass in the window of your dorm room is 2.00 ft wide, 3.00 ft high, and 0.125 in. thick. If the temperature of the outside surface of the glass is 21.0 °F and the temperature of the inside surface of the glass is 33.0 °F, what is the rate of heat loss through the window?

Need: $\dot{Q}_{cond} = ?$

Know: $T_{hot} = T_{inside} = 33.0$ °F, $T_{cold} = T_{outside} = 21.0$ °F, $A = 2.00$ ft × 3.00 ft = 6.00 ft^2, and $\Delta x = 0.125$ in. × [1.00 ft/ 12.0 in.] = 0.0104 ft.

How: The heat loss rate can be calculated from Equation (14.1).

Solve: From Table 14.1, the thermal conductivity of the glass is $k_t = 0.45$ Btu/(h·ft·R), and then Equation (14.1) gives

$$(\dot{Q}_{cond})_{loss} = k_t A \left(\frac{T_{hot} - T_{cold}}{\Delta x} \right) = 0.45 \left[\frac{\text{Btu}}{\text{h·ft·R}} \right] \times 6.00 \left[\text{ft}^2 \right] \left(\frac{33.0 - 21.0}{0.0104} \right) \left[\frac{\text{R}}{\text{ft}} \right] = 3120 \text{ Btu/h}$$

Note that the temperature difference, $T_{hot} - T_{cold}$, has the same numerical value regardless of whether Fahrenheit or Rankine temperatures are used because

$$T_{hot} - T_{cold} = (460 + 33.0R) - (460 + 21.0R) = 33.0 - 21.0 = 12.0R = 12.0°F.$$

14.3.1.2 Thermal convection

Convective heat transfer occurs whenever an object is either hotter or colder than a surrounding *fluid*.[2] Convection only occurs in a moving liquid or a gas, never in a solid. The basic equation for the rate of convection heat transfer is known as **Newton's Law of Cooling**,

$$\dot{Q}_{conv} = hA(T_\infty - T_s) \tag{14.2}$$

where $\dot{Q}_{cond}$ is the convective heat transfer rate, h is the convective heat transfer coefficient, A is the surface area of the object being cooled or heated, T_∞ is the bulk temperature of the surrounding fluid, and T_s is the surface temperature of the object (see Figures 14.4 and 14.5). In English units, h is expressed in units of Btu/(h·ft²·R), and in SI units, it is W/(m² K).[3] The algebraic sign of Newton's law of cooling is positive for $T_\infty > T_s$ (heat transfer into the object) and negative when $T_\infty < T_s$ (heat transfer out of the object). The convective heat transfer coefficient h is usually a positive, experimentally determined value.

There are two types of convection—natural convection and forced convection. **Natural convection** is produced by density differences in a fluid due to temperature differences (e.g., as in "hot air rises"). Global atmospheric circulation and local weather phenomena (including wind) are due to convective heat transfer. Figure 14.4 illustrates natural convection around a hot object.

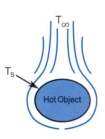

FIGURE 14.4 Natural convection around a hot object.

In **forced convection**, the fluid motion is generated by a source like a pump or a fan. It is one of the main forms of heat transfer use by engineers because large amounts of thermal energy can be transported efficiently. It is used in heating and air conditioning systems, electronics cooling, and numerous other technologies. Forced

[2] A fluid is either a liquid or a gas.

[3] Again, the "temperature" in the definition of a heat transfer coefficient is a temperature difference so that it can be expressed as Btu/h ft² °F or W/m² °C.

convection is used in designing *heat exchangers*, in which one fluid stream is used to heat or cool another fluid stream. Figure 14.5 illustrates a flat plate being cooled by forced convection from a fan.

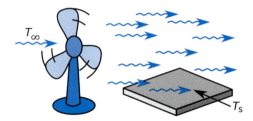

FIGURE 14.5 Forced convection over a flat plate by a cooling fan.

Table 14.2 provides values for the convection heat transfer coefficient h, under various conditions.

Table 14.2 Typical values of the convective heat transfer coefficient

Type of convection	Convective heat transfer coefficient h	
	$Btu/\left(h \cdot ft^2 \cdot R\right)$	$W/\left(m^2 \cdot K\right)$
Air, free convection	1-5	2.5-25
Air, forced convection	2-100	10-500
Liquids, forced convection	20-3,000	100-15,000
Boiling water	500-5,000	2,500-25,000
Condensing water vapor	1,000-20,000	5,000-100,000

Example 14.2

A new computer chip with a surface area of 1.00 cm^2 generates 10.0 W of heat. Determine the convective heat transfer coefficient needed to keep the temperature of the chip less than 20.0 °C above the environmental temperature. Can the chip is to be cooled with air, and will it require free or forced convection?

Need: The convective heat transfer coefficient, h.

Know: $A = 1.00$ cm^2, $\dot{Q}_{conv} = -10.0$ W (the sign is negative because the chip is losing heat), and $T_\infty - T_s = -20.0$ °C (the surface temperature, T_s, is greater than the environmental temperature T_∞ here).

How: Determine h from Equation (14.2) and compare it to the values in Table 14.2.

Solve: Equation (14.2) is $\dot{Q}_{conv} = hA(T_\infty - T_s)$, so

$$h = \dot{Q}_{conv}/A(T_\infty - T_s) = (-10.0[W])/\left\{(1.00[cm^2])(1[m]/100[cm])^2(-20.0[°C])\right\}$$
$$= 5.00 \times 10^3 \, W/m^2 \cdot K$$

Table 14.2 has this value of h in the forced convection range for liquid. Note that #/°C = #/K because the Celsius degree size is the same as the Kelvin degree size.

14.3.1.3 Thermal radiation

All electromagnetic radiation can be considered radiation transfer. Infrared, ultraviolet, visible light, radio and television waves, microwaves, X-rays, gamma rays, and so on are all forms of radiation transfer. If an object with a surface temperature T_s is emitting or receiving thermal radiating energy to or from its surroundings at temperature T_∞, then the rate of thermal radiation is expressed by the **Stefan-Boltzmann Law of Radiation**,

$$\dot{Q}_{rad} = \varepsilon A \sigma (T_\infty^4 - T_s^4) \tag{14.3}$$

where $\dot{Q}_{rad}$ is the radiation heat transfer rate, ε is the dimensionless emissivity (the hotter object is said to *emit* energy, while the colder object *absorbs* energy) of the object, A is the surface area of the object, and σ is the **Stefan-Boltzmann constant** (5.69×10^{-8} W/m$^2 \cdot$K^4 or 0.1714×10^{-8} Btu/h $\cdot$ ft$^2 \cdot$ R^4). A **black** object is defined to be any object whose emissivity is $\varepsilon = 1$, and a perfectly reflecting object has an emissivity of 0.

Note that this equation contains temperature raised to the fourth power. This means that *absolute* temperature units (Rankine or Kelvin) must always be used.

Figure 14.6 illustrates how solar radiation can be used to heat water for household use, and Table 14.3 lists values for the thermal emissivity values for various materials.

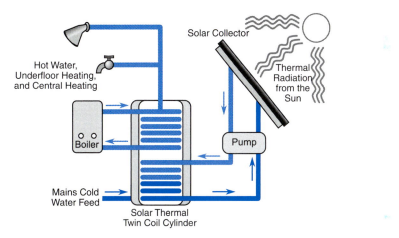

FIGURE 14.6 Solar water heater using thermal radiation from the Sun.

Table 14.3 Emissivity values of various materials

Material	Temperature		Emissivity ε (Dimensionless)
	°C	°F	
Aluminum	100	212	0.09
Concrete	21	70	0.88
Flat black paint	21	70	0.90
Flat white paint	21	70	0.88
Glass	21	70	0.95
Iron (oxidized)	100	212	0.74
Water	0-100	32-212	0.96
Wood	21	70	0.90

Example 14.3

The ancient Egyptians were said to be able to make ice in the desert by exposing a tray with a thin layer of water to a cloudless black night sky. Some say that the night sky temperature is 3 K (which is the background temperature of the universe). However, the night sky is full of air that radiates back at us, so the generally accepted value of the temperature of a cloudless night sky is about −70 °C or 203 K. The emissivity of the water is $\varepsilon=0.9$, and the surface area of the tray is 0.250 m². Determine how long it will take to freeze water initially at 0 °C if 385 kJ must be removed by radiation.

Need: Time to freeze water initially at 0 °C.
Know: $\dot{Q}=\frac{dQ}{dt}=\frac{\Delta Q}{\Delta t}$ so $\Delta t=\frac{\Delta Q}{\dot{Q}}$
How: From Equation (14.3), we know that $\dot{Q}_{rad}=\varepsilon A\sigma(T_\infty^4-T_s^4)$, $\sigma=5.69\times10^{-8}$ W/m²·K⁴, and $\Delta Q=-385$ kJ (negative because heat must be removed), $T_\infty=203$ K, and $T_s=273$ K.
Solve: $\dot{Q}_{rad}=\varepsilon A\sigma(T_\infty^4-T_s^4)=(0.9)(0.250\,[m^2])(5.69\times10^{-8}\,[W/(m^2\times K^4)])(203^4-273^4\,[K^4])=-49.4$W Then, $\Delta t=(-385\times10^3\,[J])/(-49.4\,[J/s])=7800$ s $=2.17$ h.

Note that we have to use absolute temperatures in Equation (14.3) because the temperatures are raised to a power.

14.3.2 Fluid Mechanics

A *fluid* can be either a liquid or a gas, and like solid mechanics that separates the study of statics (fixed solids) from dynamics (bodies moving under forces), fluid mechanics deals with **fluid statics** and **fluid dynamics**. Liquids are normally virtually incompressible, whereas gases are compressible. Fluids are characterized by a various physical properties, such as:

- **Density** is the mass per unit volume and is measured in kg/m³ or lbm/ft³
- **Pressure** is the force per unit area and is measured in pascals (1 Pa=1 N/m²) or psi (psi=lbf/in.²). Normal air pressure at sea level is 101,325 Pa=14.696 lbf/in.²
- **Viscosity** is the fluid's resistance to flow. Viscosity is typically measured in units of N·s/m² or lbf·s/ft².

Table 14.4 summarizes these fluid mechanic units.

Table 14.4 Fluid mechanics units

Property	Symbol	Units	
Pressure	p	N/m² (or pascal)	lbf/in.² (or psi)
Speed	V	m/s	ft/s
Density	ρ (Greek "rho")	kg/m³	lbm/ft³
Viscosity	μ (Greek "mu")	N·s/m²	lbf·s/ft²

Pressure is measured either relative to the atmospheric pressure, called *gauge* (or *gage*) pressure, or as an absolute pressure. For example, atmospheric pressure at the surface of the Earth is 0 N/m² (or 0 psi) gage, but it is 101.325 kN/m² absolute pressure. Therefore,

$$p_{absolute}=p_{gage}+p_{atmosphere} \tag{14.4}$$

Fluid flow can be either laminar or turbulent. **Laminar** flow occurs at low speeds when a fluid flows in parallel layers that slide past each another like playing cards. In laminar flow, the motion of the fluid is very orderly, moving in straight lines parallel to the walls.

Turbulent flow is characterized by chaotic fluid motion. The fluid undergoes irregular fluctuations, or mixing, and the speed of the fluid at a point is continuously changing. Common examples of turbulent flow are blood flow in arteries, fluid flow in pipelines, convective motions in the atmosphere and ocean currents, and the flow of fluids through pumps and turbines.

The factor that determines which type of flow is present is the **Reynolds number** (*Re*), defined as

$$Re = \frac{\rho V D}{\mu} \qquad (14.5)$$

where ρ and μ are the fluid's density and viscosity, respectively, and V and D are a fluid speed and a characteristic distance. For example, for fluid flowing in a pipe, V is the average fluid speed and D is the pipe diameter. The Reynolds number is dimensionless, as shown by its units:

$$\left[\frac{kg}{m^3}\right]\left[\frac{m}{s}\right][m]\left[\frac{m^2}{Ns}\right] = \left[\frac{kg}{m^3}\right]\left[\frac{m}{s}\right][m]\left[\frac{m^2 s^2}{kg \cdot m \cdot s}\right] = [0]$$

Physically, the Reynolds number is a measure of the ratio of internal forces in a fluid to its viscous forces. Fluid flow in a pipe is **laminar** for Reynolds Numbers up to 2,000. Beyond a Reynolds Number of 4000, the flow is completely **turbulent**. Between 2,000 and 4,000, the flow is in transition between laminar and turbulent. As flow is increased in laminar flow beyond $Re = 2,000$, it can suddenly become turbulent because of an otherwise very small flow perturbation.

Example 14.4

Show that the Reynolds number is dimensionless when Engineering English units are used, and calculate the Reynolds number of water flowing at an average speed of 6.75 ft/s in a pipe with an inside diameter of 3.25 in. The density of the water is 62.4 lbm/ft³, and its viscosity is 2.35×10^{-5} lbf·s/ft².

Need: The units of the Reynolds number in Engineering English units and the Reynolds number of water flowing at 3.75 ft/s inside a 3.25 in. pipe.

Know: Expression (14.5) gives the Reynolds number as: $Re = \rho V D/\mu$.

How: In Engineering English units, $[\rho] = [\text{lbm/ft}^3]$, $[V] = [\text{ft/s}]$, $[D] = [\text{ft}]$, and $[\mu] = [\text{lbf} \cdot \text{s/ft}^2]$.

Solve: Notice that since these units contain both lbm and lbf, we will need to introduce g_c at some point. Substituting into the Reynolds number equation, we get:

$$[Re] = \frac{\left[\frac{lbm}{ft^3}\right]\left[\frac{ft}{s}\right][ft]}{\left[\frac{lbf \cdot s}{ft^2}\right]} = \left[\frac{lbm}{ft^3}\right]\left[\frac{ft}{s}\right][ft]\left[\frac{ft^2}{lbf \cdot s}\right] = \left[\frac{lbm \cdot ft}{lbf \cdot s^2}\right]$$

Since this is not dimensionless, we need to divide by $g_c = 32.2 \text{ lbm} \cdot \text{ft/lbf} \cdot \text{s}^2$ to get

$$\frac{[Re]}{[g_c]} = \frac{\left[\frac{lbm \cdot ft}{lbf \cdot s^2}\right]}{\left[\frac{lbm \cdot ft}{lbf \cdot s^2}\right]} = [0]$$

For water flowing at 6.75 ft/s inside a 3.25 in. pipe, the Reynolds number is:

$Re/g_c = (62.4)(6.75)(3.25)(1/12)/[(2.35 \times 10^{-5})(32.2)] = 1.51 \times 10^5$ [dimensionless].

14.3.2.1 Fluid statics

For an incompressible fluid at rest, the fluid statics equation is just the weight of a fluid column and the pressure it exerts,

$$p_2 - p_1 = \frac{\rho g(z_2 - z_1)}{g_c} \tag{14.6}$$

where z is the height, g is the acceleration of gravity (9.81 m/s² or 32.2 ft/s²), and $g_c = 32.2$ lbm · ft/lbf · s² in the Engineering English units, and 1 [0] (an integer) in SI units. Equation (14.6) accurately predicts how the pressure varies with depth in lakes and oceans, but not the atmosphere. Figure 14.7 shows the pressure profile of the Earth's atmosphere. Note that it is not linear as predicted by Equation (14.6) because the atmosphere is compressible and Equation (14.6) is only accurate for incompressible fluids (i.e., liquids). For compressible fluids, the relationship between pressure and height is not linear.

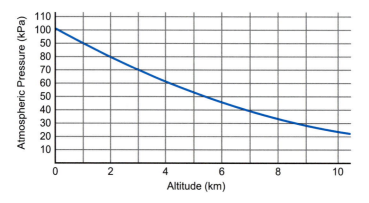

FIGURE 14.7 Atmospheric static pressure profile. We live comfortably only at the top left of this curve.

Example 14.5

Find the gage and absolute pressure under the surface of the ocean at a depth of 300. m. The density of salt water is 1025 kg/m³.

Need: $p_{absolute}$ and p_{gage} at 300. m under the sea.
Know: The density of sea water is $\rho = 1025$ kg/m³.
How: Equation (14.4) gives $p_{absolute} = p_{gage} + p_{atmosphere}$, and Equation (14.6) gives $p_{300m} - p_{atmosphere}$.
Solve: From Equation (14.6) with $g_c = 1$ [0] (an integer), we have

$$p_{300m} - p_{atmosphere} = \frac{\rho g(z_{300m} - z_{surface})}{g_c} = \frac{(1025\,[\text{kg/m}^3])(9.81\,[\text{m/s}^2])(300. - 0.\,[\text{m}])}{1}$$
$$= 3.02 \times 10^6 \text{kg}/(\text{m} \cdot \text{s}^2)\,\text{gage} = 3.02 \times 10^6 \text{N/m}^2\text{gage} = 3020\,\text{kN/m}^2\,\text{gage}$$
$$= 3020\,\text{kPa gage}$$

Then, the absolute pressure at 300. m is:

$$p_{300m} = (p_{300m} - p_{atmosphere}) + p_{atmosphere} = 3020\,[\text{kN/m}^2] + 101.324\,[\text{kN/m}^2]$$
$$= 3120\,\text{kN/m}^2\,\text{absolute (to 3 significant figures)}$$

14.3.2.2 Fluid dynamics

An incompressible fluid in steady motion can be modeled between two points 1 and 2 in a flow by the **Bernoulli Equation**:

$$\frac{p_1 - p_2}{\rho} + \frac{V_1^2 - V_2^2}{2g_c} + \frac{g}{g_c}(z_1 - z_2) = 0 \tag{14.7}$$

where p is the fluid pressure, V is the fluid speed, z is the fluid height, and ρ is the fluid density.

Also, g is the acceleration of gravity (9.81 m/s²; 32.2 ft/s²), and $g_c = 32.2$ lbm·ft/lbf·s² in the Engineering English units and 1(an integer) in SI units. This form of the equation is approximate only since it neglects any frictional losses due to the fluid's viscosity as well as pressure losses due to secondary rotational flows.

Example 14.6

Water flows through the nozzle shown below. The average water speed at the inlet is $V_1 = 20.0$ ft/s. If the pressure difference between the inlet and the outlet of the nozzle is $p_1 - p_2 = 100.$ lbf/ft², determine the outlet water speed.

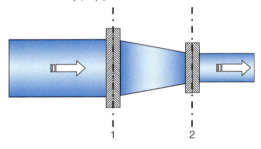

The density of the water is 62.4 lbm/ft³.

Need: The speed the nozzle's outlet air speed, $V_2 = ?$
Know: The inlet water speed is $V_1 = 20.0$ ft/s, $p_1 - p_2 = 100.$ lbf/ft², and the density of water is 62.4 lbm/ft³. The nozzle is horizontal, so $z_1 - z_2 = 0$.
How: Use Equation (14.7).
Solve: From equation (14.7), we have:

$$\frac{p_1 - p_2}{\rho} + \left(\frac{V_1^2 - V_2^2}{2g_c}\right) + \frac{g}{g_c}(z_1 - z_2) = \frac{100.[\text{lbf}/ft^2]}{62.4[\text{lbm}/ft^3]} + \left(\frac{20.0^2 - V_2^2[ft^2/s^2]}{2(32.2[\text{lbm}\cdot ft/\text{lbf}\cdot s^2])}\right) + 0 = 0$$

Then, solving for V_2 gives: $V_2 = 17.2$ ft/s. Hence the nozzle's cross-sectional area ratio (inlet to outlet) = 17.2/20.0 = 0.86 is the contraction at the nozzle's throat. This type of converging nozzle is often used for flow measurement.

14.3.3 Thermodynamics

Thermodynamics is the study of energy. The efficient use of natural and renewable energy sources is one of the most important technical, political, and environmental issues of the twenty-first century. For those reasons, thermodynamics can be used in ways that improve the lives of people around the world.

14.3.3.1 Thermodynamic system

In thermodynamics, we are interested in the energy flows of **systems** and carry out our analysis with **system diagrams**. The system is the item we are analyzing, and everything else in the universe is called the system's surroundings. This concept was introduced in Chapter 4.

There are three types of systems: isolated, closed, and open. An **isolated system** cannot exchange mass or energy with its surroundings. A **closed system** can exchange energy, but not mass, with the surroundings. An **open system** can exchange mass, energy, or both with the surroundings. Figure 14.8 illustrates these different systems.

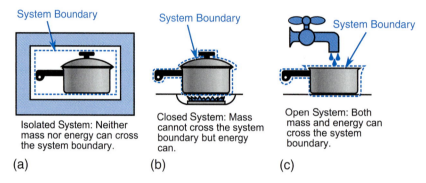

Isolated System: Neither mass nor energy can cross the system boundary.
(a)

Closed System: Mass cannot cross the system boundary but energy can.
(b)

Open System: Both mass and energy can cross the system boundary.
(c)

FIGURE 14.8 An illustration of isolated, closed, and open systems.

14.3.3.2 The first law of thermodynamics

The **first law of thermodynamics** states that energy is conserved. That is, energy cannot be created or destroyed, but it can be converted from one form of energy into another form of energy. There are many different forms of energy—kinetic, potential, thermal, magnetic, chemical, and so forth. Modern society benefits from a variety of energy conversion technologies (e.g., an automobile's engine to convert the chemical energy of gasoline into the kinetic energy of the vehicle).

Engineers have developed a way of keeping track of heat and work in thermodynamic systems. They use a concept called **internal energy**, designated by the symbol U. The first law of thermodynamics specifies that for a closed system the internal energy, U, of any system can change only if energy moves in or out of the system in the form of heat, work, or changes in the system's kinetic or potential energy. For any process that converts the system from one state (state 1) into another (state 2), the change in internal energy, $U_2 - U_1$, can be found from an energy balance on the system:

$$Q - W = U_2 - U_1 + m\left(\frac{V_2^2 - V_1^2}{2g_c}\right) + \frac{mg}{g_c}(Z_2 - Z_1) \tag{14.8}$$

where the quantity Q is the *heat added to the system*, W is *the work done by the system*,[4] the two remaining terms are the changes in kinetic and potential energy of the system, and $g_c = 32.2$ lbm·ft/lbf·s² in the Engineering English units and 1 [0] (an integer) in SI units. Equation (14.8) is the mathematical form of the first law of thermodynamics for a closed system. Note that U, Q, and W must all have the same units, Btu, ft·lbf, or joules (J).

---

[4]Note that due to historical convention, heat entering a system is positive and heat leaving a system is negative, but work has the opposite sign convention. Thus work leaving a system (i.e., the system doing work) is positive, but work entering a system (i.e., something doing work on the system) is negative. The usage developed because engineers wanted to know how much power was developed for how much coal was needed to run their steam engines, and they made these terms both positive.

The heat transfer term, Q, in Equation (14.8) consists of any of the three modes that were discussed in Section 14.3.1 of this chapter: conduction, convection, and radiation. The work term, W, in Equation (14.8) can also have several modes, mechanical work, electrical work, electrochemical work, and so forth.

Example 14.7

A 0.100 kg steel ball is dropped vertically from a height of 30.0 m and strikes the ground. The initial and final speeds of the ball are both zero. If there is no heat transfer or work involved in the process, determine the change in the ball's internal energy.

Need: $U_2 - U_1$ for the ball.
Know: Q and W for the ball are both zero, $V_1 = V_2 = 0$ and $m = 0.100$ kg.
How: Use Equation (14.8) to find $U_2 - U_1$ for the ball.
Solve: Substituting the known values into Equation (14.8), we get:

$$Q - W = 0 = U_2 - U_1 + 0 + \frac{(0.100[\text{kg}])(9.81[\text{m/s}^2])}{1}(0 - 30.0[\text{m}])$$
$$\text{or } U_2 - U_1 = 29.4\text{kg} \cdot \text{m}^2/\text{s}^2 = 29.4(\text{kg} \cdot \text{m/s}^2) \cdot m = 29.4 \ N \cdot m = 29.4\text{J}$$

The example below contains an error in the solution. The error could be in the equations, units, or computation. Can you find the error(s)?

Example 14.8: What's wrong with this solution?

A 5.00 lbm is book sitting 3.00 ft from the floor on the edge of a desk. The book is pushed off the desk and falls to the floor. Determine the change in the book's internal energy right after the book hits the floor if there is no work done or heat transfer?

Need: Where the initial potential energy of the book went.
Know: $Z_1 = 3.00$ ft and $Z_2 = 0$ ft, $W = 0$, $Q = 0$, and $V_2 = V_1 = 0$.
How: The answer can be found from Equation (14.8).
Solve: Start with Equation (14.8).

$$Q - W = (U_2 - U_1) + m\left(\frac{V_2^2 - V_1^2}{2g_c}\right) + \frac{mg}{g_c}(Z_2 - Z_1) = 0$$

Then, solve for the change in internal energy

$$U_2 - U_1 = 0 - m\left(\frac{V_2^2 - V_1^2}{2g_c}\right) - \frac{mg}{g_c}(Z_2 - Z_1) = 0 - \frac{mg}{g_c}(Z_2 - Z_1)$$
$$= -\frac{(5.00\text{lbm})(32.2\,\text{ft/s}^2)}{2(32.2\text{lbm} \cdot \text{ft/lbf} \cdot \text{s}^2)}(0 - 3.00) = 7.50\text{ft} \cdot \text{lbm}$$

Is this answer correct?[5] Why?

---

[5]The correct answer is 15.0 ft · lbf.

14.4 THE ELEMENTS OF MACHINE DESIGN

Any form of design is basically a decision-making process. The design process involves developing a plan to meet a need or solve a problem. For example, if you are designing something as simple as a chair, some of the things you will need to consider are:

1. The type of chair (e.g., an easy chair, an office chair, or a chair for a dining room table)
2. The person using the chair (e.g., an adult or a child)
3. The material used in the construction of the chair (wood, metal, plastic, or a combination of materials)
4. The strength and cost of the material used to make the chair
5. The shape and artistic appeal of the chair

In this section, we are concerned with the elements of machine design. In mechanical engineering, a "machine" is any mechanical system that transforms or transmits energy to do useful work.

14.4.1 Mechanical Behavior

When engineers select the materials that are to be used in the machines and parts they design, they need to know how these materials will behave under reasonable operating conditions. This subject is covered in detail in Chapter 13 of this book with a very short review below.

The criterion to be optimized when choosing materials for an engineering application might be cost, or weight, or performance, or some index such as minimizing weight × cost with a requirement of a particular strength. The constraints (also called design requirements) typically involve such words as **elastic modulus** (also called **stiffness), elastic limit, yield strength**, and **toughness**. While we all have loose ideas of what is meant by these terms, it is necessary to precisely express them as engineering variables. Those variables must, in turn, contain the appropriate numbers and correct units.

A design engineer must be able to

1. Define material requirements.
2. Understand classes of materials such as metals and polymers.
3. Understand the internal microstructure of materials that can be crystalline and/or amorphous.
4. Use stress-strain diagrams to express materials properties in terms of five engineering variables: stress, strain, elastic limit, yield strength, and toughness.
5. Use the results determined for those variables to carry out materials selection.

For example, if a child drops a plastic toy onto the floor, the impact should leave the toy looking like it did before it was dropped. So, as the toy's designer, how do you chose a plastic that is tough enough to survive a child's play, but not so expensive as to be unaffordable?

Example 14.9

Estimate the total energy that a 0.09 kg plastic toy car has to absorb in a 1.00 m/s collision with a perfectly rigid floor. If all of this energy is absorbed in the toy, what is the "specific" energy absorbed (i.e., energy per m^3 of volume) if the toy car has a volume of 5.69×10^{-5} m^3?

  Need: Energy absorbed and the specific energy absorbed.
  Know: Speed of impact, dimensions of toy.
  How: The conservation of energy provides the energy absorbed, and the energy absorbed divided by the volume of the toy gives the specific energy absorbed.

Solve: The toy car must **absorb** the total kinetic energy of the impact from a 1.00 m/s collision with the floor. The vehicle's kinetic energy is: $\frac{1}{2}\,mv^2 = \frac{1}{2} \times 0.09 \times 1.00^2$ [kg][m/s]$^2 = 0.045$ J.

The toy's volume $= 5.69 \times 10^{-5}$ m^3. Therefore, the

Specific energy absorbed = energy absorbed/volume of toy
$$= 0.045/(5.69 \times 10^{-5}) \text{ [J][1/m}^3] = 790 \text{ J/m}^3$$

With this information, the toy designer could survey the physical properties of various commercial plastics and choose one that meets his/her needs.

14.4.2 Machine Elements

Machine elements are basic mechanical parts used as the building blocks of most machines. They include shafts, gears, bearings, fasteners, springs, seals, couplings, and so forth. In this section, we are going to focus on the most commonly used machine element, gears.

Gears are used when engineers must deal with rotary motion and rotational speed. Rotational speed is defined in two ways, the more familiar being N, the *revolutions per minute* (RPM) of a wheel. There is also a corresponding "scientific" unit of rotational speed in terms of circular measure, radians/s. Its symbol is the Greek lowercase letter omega (ω). There are 2π radians in a complete circle. Hence, $N = 60 \times \omega/2\pi$ [s/min] [radians/s] [revolution/radian] = RPM; conversely:

$$\omega = 2\pi N/60 \,(\text{in radians/s when } N \text{ is in RPM}) \tag{14.9}$$

Angular speed ω is also directly related to **linear speed** v. Each revolution of a wheel of radius r covers $2\pi r$ in forward distance per revolution. Therefore, at N RPM, the wheel's tangential speed is $v = 2\pi rN/60 = r\omega$ (in m/s if r is in meters). Hence,

$$v = r\omega \tag{14.10}$$

Automobile engines that can have rotational speeds of 500–7000 RPM (maybe 10,000 RPM in very high-performance engines), but we have vehicles that are moving at speeds of, say, 100. km/h (62.1 mph). If the tire outer diameter is 0.80 m (radius of 0.40 m), what is the wheel's rotational speed when the vehicle is moving at 100. km/h?

In our current case, the wheels are rotating at a circular speed corresponding to the formula $r\omega = v = (100.$ km/h)$ \times (1000$ m/km) $\times (1$ h/3600 s) $= 27.8$ m/s. Hence, $\omega = v/r = 27.8/0.40 = 70.$ radians/s or $70. \times 60/2\pi = 670$ RPM. Somehow the rotational speed of the engine must be transformed into the rotational needs of the wheels. How can these two different speeds of rotation be reconciled? It is done by a mechanism called a transmission. A manual transmission is made of several gears.[6] A gear is simply a wheel with a toothed circumference (normally on the outside edge; see Figure 14.9).

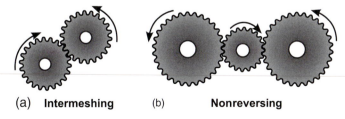

(a) **Intermeshing**   (b)   **Nonreversing**

FIGURE 14.9 (a and b) Intermeshing gears and nonreversing gears.

---

[6]In this book, all the gears are toothed and intermeshing; other types of gear such as ratchet wheels and "worm" gears are excluded from the present discussion.

A gear *set* or gear *cluster* is a collection of gears of different sizes, with each tooth on any gear having exactly the same profile as every other tooth (and each gap between the teeth being just sized to mesh). The teeth enable one gear to drive another—that is, to transmit rotation, from one gear to the other. *Note*: A simple gear pair as in Figure 14.9a *reverses* the rotational direction of the driven gear from that of the driving gear. You need at least three gears in a set of simple gears as per Figure 14.9b to transmit in the same direction as the original direction.

The **gear ratio** (GR) of a gear train is the ratio of the angular speed of the input gear to the angular speed of the output gear. It is easier to think in terms of N, the RPM, rather than in terms of angular speed ω, in radians per second, so that the GR of a simple gear train is:

$$\text{Gear Ratio (GR)} = \frac{\text{Input rotation}}{\text{Output rotation}} = \frac{N_1}{N_2} = \frac{d_2}{d_1} = \frac{t_2}{t_1} \qquad (14.11)$$

in which "d" stands for diameter and "t" stands for the number of teeth per gear. In other words, to make the output (driven gear) turn faster than the input (driving gear), we need a GR less than 1 and must choose an output (driven gear with diameter d_2) that is smaller than the input (driving gear with diameter d_1). To make the output (driven gear) turn slower than the input (driving gear), we need a GR greater than 1 and must choose an output (driven gear with diameter d_2) that is larger than the input (driving gear with diameter d_1).

Example 14.10

A 6.00 cm diameter gear is attached to a shaft turning at 2,000. RPM. That gear in turn drives a 40.0 cm diameter gear. What is the RPM of the driven gear?

Need: RPM of driven gear, N_2.
Know: Speed of driving gear $N_1 = 2,000.$ RPM. Radius of driving gear (r_1) is 3.00 cm. Radius of driven gear (r_2) is 20.0 cm. A sketch as per the inset is recommended.

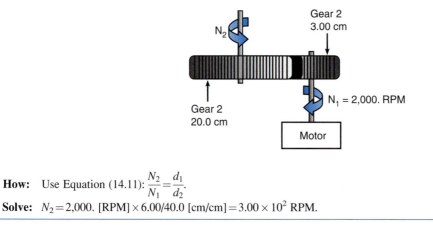

How: Use Equation (14.11): $\dfrac{N_2}{N_1} = \dfrac{d_1}{d_2}$.
Solve: $N_2 = 2,000.$ [RPM] $\times 6.00/40.0$ [cm/cm] $= 3.00 \times 10^2$ RPM.

Compound gear sets—sets of multiple interacting gears on separate shafts, as shown in Example 14.10, can be very easily treated using the gear ratio concept introduced above.

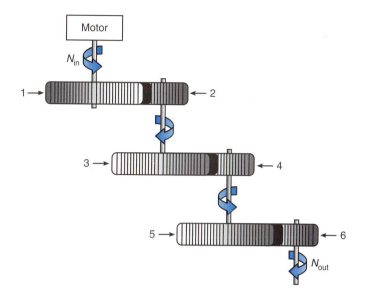

FIGURE 14.10 Compound gear train.

What is the gear ratio for the full set of compound gears? Note that some of the gears are *driven* and some are *drivers*; in addition, some of the gears are connected by internal shafts so that these turn at a common speed. These relationships make it very easy to analyze compound gear trains.

$$GR = \frac{\text{Product of diameter or number of teeth of DRIVEN gears}}{\text{Product of diameter or number of teeth of DRIVING gears}} \quad (14.12)$$

Example 14.11

A 70.0 RPM motor is connected to a 100 tooth gear that couples in turn to an 80 tooth gear that directly drives a 50 tooth gear. The 50 tooth gear drives a 200 tooth gear. If the latter is connected by a shaft to a final drive, what is its RPM? We'll go straight to solve.

Again, use a sketch to help visualize the problem.

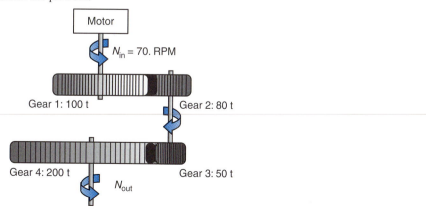

$$\text{Overall GR} = \frac{N_{in}}{N_{out}} = \frac{\text{Product of number of teeth on ''driven'' gears}}{\text{Product of number of teeth on ''driving'' gears}}$$

Hence, $N_{in}/N_{out} = (t_2 \times t_4)/(t_1 \times t_2) = (80 \times 200)/(100 \times 50) = 3.19$, and therefore,
$N_{out} = 70./3.19 = 21.9 = 22$ RPM
(Note that the number of gear teeth is a "counted" integer and thus has infinite significant figures.)

In a comprehensive analysis of gears, more than kinematics is involved. We need to consider not only the strength of the gears themselves but also how much power and force they are capable of transmitting. This subject is too complex in this introductory text, and we will just briefly introduce the idea of **torque** (or twisting moment), a central variable in gear analysis. Gears change not only the rotation speed but also the torque of the axle. Equation (14.13) shows that the torque T varies inversely with speed:

$$\frac{T_2}{T_1} = \frac{N_1}{N_2} = \frac{t_2}{t_1} = \frac{d_2}{d_1} = \frac{r_2}{r_1} \tag{14.13}$$

Therefore, to apply high torque to a shaft, you use large gears turned slowly by small intermeshing gears. However, you can use gears to achieve a desired combination of torque and RPM. For example, in a car with manual transmission, you can first use one set of gears to provide the wheels with high torque and low RPM for initial acceleration (first gear). Then, you can shift to another set of gears providing the wheels with lower torque and higher RPM as the car speeds up (second gear). Then, you can shift to a third gear combination that offers low torque and high RPM for cruising along a level highway at 65 miles per hour. In a modern automobile, there may be four, five, or six forward gears.

14.4.3 Manufacturing Processes

Virtually, everything that we use at home, at work, and at play was manufactured. Manufactured goods are everywhere—aircraft, bicycles, electronics, coat hangers, automobiles, refrigerators, toys, clothing, cans, bottles, cell phones, and so on. Manufacturing processes involve machining, drilling, milling, welding, extrusion, blow molding, and thermoforming. This subject is covered in detail in Chapter 12 of this book and thus only briefly described in this chapter.

Manufacturing plays a vital role in the U.S. economy. However, it has undergone major changes over the past several decades. Today, manufacturers use sophisticated technology to produce a variety of useful products. They have increased worker productivity, reduced the size of their workforce, and increased output. Previously, many industries required only a high school diploma for their production-related jobs, but manufacturing jobs today require workers with greater skills.

Manufacturing in the twenty-first century is far more than assembly and production. It also involves a broad spectrum of other functions such as research and development, engineering, sales and management, warehousing and storage, and shipping, just to touch on a few of modern manufacturing's key product life-cycle components (see Figure 14.11).

Life cycle manufacturing activities are critically important and add value to products by:

1) Providing a "Green" manufacturing environment.
2) Creating competitive advantage through innovation, short delivery times, customer service, and so forth.
3) Utilizing software components and communication connectivity.
4) Enabling mass customization of products for unique market segments through increased understanding of customer groups and on-demand production responses.

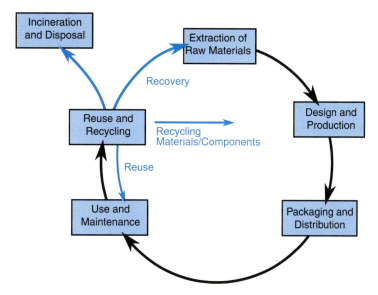

FIGURE 14.11 Life cycle manufacturing.

SUMMARY

In this chapter, we have explored the mechanical engineering fields of thermal design and machine design. Thermal design is composed of the subjects of heat transfer, fluid mechanics, and thermodynamics. Machine design is composed of the subjects of mechanical behavior, machine elements, and manufacturing processes. Each of these subjects is so important to a mechanical engineering student who will have at least one course in each of these six subjects.

EXERCISES

1. If the heat conduction rate through a 3.00 m^2 wall 1.00 cm thick is 37.9 W when the inside and outside temperatures are 20.0 and 0.00 °C, respectively, determine the thermal conductivity of the wall.

2. If a 15.0 × 3.00 m wall 0.100 m thick with a thermal conductivity of 0.500 W/m·K is losing 75.0 W of heat, determine the inside wall temperature if the outside wall temperature is 0.00 °C.

3. An engineer needs to transfer heat at a rate of 50.7 W through a wall 0.15 m thick and 3.00 m high. If the thermal conductivity of the wall is 0.085 W/m·K and the inside and outside temperatures of the wall are 20.0 and 10.0 °C, respectively, how long must the wall be in order to satisfy the heat transfer rate requirement?

4. The equation for the conduction heat transfer rate through a circular tube of length L is

$$\dot{Q} = 2\pi L k_t \frac{T_{inside} - T_{outside}}{\ln\left(r_{outside}/r_{inside}\right)}.$$

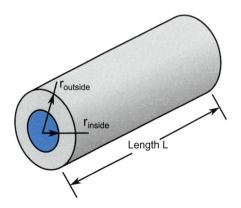

where ln() is the natural logarithm. A 35.0 ft long copper pipe with an inside diameter of 0.920 in. and an outside diameter of 1.08 in. has an inside temperature of 122 °F and an outside temperature of 118 °F. Determine the heat transfer rate through the pipe. The thermal conductivity of copper can be found in Table 14.1.

5. A cooking pot on a stove has a surface area of 0.01 m^2. The surface temperature of the pot is 70.0 °C, and the temperature of the surrounding air in the kitchen is 20 °C. The convective heat transfer coefficient is 12.0 W/($m^2 \cdot$ K). Determine the convection heat transfer rate from the pot to the kitchen air.

6. An electric heater provides 1,500. W of heat to a room. A fan blows air over the heating element with a forced convection heat transfer coefficient of 275.0 W/($m^2 \cdot$ K). The temperature of the heating element is 300.0 °C and the surface area of the heating element is 0.0195 m^2. Determine the air temperature in the room under these conditions.

7. The convective heat transfer rate from a computer chip is 3.00 W and the surface area of the chip is 1.00 cm^2. The surface temperature of the chip is 28.0 °C and the air temperature is 18.0 °C. Determine the convective heat transfer coefficient for the chip.

8. An electric circuit board is to be cooled by forced convection through cooling fins. 100. W of heat needs to be removed from the circuit board from a maximum temperature of 40.0 °C to the room air at 20.0 °C. The forced convection heat transfer coefficient is 355 W/($m^2 \cdot$ K). Determine the required surface area of the fins.

9. A flat plate at 200. °C is painted flat black. It radiates heat into a room whose walls are at 18 °C. If the surface area of the plate is 0.210 m^2, what is the rate of heat radiated into the room? The thermal emissivity of flat black paint can be found in Table 14.3.

10. The surface area of an average naked human adult is 2.2 m^2 and the average skin temperature is 33 °C (91 °F). The emissivity of human skin is 0.95. Determine the radiation heat transfer rate from the human to the walls of a room if they are at 22 °C (72 °F). This is why a naked person feels chilly at room temperature.

11. A 4.00 in. diameter glass sphere in a vacuum chamber contains a chemical reaction that produces heat at a rate of 1.00×10^5 Btu/h. The sphere radiates this to the inside walls of the vacuum chamber that are at 85.0 °F. What is the surface temperature of the glass sphere?

12. The heat exchanger on a space craft has a surface temperature of 70 °F, and emissivity of 0.83, and radiates heat to outer space at 3.00 K. If the heat exchanger must remove 375 W from the space craft, what should its surface area be?

13. What is the Reynolds number of water flowing at 3.50 ft/s through a 3.00 in. diameter pipe? The density and viscosity of the water are 62.4 lbm/ft^3 and 2.03×10^{-5} lbf·s/ft^2, respectively.

14. You are required to keep the flow of air in a 6.00 in. diameter duct in the laminar flow region. What is the maximum average speed that the air can have before it becomes turbulent? The density and viscosity of the air are 0.074 lbm/ft^3 and 3.82×10^{-7} lbf·s/ft^2.

15. Figure 14.7 is a plot of atmospheric pressure vs. height. If you draw a straight line from the starting point to the ending point of this line and determine the slope of this line, you will be able to use Equation (14.6) to determine the "average" density of the Earth's atmosphere over this height range. What is this average density?

16. If the pressure required to crush a new submarine is 11,700 kN/m^2 absolute, what is the maximum depth it can reach in the ocean? See Example 14.5 for the properties of sea water.

17. Water slowly flows through a vertical cylindrical pipe 30.0 m in height. The density of the water is 998 kg/m^3. Determine the pressure change between the inlet (bottom) and outlet (top) of the pipe.

18. What is the exit speed from the nozzle in Example 14.6 if the fluid flowing through the nozzle is air instead of water? The density of air is 1.02 kg/m^3.

19. How much heat is produced by a stationary incandescent 100. W electric light bulb that has a constant internal energy after it has been illuminated for 1.00 h?

20. A 10.00 g bullet is fired horizontally at a rigid wall. The speed of the bullet just before it hits the wall is 867 m/s, and it is completely stopped by the wall. If there was no heat transfer or work done by the bullet when it hits the wall, what was its initial kinetic energy and what happened to this kinetic energy after impact?

21. If you put 127 J of work into throwing a 0.123 kg rock vertically into the air without changing its internal energy, how high will it go?

22. If you drop a book from your desk onto the floor 0.983 m below the desk, with what speed does it hit the floor? There is no work or heat transfer done during this process, and the book's internal energy does not change during the fall.

23. If the pedal gear in the illustration below has 60 teeth and the sprocket has 30 teeth, what is the gear ratio?

Bicycle Gears

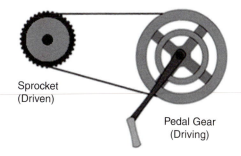

Sprocket
(Driven)

Pedal Gear
(Driving)

24. If the pedal in the illustration for Exercise 23 revolves once, how many times does the sprocket gear revolve?

25. The diameter of the sprocket in Exercise 23 is 3.25 in. What is the diameter of the pedal gear?

26. If the 6.00 cm diameter gear in Example 14.10 has 30 teeth, how many teeth does the 40.0 cm gear have?

27. If the torque produced by the motor in Example 14.11 was 875 N·m, what would be the torque at the output of the gear train?

28. What is the overall gear ratio of the compound gear train shown below, and what is the RPM of gears B and C?

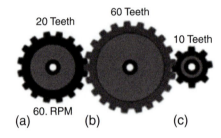

29. This concerns the case of the American Society of Mechanical Engineers (ASME) versus Hydrolevel Corporation.[7] In 1971, the engineering firm of McDonnell and Miller, Inc. requested an interpretation of the ASME Boiler and Pressure Vessel (BPV) Code from the ASME Boiler and Pressure Vessel Codes Committee. T.R. Hardin, chairman of the ASME committee and employee of the Hartford Steam Boiler Inspection and Insurance Company in Connecticut, wrote the original response to McDonnell and Miller's inquiry. ASME's interpretation was used by McDonnell and Miller salesmen as proof of Hydrolevel's noncompliance. Subsequently, Hydrolevel never acquired sufficient market penetration for sustaining business and eventually went bankrupt.

Hydrolevel then sued McDonnell and Miller, the Hartford Steam Boiler Inspection and Insurance Company, and ASME for restraint of trade. Hydrolevel's lawyers argued that two key ASME Subcommittee members acted not only in the self-interest of their companies but also in violation of the Sherman Antitrust Act. The litigation against ASME went all the way to the Supreme Court where the case was settled for $4.75 million in favor of Hydrolevel.

Please prepare a one-paragraph response to the following questions:

a. How could McDonnell and Miller have avoided the appearance of a conflict of interest?

b. What was T.R. Hardin's responsibility as the chairman of the BPV Code Heating Boiler Subcommittee? How could he have handled things differently to protect the interests of ASME?

30. You are an engineer who has experience as an expert witness.[8] You have been contacted by two attorneys working on the same case. The first attorney wants to hire you as an expert witness to support an inventor who claims to have developed a device that operates in a cycle and delivers 1000 Btu while consuming only

---

[7]Abstracted from http://ethics.tamu.edu/ethics/asme/asme1.htm.
[8]Abstracted from http://wadsworth.com/philosophy_d/templates/student_resources/0534605796_harris/cases/Cases/case64.htm.

320 Btu per cycle. The second attorney wants to hire you as an expert witness against the same inventor for allegedly defrauding investors with the invention described by the first attorney.

Please prepare a one-paragraph response to the following questions:

a. Whose expert witness would you prefer to be? Why?

b. In deciding which attorney to help as an expert witness, what ethical questions will you ask?

Nuclear Engineering

15

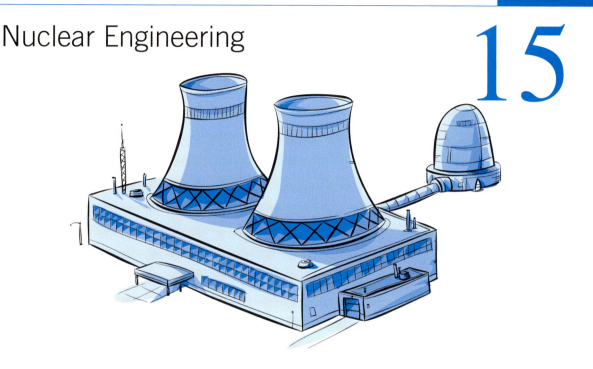

15.1 INTRODUCTION

This chapter describes one alternative to today's dominant fuels for electrical energy production, coal, oil, and natural gas, all of which produce copious amounts of CO_2, which is infamous as a greenhouse gas. We also overwhelmingly use hydrocarbon fossil fuels for our motor vehicles. Nuclear power can substitute as the source of electrical power but cannot directly make transportation fuels.

There are about a thousand nuclear reactors located around the world of which about 500 are in large electrical power generating plants. They have various sizes and perform various functions. The smallest are the research reactors with minimal power output; at the other end of the scale, the largest nuclear power reactors are designed to produce about 1300 MW[1] of electricity. The US has over 100 nuclear power reactors producing about 17% of its electrical power requirements—and without the concomitant production of CO_2.

There have been nuclear reactor incidents/accidents worldwide of varying seriousness. At least four accidents have produced significant concerns well beyond the borders of the nuclear plant that had the accident in historical order: Windscale/Sellafield (UK), Three Mile Island (U.S.), Chernobyl[2]

---

[1]If this size nuclear reactor runs for more than 300 days/year, it produces about 10^{10} kWh of electricity while producing revenue of about $2,000,000 *per day*. A reactor of this size will serve a million homes.
[2]We might ask, "Do nuclear reactors make your stomach Chernobyl?"

(Ukraine), and Fukushima (Japan). On a relative scale, these have produced total radioactive releases[3] of about (Table 15.1).

Table 15.1 Relative radioactive releases following major nuclear accidents

Reactor accident	Radioactive release relative to 3 mile island
Windscale/Sellafield	About 1500
Chernobyl	About 10,000,000
Fukushima	200,000,000-400,000,000

It is worth noting that the release from Three Mile Island (the worst nuclear accident in U.S. history) produced much less radiation release than the others because its **secondary containment** (that is, its outer steel/concrete enclosure) held. The others did not (and, in the case of Chernobyl, it had no secondary containment to begin with). The origin of the nuclear radiation will be covered later in this chapter.

This chapter is intended to inform as well as to instruct. It is organized with three objectives: (A) to present the (very) basic physics of nuclear reactions, (B) to explain what occurs in nuclear power plants, and (C) to explain the origin and implications of nuclear waste materials. The reader should then be in a position to make an *educated* pronouncement of whether possible radioactive releases from nuclear power plants under severe accidents offset the global effects of CO_2 release from fossil energy sources.

15.1.1 Nuclear Fission

An analysis starts with the basic nuclear physics of **fission**, a word meaning "splitting" of an atom by electrically neutral[4] **neutrons**. Normal chemistry depends on the motion of a local electrons belonging to atoms, but **nuclear** reactions such as fission are too energetic (perhaps several million times so) for meaningful interactions with electrons.

Fissile atoms can be fissioned, that is, split by neutrons; for reasons that will become apparent, we are particularly interested in fissile atoms that can be readily split by very low energy neutrons.

The tiny nucleus of an atom is where virtually all of its mass resides including its positively charged **protons** and its uncharged neutrons. The major atom of interest in fission is uranium-235, written as ^{235}U or as $^{235}_{92}U$; its mass is due to 235 particles consisting of 143 neutrons plus 92 protons.

An **isotope** is an atom with the same nuclear charge as another isotope; it is therefore the same chemical substance but of different mass due to more or fewer neutrons within its nucleus. U-235 is one of several uranium **isotopes** occurring to the extent of 0.72% in natural uranium, the rest being another isotope U-238.

Thus, the superscript indicates the **atomic mass** of the atom's nucleus and the subscript (which is called the **atomic number**) indicates its number of protons. Hence $^{235}_{92}U$ has 92 protons plus 143 neutrons. Similarly the isotope $^{238}_{92}U$ has and 146 neutrons but still just 92 protons.

In the nucleus, neutrons and protons are collectively called **nucleons**. For reasons that will soon become evident, in nuclear reactions, it is best for beginners to explicitly write the symbols for nuclei with their atomic numbers (even though it is redundant writing both the chemical symbol and its atomic number).

---

[3]Some of the uncertainty is due to the difficulty in accurate measurements and some due to governments/corporations wanting to minimize bad news.

[4]Charged particles such as protons are robustly repelled from the nucleus of an atom by electrostatic charges in the nucleus.

The uranium isotope $^{238}_{92}U$ is relatively stable toward neutrons but $^{235}_{92}U$ will readily absorb a low energy neutron momentarily fusing with it making its nucleus too unstable to stay together so that it will fission with a variety of possible outcomes with these typical fission fragments:

$$^{1}_{0}n + ^{235}_{92}U \Rightarrow ^{141}_{56}Ba + ^{92}_{36}Kr + 3^{1}_{0}n \tag{15.1}$$

$$^{1}_{0}n + ^{235}_{92}U \Rightarrow ^{90}_{38}Sr + ^{143}_{54}Xe + 3^{1}_{0}n \tag{15.2}$$

$$^{1}_{0}n + ^{235}_{92}U \Rightarrow ^{131}_{53}I + ^{100}_{39}Y + 5^{1}_{0}n \tag{15.3}$$

The **fission products** in these particular nuclear reactions are isotopes of barium, krypton, strontium, xenon, iodine, and yttrium, respectively, all appearing on the right-hand side of the expressions. While learning, it is best to write a neutron as $^{1}_{0}n$ instead of just n indicating it has a mass of 1 atomic unit (essentially identical with the mass of an ordinary hydrogen atom whose nucleus contains only one proton and no neutrons) and also the neutron has no charge and thus its subscript is "0".

We need to pay attention to two **conserved** quantities in nuclear reactions: the number of nuclear charges (protons) and the total number of neutrons and/or total nucleons. That sounds complicated but it is not. Table 15.2 shows how to think about nuclear reactions (it is actually quite close to how ordinary chemical reactions are balanced).

Table 15.2 Nuclear reactions

Nuclear reaction	Nuclear particle	LHS neutrons	LHS protons	Nuclear fragments	RHS neutrons	RHS protons
(15.1)	$^{235}_{92}U$	$143+1=\mathbf{144}$	**92**	$^{141}_{56}Ba/^{92}_{36}Kr$	$85+56+3=\mathbf{144}$	$56+36=\mathbf{92}$
(15.2)	$^{235}_{92}U$	$143+1=\mathbf{144}$	**92**	$^{90}_{38}Sr/^{143}_{54}Xe$	$52+89+3=\mathbf{144}$	$54+38=\mathbf{92}$
(15.3)	$^{235}_{92}U$	$143+1=\mathbf{144}$	**92**	$^{131}_{53}I/^{100}_{39}Y$	$78+61+5=\mathbf{144}$	$53+39=\mathbf{92}$

Note: LHS and RHS= left- and right-hand sides, respectively, of the reaction.

Notice that each side of the reaction balances both the number of proton charges and the number of nucleons. Any deficiency in the number of nucleons in the fissioned nuclei is balanced by the number of free neutrons produced. Be careful: these conservation laws do *not* imply that the total masses from the LHS and RHS are also exactly equal. In fact, using precision values for the masses of the fission products (including the neutrons) show that a small amount of mass is missing in each nuclear reaction (see next section).

A self-sustaining nuclear reaction occurs when sufficient neutrons are produced on the RHS of the fission reactions to supply the next round of LHS reactions with sufficient neutrons, which we here normalize to just one neutron. In practice, we need more than just enough neutrons per generation shown on the LHS because of the losses of neutrons to parasitic side reactions or other losses. In practice, in water-cooled nuclear reactors, the net number of neutrons to sustain a **chain reaction** is about 2.4 per generation. Thus, of the ~2.4 neutrons in say generation 1, just one (per reaction) must make it to be fissioned in the next generation. If we can hold it at one net neutron in generation 2 and ditto in subsequent generations, we have a self-sustaining nuclear reaction: if less than one net neutron, the nuclear reaction will die down. Of course if there is more than 1 neutron/reaction in generations 2, 3, 4 ..., we have the makings of a nuclear explosion.

Some boron compounds are added to some nuclear reactors because they have a huge ability to absorb neutrons. The favored isotope of boron is $^{10}_{5}B$; it is converted to isotopes of lithium under neutron irradiation, a strategy that is useful in the control of nuclear power reactors.

15.1.2 Nuclear Energy

Einstein's Special Relativity Theory quantitatively shows that the missing fission mass[5] in a nuclear reaction of Δm is converted to energy according to:

$$\Delta E = \Delta mc^2 \qquad (15.4)$$

where c is the speed of light, 3.00×10^8 m/s (186,000 miles/s). In a typical fission reaction, the deficiency of mass missing from the right-hand side of the reaction is about 0.2 **atomic mass units** (abbreviated as **amu**). Thus, of the $235 + 1$ amu in the reactants, the fission products are missing a mass of roughly one part in a thousand.

Example 15.1

Convert the missing mass of 0.200 amu/fission reaction into kg.

Need: 0.200 amu of mass $=$ _____ kg
Know: Avogadro's number[6] is 6.02×10^{26} particles/kmol.
How: One amu is essentially the mass of a solitary neutron or proton (or even a hydrogen atom). For example, Avogadro's number of hydrogen atoms has a mass of 1.00 kg so one hydrogen atom has a mass of $1.00/6.022 \times 10^{26} = \mathbf{1.66 \times 10^{-27}}$ kg.
Solve: Thus $\mathbf{0.200\ amu} = 0.200 \times 1.66 \times 10^{-27}$ kg $= \mathbf{3.32 \times 10^{-28}}$ kg.

Surprisingly, this is not much mass.

Energy units are much more convenient. Energy units based on a single electron's charge are useful for reactions at the single atom level. One electron charge is 1.602×10^{-19} coulombs and thus if an electron is accelerated by $+1.000$ V, its energy is 1.602×10^{-19} coulomb $\times$ volts or 1.602×10^{-19} J which is defined as 1 eV (one electron volt of energy). Mega is 10^6; so, 1 MeV $= 10^6$ eV, a convenient unit when dealing with nuclear reactions.

Example 15.2

Convert the missing mass of 0.200 amu/fission reaction to MeV.

Need: 0.200 amu of mass $=$ _____ MeV/reaction.
Know: An eV is 1.60×10^{-19} J and thus 1.00 MeV is 1.60×10^{-13} J.
How: 0.200 amu $= 3.32 \times 10^{-28}$ kg and use Einstein's equation.
Solve: Then $\Delta E = \Delta mc^2 = 3.32 \times 10^{-28} \times (3.00 \times 10^8)^2$ [kg][m/s]$^2 = \mathbf{2.99 \times 10^{-11}}$ J.

Since one MeV is 1.60×10^{-13} J, then $\Delta E = 2.99 \times 10^{-11}/1.60 \times 10^{-13}$ [J][MeV/J] $= \mathbf{187\ MeV}$.

The energy stated in joules is not a particularly convenient magnitude to remember; it is far easier to work in MeV. The result of this calculation is about right for any of the fission reactions (15.1)–(15.3) although a more accurate average number is about 200 MeV/reaction. Easy to remember but how big is it?

---

[5]The upper case Greek Δ is used to denote change of some property. For Greek letters in general see http://www.greek-language.com/Alphabet.html.
[6]This is covered in some detail in the chapter on Chemical Engineering; for the moment please either read Chapter 7, Section 7.4 or just accept Example 15.1 for now. Avogadro's number is used in several areas of engineering.

Example 15.3

Compare fission energy to a chemical reaction such as the combustion of a mole of ethanol whose heat of combustion is 1.25×10^6 kJ/kmol.

Need: Compare a typical fission reaction to the combustion of ethanol, which has a combustion heat of 1.25×10^6 kJ/kmol.
Know: Fission reactions produce about 200. MeV, which by joules is about 3.2×10^{-11} J/fission.
How: Scale to 6.02×10^{26} fissions (which is a kmol).
Solve: The fission reaction scales to $3.2 \times 10^{-11} \times 6.02 \times 10^{26}$ [J/fission][fissions/kmol] $= 1.9 \times 10^{16}$ J/kmol
   $= \mathbf{1.9 \times 10^{13}}$ **kJ/kmol.**

The energy released in this nuclear fission reaction is thus about 1.9×10^{13} kJ/kmol. It can be compared to a heat of combustion of ethanol of about 1.25×10^6 kJ/kmol—*a difference of about 10 million.* This is the energy source for both nuclear reactors and nuclear bombs.[7]

Most of the 200. MeV/fission energy ends up as heat inside a nuclear reactor. The reactor works by directing all of this fission energy into a form that can be extracted from the reactor.

15.2 NUCLEAR POWER REACTORS

The way a nuclear reactor works starts with a fuel pellet.[8] Most reactors use uranium oxide pellets although some countries use a mixture of uranium oxide and plutonium oxide, the latter containing fissile $^{239}_{93}Pu$. These pellets fit tightly into zirconium[9] metal tubes; fuel rods are these tubes filled with fuel pellets and which are then fitted into subassemblies (Figure 15.1).

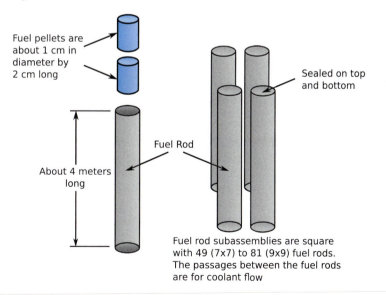

Fuel pellets are about 1 cm in diameter by 2 cm long

Sealed on top and bottom

About 4 meters long

Fuel Rod

Fuel rod subassemblies are square with 49 (7x7) to 81 (9x9) fuel rods. The passages between the fuel rods are for coolant flow

FIGURE 15.1 Nuclear fuel rods.

---

[7]It took the annihilation of about one gram of U-235 and Pu-239, respectively to vaporize the cities (and populations) of Hiroshima and Nagasaki in 1945, thus ending WWII.
[8]The fuel contains only about 2–4% of $^{235}_{92}U$ with the rest being the much more stable $^{238}_{92}U$. The fuel has been enriched in ^{235}U from its natural extent of about 0.7%. That step is perhaps the technically most challenging in the whole of nuclear power production.
[9]Zirconium metal is resistant to damage by nuclear bombardment inside the reactor.

The picture below shows a completed subassembly. Notice the technician has minimal protective wear because non-irradiated nuclear fuel is only very slightly radioactive (Figure 15.2).

FIGURE 15.2 Nuclear fuel subassemblies. *Source*: US Nuclear Regulatory Commission, http://www.nrc.gov/images/reading-rm/photo-gallery/20071114-045.jpg.

A rod bundle is then inserted into a rectangular metal sleeve. Finally, coolant such as water is forced into the sleeve from the bottom and removes the heat generated by nuclear fission. The coolant also serves to **generate** the nuclear heat by the process of neutron moderation as described in Section 15.3.

The next step is to insert the subassemblies into the shell of the nuclear reactor. The shell is a massive pressure vessel that contains the hot water and steam circulating within reactor. Generic reactor flow circuits are shown in Figure 15.3.

Two of the most common types of reactors are schematically shown in Figure 15.3: they are a pressurized water reactor (PWR) and a boiling water reactor (BWR) that are collectively known as "light water reactors (LWR)," light water being ordinary water. In nuclear reactors, water is boiled to steam either directly or indirectly in a steam generator and then the steam is passed through a steam turbine that turns an electrical generator.

PWRs operate at a substantial pressure of about $160 \times$ atmosphere and BWRs at approximately one half that. They both have efficiencies of about 33% meaning that about 33% of the heat from the nuclear fuel is captured into electricity and the remaining 67% is lost as waste heat in the steam condenser. An important issue therefore is that nuclear reactors are often located near large sources of water so they can condense the turbine exhaust steam. Note that the balance of the plant (for producing electric power) is similar for either kind of LWR.

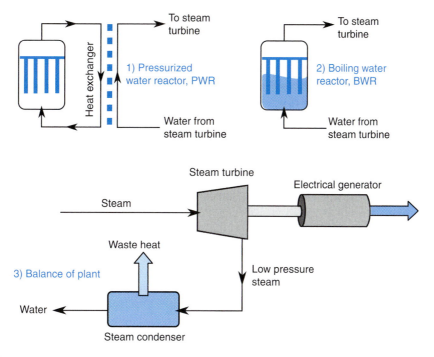

FIGURE 15.3 Nuclear power reactors.

Nuclear power plants are used as "base load," meaning they are run flat out for as long as possible because they are expensive to install (so you want to run them 24/7 to spread their capital costs as much as possible) but they are cheap to run primarily because so little fuel[10] is used.

15.3 NEUTRON MODERATION

Water in light water reactors has two purposes. The first is the more obvious—removing the nuclear energy produced as a result of fission as heat and delivering that heat in steam to a steam turbine-generator set. But there is also a much more compelling nuclear effect of water that is responsible for the ultimate reactivity of the nuclear chain reaction. When a neutron collides with a proton (i.e., a hydrogen atom less its electron), kinetic energy is efficiently exchanged between them. This is important because fresh neutrons from the fission process called **fast neutrons** have an average energy of about 2 MeV (so they move at about 5% the speed of light) but they are slowed by collisions with protons to "**thermal energies**" or as "**thermalized neutrons**" and

---

[10]A 1000 MW power reactor produces about 100 tons/*year* of (highly radioactive!) spent fuel; a coal-fired power plant of the same capacity needs about 8000 tons/*day* of coal producing about 800 tons/*day* of solid waste and about 9000 tons/*day* of CO_2.

behave rather like atoms in an ideal gas. Their energy is in the form of kinetic energy, $\frac{1}{2}mV^2$. Neutron thermal energies are only a few hundredths of an eV compared to ~2 MeV for fast neutrons.

Example 15.4

What is the average energy of an atom in an ideal monatomic gas at 20.0 and 350. °C? Assume the average energy of an ideal gas is kT in which T is the absolute temperature and k is the Boltzmann's constant, 8.617×10^{-5} eV/K.

> **Need:** Average energy of an atom in a gas at 20.0 and 350. °C.
> **Know:** A gas atom's or molecule's average energy is kT and $k = 8.617 \times 10^{-5}$ eV/K.
> **How:** Convert temperature to Kelvin. $T = 20.0 + 273 = 293$ K and $T = 350. + 273 = 623$ K.
> **Solve:** Atom's energy $= kT = 8.617 \times 10^{-5} \times 293 = \mathbf{0.0252\ eV}$ **at 20.0 °C** and at **350. °C**,
> $8.617 \times 10^{-5} \times 623 = \mathbf{0.0537\ eV}$.

Which kind of neutron causes the faster rate of nuclear fissions? It turns out that the probability of a fission reaction is about 500–1000 × greater for thermal neutrons than for fast ones. For this reason, we need only modest enrichment of the fissile components in the fuel[11] so water-moderated nuclear reactors need only 2–4% enrichment of ^{235}U compared to its naturally occurring concentration of about 0.7%.

A good demonstration of the effects of collisions between neutrons and protons is demonstrated by Newton's Cradle. A video[12] is immediately convincing and the principle is shown in Figure 15.4. Two steel balls of the same mass are suspended on hooks by a piece of string, nylon, chain, etc. One is hanging vertically and the other is pulled away and released so that it hits the stationary ball full on. If the two balls are both hard and 100% elastic (see Chapter 13 for what is meant by "elastic"), *all* of the kinetic energy of the swing ball is transferred to the previously stationary one.[13] In Figure 15.4, ball A hits ball B and it swings exactly to a position mirroring ball A's initial condition while ball B stops completely. Of course, ball B is then free to repeat the motions with the ultimate number of collisions being determined by various small sources of friction.

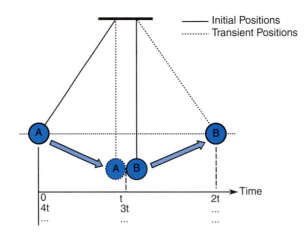

FIGURE 15.4 Newton's cradle.

---

[11]For this reason that you can't make a nuclear bomb without considerably further enriching ^{235}U.
[12]See http://www.youtube.com/watch?v=DJ_lulkpL1s
[13]The analysis requires both that energy and linear momentum are conserved.

In a perfectly elastic system (one that loses no kinetic energy after the rebound), the swing amplitude and the incremental time to swing between collisions remain fixed. The time intervals are given as successive multiples of "t" in Figure 15.4.

When a neutron hits a proton head on, the neutron will behave the same way—it will lose all of its kinetic energy and come to a stop with the proton taking all of the original kinetic energy. The process is called **elastic scattering**. Of course, in practice, only a few neutrons will hit protons head on; more likely they will give a glancing blow. A fuller analysis of the scattering process shows about 17 collisions with protons is enough to slow the neutron to thermal energies. The overall process is known as **neutron moderation** and the substance doing it (in this case, the hydrogen nuclei of ordinary water) is the **moderator**.

When neutrons hit larger nuclei than protons, say the nucleus of a carbon atom (which contains six neutrons and six protons), it requires many more collisions (say a hundred or so) to slow down a fast neutron. If a neutron elastically hits a still heavier nucleus, it simply bounces off.

This process is known as "thermalization" because it ultimately slows the 2 MeV fast neutrons on fissioning to thermal speeds characteristic of the protons in the moderator.

Example 15.5

Suppose a fast neutron in a nuclear reaction has energy E_{init} before thermalization. After thermalization, it has energy E_{final}. The number of collisions to thermalize the fast neutron p is given by:

$$p = \frac{\ln\left(\frac{E_{init}}{E_{final}}\right)}{\xi} \qquad (15.5)$$

in which ξ (pronounced "xi") $= 1.00$, otherwise $\xi = \frac{2}{A+2/3}$ where $A = $ mass of struck nucleus/mass of neutron but if $A = 1.00$ then $\xi = 1.00$. The logarithmic term in Equation (15.5) is called the logarithmic energy decrement. How many collisions does it take for neutrons in a water-moderated reactor to slow to thermal energies at 20.0 °C and at 350. °C?

Need: The number of collisions to thermalize a fast neutron given its initial e energy is ~2 MeV

Know: At 20.0 °C, a thermal neutron has energy of 0.0252 eV and at 350. °C, energy of 0.0537 eV. For a hydrogen nucleus relative to a neutron, $A = 1.00/1.00 = 1.00$ and also $\xi = 1.00$.

How: Use the logarithmic energy decrement method, Equation (15.5).

Solve: At 20.0 °C $(E_{init}/E_{final}) = 2.0 \times 10^6/0.025 = 8.0 \times 10^7$ and at 350. °C, $(E_{init}/E_{final}) = 2.0 \times 10^6/0.0537 = 3.7 \times 10^7$.

Thus at **20.0 °C**, $p = \frac{\ln\left(\frac{E_{init}}{E_{final}}\right)}{\xi}$

Therefore, $p = \frac{\ln\left(8.0 \times 10^7\right)}{1.00} = $ **19 collisions at 20 °C.**

At **350. °C**, $p = 17$ **collisions**.

15.4 HOW DOES A NUCLEAR REACTOR WORK?

In order to calculate how a nuclear reactor manages to balance a lot of disparate factors, first break down the reactor core physics into several manageable pieces. The object is getting a nuclear reactor core that achieves self-perpetuating "criticality," the state in which a reactor exactly produces the correct amount of neutrons from one generation to the next generation.

15.4.1 Fissile Nuclei

As we have already seen, the collisions of fissile nuclei with neutrons break down certain heavy and unstable atoms destroying a small amount of mass in each reaction and also leaving two fragments of lighter atoms/fission (see Figure 15.5). Fissile atoms include U^{233}, U^{235}, Pu^{239}, and Pu^{241} with the prime example being the fissioning of U-235. This chart is very important in how we treat the waste products of nuclear reactions (because they may be very radioactive and some are very abundant).

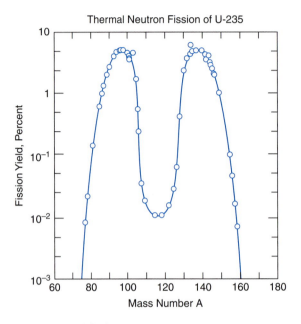

FIGURE 15.5 Fission product yields by thermal fission.

Notice that there are two major peaks in the fission product spectrum. We will return to Figure 15.5 later in this chapter.

15.4.2 Fertile Nuclei

Fertile atoms will convert to fissile atoms by neutron collisions followed by a radioactive decay sequence of which the most important product is plutonium, for example,

$$^{238}_{92}U \rightarrow {}^{239}_{94}P$$

How did two additional positive charges in the nuclei (see subscripts) come about? The radioactive decay of $^{239}_{92}U$ to $^{239}_{93}Np$ and subsequent decay of $^{239}_{94}Pu$ each produce a by-product of negatively charged electrons, which accordingly raises the number of positively charged protons by two. A simple view is that neutrons break down to produce a proton and an electron, the latter being called beta decay.

Pu-239 is an important atom because it is a fissile atom and not simply a fertile one. It is used in nuclear reactors as an additive to uranium fuel and also it is formed internally from fertile U-238, the latter being far and away the most abundant atom in a reactor core.

There is only 2–4% directly fissionable U-235 in the reactor core. Fortunately the remaining ~97% of the core uranium is fertile U-238. As such, it provides extra fuel beyond that loaded by fissioning of the initial U-235 so that the conversion of fertile atoms is a significant contributor to the performance of nuclear reactors.

Example 15.6

$^{232}_{90}$Th has an even number of nucleons. This confers stability against fissioning. But it is fertile. What fissionable isotope can it form? (Hint: Fissionable atoms have odd numbers of nucleons. Assume also our fissile product is the result of two beta decays.)

Need: Fissionable isotope from Th-232.

Know-How: Th-232 is fertile so it can absorb a neutron. Th-233 is likely unstable and will beta decay rapidly.

Solve: The double beta decay will raise the atomic number by two; $^{232}_{90}$Th has 90 protons so our new element will have 92 protons. This must be an isotope of uranium and it must have mass 233 after absorbing a neutron. The **fissionable isotope is** $^{233}_{92}$U.

15.5 THE FOUR FACTOR FORMULA

For a nuclear reactor to work continuously, you must make the proper number of neutrons during fission reactions—not too many and not too few as Goldilocks would say. To understand what is in that statement, we will consider how to calculate the number of neutrons by breaking down the problem into simpler pieces.

The factors to consider sustaining a self-perpetuating neutron cycle are:

- The fast fission factor ε

$$\varepsilon = \frac{\text{Total neutrons from fast and thermal fission neutrons}}{\text{Fission neutrons produced from thermal fissions alone}}$$

- Resonance escape probability, p

$$p = \frac{\text{Number fission neutrons slowing to thermal energies}}{\text{Total of fast neutrons}}$$

- The thermal utilization factor f

$$f = \frac{\text{Thermal neutrons absorbed into fuel}}{\text{Total of thermal neutrons absorbed}}$$

- The thermal fission factor (aka the fast fission reproduction factor), η

$$\eta = \frac{\text{Neutrons produced by fission}}{\text{Neutrons absorbed by fuel etc}}$$

When you multiple these factors together, they will give the infinite multiplication factor, k_∞.

$$k_\infty = \frac{\text{Neutrons in current generation}}{\text{Neutrons in preceding generation}} = \eta f p \varepsilon \qquad (15.6)$$

If $k_\infty = 1$ in an infinitely large reactor, it is critical; if $k_\infty \leq 1$ it is subcritical; and if $k_\infty \geq 1$, it is supercritical. For a finite-sized reactor, you need $k_\infty > 1$ to account for leakages. We still need to adjust this prediction for a finite reactor core that suffers neutron leakage.

Equation (15.6) is called the Fermi Four Factor equation after the great Italian physicist Enrico Fermi, who built the world's first nuclear reactor (under the sports stadium at the University of Chicago). The equation can be derived by chasing the fate of the first round of neutrons through the next generation as in Figure 15.6.

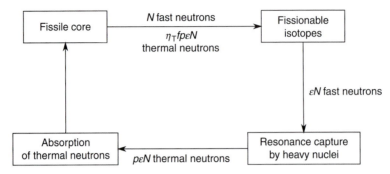

FIGURE 15.6 Neutron cycle in a nuclear reactor.

These factors can be calculated using concepts that are not covered in this book so our calculations will be accordingly oversimplified. Two of these factors, η and ε, deal with how the neutron population is increased and two, p and f, deal with how neutrons are absorbed.

At this stage, in our knowledge of nuclear reactor physics, you do not have the tools to deduce the values of these four factors by their fundamental physics so we will just assume their numerical values as needed.

Equation (15.6) is only a guide that strictly works only for an infinitely large reactor; it is a good first estimator of a nuclear reactor's behavior as to whether you have a viable design.

Example 15.7

Consider a homogeneous[14] nuclear reactor containing 95.00% of a graphite moderator and 5.00% fissile fuel $^{235}_{92}U$.

a. What is the value of k_∞?

b. How much neutron leakage can you afford with this design? These are the pertinent factors to be used in Equation (15.6): $\eta = 1.931$, $\varepsilon = 1.000$, $f = 0.982$, and $p = 0.648$.

c. Why is $\varepsilon = 1.000$ for this design?

> **Need:** The value of k_∞, the percent of neutrons lost to leakage, and why $\varepsilon = 1.000$.
>
> **Know:** Reactor core arranged homogeneously, $\eta = 1.93$, $\varepsilon = 1.000$, $f = 0.982$, and $p = 0.648$.
>
> **How:** Use the Four Factor Formula to calculate k_∞ and, from that, the loss of neutrons.
>
> **Solve: 1.** $k_\infty = \eta f p \varepsilon = 1.931 \times 0.982 \times 0.648 \times 1.000 = 1.229$.
>
> **2.** To get $k = 1.000$ for a critical finite core, $k/k_\infty = 1.000/1.229 = 0.814$ and thus **leakage = 0.186**. (18+percent losses of neutrons is probably larger than would be tolerated.) You could add a reflector such as a pure graphite shield to reduce the losses.
>
> **3.** Why is $\varepsilon = 1.000$ for a homogeneous reactor? Look at the core design—95.00% moderator and 5.00% fissile fuel. In the immediate vicinity of the thermal fission, there are 19 moderator nuclei for each fissionable one so the fast neutron that is formed will immediately slow to thermal speeds and **the fast fission factor $\varepsilon = 1.000$**.

---

[14]"Homogeneous" means uniform properties in all directions as when two substances are crushed to a powder and then very well mixed.

15.6 FISSION PRODUCTS AND NUCLEAR WASTE

All three nuclear reactions shown in Equations (15.1)–(15.3) produce fission products. Of these, iodine-131 and strontium-90 are notorious because they are dangerous radioactive elements that can enter the food chain.

Without studying the details of radioactive processes, let us note that nuclear radiation can negatively affect living creatures. Iodine-131, for example, behaves chemically and biochemically as ordinary iodine, natural iodine being entirely composed of a single nonradioactive isotope, I-127. Biochemically, any isotope of iodine has the same affinity for the thyroid gland. Radioactive I-131 can thus move through the food chain into the human thyroid and cause throat cancer. But it has a short **half life**[15] of only 8 days.[16] So, if milk is exposed to I-131, it will become radioactively contaminated. But it can be made into cheese and aged for a few months to make it wholesome again.

Radioactive strontium-90 also behaves chemically and biochemically as its nonradioactive cousin Sr-88. Further, since Sr is in the same Periodic Table group as calcium, it can fool the body into absorbing it into bone; if Sr-90 is absorbed into the body, it can produce serious illness (including fatal bone cancers). Sr-90 has a half life of 29 years and represents a serious hazard if it leaks from its containment.

Unfortunately, the above two radioactive isotopes are not the only radioactive species generated in nuclear reactions. In the fissioning process, hundreds of nuclear reactions occur beyond those expressed as Equations (15.1)–(15.3). The fissioning process does not split atoms uniformly into equal amounts of radioactive nuclei, but forms a broad spectrum of fission products.

What is happening during fissioning is that the uranium nucleus becomes unstable when it absorbs an extra neutron. It flies apart as quickly as it can and forms a smorgasbord of fission products. Figure 15.5 neatly summarizes the amounts of fission products that are formed as well as the concentrations of essentially all of their radioactive radioisotopes.

Notice the ordinate, yield percent, is logarithmic and covers four orders of magnitude. Smaller amounts of other fission products species are not shown because they are only produced in parts per million (or even much less) and are too dilute to show. Still some of the most persistent radioactive isotopes are technetium-99, neptunium-237, and plutonium-242, all with half lives are measured in hundreds of thousands of years.

In the fission products' curve, there are two prominent peaks with concentrations between about 1% and 8%. These peaks contain the most abundant of fission products. One broad peak occurs at a mass number of about 95 ± 10 (which includes Sr-90) and the other at about 140 ± 10 (ditto for I-131).

Example 15.8

How many years does it take to reduce the initial activity of Sr-90 to 1.0% of its original radioactivity? Assume the rate of radioactive decay of an isotope follows an exponential decay in time: $C = C_0 \exp\left[-\left(\frac{0.693}{t_{1/2}}\right)t\right]$. The half-life $t_{1/2}$ is defined when 50.% of the isotope has decayed.

Need: Time in years to reduce the radioactivity of Sr-90 to 1.0%.
Know: Half-life, $t_{1/2} = 29.$ years.
How: The rate of radioactive decay follows an exponential decay in time:

$$C = C_0 \exp\left[-\left(\frac{0.693}{t_{1/2}}\right)t\right] \tag{15.7}$$

---

[15]Meaning 50% of the atoms of the iodine-131 decay after just 8 days and 75% after 16 days etc.
[16]It loses 50% of its activity every 8 days.

where C is the final number of atoms of Sr-90 and C_0 is the original amount written in the same units.[17]

Solve: $C = C_0 \exp\left[-\left(\dfrac{0.693}{t_{1/2}}\right)t\right]$ in which $C/C_0 = 0.01$.

Therefore $\ln(0.01) = -\dfrac{0.693t}{t_{1/2}}$ and $t = \dfrac{-4.605 \times 29}{-0.693} = 193\,\text{years}$.

One can see that management of a radioactive accident site (as has befallen the Japanese at Fukushima) is a long-term affair. Most of the high activity fission products have half-lives of 30–50 years. Chernobyl in the Ukraine, the site of one of the worst nuclear accidents in history, has been totally abandoned due to radiation contamination.

15.6.1 Nuclear Waste, the Achilles Heel of Nuclear Power

The spectrum of fission products is the ultimate weakness of nuclear power because so many of these products are highly radioactive and many are biologically active as well. Fission products are an enormous burden that, if and when they leak, can cause expense (clean up costs), illness, and death. The effects of nuclear radiation depend on the types of radiation emitted as well as on the chemical and biochemical properties of the materials. For example, radioactive xenon Xe-133 has a half-life of 5 days but, importantly, xenon is a rare gas meaning that it does not form any chemical bonds and is chemically and biochemically inert. In general, it is less dangerous whereas I-131, with a similar half life, can enter the human thyroid gland with devastating consequences.

15.6.2 Nuclear Accidents

When a 1000-MW nuclear reactor is operating at full power, it will produce the entire radioactive spectrum described in Figure 15.5. If such a reactor shuts down, intentionally or not, the radioactive burden of this reactor at shut down is about 10^{20} Bq (radioactive disintegrations/s).[18] The accident at Chernobyl *released* somewhere in the range 5×10^{18} Bq and Three Mile Island about 5×10^{11} Bq. The releases at Fukushima are still in flux by a factor of more than 10.

There is another very important phenomenon that accompanies fission products. It is that radioactive decay is an exothermic event, which is harder to deal with as the reactors get larger. When a nuclear power reactor is shut down by inserting neutron absorbing rods into the reactor's core, it does not *fully* shut down. For example, a 1000 MW[19] nuclear reactor at the instant of shutdown, still continues immediately after shutdown to produce about 180 MW of heat from the radioactive decay of the fission products. After 1 h of shutdown this falls to about 45 MW of heat and after a day to 15 MW, which is still a large amount of heat. This radioactive heat is not just a minor nuisance; if not continuously cooled, the cladding of the fuel rods will melt and red hot

---

[17]The rate of radioactive decay is proportional to the amount of remaining radioactive isotopes. It follows the relationship, $\frac{dC}{dt} = -\left(\frac{0.693}{t_{1/2}}\right)C$ with $C = C_0$ at $t=0$. Read dC/dt as the rate of change of C with time. It's negative because the radioactive atoms are decaying to other atoms. Starting with the rate of decay relationship, Equation (15.7) is then easy to derive. Almost any atomic or molecular units can be used for C including atoms/s or mols/year etc. The numerical factor 0.693 is merely $\ln(2)$ that originates when $C/C_0 = \frac{1}{2}$ for a half life calculation. The decay equation can also be written, $C = C_0 \exp\left[-\left(\frac{0.693}{t_{1/2}}\right)t\right]$. The differential form $\frac{dC}{dt} = -\left(\frac{0.693}{t_{1/2}}\right)C$ is convenient when you want the rate of radioactive decay rather than the instantaneous amount of radioactive species.

[18]One disintegration/s is called a Becquerel (Bq) after the French discoverer of radioactivity; an older unit is a Curie (Ci) equal to 3.7×10^{10} Bq.

[19]It actually produces 3000 MW of thermal power, of which only about 1/3 can be usefully extracted.

cladding metal[20] will chemically react with water producing explosive hydrogen. In the absence of cooling, the fuel pellets will also melt (at ~2860 °C) and destroy the pressure vessel that had now contained volatile fission products. Most of these things happened during the accidents at Three Mile Island in the United States, at Chernobyl in the Ukraine, and at Fukushima in Japan.

Example 15.9

Statistics of ^{90}Sr fission product in a large power reactor core:

Breakdown Accounting	Result
Fuel in nuclear core as UO_2	More than 100 tonnes (1T = 1000 kg)
Enrichment of ^{235}U	3% or 3 tonnes = 3000 kg
At reactor shutdown, assume total fission products ~6% of ^{235}U	180 kg
Yield of ^{90}Sr is about 4.5% at shutdown	8. kg of ^{90}Sr

How many disintegrations/s will the Sr initially undergo? What about after 193 years of storage?

Need: ^{90}Sr disintegrations/s: (A) On shutdown and (B) after 193 years of storage.

Know: $t_{1/2} = 29.$ years and radioactivity drops to 1% after 193 years (see Example 15.8). The atomic mass of ^{90}Sr = 89.9 kg/kmol. Avogadro's number is 6.02×10^{26} particles/kmol.

How: Radioactive decay law for $\frac{dC}{dt} = -\frac{0.693C}{t_{1/2}}$ with $C = C_0$ at $t = 0$. The required units are disintegrations of atoms of ^{90}Sr per second. Convert $t_{1/2} = 29.$ years to seconds $= 9.1 \times 10^8$.

Solve: 8 kg of ^{90}Sr converted into atoms $= (8./89.9) \times 6.02 \times 10^{26}$ [kg][kmol/kg][atoms/kmol] $= 5.4 \times 10^{25}$ atoms of ^{90}Sr.

Rate of disintegrations of ^{90}Sr atoms/s $= 0.693 \times 5.4 \times 10^{25}/9.1 \times 10^8 = $ **4.** $\times$ **10^{16} atoms of ^{90}Sr** that decay every second. After 193 years, this falls to 1% of the original activity or **4.** $\times$ **10^{14} atoms/s**, which still is not small!

15.7 IS NUCLEAR POWER A VIABLE RENEWABLE ENERGY SOURCE?

Uranium fuel is assuredly a renewable energy source. Not only are there significant reserves of the ore (known as "pitchblende") but, with better fuel cycles than the common ^{235}U/LWR combination, there is a large potential supply of fissile materials. Uranium-235 is used in conjunction with the ^{238}U, which is roughly 97-98% of the total uranium in nuclear reactor fuel. Some neutrons are absorbed in this fertile material eventually making plutonium-239, which is also fissile, and which will sustain additional nuclear reactions much as U-235 does. If this ^{239}Pu is chemically separated from the rest of the fission products, it can be used as additional nuclear fuel when it is reloaded into a fresh batch of fuel. In fact one can, in principle, extract the ^{239}Pu repeatedly and end up with 250% of the original ^{235}U fuel value!

In addition, light water reactors are neutron hogs because there is a significant likelihood that protons in water will parasitically absorb some of the spare neutrons making **heavy hydrogen** (also called deuterium) by neutron absorption, that is,

$$^1_0n + ^1_1p \Rightarrow ^2_1d \tag{15.8}$$

where d is its nucleus with twice the mass of ordinary hydrogen.

---

[20]Usually a zirconium alloy.

There are other reactor schemes such as graphite moderated reactors that are so sparing with neutrons you can recover up to 95% of the original fuel as ^{239}Pu in each complete fuel cycle. Further, this fuel can be chemically recovered and it will deliver $20 \times$ the original ^{235}U fuel worth by reprocessing. **Heavy water** (made from deuterium) is used to moderate Canadian reactors. Its advantage is that it is parsimonious with parasitic neutron absorption compared to ordinary water.

There are also "fast" reactors that produce more uranium than they consume. They have no moderator but rely on an efficient capture process for fast neutrons. These can in principle recycle all of the fertile ^{238}U as ^{239}Pu (^{238}U being $\sim 99 \times$ as abundant as ^{235}U).

Finally, thorium 232 and uranium 233 can be used as a nuclear fuel (although no one has done this commercially). Thorium is available as beach sand in India!

Levels of uranium in the range of 1–10 ppm are widely scattered in topsoil. In fact, soil is slightly radioactive due to the presence of both uranium and thorium. Uranium is also present in coal ash so that a coal-fired power plant will emit thousands of times more radioactivity in particulates up their smoke stacks than will a nuclear power plant (providing of course that the nuclear power plant does not have an accidental radioactive leak)!

SUMMARY

Nuclear fission reactions can sustain a nuclear **controlled chain reaction** and will produce both large quantities of thermal energy and **neutrons**. The key step is **moderation** of **fast** neutrons produced by **fission**. This occurs in light water reactors when neutrons collide with protons in water molecules that are there for the dual purposes of providing a coolant and a moderator. A necessary variable in the design of a nuclear reactor is that the **infinite multiplication factor** is >1.000 and, with neutron losses accounted for in a real reactor, $k=1$ to maintain the proper cycling of neutrons.

An important and unfortunate consequence of fission reactions are **fission products** that are not only dangerously **radioactive** (and for a long time) but also produce **decay heat**, the latter capable of causing meltdown of both the nuclear fuel and the nuclear containment system. Loss of containment has the potential of releasing large quantities of radioactivity into the biosphere.

The prime advantage of nuclear power is that it does *not* produce a CO_2 effluent, thus not tipping the atmosphere into an imputed long warming period. Does this outweigh its disadvantages? These include long-term spent fuel custodianship and the possibility of more nuclear meltdowns. The jury is still out.

EXERCISES

SOME USEFUL CONVERSIONS:

1.00 MeV is 1.60×10^{-13} J.
1 fission reaction produces 2.0×10^2 MeV.
Boltzmann's constant, 8.617×10^{-5} eV/K.
Avogadro's number $= 6.02 \times 10^{23}$ entities per mol or 6.02×10^{26} entities per kmol.

1. Write down the possible fission reactions (conserving charge and nucleons) for $^{92}_{36}$Kr, $^{142}_{56}$Ba, $^{92}_{38}$Sr and $^{140}_{54}$Xe (krypton, barium, strontium, and xenon, respectively).

2. Find the error in this nuclear reaction: It describes a process for nuclear **fusion**. (Fusion is the opposite of fission, because it is the addition of a nuclear particle to a nucleus to make a heavier element than the starting materials.[21])

$$^2_1H + ^3_1H \Rightarrow ^4_2He + 2^1_0n$$

Why is tritium (see footnote) a *particularly* dangerous radioisotope?

3. How much uranium-235 is used in grams in producing 1.0 MWD (i.e., a megawatt-day) of energy given in 2.0×10^2 MeV/fission?

4. In the previous question, approximately how much of this uranium has been *annihilated* (in mg).

5. **Find the Error in This Solution**: What is the average energy for an ideal gas of neutrons in a graphite-moderated reactor at 750 °C? Note: An ideal gas's average energy is kT with $k = 8.617 \times 10^{-5}$ eV/K.

   Need: Average energy for an ideal gas of neutrons in a graphite-moderated reactor at 750 °C.
   Know-How: An ideal gas's average energy is kT with $k = 8.617 \times 10^{-5}$ eV/K.
   Solve: $E = kT = 8.617 \times 10^{-5} \times 750 = 0.065 \text{eV}$

6. **Find the Error in This Solution**: What is the logarithmic energy decrement for a neutron that thermalizes from 2.0 MeV to 0.1 eV?

   Need: The logarithmic energy decrement for a neutron between 2. MeV and 0.1 eV
   Know-How: The logarithmic energy decrement is defined as $\text{LED} = \log\left(\frac{E_{\text{Init}}}{E_{\text{Final}}}\right)$
   Solve: $\text{LED} = \log\left(\frac{2. \times 10^6}{0.1}\right) = 7.3$

7. How many collisions does it take for neutrons in a graphite-moderated reactor[22] to slow to thermal energies at 20 °C and at 350 °C?

8. How many times can you reuse a charge of $^{235}U/^{238}U$ fissile fuel if you reprocess the fuel when it is spent and recover all of the fissile Pu-239 there? Assume that 60% of the original fissile charge of U-235 was indirectly converted to Pu-239 from U-238. (**Hint**: It is a geometric series if all the Pu-239 converted from the original U-235 is reused.)

9. What is the ultimate fuel value of ^{235}U in a light water reactor ($x = 60\%$) and in a graphite-moderated reactor ($x = 95\%$)? (**Hint**: See Problem 8).

10. How much heat in watts is generated by radioactive ^{90}Sr in a nuclear core at shutdown and at 193 years later? Assume that each radioactive decay of ^{90}Sr produces 0.545 MeV/disintegration and that the initial amount of ^{90}Sr is 8.0 kg (see Example 15.9). Does ^{90}Sr meaningfully add to the decay heat in a nuclear meltdown?

---

[21]The stable hydrogen isotope ^2_1H is called "heavy hydrogen" or "deuterium", and its atom is often written as D; its nucleus commonly written as d. A ^3_1H is a radioactive isotope of hydrogen with a 12.5-year half life called "tritium", T. Its nucleus is written as t. Finally ^4_2He is the nucleus of a helium atom called an alpha particle, symbol α. It's the second element in the Periodic Table (having just two protons and two neutrons in its nucleus). An alpha particle is a fairly common radioactive decay product and possibly dangerous in its own right.
[22]High temperature graphite-moderated reactors are rare. Here are two: (1) a fixed body of graphite with hexagonal channels containing highly enriched uranium fuel (known as a "Prismatic Reactor") and (2) many thousands of ceramic-coated tennis ball-sized pellets containing graphite and moderately enriched fuel that continuously moves through the nuclear core zone (this is known as a "Pebble Bed Reactor"). Many experts think this is the safest kind of nuclear power plant but two of the worst nuclear accidents (Windscale/Sellafield in the UK and Chernobyl in the Ukraine) had disastrous graphite fires that dispersed radioactive fallout over wide areas.

11. The total decay heat on shutdown is about 180. MW for a large power reactor. What is the average fission product half-life *on shutdown* if there are 500. kg of fission products and each radioactive decay releases 0.500 MeV? Assume the average molecular mass of the fission products is 130. kg/kmol.

12. A heterogeneous reactor, as the name suggests, has non-uniform properties, for example, a graphite moderator slab adjacent to a fissile $^{235}_{92}$U slab. Without doing the arithmetic can you deduce arguments why this might increase the net thermal neutron flux? **Hint:** Sketch the expected neutron distribution, both fast and thermal, under the sketch below. What might this do to the resonance escape probability?

U235 Fissile Fuel

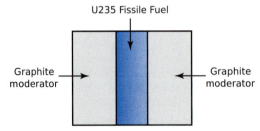

Graphite moderator   Graphite moderator

13. The world's first nuclear reactor was designed and built by Enrico Fermi in 1942. It consisted of a checkerboard of three-dimensional alternating blocks of a graphite moderator and fissile U-235. The fuel was natural grade uranium containing just 0.7% U-235 because no one had yet solved the problem of how to enrich it beyond its natural state.

The resonance escape probability for an infinite (no neutron loss) homogeneous core is about 0.648. Predict k_∞ if $\varepsilon = 1.000$, $\eta = 1.931$, $f = 0.750$, and $p = 0.648$. If you calculate $k_\infty \leq 1.000$, how did Fermi get it achieve criticality?

14. Find the Error in This Solution: You have a radioactive sample consisting of 10^{25} atoms. It is decaying at a rate of 10^{21} per second. What is the residual activity after 24 h?
    Need: Residual activity of a radioisotope after 24 h.
    Know: Original amount is 10^{25} atoms decaying at a rate of 10^{21} per second.
    How: The original amount present divided by the rate gives the time to complete the decay.
    Solve: Time to decay to zero is $10^{25}/10^{21}$ [atoms][atoms/s] $= 10^4$ s or **2.8 h**. Since it was all decayed after 2.8 h, it was also completely decayed after 24 h as none was left.

15. How much radiation in Becquerel's[23] (radiation particles/s) is released by the burning of 8000 metric tons/day of coal in a power plant? (The low-level radiation is there because uranium is slightly radioactive and widespread in minerals[24] of all types including coal.)

Assume coal contains 10.0% by weight of ash. If the coal contains 10.0 parts per million (ppm) of $^{238}_{92}$U the ash contains 100. ppm of $^{238}_{92}$U. The half-life of $^{238}_{92}$U is 4.468 billion years. There are eight other radioactive isotopes per disintegration of $^{238}_{92}$U.

16. Milk, which is easily contaminated by I-131, can be made into cheese[25] and aged for a few months so it will become wholesome again. How long should you wait?

---

[23]Symbol Bq. The radiation is mainly in the form of alpha rays or nuclei of helium atoms, $^4_2$He.
[24]More likely, present as its oxide U_3O_8. For extra credit, convert to this form of uranium.
[25]After contamination of central European farms by the Chernobyl nuclear disaster in the Ukraine in 1986, the Poles did indeed convert their contaminated milk into cheese and then aged it for several months.

17. A particularly long lived isotope is ^{237}Np with a half life of 2.2 million years so that its radioactivity is long lasting.[26] How many Bq does 1.00 mg of it produce?

18. What does it mean when a nuclear reactor is to be moderated by graphite and cooled by helium? Can you *sketch* such a reactor?

19. While not concerned with nuclear power, the analytical tools you have mastered can be used to give a lower limit on *the age of the Earth*. U-238 decays through multistep radioactive cascade to Pb-206, a stable isotope of lead. It is dominated by the first step in the decay which has a half-life of 4.47 billion years.

   The measurement depends on small zircon crystals (ZrSiO$_4$), which can dissolve uranium compounds (but they will not dissolve lead compounds). Liquid zircons will freeze at about 2500 °C trapping any available uranium in them. This uranium will eventually decay to stable Pb-206, which is then trapped within the zircon crystal. In one zircon crystal, the ratio of U-238/Pb206 was determined to be 0.888. When the Earth was last molten?

20. You are a nuclear engineer in the now defunct Soviet Union, which was a heavy-handed dictatorship tolerating neither criticism nor behavior contrary to the official line. You are shown a proposed design for the Chernobyl reactor that is destined to explode in 1986 causing about 30 immediate deaths and inflicting thyroid damage on thousands (perhaps tens of thousands) of other victims (particularly the young). You immediately see that the reactor has no secondary containment and, perhaps as bad, the last 60 cm of the control rods are coated with a heavy layer of a graphite moderator. You point out to the engineer-in-chief that a sudden insertion of withdrawn control rods all the way into the core of the reactor will not shut it down as expected, but will *increase* the nuclear reactions by virtue of the extra moderation via the graphite coating. Furthermore, lacking a secondary containment vessel, there could be disastrous widespread consequences. The engineer-in-chief takes you aside and says that the reactor is designed to be economical.[27] The design must stand essentially as is. He warns you to be careful because, if you bring it up again, the Soviet oversight authorities may send you to jail. What should the engineer do? (Use the Engineering Ethics Matrix and the NSPE Code of Ethics for Engineers.)
   a. Work within the system to get the best reactor he can under the circumstances.
   b. Still insist on changes to make the reactor safer including changes he has been warned to ignore.
   c. Resign and seek a new job, a job that will surely be a serious demotion given the Soviets' aversion to explicit or implicit criticism.

21. You are a nuclear engineer designing the nuclear reactor complex in Fukushima, Japan. It contains six virtually identical reactors. Japan is of course a democracy but one imbued with a culture of respect for one's elders. During a design review for the nuclear complex, you notice that all the secondary cooling systems have a common design based on diesel-electric generators to run the cooling pumps. You realize that a common failure mode exists. Anything that knocks out one generator could do the same to all of the reactors. Very politely you explain your concerns to the engineer-in-chief who tells you the design has been fixed as a compromise between absolute safety and safety based on an extremely unlikely event. And, based on the virtually zero likelihood of a common mode failure, the design stands. Remember, in Japan, consensus is very important. Are you OK with the engineer-in-chief's decision? What should the engineer do? (Use the Engineering Ethics Matrix and the NSPE Code of Ethics for Engineers.)

---

[26]Radioactive materials such as these are used in smoke alarms because they can produce electric charges that attach to smoke particles, which then can be easily detected. They are considered non hazardous provided the radioactive source stays confined.
[27]It was also designed to produce weapon's grade lithium isotopes for H-bombs!

    a. Work within the system to get the best reactor he can under the circumstances.

    b. Still insist on changes to make the reactor complex safer including changes he has been warned to ignore.

    c. Resign and seek a new job, a job that could be a demotion given the Japanese aversion to explicit or implicit criticism of their collective decision making.

Emerging Engineering Fields

2.1

We can expect many new fields of engineering to emerge as technology advances in the twenty-first century. Some of these fields are apparent today. Fields such as micro-electro-mechanical systems (MEMS), mechatronics (robotics), nanotechnology, synthetic biology, and smart technologies (cars, structures, buildings, etc.) are all emerging engineering fields. New engineering fields are often multidisciplinary and hold enormous opportunities for future engineers. The following chapters are devoted to three specific emerging engineering fields.

Bioengineering combines the analytical and experimental methods of engineering with biology and medicine to produce a better understanding of biological phenomena and develop new therapeutic techniques and devices. Students with bioengineering degrees may work as biomedical engineers to develop new medical techniques, medical devices, and instrumentation for hospitals or individual patients.

Electrochemical engineering is a combination of electrical engineering and chemical engineering. While the subject involves some mature technologies, such as the synthesis of chemicals, refining metals, the development of new batteries and fuel cells, electroplating, etching, and corrosion resistance, it also holds a central position in the development of new generations of hybrid cars as well as being the core technology for all-electric vehicles of the future.

Green energy engineering is energy produced from sources that are environmentally more friendly (or "greener") than fossil fuels (coal, oil, and natural gas). Green energy includes all renewable energy sources (including solar, wind, geothermal, biofuels, and hydropower), and by definition, should also include nuclear energy (though there are many environmentalists who oppose the idea of nuclear energy as green energy because of the nuclear waste issues, accidents, and its harmful environmental effects).

Bioengineering

16

Source: © iStockphoto.com/Antonis Papantoniou

16.1 INTRODUCTION

Bioengineering is the application of engineering, life sciences, and applied mathematics to define and solve problems in biology, medicine, health care, and other fields that deal with living systems. Some examples of bioengineering include the design and development of:

- Devices that substitute for damaged body parts such as hearing aids, cardiac pacemakers, and synthetic bone and teeth.
- Artificial kidneys, hearts, heart valves, blood vessels, arms, legs, hips, knees, and other joints.
- Medical imaging techniques (ultrasound, MRI, CT, and others).
- Engineered organisms for chemical and pharmaceutical manufacturing.
- Blood oxygenators, dialysis machines, and diagnostic equipment.

Bioengineering is a relatively new discipline that combines many aspects of traditional engineering fields, such as chemical, electrical, and mechanical engineering as well as a number of subspecialties, including biomedical engineering,[1] biotechnology, biological engineering, biomechanics, biochemical, and clinical engineering. Each of these fields may differ slightly in its focus of interest, but all are concerned with the improvement of human life.

16.2 WHAT DO BIOENGINEERS DO?

Bioengineers have a wide variety of career choices. Some work alongside medical practitioners developing new medical techniques, medical devices, and instrumentation for manufacturing companies. Hospitals and clinics employ clinical engineers to maintain and improve the technological support systems used for patient care. Engineers with advanced bioengineering degrees can perform biological and medical research in educational and government research laboratories.

Many bioengineers help people by solving complex problems in medicine and health care. Some bioengineering areas combine several disciplines, requiring a diverse array of skills. Digital hearing aids, implantable

---

[1]The term *biomedical engineering* is sometimes used synonymously with *bioengineering*.

defibrillators, artificial heart valves, and pacemakers are all bioengineering products that help people combat disease and disability. Bioengineers develop advanced therapeutic and surgical devices, such as a laser system for eye surgery and a pump that regulates automated delivery of insulin.

In genetics, bioengineers try to detect, prevent, and treat genetic diseases. In sports medicine, bioengineers develop rehabilitation and external support devices. In industry, bioengineers work to understand the interaction between living systems and technology. Government bioengineers often work in product testing and safety, where they establish safety standards for medical devices and other consumer products. A biomedical engineer employed in a hospital might advise on the selection and use of medical equipment or supervise performance testing and maintenance.

In biocommunication, bioengineers develop new communication systems that enable paralyzed people to communicate directly with computers using brain waves. Through bioinformation engineering, they are exploring the remarkable properties of the human brain in pattern recognition and as a learning computer. Through biomimetics, bioengineers try to mimic living systems to create efficient designs. These areas have already developed far beyond the material in this chapter, but their basic themes are similar.

By engineering analysis of the situations to which living matter might be exposed and by characterizing, in engineering terms, the remarkable properties of living matter, knowledge is gained that improves safety and health. That understanding can be used to make life safer and healthier. Among the tasks undertaken by bioengineers is the design of safety devices, ranging from football helmets to seat belts to air bags; the development of prosthetic devices for use in the human body, such as artificial hip joints; the application of powerful methods for imaging the human body, such as computed axial tomography (CT/CAT scanning) and magnetic resonance imaging (MRI); and the analysis and mitigation of possible harmful health effects on humans subject to extreme environments, such as the deep sea and outer space.

In this chapter, we introduce a few simple descriptions of human anatomy and the effects of large forces on hard and soft human tissues. We also learn (1) why vehicle collisions can kill, (2) how to make a first approximation of the likelihood of damage during collisions to the human body using a fracture criterion, (3) how to predict the injury potential of a possible accident using a criterion that we call "stress-speed-stopping distance-area" (**SSSA**), and (4) how to apply two other criteria for the effect of deceleration on the human body (the 30. g limit and the Gadd Severity Impact parameter).

We investigate just two areas of human anatomy: the first is to understand how serious blunt force trauma can affect the operation of the brain and other neurological tissue; the second is to understand how bone protects internal soft tissues.

16.3 BIOLOGICAL IMPLICATIONS OF INJURIES TO THE HEAD

In many automobile accidents, the victims suffer severe head and neck injuries. Some of the accidents directly cause brain trauma, and others cause neck injuries. The kind of accident caused by severe overextension of the neck and associated tissues is called **whiplash**. Head and neck injuries are all too common in accidents where people suffer **high g decelerations** (meaning decelerations of many times that of gravity). What this means is that the victim absorbs very high decelerations on impact with another vehicle or with a stationary object. Still other victims are injured within the vehicle when they contact interior components of the passenger compartment, such as the dashboard and the windshield.

We first take a brief look at human anatomy from the neck up to understand what forces can do to neurological function. Figure 16.1 shows the essentials of the skull, the brain, the spinal column, and the spinal cord.

The skull protects the brain, which floats inside the skull in a fluid-like layer. The base of the brain connects to the spinal cord through the spinal column, which provides protection for the spinal cord. The spinal column consists of individual bony vertebras that surround the all-important spinal cord. The spinal cord is the essential

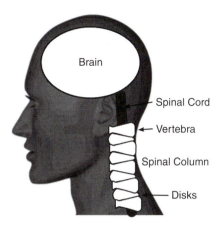

FIGURE 16.1 Part of the human nervous system.

"wiring" that takes instructions from the brain to the various bodily functions. The vertebrae are separated by cartilaginous matter known as *disks* to provide flexibility and motion. Injury to the disks can result in severe pain and, in the case of injuries to the spinal cord per se, in paralysis. Injury to the spinal cord can lead to very severe bodily malfunctions, while injury to the brain can cause a number of physical deteriorations or death. The brain, protected within the hard shell of the skull, floats in a fluid-like environment. The system, as efficient as it is, can suffer a number of possible injuries during high *g* decelerations.

Consider the injuries that are illustrated in Figure 16.2. The brain moves relative to the skull in its fluid-like layer when experiencing high *g*'s. In **hyperflexion**, the skull moves forward relative to the brain, causing damage to the occipital lobe (the back of the brain), and in **hyperextension**, it moves in the opposite direction relative to the brain, causing damage to the frontal lobe. Further, damage can occur to the basal brain, a potentially devastating injury since it may interrupt the nerve connections to the spinal cord. Also, there can be fractures to the vertebrae and extrusion or rupture of the protective disks.

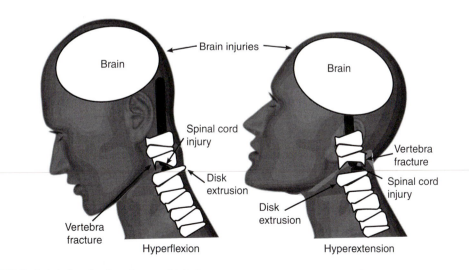

FIGURE 16.2 Brain injuries due to extreme whiplash.

16.4 WHY COLLISIONS CAN KILL

What is it about collisions that can kill or severely injure people? The challenge of this chapter is to express the answer using engineering variables. By using the right numbers and units for the values of those variables, criteria can be identified that distinguish a potentially fatal crash from one where the victims walk away.

One key variable in the bioengineering analysis of automobile safety is deceleration. Just as the "zero-to-sixty"[2] time of acceleration is a measure of a car's potential for performance, the "sixty-to-zero" time of deceleration is a first measure of an accident's potential for injury. But, while the zero-to-sixty time is typically measured in seconds, the sixty-to-zero time may be measured in milliseconds, that is, in thousandths of a second.

Rapid deceleration causes injury because of its direct relation to force. As discussed in the section on Newton's Second Law (Chapter 3), force is directly proportional to positive or negative acceleration (i.e., deceleration). And, as discussed in Chapter 13, force divided by area is stress. As further discussed in that chapter, the stress-strain curve gives the yield strength of a material. Suppose that the biologic material is flesh, bone, or neurons (nerve cells found in the brain, spinal column, and local nerves). If their yield strength is exceeded, serious injury or death follows.

An engineering forensic analysis of collision injury combines the effects of acceleration with the properties of materials. So, a first step in characterizing collisions is to determine the deceleration that occurs. We already have a tool for visualizing this: the $v-t$ graph.

Example 16.1

A car traveling at 30.0 miles per hour (mph) runs into a sturdy stone wall. Assume the car is a totally rigid body that neither compresses nor crumples during the collision. The wall "gives" a distance of $D_S=0.030$ m in the direction of the collision as the car is brought to a halt. Assuming constant deceleration, calculate that deceleration.

Need: Deceleration = _____ m/s^2.

Know-How: First sketch the situation to clarify what is occurring at the impact. Then, use the $v-t$ graph to describe the collision. A positive slope of the $v-t$ graph indicates "acceleration," and a negative slope indicates "deceleration."

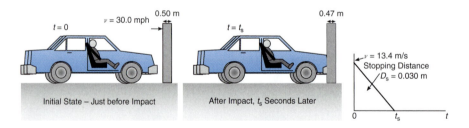

Solve: We first need to calculate the stopping time, t_s, to decelerate from 30.0 mph (13.4 m/s) to zero. If the deceleration is constant, we know that stopping distance is equal to the area under the $v-t$ graph or $D_S=\frac{1}{2}\times v\times t_s$, where t_s is the stopping time. So, $D_S=\frac{1}{2}\times 13.4\,[\text{m/s}]\times t_s\,[\text{s}]=0.030$ m.

Therefore, $t_s=0.06/13.4=0.0045$ s, and the slope of the $v-t$ graph is:
$\Delta v/\Delta t=(0-13.4)/(0.0045-0)=-2980=\mathbf{-30.\times 10^2\ m/s^2}$.
The negative sign indicates this is the **deceleration** rate.

---

[2]The automotive term for the time to go from a complete standstill to 60 mph.

What does this deceleration mean? How high is it? One good way to understand its damaging potential is to compare it to the acceleration due to gravity. This nondimensional ratio is $30. \times 10^2/9.8 = 310g$'s (to two significant figures).

The use of acceleration in terms of g can be quite helpful. Referring to Chapter 3, what is the weight of a mass m on earth? It is mg. What is the force on a mass accelerating by an amount a? It is ma. But, let us simultaneously multiply and divide that force by g, that is,

$$F = ma \times (g/g) = mg \times (a/g) = \text{the body's weight } (mg) \times \text{number of } g\text{'s } (a/g) \qquad (16.1)$$

In the impact described in Example 16.1, your head would weigh 310 times its usual weight (and the same for the rest of your body). An average human head has a mass of about 4.8 kg, so it will experience a force of about $4.8 \, [\text{kg}] \times 9.8 \, [\text{m/s}^2] \times 310 = 15{,}000$ N or about 3300 lbf! Your entire body will experience a similar force 310 times its normal weight.

You can also calculate the forces involved by kinetic considerations as opposed to these basically kinematic considerations. The force $\times$ distance to stop is equal to the kinetic energy that is to be dissipated. Let's assume the driver is belted into the car and therefore decelerates at the same rate as the car.

Example 16.2

The car and its belted driver of Example 16.1 suffer the same constant deceleration of $310g$ or $30. \times 10^2$ m/s^2. What is the force the driver experiences during the collision if his mass is 75 kg? Assume the car is a totally rigid body that neither compresses nor crumples during the collision.

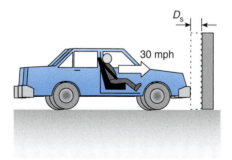

Need: Force stopping 75 kg driver on sudden impact $=$ _____ N.
Know: The driver's weight is $75 \, [\text{kg}] \times 9.8 \, [\text{m/s}^2] = 740$ N and deceleration rate is $310g$.
How: $F = ma = mg \times a/g$.
Solve: $F = 740 \times 310 = \mathbf{2.3 \times 10^5}$ **N** (or about 52,000 lbf!).

16.5 THE FRACTURE CRITERION

Under what conditions does a force cause living material to fail, that is, to crack, or break into pieces, or lose its ability to contain or protect fluid or soft structures? As we saw in Chapter 13, a material breaks when the stress on it exceeds its ultimate strength. Stress and strength are variables with units (N/m^2) that can be calculated from a description of the loading and the material's stress-strain diagram.

One aspect of the loading is the applied force, which was found in Example 16.2. Another aspect is the area over which that force is applied. This is needed to calculate the stress on the material in question.

Figure 16.3 shows a crude illustration of a bone. Bone is a natural composite material that consists of a porous framework made of the mineral calcium phosphate, interspersed with fibers of the polymeric material collagen. The calcium phosphate gives bone its stiffness, and the collagen provides flexibility. Figure 16.3 shows the internal structure of the body's all-important long bones, which have a hard outer layer, spongy interior layers with axial channels built from bone cells known as *osteoblasts*, and a central core of marrow that is responsible for renewing blood supply and other important cells. Bone is "anisotropic," which means that its properties are not the same in all directions. For example, it has different mechanical properties along its axis than it has perpendicular to the axis.[3]

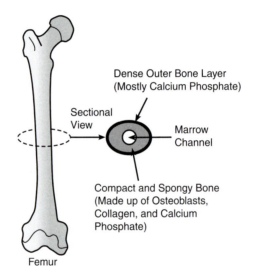

FIGURE 16.3 Structure of the body's long bones.

The properties of bone can be approximated by a stress-strain diagram as in Figure 16.4.[4] Notice bone is more elastic along (that is, in the longitudinal direction) the bone's length than crosswise (that is, in its transverse direction). These properties are due to the dense outer bone layer being tough with the compact and spongy interior bone providing surprisingly high flexibility. This model is a simplification of real bone behavior and is modeled here as if it were a perfectly elastic and perfectly plastic material. The corresponding data are shown in Table 16.1.

---

[3]Wood has similar anisotropy; you can split wood rather easily along the grain, but it is hard to break cross grain. The cellular patterns in wood are responsible for this useful behavior (for trees as well as for people).

[4]Adapted from A. H. Burstein and T. M. Wright, *Fundamentals of Orthopedic Biomechanics* (Baltimore, MD: Williams and Wilkins, 1994), p. 116.

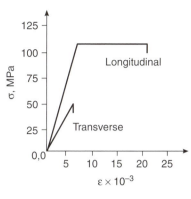

FIGURE 16.4 Assumed bone mechanical properties.

Table 16.1 Approximate Mechanical Properties of Human Bone

Direction	Elastic yield, MPa	Young's modulus, GPa	Strain at yield, %	Strain at fracture, %	Toughness at yield and fracture, MJ/m^3
Longitudinal	110	16.	0.7	2.0	0.4 at yield and 1.8 at fracture
Transverse	60.	9.	0.7	0.7	0.2 at yield and fracture

Bone, like other materials, may fail in several ways:

1. If bone is subjected to a local stress greater than its yield strength, it takes on a deformation.
2. If that stress exceeds the bone's ultimate stress, the bone will break.
3. If the energy absorbed exceeds the toughness at yield × volume of affected bone, the bone takes a permanent set.
4. If the energy absorbed exceeds the toughness at fracture × volume of affected bone, the bone breaks.

In general, bone failure modes 1 and 2 apply most to the situation in which the load is applied slowly. For a suddenly applied load, it is usually the toughness criterion that applies, given that bone toughness is often measured by the strike of a heavy hammer in a sudden blow.

Most bones that fail in vehicular accidents fail due to transverse stress rather than longitudinal stress[5] (for example, the ribs hitting the steering wheel if their owner is not protected by a seat belt or air bag). However, whiplash injuries may produce neck bone failure in the longitudinal direction. In general, we prefer not to permanently distort bone. Therefore, the elastic yield condition is the more conservative situation for a victim of an accident.

In another version of the accident described in Examples 16.1 and 16.2, the driver is unbelted, so the skull hits and dents the dashboard. How does the stress on the skull compare with the elastic yield point of the bones in the skull? That depends on the area of the skull over which the force is applied. For a given force, the smaller the area, the higher is the stress.

---

[5]Think of snapping a dead tree limb across your knee. The break occurs on the side away from your knee as fulcrum. That layer of branch is being *stretched*, and it fails mostly in tension.

Example 16.3

This is an accident similar to that in Examples 16.1 and 16.2, except the driver is not wearing a seat belt. Assume the driver's skull dents the dashboard to a depth of 0.030 m. The area of contact between the driver's forehead and the dashboard is $3.0 \text{ cm} \times 3.0 \text{ cm} = 9.0 \times 10^{-4} \text{ m}^2$, and the properties of bone are those given in Table 16.1. Will the collision *fracture* the skull?

Need: Is the yield strength of the skull exceeded? (Yes/No)

Know: Since the head suffers the same deceleration of $310g$ given $D_S = 0.03$ m, we can use the results of Examples 16.1 and 16.2. Consequently, the force on impact is 2.3×10^5 N.

How: Stress is force per unit area, and from Table 16.1, the yield strength of the bone is 60. MPa.

Solve: Compare $60. \times 10^6 \text{ [N/m}^2]$ with the applied stress on impact. That stress is $(2.3 \times 10^5)/(9.0 \times 10^{-4}) \text{ [N]/[m}^2] = 2.6 \times 10^8 \text{ N/m}^2$.

which is considerably greater than $60. \times 10^6 \text{ N/m}^2$. So, yes, the bone fractures, and the brain is seriously damaged. Unfortunately, this driver is fatally injured.

16.6 THE SSSA CRITERION

Let's generalize the results of Example 16.3. Collisions kill or injure because they involve a high rate of deceleration. That deceleration causes bodily contact with inflexible material, and the result is experienced as a force. The smaller the area over which that force is applied, the higher is the stress on the bone or tissue. The higher the localized stress, the greater is the likelihood that the material fails in that region.

We can express this insight as a relationship among the variables. Consider first the generalized form of the $v - t$ diagram of a collision with a constant deceleration rate. Suppose a body of mass m is subjected to a constant deceleration a and stops in a distance D_S. Then, from Newton's Second Law, the force experienced by the body is $F = ma$, and $a = v/t$, so $F = m \, v/t$. For a constant deceleration, the stopping distance is $D_S = \frac{1}{2}vt$, so $t = 2D_S/v$. Then, the force experienced by the body is:

$$F = mv^2/2D_S \tag{16.2}$$

Now suppose that this force is experienced on an area A of the head or body with a hard surface. Then, the stress can be calculated from Equation (16.2) as

$$\text{Stress} = \sigma = F/A = mv^2/2AD_S \tag{16.3}$$

We call this relationship the **stress-speed-stopping distance-area (SSSA) criterion**. It states that, technically, it's not enough to say that "speed kills." What kills is the combination of high speeds, short stopping distances, *and* small contact areas!

Strategies for reducing the severity of a collision follow from these insights. By decreasing speed, increasing the stopping distance, and increasing the area of application of the force, the effects of the collision on living tissue can be reduced.

The presence of the v^2 term in the numerator of Equation (16.3) indicates that decreasing speed is a highly effective way of decreasing collision severity. Cutting speed in half reduces the stress of a collision to one-fourth its previous value (if everything else remains the same).

How might the other two terms in the SSSA criterion be brought into play? Application of the area term is simple in concept, though more difficult in practice. The larger the area of contact between body and surroundings during a collision, the smaller is the stress. This is one motivation behind the air bag. It is a big gas-filled cushion that can increase the area of contact by a large factor. Of course, that cushion must be deployed rapidly enough to be effective in a collision.[6] Only by conquering these two technical challenges (deployment in a few milliseconds when appropriate, nondeployment at all other times) did engineers turn the air bag into a valuable safety tool.

However, the area portion of the SSSA criterion is only part of the air bag story. A seat belt does, indeed, increase the area over which force is applied (compared to the forehead) and, as important, reduces the probability of an impact of the head with the dashboard. But that increase would not by itself account for the great life-saving and injury-reducing potential of seat belts. A more important part of that potential involves that other term in the denominator, the stopping distance, D_S.

A more practically important contribution of seat belts is to attach the driver to a rigid internal passenger shell, while the rest of the car shortens by crumpling. The high effectiveness of the seat belt as a safety device has been achieved only in combination with the design of a car, which significantly crumples during a collision. Let's now assume that the car has such a *crumple zone*. This is illustrated in Figure 16.5.

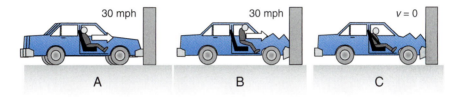

FIGURE 16.5 Seat belts and crumple zones.

Illustration A shows the situation of the driver, car, and wall during a 30 mph collision. The car rapidly crumples to a stop several milliseconds later, as shown in picture B. But the driver is still moving forward at the speed of 30. miles an hour. It is the collision of the driver's head and the windshield that brings the driver to a stop in a short distance, on the order of a few hundredths of a meter. By contrast, when the driver is wearing a seat belt, the body stops with the car, as shown in picture C.

In the last case, the driver gets the full benefit of the crumple of the car, which is on the order of a half to 1 m. This hundredfold increase in stopping distance results, according to the SSSA criterion, in a reduction in stress to one-hundredth of its previous value. This can be the difference between life and death (Example 16.4).

Example 16.4

A car traveling at 30.0 miles per hour (13.4 m/s) runs into a rigid stone wall. Assume the car's crumple zone results in a stopping distance of 0.60 m. Assume that the 75 kg driver is wearing a seat belt. Assume that the area of contact of seat belt and body is 4.0 cm × 30. cm $= 0.012$ m^2. Determine the maximum stress and the value of g the driver's body experiences.

Need: Maximum stress on the driver's body _____ N/m^2 and the number of g's.

Know-How: We could simply repeat the analysis of the previous examples, but the SSSA criterion provides a shortcut.

Solve: Stress, $\sigma = mv^2/2aD_S$. Substitute into the SSSA formula:

$\sigma = 75 \times 13.4^2/(2 \times 0.012 \times 0.60)$ [kg] [m/s]^2[1/m^2][1/m] $= \mathbf{9.4 \times 10^5}$ **N/m^2** $\approx \mathbf{1.0\ MPa}$ (which is less than the elastic yield of bone from Table 16.1). The value of g is $9.4 \times 10^5 \times 0.012/(75. \times 9.8)$ [N/m^2][m^2][1/kg][s^2/m] $= 15$. (Previously, it was 310, so, survivability has been greatly enhanced.)

[6]And, it must also *not* be deployed unless a collision has just occurred.

To sum up, here is an answer to our original question, expressed in engineering variables and units. The principal way seat belts save lives is by attaching the driver securely to the inner shell of the car, enabling the driver to take advantage of the car's crumple zone of about 0.5–1 m. During a collision from 30. miles an hour, that strategy restricts deceleration to less than about 150 m/s^2, resulting in stresses on the body that are less than about 1.0 MPa and almost surely less than enough to break bones.

16.7 CRITERIA FOR PREDICTING EFFECTS OF POTENTIAL ACCIDENTS

The preceding sections established that the effects of rapid deceleration on human bone and tissue may produce serious injury. We just indicated one way to address that issue by translating deceleration into force and force into stress. A comparison of maximum stress experienced by the body with the strength of bone provides a first criterion for the prediction of accident severity.

The effects of acceleration and deceleration on the human body go far beyond the potential to break bones. These effects range from the danger of blackouts for pilots experiencing very high acceleration or deceleration to the bone loss and heart arrhythmia experienced by astronauts exposed to the microgravitational forces of space flight for extended periods of time.

We continue to focus on the effects of high accelerations/decelerations. Just how much g can the human body stand? As a first approximation, engineers drew on a wide range of experiences, such as those of pilots and accident victims, to arrive at the following criterion.

The human body should not be subjected to more than 30.g's.

At 30.g's and above, the damaging effects of acceleration or deceleration on the human body can range from loss of consciousness to ruptured blood vessels to concussion to the breaking of bone to trauma or death. This criterion still serves as an initial rule of thumb for the design of safety devices (Example 16.5).

Example 16.5

Does a driver without a seat belt who experiences a 30.0 mile per hour (13.4 m/s) collision and is stopped in 0.10 m by a padded dashboard exceed the 30.g criterion? Does the role of a seat belt in taking advantage of the crumple zone to increase collision distance to 0.60 m meet the 30.g criterion?

Need: Deceleration without seat belt = _____ (less than/equal/more than 30.g?)
             Deceleration with seat belt = _____ (less than/equal/more than 30.g?)

Know-How: From Equation (16.2), $F = ma = mv^2/(2D_S)$; therefore, in g,

$$F = mg = ma/mg = a/g = mv^2/(2mgD_S) = v^2/(2gD_S)$$

Solve: Case 1: $D_S = 0.10$ m, $F/mg = (13.4^2)/(2 \times 9.8 \times 0.10) = 92g$, which is **more than 30.g**.
             Case 2: $D_S = 0.60$ m, $F/mg = (13.4^2)/(2 \times 9.8 \times 0.60) = 15g$, which is **less than 30.g**.

Consistent with our previous analysis, the combination of seat belt and vehicle crumple zone reduced the driver's deceleration from a probably fatal 92g to a probably survivable 15g.

However, experiences under extreme conditions, such as high-speed flight, soon revealed that the 30.g criterion is seriously incomplete. In some cases, humans have survived accelerations far above 30.g. In other cases, accelerations significantly below 30.g caused serious injury.

The missing element in the simple 30.g criterion is **time.** Deceleration has many effects on human tissue, ranging from destruction at one extreme to barely noticeable restriction of blood flow at the other. Each of these effects has a characteristic time interval needed to take effect. Very high decelerations, measured in hundreds of

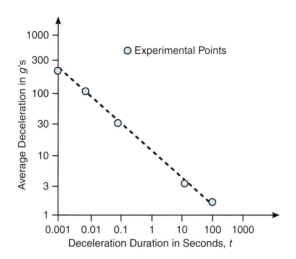

FIGURE 16.6 Injury criterion based on the deceleration in g's and the deceleration duration time, t. S.A. Berger et al., Introduction to Bioengineering (Oxford: Oxford University Press, 1996)

g, can be survived *if* the exposure is short enough. Deceleration at moderate g levels, on the other hand, can prove fatal if the exposure is long enough (see Figure 16.6).

In Figure 16.6, the axes are expressed in terms of the **$\log_{10}$** of the variables. Not only can we get a scale covering several orders of magnitude by this neat trick, but we can also infer something important about the relationship. In fact, it is linear, not between the actual variables g and t_S, but between $\log_{10}g$ and $\log_{10}t_S$.

The dotted line represents a *rough boundary* between accelerations likely to cause serious injury or death (above and to the right of the line) and survivable accelerations (below and to the left of the line). Notice that the previous "30.g" criterion is the point on that line corresponding to a duration of 0.1 s (the right order of magnitude for the duration of an automobile accident).

Figure 16.6 can also be expressed as a formula. A straight line on a log-log plot corresponds to the number on the x-axis raised to a certain power. In this case, the formula for the line dividing serious injury from survivability is

$$a = (0.002\,t_S)^{-0.4} \tag{16.4}$$

with a expressed in the nondimensional units of g's and with t_S expressed in seconds. The slope of the log/log graph is -0.4. This is *not* a basic scientific law. Rather, it is an empirical relationship summing up the net effects of many physical and biological properties. It is, however, a useful guide. It is more convenient by rearranging it into the form

$$a^{2.5}t_S = 500 \tag{16.5}$$

The quantity on the left of Equation (16.5) is the **Gadd Severity Index** (GSI).[7]

---

[7]The Gadd Severity Index was introduced in C. W. Gadd, in 1961. "Criteria for Injury Potential." National Research Council Publication #977 (Washington, DC: National Academy of Sciences, 1961), pp. 141–145.

$$GSI = a^{2.5}t_S \tag{16.6}$$

(where t_S is in seconds and a in g's)

The GSI in Equation (16.6) is a simplified numerical index that comes from assuming either constant deceleration or that a meaningful average deceleration can be measured. To use this equation, first calculate the GSI and then compare the result to the number 500. If the GSI is greater than or equal to 500, there is a serious danger of injury or death (Examples 16.6 and 16.7).

Example 16.6

Calculate the GSI for a driver without a seat belt who experiences a constant deceleration in a 30.0 miles per hour (13.4 m/s) collision and is stopped in 0.050 m by the dashboard.

Need: $GSI =$ _____ (a number).

Know-How: From Example 16.1, we know that, if deceleration is constant, the stopping distance is equal to the area under the $v - t$ graph or $D_S = \frac{1}{2} \times v \times t_s$, where t_s is the stopping time. So, then, $t_s = 2D_s/v = 2 \times 0.050/13.4 = 0.0074$ s. Therefore, the deceleration is: $a = \Delta v/\Delta t = (0 - 13.4)/(0.0074 - 0) = -1800$ m/s^2, or $a = 180g$'s (neglecting the negative sign indicating it's a deceleration).

Solve: $GSI = a^{2.5} t_S = 180^{2.5} \times 0.0074 = 3200$, which is much greater than 500. This suggests that severe injury or death is likely.

Example 16.7

Calculate the GSI for a driver with a fastened seat belt who experiences a constant deceleration in a 30.0 miles per hour (13.4 m/s) collision and is stopped by a 1.0 m crumple zone.

Need: $GSI =$ _____ (a number).

Know-How: From the development of Equation (16.2), the stopping time can be calculated from $t_S = 2D_S/v = 2(1.0)/13.4 = 0.15$ s, and the deceleration is $a = \Delta v/\Delta t = (0 - 13.4)/0.15 = -89.$ m/s^2, or $a = 9.1g$.

Solve: $GSI = a^{2.5} t_S = 9.1^{2.5} \times 0.15 = 37$, which is much less than 500. This suggests that severe injury or death is very unlikely.

SUMMARY

Bioengineering applies the methods of engineering to living bodies, organs, and systems. This chapter illustrated just one aspect of bioengineering, the use of biomechanics and engineering analysis of biomaterials to improve safety. In this chapter, we learned why collisions can kill, how to make a first approximation of the likelihood of collision damage to the human body using a fracture criterion, how to predict the injury potential of a possible accident using the **stress-speed-stopping distance-area (SSSA)** criterion, how to apply two criteria for the effect of deceleration on the human body (the **30.g limit** and the **Gadd Severity Impact** parameter), and how to analyze bioengineering problems.

EXERCISES

Exercises 1 and 2 concern the following situation: A car is traveling 30.0 mph and hits a wall. The car has a crumple zone of zero, and the passenger is not wearing a seat belt. The passenger's head hits the windshield and is stopped in the distance of 0.10 m. The skull mass is 5.0 kg. The area of contact of the head and the windshield is 0.010 m^2. Assume direct contact and ignore the time it takes the passenger to reach the windshield.

1. Provide a graph of $v - t$ of the collision of the skull and the windshield and then graph the force experienced by the skull as a function of time. (**A: See the following figure.**)

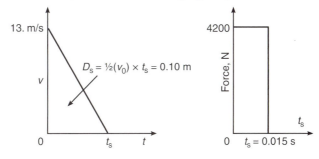

2. If the compressive strength of bone is 3.0×10^6 N/m², does the collision in the previous exercise break the skull?

Exercises 3 and 4 concern an experiment in the 1950s, when Air Force Colonel John Paul Stapp volunteered to ride a rocket sled to test the resistance of the human body to "g forces." The sled accelerated from 0 to 625 miles per hour in 5.0 s. Then, the sled hit a water brake and decelerated in 3.0 s to a standstill. Assume that Stapp was rigidly strapped into the sled and that he had a mass of 75 kg.

3. Prepare $v - t$ and $F - t$ graphs of Stapp's trip and compute the "g forces" he experienced in the course of acceleration and deceleration. (**A: 5.7g, $-9.5g$., see the following figure.**)

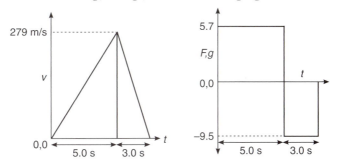

4. Using the force vs. time graph (Figure 16.6) for human resistance to g forces, predict whether Stapp suffered serious injury in the course of his record-breaking trip.

5. A tall person sits down on a sofa to watch TV. Assume that the center of gravity of the person falls 1.0 m with constant gravitational acceleration in the course of sitting down. The sofa compresses by 0.05 m. Assume constant deceleration. Determine the g forces experienced by the person in the course of this sitting down. (**A: 20.g**, so be kinder to couch potatoes!)

Exercises **6 and 7** concern an infant's rear-facing safety seat as illustrated next.

6. A rear facing child safety seat holds a child of mass 12. kg rigidly within the interior of a car. The area of contact between the seat and the child is 0.10 m². The car undergoes a 30. mph collision. The car's crumple zone causes the distance traveled by the rigid interior to be 1.0 m. Give the stress experienced by the child's body in terms of a fraction of the breaking strength of bone assuming an infant's bone breaks at a stress of 10. MN/m². (**A:** As a fraction of breaking stress, $1.0^8 \times 10^3$, the bones should hold and the infant should be safe.)

7. A rear-facing child safety seat holds a child of mass 25. kg rigidly within the rigid interior of a car. The area of contact between the seat and the child is 0.10 m². The car undergoes a 30. mph collision. The car has no crumple zone, but a harness attached to the car seat stops it uniformly within a distance of 0.30 m. According to the Gadd Severity Index, will the child sustain serious injury or death?

8. Consider a parachute as a safety device. When the parachute opens, the previously freely falling person has typically reached a speed of about 50. m/s. The parachute slows to a terminal speed of about 10. m/s in 1.3 s. Approximating this set of motions by a constant deceleration, what is the maximum g experienced by the parachutist? (**A: −3.1g.**)

9. In the previous exercise, the force exerted by the parachute is spread by a harness in contact with 0.50 m² of the parachutist, and the parachutist has a mass of 75 kg. What is the force per unit area (stress) experienced by the person during the deceleration?

10. The parachutist in the previous two exercises hits the ground (still wearing the parachute!) and is stopped in a distance of 0.10 m. If this final deceleration is constant, calculate the Gadd Severity Impact of the landing. (**A: GSI = 370.**)

11. A 75 kg person jumping from a 1.00×10^3 m cliff will reach a terminal speed of 50. m/s and uses a 1.00×10^2 m bungee cord to slow the descent. The bungee cord exerts a force F proportional to its extension, where F (in newtons) $= (5.0 \text{ N/m}) \times$ (extension in m) and is designed to extend by 5.00×10^1 m in the course of bringing the user to a stop just above the ground. Is the maximum deceleration in g experienced by the falling person more or less than the maximum deceleration experienced by a parachutist undertaking the same leap (excluding landing forces)?

12. An air bag is designed to inflate very quickly and subsequently compress if the driver hits it. A collision uniformly stops a car from 30.0 mph to 0.0 and triggers the air bag. The driver is not seat belted and so hits the inflated air bag. This acts as a local "crumple zone" and consequently compresses by 0.20 m as his head is brought to rest. According to the Gadd Severity Index, does the driver suffer serious injury? Assume constant deceleration of the driver's head after hitting the air bag. (**A: GSI = 420**, and no, the driver should not suffer serious injury or death.)

13. Two 100. kg football players wearing regulation helmets collide helmet to helmet while each is moving directly at each other at 10.0 m/s and come to a near instantaneous (<1.0 ms) stop. The area of contact is 0.01 m², and the helmets are each designed to provide a crumple zone of 0.025 m. What is the maximum stress exerted on each player? (**A: 4.0 × 10^7 N/m².**)

14. A designer of football helmets has two options for increasing the safety of helmets but, for economic reasons, can implement only one. One option is to double the area of contact experienced in a helmet-to-helmet collision. The other is to double the crumple distance experienced in a helmet-to-helmet collision. Which will be more effective in reducing the maximum stress? (Hint: Try the previous exercise first.)

15. A soccer player "heads" a wet 0.50 kg soccer ball moving directly toward him by striking it with his forehead. Assume the player initially moves his head forward to meet the ball at 5.0 m/s, and the head stops after the ball compresses by 0.05 m during impact. Assume the deceleration of the head is constant during impact. Compute and comment on the calculated Gadd Severity Impact of heading a soccer ball under these conditions.

16. A car strikes a wall traveling 30.0 mph. The driver's cervical spine (basically, the neck) first stretches forward relative to the rest of the body by 0.01 m and then recoils backward by 0.02 m as shown in the following figure. Assume the spine can be modeled by a material with a modulus $E = 10.$ GPa and a strength of 1.00×10^2 MPa. Does the maximum stress on the cervical spine during this "whiplash" portion of the accident exceed the strength of the spine? Assume a 0.15 m length of the cervical spine. (**A:** Stress on cervical spine is **greater than** its tensile strength.)

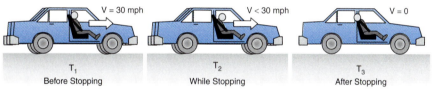

17. Which do you think has been more effective in reducing fatalities on American highways, seat belts or air bags? Give an engineering reason for your answer, containing variables, numbers, and units. (Hint: Recall the SSSA formula previously developed and consider what other safety element is designed into a modern automobile.) Then, go on the Web and see if you were right.

18. As a bioengineer at the Crash Safety Test Facility of a major automobile company, you are asked to provide more data for the Gadd Severity Index (Figure 16.6). Your boss suggests using live animals, dogs and cats from the local pound, in hard impact tests and inspecting them for injury. You know their injuries will be severe or fatal, and using dogs and cats seems cruel. What do you do? (Use the Engineering Ethics Matrix format to summarize your conclusions.)
    a. Nothing. Live animals are used regularly in product testing, and besides, they will probably be killed in the pound anyway.
    b. Suggest using dead animals from the pound, since their impact injuries probably don't depend on whether or not they are alive.
    c. Suggest using human cadavers since you really want data on humans anyway.
    d. Suggest developing an instrumented human mannequin for these tests.

19. You are now a supervisor in the bioengineering department of a major motorcycle helmet manufacturer. Your engineers are testing motorcycle helmets manufactured by a variety of your competitors. Motorcycle

helmets contain an inner liner that crushes on impact to decrease the deceleration of the head. This liner material is very expensive and can be used only once (i.e., once the helmet sustains a single impact it must be replaced). Your company has developed an inexpensive liner that withstands multiple impacts but is less effective on the initial impact than any of your competitors. The vice president for sales is anxious to get this new helmet on the market and is threatening to fire you if you do not release it to the manufacturing division. What do you do? (Use the Engineering Ethics Matrix format to summarize your conclusions.)

a. Since your company has invested a lot of money in the development of this helmet, you should release it, and besides, if you don't, someone else will.

b. Recommend continued testing until your company's product is at least as good as the worst competitor's product.

c. Contact your company's legal department to warn them of a potential product liability problem and ask for their advice.

d. Go over the vice president's head and explain the problem to the company's president.

20. During World War II, Nazi Germany conducted human medical experimentation on large numbers of people held in its concentration camps. Because many German aircraft were shot down over the North Sea, they wanted to determine the survival time of pilots downed in the cold waters before they died of hypothermia (exposure to cold temperatures). German U-boat crew faced similar problems. In 1942, prisoners at the concentration camp in Dachau were exposed to hypothermia and hypoxia[8] experiments designed to help Luftwaffe pilots. The research involved putting prisoners in a tank of ice water for hours (and others were forced to stand naked for hours at subfreezing temperatures) often causing death.

Research in the pursuit of national interests using available human subjects is the ultimate example of unethical bioengineering. Since the Nazi scientific data were carefully recorded, this produces a dilemma that continues to confront researchers. As a bioengineer today, should you use these data in the design of any product (such as cold-weather clothing or hypothermia apparatus for open heart surgery)? (Use the Engineering Ethics Matrix format to summarize your conclusions.)

a. Since these experiments had government support and were of national interest at the time, they should be considered valid and available for scientific use now.

b. Should you use these data, since similar scientific experiments have been conducted in other countries during periods in which national security was threatened and these data are not questioned today? Even the United States conducted plutonium experiments on unsuspecting and supposedly terminally ill patients (some of whom survived to old age!).

c. This is just history and should have no bearing on the value or subsequent use of the data obtained.

d. Do not use the data.

---

[8]Deprived of sufficient oxygen.

Electrochemical Engineering

17

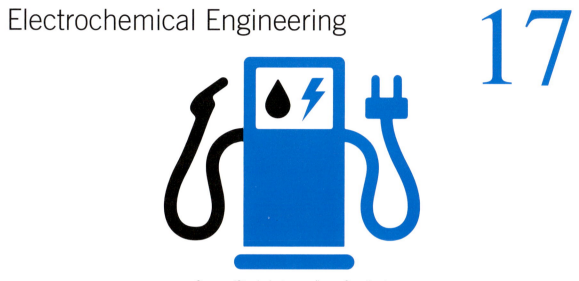

Source: iStockphoto.com/Leon Goedhart.

17.1 INTRODUCTION

Alternate energy sources are the hope for a tomorrow where we eliminate or to relieve the heavy dependence we have today on petrochemical fuel sources. One large vision, still under development, is the use of electrochemical engineering systems to propel electric cars, improve electronic devices, and supplement solar energy systems.

But, what is electrochemical engineering and why do engineers care about it? And, as a follow-on question, what kinds of engineers are engaged in the study and development of electrochemical engineering? If we answer the first of these questions, then the answers to the subsequent one makes logical sense.

17.2 ELECTROCHEMISTRY

Electrochemistry is the science behind **batteries** and **fuel cells** and, as such, plays an essential role in the development of novel power sources. We will get to fuel cells in good time, but batteries are common enough, so we will start there.

You might be aware there is a heavy-duty battery in every automobile (to activate the starter at the twist of the ignition key as well as for auxiliary uses in headlights, car radios, etc.). But, there are also small, familiar batteries that power flashlights, electronic devices, computers, and much more. In addition, today, special batteries are an integral part of a "hybrid" automobile. (*Hybrid* here means that an automobile has two sources of motive power: an electric motor and a relatively small assisting gasoline engine.)

Batteries are intimately involved in various alternate power concepts. For example, battery electrical storage units may be used in solar energy systems, be they small systems for individual houses or large ones for central power stations (Figure 17.1).

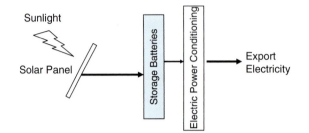

FIGURE 17.1 Solar power schematic.

A problem posed by such schemes is that they are subject to the vagaries of the weather as well as fluctuations in the demand for the electricity, especially because electricity is the ultimate "consumable." It has to be used immediately because it can't be stored on the power grid. In other words, you can make electricity when the sun is shining (or the wind is blowing if you use windmill power). But what happens if you don't have an immediate use for that power? Do you somehow discard it even though this is a very wasteful thing to do?

And, what happens when you do need more power than you are producing at that moment? Suppose you use electric power for cooking the morning and evening meals. Your highest demand is probably in the early morning and evening respectively before the sunrise and after the sunset. Figure 17.2 shows the problem.

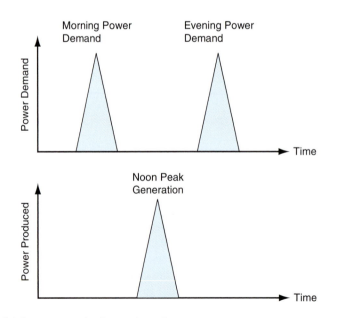

FIGURE 17.2 Mismatch of daily power production and need.

Obviously, what you want to do is to store the excess energy when it is abundant (noon in Figure 17.2) and use it when it is required (early morning and evening in Figure 17.2). Batteries are a relatively simple way to achieve this harmony and, indeed, individual house solar installations may rely on this tactic, known as **load leveling**, as shown in Figure 17.3.

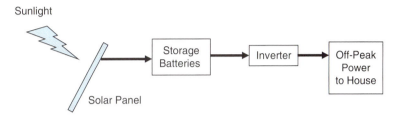

Sunlight

Solar Panel

FIGURE 17.3 Load leveling for solar energy. An inverter takes direct current (DC) and converts it to alternating current (AC).

The economics of solar power depends on the efficiency of the battery storage units that save the excess electrical energy for when it is needed. Consequently, batteries are a main technological and economic element of alternative energy practices.

The other major topic of this chapter related to batteries is fuel cells. What are they, and why do we need them? Simply put, fuel cells are continuously refueled batteries. Normal batteries, such as the battery in a car, or a flashlight battery, or a rechargeable computer battery, have a finite amount of chemical energy stored in them. That's why batteries go "dead" after extended use without recharging—the potential energy stored in their electrodes/electrolyte (the chemical heart of a battery) has been extracted, and the equivalent amount of electrical energy has been consumed. A fuel cell does not suffer this limitation. It is constantly replenished with fresh chemicals and therefore with fresh chemical energy as more electricity is demanded. The high-energy feed chemicals (e.g., hydrogen gas or methanol) produce effluents that are low energy, such as water or carbon dioxide.

We can now answer the initially posed question: What kind of engineers are interested in electro-chemical engineering? Since chemistry is heavily involved, chemical engineers have been at the vanguard along with mechanical and electrical engineers who integrate such systems into useable energy resources.

Example 17.1

Between 10:00 am and 2:00 pm on an otherwise cloudy day, the sun comes out and a solar-powered house produces and stores 20. kWh of electric energy. (Note: 1 kWh is the energy produced by 1 kW for 1 h and is equal to 3600 kJ.) The rest of the day it produces nothing. The house requires 1.0 kW from 10:00 am to 6:00 pm, and 2.5 kW between 6:00 pm and 8:00 pm. Is a 10. kWh storage battery large enough to supply the house's electrical needs between 6:00 and 8:00 pm?

Need: Size of storage battery for load leveling.

Know: The house uses 1.0 kW except between 6:00 pm and 8:00 pm when it requires 2.5 kW. It stores 20. kWh between 10:00 am and 2:00 pm.

How: We need an energy balance relating the energy flow and the net energy stored. We make the assumption that no energy is stored until 10:00 am.

Solve: The easiest way to solve this kind of problem is to graph it.

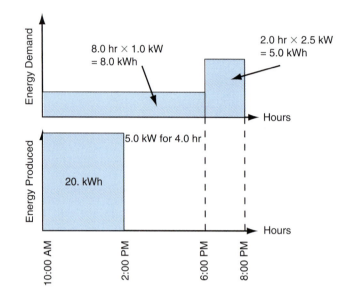

As the schematic shows, our house needs electricity at a rate of 1.0 kW for 4.0 h (4.0 kWh) from 2:00 pm to 6:00 pm and another 5.0 kWh between 6:00 pm and 8:00 pm for a total of **9.0 kWh**.

We store a net 20. − 4.0 = 16. kWh of electrical energy between 10:00 am and 2:00 pm, much of which we have to dispose of, but our **10.0 kWh battery storage is large enough** for our needs. The additional 7.0 kWh may be sold to the electric power grid if the local utility allows it.

17.3 PRINCIPLES OF ELECTROCHEMICAL ENGINEERING

Why does a battery work at all? Where do the electrons come from that carry the current in the external wires? To partially answer these questions, we briefly digress into elementary chemistry. First, many simple compounds are **ionic**, meaning they are held together by the balancing of positive and negative electrical charges. The perfect example is common salt, which can be symbolically written as either NaCl or Na^+Cl^-. Ionic compounds are formed by the transfer of an electron from an electrically neutral atom or molecule to another electrically neutral atom or molecule. In a *crystal* of common salt, these charges hold the crystal lattice together and are called Columbic forces or ionic forces, which are nothing but the forces of electrostatics governed by Coulomb's Law, expressed as follows:

$$F = \frac{e^2}{kr^2} \tag{17.1}$$

Here, F is the attractive force between two electrostatically opposite charges each of magnitude $e = 1.60 \times 10^{-19}$ coulombs and held apart at some distance of r. The constant k is the **dielectric constant** (or **permittivity**) of the medium in which the compound is immersed. This force is strongly attractive and, as noted above, is responsible for the crystalline structure of common table salt.

If an ionic crystal is immersed in water instead of in air, the relative value of k increases by a factor of ~ 80 and the corresponding force between atoms in the NaCl crystal is loosened by the same factor of ~ 80. The result is that positive ions of Na^+ and negative ions of Cl^- dissolve in water, and the ions drift apart. Thus, a solution of NaCl becomes a solution of two kinds of ions, Na^+ and Cl^- in equal proportion (and thus neutral overall). The trick in any electrochemical device is to physically separate these charges enough and create a voltage potential; this potential can then drive electrical charges through the external circuit.

Before we describe any electrochemical applications, it is necessary to define the terms **electrolyte** and **electrodes** (which are further subdivided into **anodes** and **cathodes**). Figure 17.4 illustrates these concepts.

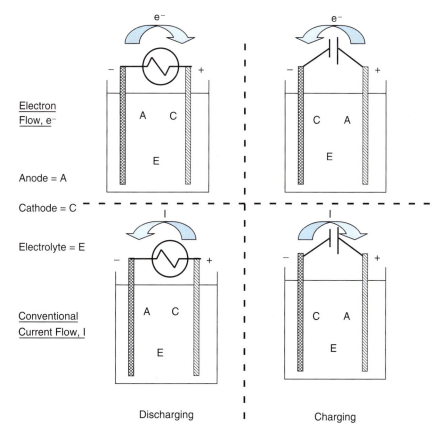

Electron
Flow, e^-

Anode = A

Cathode = C

Electrolyte = E

Conventional
Current Flow, I

Discharging   Charging

FIGURE 17.4 Some electrochemical terms.

This has been confusing ever since Ben Franklin arbitrarily decided that electric current flowed from the positive to the negative terminal. Much later, the British physicist J. J. Thompson discovered that negatively charged electrons carry the current in the opposite direction assumed by Franklin, that is, from negative to positive. It's much easier to think in terms of electrons than Franklin's current flows. For a discharging battery, the key concept is that the anodes churn out electrons and cathodes consume them. Conventional current during discharge flows into the anode (so, it is marked as negative) and away from the cathode (so, it is marked as positive).

17.4 LEAD-ACID BATTERIES

As our first electrochemical example, consider today's lead-acid car batteries. The principles of their operation are conceptually similar to those for most electrochemical devices. The lead-acid battery was invented more than 150 years ago and has been much improved with detailed understanding of its internal chemistry.

One electrode is basically lead (chemical symbol Pb) plus several other metals added to improve its performance. The other electrode is lead coated with a lead oxide, known as litharge. The electrolyte is about ⅓ sulfuric acid (H_2SO_4) in water. The sulfuric acid solution in water may be thought of as dissociating into its constituent ions $2H^+$ and $SO_4^=$.

The discharge chemistry at the anode dissolves some lead and puts the subsequent positive lead ions into solution (since electrons are produced in the anode and flow into the external circuit).

Anode: The anode is made of "pure" lead. The principle anodic reaction is

$$Pb = Pb^{++} + 2e^- \tag{17.2}$$

The lead ions Pb^{++} are doubly charged, combine with the sulfate ion from sulfuric acid, and mostly precipitate, since lead sulfate is only sparingly soluble in water:

$$Pb^{++} + SO_4^= = PbSO_4 \tag{17.3}$$

The removal of sulfate ions from solution means they no longer exactly match the hydrogen ion concentration and the excess H^+ ions migrate across the electrolyte toward the cathode.

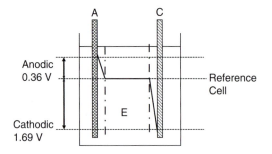

FIGURE 17.5 Cell potentials for lead-acid half-cells.

In the case of a lead-acid cell, the reactions near the anode produce a voltage of about 0.36 V and, near the cathode, another 1.69 V for a total cell voltage of about 2.0 V (see Figure 17.5). There is also a voltage loss across the electrolyte, which we ignored, and is referred to as *Ohmic losses* because it is explained by Ohm's Law. It occurs only when current is being drawn from the battery.

Cathode: The cathode is mostly made of a sheet of lead coated with a paste of lead oxide (PbO_2). During discharge, the cathode receives electrons from the external circuit that, in turn, help to convert the lead oxide to its sparingly soluble sulfate, most of which stays on the cathode's surface. Overall, this represents the primary reactions:

$$PbO_2 + 4H^+ + SO_4^= + 2e^- = PbSO_4 + 2H_2O \tag{17.4}$$

Both cathodic and anodic reactions remove lead ions and sulfate ions from the electrolyte and thus deplete the charge. The battery can be recharged several hundred times by applying a reverse voltage in excess of 2.0 V per cell. Eventually, the repeated solution/dissolution of the lead sulfate produces debris that falls to the bottom of the cell and fatally shorts it.

Table 17.1 Measures of Lead-Acid Battery Efficiency

Battery mass	25 kg
Energy content	3,000 kJ
Mass energy storage density	120 kJ/kg
Volumetric energy storage density	250 kJ/L
Power	5 kW

Source: Derived in part from: http://en.wikipedia.org/wiki/Lead-acid_battery and http://www.wdv.com/Hypercars/EAATalk.html

A car battery has six lead-acid cells in series and weighs about 25 kg. It can deliver 200-300 A at about 12 V for several minutes. The principal reason for its heavy weight is that it contains lead whose density is more than 11 times that of water. This fact is important for the future use of lead-acid batteries as energy storage devices.

Example 17.2

Compare the statistics in Table 17.1 to that in 25. kg of gasoline. Comment on the prospects for all-battery driven cars in the future.

Need: Energy in 25. kg of gasoline and compare to Table 17.1.

Know: Combustion energy of gasoline is the same as its mass energy storage density, or about 46,500 kJ/kg (see Chapter 7).

How: We need the mass density of gasoline to compute its volumetric energy storage density. It is 740. kg/m^3 or 0.740 kg/L.

Solve: The energy content is $46{,}500 \times 25.$ [kJ/kg][kg] $= \mathbf{1.2 \times 10^6\ kJ}$. For the volumetric energy storage density, $46{,}500 \times 0.740$ [kJ/kg][kg/L] $= \mathbf{34{,}400\ kJ/L}$.

Property	Lead-Acid Battery	Gasoline
Mass	25. kg	25. kg
Energy	3000 kJ	1.2×10^5 kJ
Mass energy storage density	120 kJ/kg	46,500 kJ/kg
Volumetric energy storage density	250 kJ/L	34,400 kJ/L
Power	5 kW	Typically >100 kW

In the tabular form, you can easily see the challenge for the all-electric car that uses conventional batteries. Basically, a lead-acid battery is too heavy and has too little stored energy to compete with gasoline (hence, the compromise solution of hybrid gasoline/electric cars).

17.5 THE RAGONE CHART

A neat way to compare the energy storage capability of an electrical storage device with its power producing capability is known as the Ragone[1] plot. It visually shows the energy storage and power producing characteristics of electrical storage devices. Figure 17.6 is a pictorial representation of the logarithm[2] of energy storage density (expressed as Wh/kg) against the logarithm of power producing density (expressed as W/kg).

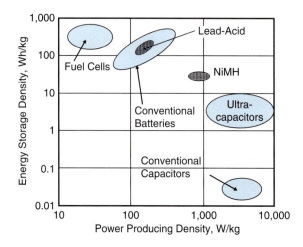

FIGURE 17.6 Ragone plot showing characteristics of some batteries and other electrochemical devices. *Source:* Modified from a graphic of Maxwell Technologies: http://www.maxwell.com.

We want to create an electrochemical device that operates in the upper right-hand corner, so we can simultaneously maximize the power producing mass density and the energy storage mass density and make the smallest possible battery with the least materials. For example, the Ragone plot indicates that a lead-acid battery is more competitive in the area of energy storage capacity than for its power producing density compared to the NiMH (nickel metal hydride) battery. There are also fuel cells and ultracapacitors on this plot, the last that we will investigate later and in the exercises.

17.6 ELECTROCHEMICAL SERIES

To introduce other batteries, we bring in one other useful concept arising from what we have already seen. The *electrochemical series* is defined by the relationship among half-cell reactions (such as developed for the lead-acid battery at each electrode) and put in order of their potentials measured against a standard cell called the hydrogen electrochemical half-cell.

---

[1]Pronounced "ra-GOH-nee" after David Ragone, who introduced the plot in 1962 (http://en.wikipedia.org/wiki/Ragone_chart).
[2]All logarithms are base 10 in this chapter.

Table 17.2 The Electrochemical Series

Half-Cell Chemistry	Potential in Volts
$Li^+ + e^- \leftrightarrow Li(s)$	-3.05 V
$Na^+ + e^- \leftrightarrow Na(s)$	-2.71 V
$Mg^{++} + 2e^- \leftrightarrow Mg(s)$	-2.37 V
$Zn^{++} + 2e^- \leftrightarrow Zn(s)$	-0.76 V
$Fe^{++} + 2e^- \leftrightarrow Fe(s)$	-0.44 V
$Ni^{++} + 2e^- \leftrightarrow Ni(s)$	-0.25 V
$2H^+ + 2e^- \leftrightarrow H_2(g)$	0.00 V (hydrogen half-cell is defined as zero)
$Cu^{++} + 2e^- \leftrightarrow Cu(s)$	0.34 V
$Cu^+ + e^- \leftrightarrow Cu(s)$	0.52 V
$Ag^+ + e^- \leftrightarrow Ag(s)$	0.80 V
$Pd^{++} + 2e^- \leftrightarrow Pd(s)$	0.95 V

Note: In principle, these reactions are all reversible. All ions are in aqueous solution, (s) means as a solid and (g) means as a gas.

Simple cells operate on principles that use this electrochemical series.

Example 17.3

In a Daniell cell, the electrolytes are $ZnSO_4$(aq) with a Zn anode in its half-cell and $CuSO_4$(aq) with a copper cathode in its half-cell. The two electrolytes are separated by an electro conducting porous and inert "salt bridge" that completes the path for the current flow while allowing the separation of ions from each electrolyte. Write down the reactions in each electrolyte and explain what happens in the salt bridge. Finally, what is the voltage produced by the Daniell cell?

Need: Explanation of half-cell reactions in Daniell cell and the voltage produced therein.
Know: The electrochemical series table shown in Table 17.2.
How: Draw the cell and use the electrochemical series, knowing that the Zn half-cell is the anode.

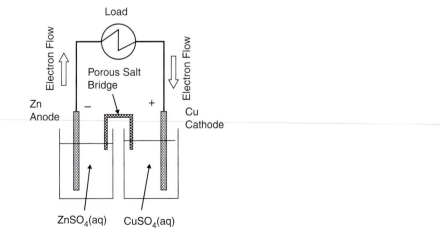

Solve: In the anodic electrolyte, Zn^{++} and $SO_4^=$ must be in aqueous solution in balance with each other:

$$ZnSO_4(aq) \leftrightarrow Zn^{++} + SO_4^=$$

But the anode is also dissolving and thus yields some locally extra Zn^{++} ions according to $Zn(s) \leftrightarrow Zn^{++} + 2e^-$ $(V = +0.76 \text{ V})$

Note that this is the reverse reaction of that in Table 17.2, so this half-cell reaction produces a positive voltage. We have also produced some additional Zn^{++} ions at a rate of one ion for every two electrons that leave the half-cell. A corresponding number of Zn^{++} ions must also move across the salt bridge and be "neutralized" by a corresponding number of $SO_4^=$ ions.

The cathodic electrolyte must be electrically neutral, containing equal numbers of Cu^{++} and $SO_4^=$ ions; the cathodic reaction must be written just as in Table 17.2:

$$Cu^{++} + 2e^- \leftrightarrow Cu(s) + 0.34 \text{ V}$$

Notice that we have removed copper ions from solution; therefore, there must be a corresponding reduction in $SO_4^=$ ions in the electrolyte. They must move into the salt bridge to exactly counteract the Zn^{++} ions from the anodic side. Therefore, the cell potential is equal to $(+0.76 \text{ V}) + (+0.34 \text{ V}) = \textbf{+1.10 V}$.

Something else is going on that you might have spotted. In the anode, we dissolved one unit of $Zn(s)$ that became Zn^{++} ions; and in the cathode, we precipitated the corresponding number of ions of Cu^{++} ions as $Cu(s)$. This is the type of reaction used in the electroplating industry. In that industry, solutions of metal ions are plated out to produce gold-, silver-, chromium-, copper-, and nickel-plated objects. In addition, corrosion engineers are interested in the opposite of plating—the dissolution of bulk metals. For example, the rusting of steel occurs in part because of the half-cell reactions $Fe(s) \leftrightarrow Fe^{2+} + 2e^-$ and $Fe(s) \leftrightarrow Fe^{3+} + 3e^-$.

Example 17.4

An electroplater wants to coat a 10.0 by 10.0 cm copper plate with 12.5 μm of silver. How many electrons must pass in the external circuit? How many coulombs are passed? If the plating takes 1200. s, what's the electrical current in amperes in the external circuit?

Need: Number of electrons and the current flow to deposit 12.5 μm of silver onto 100. cm^2 from a solution containing Ag^+ ions.

Know: Atomic mass of Ag is 108 kg/kmol. Its density is 10,500 kg/m^3. Avogadro's number (N_{Av}) is 6.02×10^{23} atoms/mol or 6.02×10^{26} atoms/kmol. What we call current is nothing but the rate of flow of electrons, so 1.00 A = 1.00 coulomb/s and one electron carries 1.60×10^{-19} C.

How: Sketch a credible electrochemical circuit; use an electrolyte of silver nitrate (which is soluble in water producing equal numbers of Ag^+ and NO_3^- ions). Use a silver anode and the copper plate as a cathode. Finally, connect a battery with the polarity chosen as shown in the following figure.

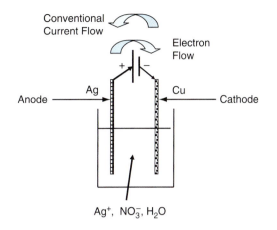

Solve: The reaction at the anode is $Ag(s) \rightarrow Ag^+ + e^-$, and the reaction at the cathode is $Ag^+ + e^- \rightarrow Ag(s)$; hence, one atom of silver dissolves at the anode and one atom of silver is deposited at the cathode. For each atom of silver dissolving at the anode and depositing at the cathode, one electron must circulate in the external circuit.

First, we need to know what the mass of silver is in the coating. We assume the deposited silver atoms fully pack with no internal pores. Then the mass of deposited silver is

$$\text{Mass} = \frac{12.5 \times 10^{-6} \times 100. \times 10{,}500}{1.00 \times 10^4} [\mu m][m/\mu m][cm^2][kg/m^3]/[cm^2/m^2] = 1.31 \times 10^{-3} \text{ kg}$$

Next, convert to kmol:

$$\text{kmol} = \frac{1.31 \times 10^{-3}}{108} [kg][kmol/kg] = 1.22 \times 10^{-5} \text{kmol}$$

Now, convert into the number of silver atoms using Avogadro's number:

$$\text{Number of atoms of Ag(s) deposited} = 1.22 \times 10^{-5} \times 6.02 \times 10^{26} [kg][atoms/kg] = 7.32 \times 10^{21}$$

This is equal to the number of electrons that flowed in the external circuit. Hence, we have used the services of **7.32×10^{21} electrons**.

We know that an ampere is the charge flowing/time; the electrical charge in this case is $7.32 \times 10^{21} \times 1.60 \times 10^{-19}$ [e$^-$] [coulombs/e$^-$] $= 1.17 \times 10^3$ coulombs. This charge flows for 1200. s, and hence, the **current I** $= 1.17 \times 10^3/1200.$ [coulombs]/[s] $= $ **0.975 A**.

17.7 ADVANCED BATTERIES

Many batteries, such as the common "alkaline" battery, use a solid electrolyte rather than a liquid, but all operate on similar principles. So we must ask, "Are there higher-performance batteries available?" And, of course, there are. A widely available battery today is the **nickel-metal-hydride** (NiMH) battery. They are used both to propel hybrid cars and to start their auxiliary engines. Smaller versions of them are also used for long-lasting electronic devices.

In the anode, the overall principle reaction that occurs is[3]

$$Ni(OH)_2 + OH^- \leftrightarrow NiO(OH) + H_2O + e^- \tag{17.5}$$

and the corresponding overall cathodic reaction is

$$M + H_2O + e^- = MH + OH^- \tag{17.6}$$

These reactions are not fundamental but represent several internal reactions; the M is a metal and MH is a metal hydride. The "metal" is an intermetallic compound of two metallic elements, one of which is called a *rare earth* element. It is any of elements of atomic number 58 through 71, characterized by having very similar chemistry because their chemistry is mediated by inner shell electrons instead of their outermost electrons. The other metal is one or more of Ni, Co, Mn, or Al. The electrolyte is mostly KOH, a strong alkali.

The basic advantage of this kind of battery is that it has both relatively high energy storage density and power producing density so, when packaged as a common D-size flashlight battery, it can supply more than 10 A for 1 h (which is more than 50 kJ). The Ragone plot, Figure 17.6, shows its power performance is considerably better than a lead-acid battery. No wonder these batteries in various sizes are preferred for rechargeable use in higher-end electronic devices. For automotive hybrid use, similar sizes to D cells are packaged in a large array, thus simultaneously providing sufficient voltage and current capacity. But, the NiMH is being overtaken by the lithium-ion battery, whose overall chemistry is a shift of a Li ion, Li^+ from a compound of lithium, cobalt, and oxygen to lithium carbide. Li-ion batteries have "high" voltages (approaching 4 V compared to most batteries of 1.5–2 V) and still have a high energy storage density.

Whether NiMH or Li-ion (or the more common NiCd, pronounced "nyecad"), these batteries are two or three times better than lead-acid batteries in capacity and therefore will continue to find use in high-grade electronics and hybrid cars. None is likely to completely replace the gasoline engine because the mass and volumetric energy densities are simply too low.

17.8 FUEL CELLS

If batteries with stationary solid or liquid electrolytes fail to completely replace gasoline engines, then why not use a constant flow of fresh replacement electrochemical fuel and continuously purge the used material? This is what a fuel cell does. It is not a recent invention: a fuel cell based on phosphoric acid was invented about 175 years ago. Since then, there have been many variations. One important class of fuel cells was invented after fundamental research by two GE scientists, Thomas Grubb and Leonard Niedrach, in the 1950s. The NASA moon program in the 1970s used these fuel cells. They almost led to disaster during the Apollo 13 moon shot, when internal poorly insulated electric wires shorted and sparked, resulting in a hydrogen/oxygen explosion.

The hydrogen fuel cell is the most promoted of all fuel cells for future transportation use, since it converts H_2 to H_2O *and nothing else* and thus would be a welcome relief to the air quality of urban areas currently relying on gasoline vehicles.[4]

---

[3]http://en.wikipedia.org/wiki/Nickel_metal_hydride.
[4]This is a bit misleading, since hydrogen must come from somewhere; its commercial manufacture consists of burning hydrocarbons in the presence of air and steam, thus venting significant quantities of CO_2 waste product into the atmosphere at its point of manufacture.

A schematic of the central feature of a fuel cell is shown in Figure 17.7. The reactants, hydrogen and oxygen (actually as air), flow to the anode and cathode, respectively, and hot water or steam, the sole product of reaction, is removed.

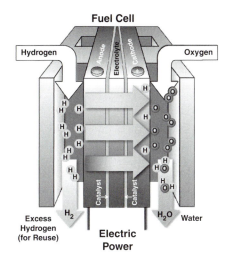

FIGURE 17.7 Principle of a proton exchange membrane.

The key step of Grubb and Niedrach was to introduce the proton exchange membrane (PEM) based on a fluorinated polymer membrane. Niedrach showed how to impregnate the membrane with platinum, which is a catalyst for the oxidation of hydrogen by air at ambient temperatures. Hydrogen diffuses into the membrane, where it is ionized. The resulting electrons flow from the anode, as shown in Equation (17.7):

$$2H_2 \rightarrow 4H \rightarrow 4H^+ + 4e^-$$ (17.7)

The oxygen gas counter diffuses in the PEM and reacts to form water while absorbing electrons:

$$4H^+ + 4e^- + O_2 \rightarrow 2H^+ + 2OH^- \rightarrow 2H_2O$$ (17.8)

The actual construction of a fuel cell is quite complex (as schematically shown in Figure 17.7), but this is not its principal challenge. The biggest problem for a hydrogen fuel cell is hydrogen.[5]

---

[5]It's a very flammable and explosive gas. Further, as a gas, the amount of hydrogen that can be stored is limited. You can compress hydrogen in pressure vessels, reversibly absorb it in a solid, or liquefy it. As a gas at 200 atmospheres pressure (about 3000 lbf/in^2), its volumetric energy storage density is only 2250 kJ/liter compared to gasoline's 34,000 kJ/liter. The highest *possible* energy storage density for hydrogen of 9700 kJ/liter can be achieved by its liquefaction; unfortunately, the liquefaction process is energy intensive (and thus occurs with a CO_2 effluent). Liquid hydrogen boils at -253 °C, which is another engineering obstacle for vehicular use. There is also the practical matter of replacing gas stations with hydrogen stations all across the world. One last problem is that a hydrogen-fuel-celled car is not truly "green," because the hydrogen has to be produced somewhere. That process is also net CO_2 generating, although, as previously noted, it is vented far away from where the hydrogen is used.

17.8.1 Fuel Cells Using Novel Fuels

While lots of electrochemical schemes function as a fuel cell, one other shows some promise because it uses a fuel that can be stored as a liquid—methanol, CH_3OH. The anode feed for such a cell is a mixture of methanol and steam onto a PEM membrane. With the appropriate catalyst on the PEM, the methanol is broken down, producing electrons and protons (H^+) as shown in Equation (17.9):

$$CH_3OH + H_2O \rightarrow CO_2 + 6H^+ + 6e^- \tag{17.9}$$

The electrons are removed via the anode, and the protons transport across the PEM as in a hydrogen cell. CO_2 is vented, a serious disadvantage for large-scale applications. At the cathode, the reaction is

$$^3/_2O_2 + 6H^+ + 6e^- \rightarrow 3H_2O \tag{17.10}$$

Overall, the reaction is

$$CH_3OH + ^3/_2O_2 \rightarrow CO_2 + 2H_2O \tag{17.11}$$

This cell operates at reasonable temperatures from 60 to 120 °C. Its biggest disadvantage is that it produces CO_2; this is offset by the fact that an all-electric drive is roughly twice the efficiency of a gasoline powered car (thus about doubling its mpg). As a further use of these fuel cells, Figure 17.8 shows a prototype methanol fuel cell designed to boost the recharging interval of laptop computers to 10 or more hours.

FIGURE 17.8 Fuel cells for laptops. *Source:* http://science.nasa.gov/headlines/y2003/images/fuelcell/notebook_med.jpg.

If we plot the electrical characteristics of fuel cells on a Ragone plot, they lie toward the upper left-hand corner, because their energy storage density is high (perhaps, 500 Wh/kg) while their power producing density is low (perhaps, 20–40 W/kg).

Example 17.5

A PEM cell is fed with 100. standard[6] ml/minute of $H_2(g)$ and its stoichiometric equivalent of air. What's the cell voltage at zero current given the half cathodic cell produces 1.23 V, and what's the maximum possible electrical output of the cell?

Need: Hydrogen cell voltage and current capability.

Know: Anode half-cell: $2H_2 \rightarrow 4H \rightarrow 4H^+ + 4e^-$ for which $V = 0.00$ V (see Table 17.2); and the cathode half-cell: $4H^+ + 4e^- + O_2 = 2H_2O$, for which $V = +1.23$ V

$N_{Av} = 6.02 \times 10^{26}$ molecules/kmol and the electronic charge $= 1.60 \times 10^{-19}$ C.

How: Find the voltage from the half-cell potentials and current from the assumed 100% conversion of the energy in 100. standard ml/min of $H_2(g)$. Convert standard ml to moles using $pV = nR_uT$, where R_u is the **universal** gas constant, 8.31×10^3 J/kmol K.

Solve: The **voltage** for the cell is found by algebraically adding the half-cell potentials $= 0.00 + 1.23$ V $= $ **1.23 V.**

So, **100.** standard ml/min of hydrogen requires **50.0** standard ml/min of O_2 for the reaction $H_2 + \frac{1}{2}O_2 = H_2O$. The **amount of air** needed is 4.76×50.0 standard ml (see Chapter 7) $= $ **238 standard ml/min.**

The net reaction uses 100. standard ml/min of H_2. The number of kmol/min of H_2 consumed is $\dot{N} = p\dot{V}/R_uT = (1.00 \times 10^5) \times (100.)(10^{-6})/(8.31 \times 10^3 \times 273)$ [N/m^2][ml/min][m^3/ml]/[J/kmol·K][K] $= 4.41 \times 10^{-6}$ kmols/min.[7]

This contains $4.41 \times 10^{-6} \times 6.02 \times 10^{26} = 2.65 \times 10^{21}$ molecules/min of $H_2(g)$ or 4.42×10^{19} molecules/s flowing in. The maximum current is:

$$2.00 \times 4.42 \times 10^{19} \times 1.60 \times 10^{-19} \ [e^-/\text{molecule}] \ [\text{molecule/s}][C/e^-] = \textbf{14.1 A.}$$

Note the factor of 2, since two electrons are needed for each molecule of H_2.

If methanol is an "interesting" fuel for a fuel cell, can we use a hydrocarbon such as gasoline directly in a fuel cell? In due course, a fuel cell that could use gasoline directly could be advantageous, since the present infrastructure for gasoline could be used; an all-electric output all-gasoline fueled vehicle has the potential to double today's gasoline combustion engine mileage. A major scientific and engineering obstacle is to develop a low-temperature catalyst that readily breaks the sturdy chemical barriers between the C-C atoms that form most of the compounds in gasoline.

That this is a complicated high-pressure/high-temperature affair is shown by a state-of-the-art reformer that produces 250 kW (335 HP) or enough for a large truck (Figure 17.9). The final module is large compared to a diesel fuel engine and fuel tank for the same purpose.

FIGURE 17.9 A 250 kW fuel cell with a built-in natural gas reformer (Ballard Power Systems).

---

[6]Meaning measured at atmospheric pressure and 0 °C/273 K temperature.
[7]The over-dot in $\dot{N}$ is used to denote rate, dN/dt.

17.9 ULTRACAPACITORS

The common **electrical capacitor** uses physical principles (instead of chemical ones) to store electrical energy in the form of electric charges. It works by charging a **dielectric** (which is essentially a nonconductor of electricity, such as a polymer or glass) by applying a voltage across it by sandwiching the dielectric material between two metal foils or plates. If the dielectric material has the appropriate structure, electrons can be deformed out of their normal orbits to induce net charges at the interface between the metal and the dielectric material. Common electrical capacitors have an energy storage density of less than 0.1 Wh/kg, which is too small for virtually all battery storage operations but is often useful for timing applications in electric circuits.

Ultracapacitors (sometimes called *supercapacitors*) are a cross between a common electrical capacitor and a battery. Like common capacitors, they store electrical charges on surfaces between charged electrical conductors. Unlike regular capacitors, the charges accumulate in porous carbon immersed in oil, see Figure 17.10.

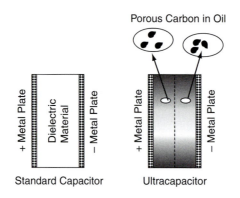

FIGURE 17.10 Comparison of conventional capacitor and ultracapacitor.

A detailed description of ultracapacitors reveals that their charges are in double layers at interfaces, and so they are sometimes called *electric double-layer capacitors*. In ultracapacitors, porous carbon particles are the electrical storage medium because they can have huge internal surfaces of $100–300$ m^2/g. If the carbon particles were nonporous, the carbon's external surfaces would total only a few cm^2/g.

Ultracapacitors are constructed with a separator and two compartments, so + and − charges can be separated and collected on their external plates. There is a factor of about 10,000 in total charge if the charges can fill the internal pores of the carbon instead of just its external surface areas. This huge multiplier, compared to a standard capacitor, means the ultracapacitor has a correspondingly large energy storage capacity. Indeed, ultracapacitors have energy storage densities of 1–10 Wh/kg, but their capability to deliver this energy as power can be as high as several thousand W/kg. Unlike batteries, the charge/discharge cycle has no *net* chemical changes and thus ultracapacitors should be capable of thousands of charge/discharge cycles.

The Ragone chart, Figure 17.6, shows the areas of applicability of electrochemical batteries, fuel cells, and ultracapacitors. Ultracapacitors can deliver short pulses of high power at reasonable energy storage densities. The region of applicability for ultracapacitors is unique; the ability to deliver short bursts of energy for high-power applications coupled with their energy storage capability makes them suitable for small power demands, such as personal electronics.

SUMMARY

Electrochemical systems are a part of most alternate energy schemes to store electrical energy for **load leveling** purposes. Batteries work because of the electrical nature of matter. Batteries all have **anodes** (as a source of electrons) and **cathodes** (a sink for them). The lead-acid battery is familiar because of its automotive uses but is limited by weight and its energy storage capacity.

Batteries are made up of two **half-cells**, the polarity of which is a result of the position of the half-cells in an electrochemical series. Such half-cells are also important in electroplating and metal corrosion.

Modern battery systems, for example, NiCd, MNiH, and Li-ion, have higher energy storage density than traditional batteries but are inadequate for a purely electric vehicle. They find application in electronic equipment that needs longer intervals between recharging. For longer-term or continuous use, a fuel cell is required. Practical fuel cells include the hydrogen **PEM** cell and, to a lesser degree, the PEM methanol fuel cell.

Both ultracapacitors and fuel cells fill power-energy niches not satisfied by standard battery systems. A **Ragone plot** offers a quick assessment of the region that a given electrochemical engineering technology best fills.

EXERCISES

Some useful unit conversions: 1.00 kJ=0.278 Wh; 1.00 Wh=3600 kJ; 1.00 U.S. gallon=3.79 L; 1.000 metric ton (tonne)=1000. kg; one electron charge=1.60×10^{-19} coulombs; Avogadro's number=6.022×10^{26}/ kmol; standard molar volume=22.4 m^3/kmol, or *equivalently* R_u is the universal (i.e., per kmol) gas constant, 8.31×10^3 J/K kmol.

1. A vehicle has a 15. U.S. gallon gas tank and can be filled from empty in 60. s.
   a. What is the rate that power is transferred to the vehicle?
   b. If the vehicle is converted to an all-battery system, using a battery pack to replace the gas tank, what is the rate of power transferred to the batteries if it takes 4.0 h to charge?
   Assume the density of gasoline of 740 kg/m^3, its heat of combustion is 46,500 kJ/kg, and the battery's energy storage density is 525 kJ/L. (**A: a. 33. MW;** b. **2.1 kW**.)

2. A windmill produces mechanical power according to this formula: $P=\frac{1}{2}\eta\rho V^3 A$, where η is its efficiency (assume $\eta=60.\%$), $\rho=$density of air (1.00 kg/m^3), $V=$wind speed in m/s (assume 5.0 m/s), and A is the cross section of the mill that faces the wind (assume it is circular with a radius of 35.0 m).
   a. How many kW does the windmill produce?
   b. To load level, we have a battery storage device that can store 2.00 MWh. How long will we be charging it?
   c. If the energy storage density we can achieve is 125 Wh/kg, how big is this storage battery in tons?

3. We wish to store 2.00 MWh as an emergency power supply for a "big-box" store. If the gross energy storage density of the battery is 425 kJ/L, how big is the storage battery in m^3? If the density of the battery averages 2.5×10^3 kg/m^3, is this reasonable compared to other batteries on the market? (Hint: See Figure 17.6). (**A: 2.60 meter cube, yes.**)

4. In a proposed hybrid car, the battery is "reinforced" by an ultracapacitor with the following characteristics: 5.0 Wh/kg and 5.0 kW/kg. If the car wants to draw on its ultracapacitor for 10. kW for 10.0 s, is it power or energy limited?

5. Will a Ragone plot confirm the conclusions of Exercise 17.4? Describe how by referencing Figure 17.6. (Hint: How long does it take to discharge the ultracapacitor?)

6. Draw lines of constant discharge time on the Ragone plot. Interpret these lines. (Hint: Energy/power=time. On a log-log plot, since time is represented only to the first power, its slope is 1, which is represented as one ordinate decade to one abscissa decade.)

The figure that follows applies to **Exercises 7, 8, and 9.**

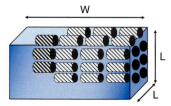

7. Design a battery pack composed of D cells (6.35 cm long × 3.18 cm diameter, each weighing 0.100 kg). The final package (see figure) must supply 42.0 V (a coming standard for all cars) at 30.0 A for 2.00 h. Each cell produces 3.0 V and has a *power producing density* of 125 W/kg. Estimate W and L for the package (ignore its energy storage density for now).

8. Design a battery pack composed of D cells (6.35 cm long × 3.18 cm diameter, each weighing 0.100 kg). The final package (see figure) must supply 42.0 V (a coming standard for all cars) at 30.0 A for 2.00 h. Each cell produces 3.0 V and has an *energy storage density* of 125 Wh/kg. Estimate W and L for the package (ignore its power producing density for now).

9. Design a battery pack composed of D cells (6.35 cm long × 3.18 cm diameter, each weighing 0.100 kg). The final package (see figure) must supply 42.0 V (a coming standard for all cars) at 30.0 A for 2.00 h. Each cell produces 3.0 V and has a *power producing density* of 125 W/kg **and** an *energy storage density* of 125 Wh/kg. Estimate W and L for the package.

10. Compare a battery made from the aqueous half-cells: $Mg^{++}+2e^- \leftrightarrow Mg(s)$ and $Cu^{++}+2e^- \leftrightarrow Cu(s)$ with one made from $Fe^{++}+2e^- \leftrightarrow Fe(s)$ and $Ag^++e^- \leftrightarrow Ag(s)$. What are the voltages and can you speculate on the relative weights of the batteries if the densities of Mg(s), Cu(s), Fe(s), and Ag(s) are 1740, 7190, 7780, and 10,500 kg/m^3, respectively? (Hint: A significant part of the weight of an electrochemical cell is the weight of its electrodes.)

11. Consider a graph of the half-cell potentials in Table 17.2 versus the density of the electrode metal in the half-cell. Explain whether or not it is possible to design a cell that is both light and high voltage (say, more than 3 V).

12. An industrial fuel cell has to supply a continuous 20. MW to a perfectly efficient inverter at 123 V. Assuming it uses a hydrogen gas/air PEM system, what are the needed flows of H_2 and air to the cells in standard m^3/s? See the simplified schematic that follows. Assume each fuel cell produces 1.23 V.

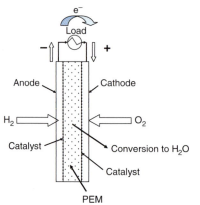

13. In the fuel cell configuration in Exercise 12, the actual voltage delivered by the cell varies with load; assume the voltage measured at full load is 0.75 V so that the cell electrolytic efficiency is $0.75/1.23 = 61.\%$. How many cells do we need in series and what is the fate of this inefficiency?

14. If you run a fuel cell in reverse as an electrolysis cell, H_2O is split into its elementary components, $H_2(g)$ and $\frac{1}{2} O_2(g)$. This way we can make hydrogen and have it stored in case of the loss of outside electric power. The stored hydrogen can then be fed to our electrolysis unit acting as a fuel cell. If

   a. it takes 1.23 V to electrolyze water, what is the overall efficiency, defined as (Theoretical Power Required in MW)/(Actual Power Required in MW), of this scheme?

   b. the electrolysis requires 1.50 V, what then is its overall efficiency?

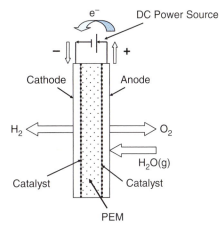

15. Solid oxide fuel cells use a unique electrolyte—a hard ceramic, based on a solid-state solution of zirconium oxide "stabilized" by the element yttrium—which conducts oxygen ions when hot enough (about 1000 °C). The anode is a Ni/ceramic material and the cathode is an exotic material, such as lanthanum strontium manganite. It works similar to a PEM fuel cell. Air is applied to the cathode and hydrogen to the anode. It is about 60% efficient, and if it has an application, it will be to sophisticated central stationary power plants. Write down the anode and cathodic reactions; what is the theoretical voltage of such a cell?

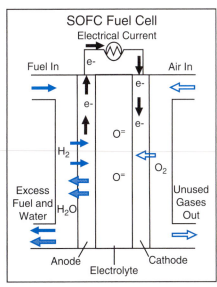

16. A methanol fuel cell is used as a battery for a laptop whose average electrical demand is 20. W at 10.0 V. Each cell produces 0.50 V. How much methanol in ml is needed each day? Methanol density is 792 kg/m³.

17. A steam reformer is fed 2.00 L/s of methane, 0.952 L/s of air, and 1.00 L/s of steam. If it reaches its stoichiometric maximum extent of reaction, what is the composition of its product gases? Assume all quantities given in standard L/s (1.00 bar, 0 °C basis). (**A: CO = 0.600 L/s, CO₂ = 0.400 L/s, H₂ = 5.00 L/s, and N₂ = 0.752 L/s.**)

18. The same reformer as in the previous problem has a problem: the 3.00 L/s of H_2 is contaminated with 0.600 L/min of CO. Assume you can remove this contaminant down to 10.0 ppm of CO. What is the final composition of its product gases?

19. For the reformer in the two previous questions, what is its maximum electrical power production?

Comment: The next two problems are loosely based on experiences of Thomas Edison in developing a new battery, initially invented for use in automobiles and submarines. For details of the actual case, see Byron Vanderbilt, *Thomas Edison, Chemist.*

20. You are an engineer on a team developing a radically new battery invented by your immediate supervisor, a successful and famous inventor. Preliminary tests suggest that the new battery, if used in automobiles, will triple the range over the best existing battery. Detailed tests have not been made as to the safety or lifetime of the battery. But, the inventor urges immediate introduction to the market, using his personal reputation as a guarantee to customers of the value of the product. Only by testing it in actual use, the inventor argues, can any remaining "bugs" be identified and removed. You are concerned that customers will regard the inventor's confident statements as an implied warranty. You urge delay until safety and life tests can be done. The inventor, your immediate supervisor, rejects this approach. What do you do? (Use the Engineering Ethics Matrix.)

21. You are the CEO of a company that has developed a new fuel cell invented by one of your company's most creative engineers. After thoroughly testing the system for safety and reliability under a wide range of ordinary conditions, you arrange a sale to a customer operating under unusually stressful conditions (undersea exploration to unprecedented depths and pressures) not covered by your tests. As part of the contract, the customer takes full responsibility for any difficulties that arise under these conditions. In use, the fuel cell fails miserably in the application due to an unanticipated pressure effect. The inventor, embarrassed by the failure, urges that you fully reimburse the customer for the sale price, keep the failure secret, and allow him to correct the problem at company expense. He argues that you are ethically obligated to do this to uphold the company's good name. What is the ethical way to respond to the inventor's request? (Use the Engineering Ethics Matrix.)

Green Energy Engineering

Earth Rise as Seen by the Apollo 16 Astronauts. *Source:* Courtesy of NASA.

18.1 INTRODUCTION

This chapter describes how nonfossil sources of energy may be substituted for today's dominant fossil fuels, which are coal, oil, and natural gas. Principally, we use these fuels for electric power production and for our motor vehicles.

We already looked at some conventional energy sources based on these fuels in Chapter 7 as well as dependent electrochemical sources of stored energy and power in Chapter 17. One of the things we learned is that a major effluent from fossil fuel sources is carbon dioxide, a gas implicated in global warming. The subject is contentious because short-term weather fluctuations are highly variable. On the other hand, climate studies look at long range changes in weather patterns, such as previous historical earth warming and ice age cooling periods, of which there have been many in Earth's history. If and when either returns, the survival of humankind will be sorely tested.

There are both theoretical and experimental reasons to expect global warming due to CO_2 accumulation in the atmosphere. Of the radiation reaching us from the Sun, a small fraction has some high-energy ultraviolet light (UV) content. While this is a small part of the total sunlight spectrum, it is absorbed by the atmosphere, preferentially by CO_2 molecules (also by water vapor). A measured solar spectrum is shown as Figure 18.1. The dips in the curve are due to absorptions by CO_2, H_2O, O_3 (ozone), and other constituents of the atmosphere, and the high-energy solar radiation occurs at short wavelengths (i.e., near the origin of this graph).

The incoming photons (packets of light) from sunlight are measured as a function of **wavelength**[1] (usually in nanometers, abbreviated nm); the longer the wavelength, the redder is the light and the lower its energy. The ordinate is in watts/m^2 per nanometer. To get the total solar insolation (that is, the amount of incoming sunlight), we must integrate for the area under the curve. In practice, we need to integrate only over the visible

---

[1]Individual light waves are sinusoidal. Their frequency (which is inverse to their wavelength) is proportional to their energy. High frequency (or short wavelengths) indicates energetic waves and vice versa.

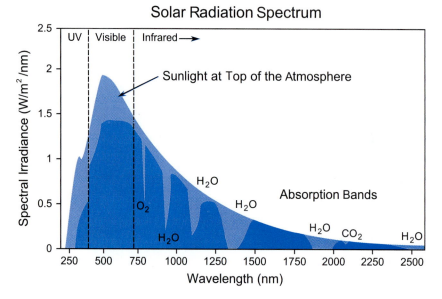

FIGURE 18.1 Solar irradiation as a function of wavelength.

spectrum from its low-energy red end (or infrared (IR)) to its high-energy blue end of the spectrum, the ultraviolet (UV). Integrating the spectrum confirms the Earth receives an average of about $1.3 \, \text{kW/m}^2$ of its surface area.

If none of the incident sunlight on Earth were absorbed in the atmosphere, we would not have very many warm evenings because they are due, in part, to the retention of the absorbed heat in the atmosphere that accumulated during the sunny hours of the day. In addition, after striking the Earth, the resulting relatively short wave ultraviolet radiation is degraded to long wavelength infrared radiation, which is trapped by the lower atmosphere (hence, the term *greenhouse gas*).

Over a century ago, the eminent Swedish chemist Svante Arrhenius first recognized the atmosphere could heat up due to CO_2 accumulation. The amount of CO_2 in the atmosphere has increased from the pre-Industrial Age value of about 280 parts per million (ppm) to a current near 400 ppm, quite probably because worldwide we continue to emit some 8 billion tons of CO_2 into our atmosphere every year.

Many mathematical studies of the world's climate do show a long-term heating trend in our atmosphere; however, global warming due to CO_2 retention is a hotly contested subject because the calculations are sensitive to small changes in input data and our observed weather patterns are anything but smooth and uniform.

In this text, we take the position that it is better not to gamble that an increase in CO_2 in the atmosphere is benign. This proposition leads to a major conclusion: we should consider ways to produce both electric power and motor fuel in ways that do not increase the CO_2 concentration in the atmosphere. Figure 18.2 shows the overall pool of available energy resources.

Renewable energy resources are only 8% of the total available energy resources according to Figure 18.2. These include solar, geothermal, wind, and hydroelectric-generated power but we will study only direct solar- and indirect solar-derived methods of power generation, as shown in Table 18.1.

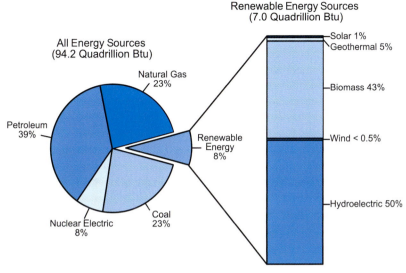

Totals may not equal sum of components due to independent rounding.

FIGURE 18.2 Current world energy resources. A Btu (British thermal unit) is the heat required to heat 1 lbm of water 1 °F.

Table 18.1 Solar Alternatives to Fossil Fuels

Power Methods	Ultimate Source of Power	Comments and Rough Capital* Costs per kW
Photovoltaics	Direct Sun	Relatively expensive, large potential for small applications (**$4750/kW**)
Solar thermal electric power plants	Direct Sun	May use the Sun directly to produce steam; expensive per unit of power (**$3150/kW**)
Solar thermal heat	Direct Sun	Direct household use; costs depend on specifics
Windmills	Wind (itself dependent on the Sun)	Large installations are economical; limited sites available (**$1200/kW**)
Hydropower	Hydroelectric river dams and tidal power generators (which are really another effect of solar energy moving water)	Many large facilities in the world; number of acceptable sites declining (**$1500/kW**)

*Capital costs refer to the hardware and are exclusive of additional costs of running the power plant (such as labor and fuel costs). If the capital cost is $2000/kW, the total plant cost for a 1000 MW power plant is $2 billion. The price of electricity, normally quoted in cents/kWh, depends on capital costs, interest rates, cost of fuel, and other plant operating costs. Electric power in the United States generally costs about 5–10 cents/kWh.

The first two kinds of green power plants in the table rely on immediate, bright sunlight. When the Sun goes down, the generated power drops to zero. In Chapter 17, we raised the question of how to deal with synchronizing solar output with immediate demand for such systems, since load leveling by batteries is expensive and inconvenient.

As indicated in Table 18.1, wind is a form of solar energy because it is generated by air heated by the Sun. For example, hot air rises and cool air sinks, and the interaction of these vertical winds with the Earth's rotation can transform them into horizontal winds.[2] The energy of these winds can be captured by the blades of large windmills.

Hydropower is another form of solar energy. Water from the oceans evaporates into the atmosphere. Eventually, this water rains or snows out in the mountains due to atmospheric cooling as it is lofted. If we are clever enough to catch the rain or the snowmelt behind a dam, we can release water from the dam when it is needed and convert its potential energy into electricity. Tidal flows can also be used in hydroelectric schemes.

Did you wonder why hydrogen has not been included in renewable green sources? After all, the by-product of the combustion of hydrogen is water *and only water*. It seems to be the perfect substitute for fossil fuels since hydrogen-fueled motor cars and fuels cells (Chapter 17) are well within our technological capabilities. The question is this: What is a viable source of hydrogen?

There are two industrial methods for making hydrogen: (1) by adding steam to partially burned hydrocarbons and, as such, is a copious producer of CO_2 and (2) electrolysis, which is only used to make relatively small quantities of hydrogen, and since it uses electricity, it too is an indirect producer of CO_2.

18.2 SOLAR ENERGY

Solar energy is the light and heat we receive from the Sun. The Earth receives 1.74×10^{17} watts of solar radiation at the upper atmosphere. Approximately 30% is reflected back to space, while the rest is absorbed by clouds, oceans, and land masses.

Solar energy technologies include solar heating, solar photovoltaics, solar thermal electricity, and solar architecture, which can make considerable contributions to solving some of the energy problems the world now faces.

18.2.1 Photovoltaic Power

Sunlight falls onto certain materials and electricity is produced. How? What is the basic process? and Does it infer limitations to this method of producing electric power from photovoltaics? To answer these questions, we need to delve a little into quantum mechanics.

You may already know that classical mechanics was unable to explain the photoelectric effect, which occurred when light shining on a heated electric wire produced a small current of electrons. More light did not necessarily help. What did the trick, as Einstein explained, was to shine a light of sufficient energy (equivalently blue color or high **frequency** or short wavelength) onto the wire. In other words, UV light worked well, but longer wavelength (redder light) less so.

Einstein showed that an electron was produced only when the light exceeded a certain frequency. At longer wavelengths, there was zero current. This effect was immediately associated with Planck's postulate that sources of light emit in discrete frequencies, as measured by the famous Planck equation:

$$E = hv \tag{18.1}$$

where E is the photon's (i.e., light wave's) energy, v is its frequency (in cycles/s), and h was originally thought of as an empirical proportionality constant naturally called *Planck's constant*, which, in MKS units, is 6.626×10^{-34} J·s.

We can also express the energy of light by other variables (wavelength and frequency), using the traveling wave equation in which the speed of light is $c = 3.00 \times 10^8$ m/s:

$$\lambda v = c \tag{18.2}$$

---

[2]In Chapter 5, we saw that a crude estimate of wind energy over the whole Earth is about 0.002% of the total received solar energy.

Example 18.1

The wavelength of the visible light spectrum is from (a) 0.4 μm to (b) 0.70 μm. What are the corresponding energies and frequencies in joules and hertz (cycles/s), respectively, of near-IR and near-UV photons?

Need: Frequencies and energies of near-IR and near-UV photons.

Know: Wavelengths of near-IR and near-UV photons, respectively, are 0.70 and 0.40 μm.

How: Use Planck's Law, Equation (18.1) with $h = 6.626 \times 10^{-34}$ J·s, and convert to frequency from Equation (18.2) with $c = 3.00 \times 10^8$ m/s.

Solve: $\lambda\nu = c$ gives $\quad \nu = c/\lambda = 3.00 \times 10^8 \times 10^6/0.70\,[m/s]/[\mu m] \cdot [\mu m/m]$
$$= 4.3 \times 10^{14}\text{cycle/s} = 4.3 \times 10^{14}\,\text{Hz} \quad \text{for near} - \text{IR photons}$$
and $\quad \nu = c/\lambda = 3.00 \times 10^8 \times 10^6/0.40\,[m/s]/[\mu m] \cdot [\mu m/m]$
$$= 7.5 \times 10^{14}\text{cycle/s} = 7.5 \times 10^{14}\,\text{Hz} \quad \text{for near} - \text{UV photons}.$$

Then,
$$E = h\nu \text{ or } E = 6.626 \times 10^{-34} \times 4.3 \times 10^{14}\,[J \cdot s][1/s] = 2.9 \times 10^{-19}\,\text{J for near-IR photons, and for near-UV photons,}$$
$$E = 4.1 \times 10^{-19}\,\text{J}.$$

While the numbers are correct, answers in these units are not very intuitive.

In the photoelectric effect, a photon of light with energy less than a given frequency gives zero current, but at another incrementally higher frequency, the current flows. A sufficiently energetic photon can thus produce a detectible electron. This is the origin of photovoltaic power.

A complication arises when we are not dealing with the photoelectric effect *per se* but with semiconductor solids that are relatively poor conductors of electricity (see Chapter 10). The situation is displayed in Figure 18.3, which is known as an **energy level diagram**. On the left is a sketch of the basic photoelectric effect; an incoming photon stimulates the release of an electron only if its energy (i.e., frequency) is large enough. The analogous situation for photovoltaics is shown on the right. In photovoltaic solids, the number of energy levels is huge because of the vast number of interconnected atoms. The energy levels smear into regions in which the electrons can exist. These regions are known as *valence* and *conduction bands*.[3] Electrons exist in conduction and valence bands and not in between them. If the bands overlap, as in metal conductors, electrons spontaneously flood the conduction band, leading to easy electricity flow. If the bands are so far apart electrons cannot jump to the conduction band, the material is an insulator. In between are the **semiconductors**, whose properties are dominated by the properties of the band gap.

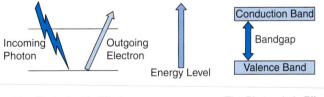

The Photoelectric Effect   The Photovoltaic Effect

FIGURE 18.3 Photovoltaic basics.

[3]These are so-called because electrons trapped in these bands do what the names suggest, make chemical bonds (the valence band), and conduct electricity (the conduction band).

Band gap values for four common semiconductors are given in Table 18.2, of which the most important is silicon. It is convenient to express the energy in the unfamiliar but useful units of the electron volt (eV).[4] Crystalline silicon's band gap is 1.11 eV, meaning, among other things, that we would need to connect many such cells in series to reach a useful household level of 115 volts.

Table 18.2 Band Gaps for Some Semiconductors

Material	Symbol	Band gap (eV)
Crystalline silicon	c-Si	1.11
Amorphous silicon	a-Si	1.7
Germanium	Ge	0.67
Indium antimonide	InSb	0.17
Diamond	C	5.5

Note: See also Chapter 13 for a discussion of crystalline versus amorphous materials.

To promote a free electron in c-Si, we need incoming photons with energies ≥ 1.11 eV. Using Equation (18.1) and Planck's constant in eV units, we can directly compare electron energies to the photon energies that produce them. In these units, Planck's constant is $h = 4.135 \times 10^{-15}$ eV·s (Example 18.2).

Example 18.2

What are the near-IR and near-UV photon energies in eV corresponding to wavelengths of (a) 0.40 and (b) 0.70 μm?

Need: The near-IR and near-UV photon energies in eV.
Know: Planck's law $E = h\nu$ and wave speed equation $\lambda \nu = c$, with $h = 4.135 \times e^{-15}$ eV·s.
How: Combine the traveling wave equation with Planck's Law, $E = hc/\lambda$.
Solve: $E = hc/\lambda = 4.135 \times e^{-15} \times 3.00 \times 10^8 \times 10^6/0.4$ [eV·s][m/s][μm/m]/[μm] $= 1.240/0.4$ eV $= $ **3.1 eV** for near-UV photons and **1.8 eV** for near-IR photons.

Figure 18.4 shows the band gap of amorphous silicon in eV relative to the Sun's spectral energy (smoothed to ignore extraneous atmospheric absorbencies).

---

[4]An electron volt (eV) is a convenient unit of energy for atomic and subatomic particles. It's the energy required to push one electron up a gradient of 1 volt. Since an electron's charge is 1.60×10^{-19} coulombs, **1 eV $= 1.60 \times 10^{-19}$ C $\times 1.0$ V $= $ 1.60 $\times 10^{-19}$ joules.** (Note that 1 [C $\times$ V] $= 1$ [A·s] $\times$ [V] $= 1$ J.)

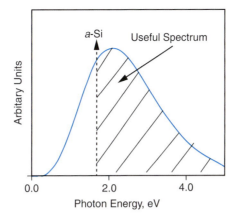

FIGURE 18.4 Band gap limitation on photon efficiency.

Since only photons with energy above the band gap contribute to the electric current, the smaller band gap in a germanium crystal theoretically has a larger efficiency than c-Si. Integration of the area under the spectral curve for energies larger than the band gap determines the possible photovoltaic efficiency (although other limitations are imposed by electrical, thermal, and optical properties in a practical solar device).

Real photovoltaic devices have many factors that enter into their efficiencies. Even the "contamination" of c-Si by a-Si reduces the cell's efficiency given that c-Si is considerably harder to produce and thus more expensive than a-Si. A possible solar cell installation is shown in Figure 18.5.

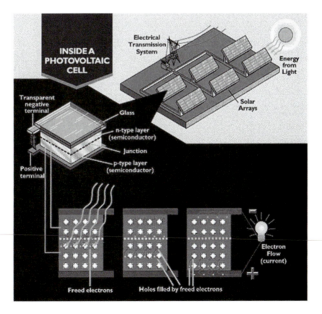

FIGURE 18.5 Photovoltaic installation. Source: Office of Energy Efficiency and Renewable Energy, U.S. Department of Energy.

This figure shows several factors other than band gap energy that are relevant to photovoltaic efficiency. Within the cell itself we have to consider diffusion of the photon-produced electrons,[5] as they flow toward the electrical collector plates on either side of the semiconductor's surface. Glass plates are also needed to protect the cells. Dopants (small quantities of active elements) are added to the semiconductors to modify their band structure. The cells may have many layers of semiconductor to optimize their efficiency. Actual *a*-Si photovoltaic cells have overall efficiencies of 10-20%. Germany is the current leader in solar photovoltaics, with an installed capacity in 2010 of almost 17,000 MW.[6]

Example 18.3

If a single 100. cm^2 silicon cell produces a voltage of 0.50 V and a current of 2.0 A when exposed to sunlight with an intensity of 1000. W/m^2, determine the total number and area of these silicon cells required to produce 17,000 MW.

Need: The total number of 100. cm^2 silicon cells and their total area.
Know: The power from a single cell is 0.500×2.00 [V][A] $= 1.00$ [VA] $= 1.00$ W.
How: The number of cells needed is the total power required divided by the power produced by a single cell.
Solve: Therefore, to produce 17,000 MW $= 17 \times 10^9$ W, we would need **17×10^9 cells**, which would have a total area of Area $= 17 \times 10^9$ [cells] [1 cm^2/cell][m/100 cm]$^2 = $ **17×10^7 m^2** or about 66 square miles.

18.2.2 Solar Thermal Power Plants

Standard fossil power plants use combustion heat to create high-pressure steam. The steam then drives a turbine connected to an electric generator that, in turn, produces the electricity. A similar conceptual solar thermal design is shown in Figure 18.6. It concentrates focused sunlight onto tubes containing hot oil, which then produces high-pressure steam. The solar heat is concentrated about 75 times as the parabolic mirrors swivel to follow the maximum sunlight.

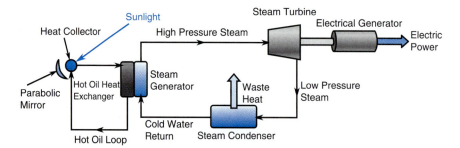

FIGURE 18.6 Solar thermal power plant.

As in fossil power plants, the high-pressure steam drives a turbine that then drives an electrical generator. The steam exiting the turbine is condensed back to liquid water, which is pumped back into the heat exchanger to be reconverted into high-pressure steam.

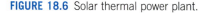

[5]Electrons are, of course, electrically negative, and "holes" are regions that have been depleted of electrons; therefore, they are relatively positive. Both holes and electrons are collected at their respective electrodes as the product of the photoionization process.
[6]http://en.wikipedia.org/wiki/Solar_power_in_Germany.

The theoretical amount of power produced by the turbine is the mass flow rate of the steam multiplied by the difference between the energy in the steam at the turbine's inlet and the turbine's exhaust. State-of-the-art[7] solar thermal power plants are located in desert regions. An existing solar plant heats thermally stable synthetic oil to 400 °C and produces up to 80 MW of electricity from a mirror field of nearly ½ million square meters.

Example 18.4

What is the gross efficiency of a solar power station that produces 80.0 MW for 10.0 hours/day from an array field of parabolic mirrors covering 5.00×10^5 m^2 if the solar thermal flux is 1.30 kW/m^2?

Need: The efficiency of a solar thermal power plant.

Know: The power output is 80.0 MW for 10.0 hours/day. Solar power is 1.30 kW/m^2, and the array field is 5.00×10^5 m^2.

How: Compare the power produced with the solar insolation of 5.00×10^5 m$^2 \times 1.30$ kW/m$^2 = 6.50 \times 10^5$ kW.

Solve:

Efficiency $= 100 \times 80.0 \times 1000/(6.50 \times 10^5)$ [%][MW][kW/MW]/[kW] $= \textbf{12.3\%}$.

Note that the United States can afford to use such large land areas in underpopulated southwestern deserts. But, Europe and Asia have larger population densities than the United States, and this power plant could be considered an underutilization of land.

18.2.3 Solar Thermal Heating

Certainly, those of you who live east of the Rocky Mountains are familiar with the cold that winter brings. Do you know how your house is heated in winter? Do you know what fuel is used to keep your house warm? Did you realize your heating system has a CO_2 effluent[8]?

Household heating is an expensive and reoccurring expense. A typical northern suburban house consumes about 100×10^6 kJ per heating season. Provided that a typical winter's day is not too overcast, solar heating to supply this household heat would seem to be an obvious application. All that is needed is a strategically placed solar collector mounted perpendicular to the rays of the winter Sun as in Figure 18.7 (usually on a roof with a southern[9] exposure). Because we need only relatively warm water for household heating, we can use simple tubes with flowing water inside[10] to collect some of the incident solar radiation. Since black surfaces are very efficient at collecting solar radiation, the tubes are often blackened as well as being insulated with a glass cover.

Inside the house, we need a pump, thermal storage (such as drums of warm water to be called on when the Sun is not shining), and standard heating coils to distribute the warm water throughout the house (see Figure 18.8).

---

[7]http://en.wikipedia.org/wiki/Solar_Energy_Generating_Systems.
[8]If you have an electrically heated home, it too emits CO_2 but remotely at the power station.
[9]This maximizes the collection time for sunlight as the sun sweeps from east to west.
[10]In cold climates, the tubes are usually filled with antifreeze, which then passes through a heat exchanger to produce warm water for household heating.

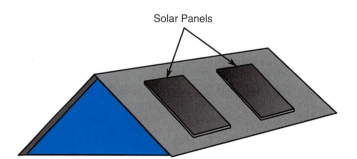

FIGURE 18.7 Solar heating.

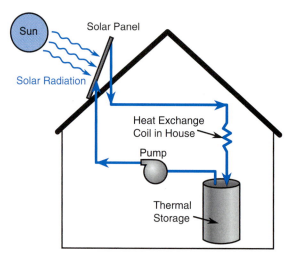

FIGURE 18.8 Domestic heating systems.

Example 18.5 Design of a Solar Thermal Heating House

A typical suburban house in midwinter in a moderate climatic zone has a furnace capacity of about 20 kW that runs on a 50% cycle (meaning that it is on half the time). This means that, on average, about 10 kW is delivered to the house, and this is also the heat rate that needs to be supplied by solar heating.

If the solar irradiation is 0.50 kW/m^2 and is directly delivered to the collection fluid in the roof panel, and the heat capacity of the storage fluid is 1.10 kJ/kg°C and its density is 1100 kg/m^3, determine the following items.

a. The size of the required roof panels and is it possible to put panels of this size on a typical house roof?

b. The storage size required for the hot fluid for a winter night of 14 hours if the house is maintained at 22 °C and the hot fluid collected in the tank is at 45 °C.

c. How much heat can be stored in the storage tank?

d. Is this a reasonable tank size?

Need: **a.** Size of the solar collector = _____ m².
  b. Solar heat stored in thermal storage tank = _____kJ.
  c. Size of solar thermal storage tank = _____m³.
  d. Is this tank reasonable?

Know: Required heat rate is 10. kW, solar insolation rate is 0.5 kW/m², inlet temperature to thermal storage tank is 22 °C, and outlet is 45 °C. Heat capacity of coolant is 1.10 kJ/kg°C, and its density is 1100 kg/m³.

How: Equate the solar insolation to the house's needs as measured by the furnace capacity.

Solve: **a.** Assume the solar panel area is A m². Hence, collected heat rate $= 0.50 \times A$ [kW/m²][m²] $= 10$ kW. Therefore, $A = 10.0/0.50$ m² $= \mathbf{20.\ m^2}$, and if the panels are twice as wide as high, their size is $A = HW = 2\,W^2$, so $W = \sqrt{10.} = \mathbf{3.2\ m}$ and $H = \mathbf{1.6\ m}$. These will fit on a typical house roof.

  b. After dusk, all of the energy needed by the house is to be recovered by the heat in the storage tank(s). This amount is 10. [kW] $\times$ 14. [hr] $= 140$ kWh $= 140 \times 3600$. [kWh][kJ/kWh] $= 5.0 \times 10^5$ kJ. Therefore, the solar storage capacity of the reservoir tank $= \mathbf{5.0 \times 10^5\ kJ}$.

  c. Let the tank's volume $= V$ m³ so that the delivered heat from the tank is $Q = \rho V C \Delta T$, where Q is the stored heat, ρ is the density of the thermal fluid, C is its heat capacity, and ΔT is the difference in temperature of the stored fluid minus the delivered temperature. Therefore, $Q = \rho V C \Delta T = 1100 \times V \times 1.10 \times (45 - 22)$ [kg/m³] [m³][kJ/kg°C] [°C] and $Q = 2.8 \times 10^4\ V$ kJ $= 5.0 \times 10^5$ kJ so that $V = 5.0 \times 10^5/2.8 \times 10^4 = \mathbf{18.\ m^3}$.

  d. A standard metal oil drum has a capacity of 0.16 m³, so we would need 112 drums, which is a lot for most houses! If the stored thermal energy is too small, we would also need a supplementary house furnace to supply additional heat during the dark hours. In other words, we may need two complete heating systems: the solar one and a conventional one when the thermal storage has been exhausted. As a quick rule of thumb, the cost of a solar-assisted unit is about twice that of a conventional furnace.

18.2.4 Windmills

Terrestrial wind is driven by solar heating in the atmosphere. Once winds are in accessible regions, meaning the atmospheric boundary layer we all live in, their kinetic energy can be harvested, as it has been for centuries by windmills.

FIGURE 18.9 A modern windmill. *Source:* Photograph by Wagner Christian, Creative Commons CC-BY-SA-2.5.

The basic analysis is easy. The power output of a windmill is directly proportional to the cross-sectional area of its rotors, A (because that multiplied by the axial wind speed v determines the volumetric flow rate of air through the rotors, $\dot{Q}$).

$$\dot{Q} = Av \tag{18.3}$$

After the passage of the wind through the windmill's rotors, the wind spreads out behind the windmill, since it has slowed after having some of its energy removed (see Figure 18.10).

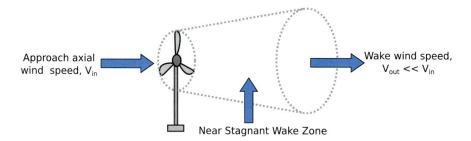

FIGURE 18.10 Fluid mechanics of a windmill. The air flow is somewhat different for an isolated windmill and an array of them; carefully designed placing of many windmills can enhance their interactions to increase the overall efficiency.

Ideally, all the wind's power is converted to mechanical power, but stationary wake air behind the windmill would stop the flow through the windmill. An ideal power calculation ignores this:

$$P_{\text{ideal}} = \dot{Q} \times \frac{1}{2}\rho v^2 = \frac{1}{2}\rho v^3 A \tag{18.4}$$

Since the ideal lossless state cannot be achieved, this equation must be corrected for the finite flow behind the windmill's rotors. This detailed calculation shows that there is a maximum in the wind recovery behind the windmill, which balances the flow through the rotors with the need to purge the wake airflow:

$$P_{\text{actual}} = \frac{1}{2}C_p\rho v^3 A \tag{18.5}$$

The term C_p is called a **power coefficient**, which has a theoretical maximum of $16/27 = 0.593$ (or 59%). In this form, Equation (18.5) is called **Betz's Law** and the power coefficient is called the **Betz's limit**.[11] In practice, the efficiency limit is about 0.4-0.5 (or 40-50%).

Example 18.6

A modern commercial windmill has a diameter of 100. meters and operates in winds of just 5.00 m/s. If the Betz limit is 0.50, what is the power produced by this windmill? Assume that the air density is 1.05 kg/m³. What wind speed is needed to achieve the windmill's design power output of 2.5 MW?

  Need: The power produced by a windmill of 100. meters diameter in a wind of 5.00 m/s with a power coefficient of 0.50. What wind speed corresponds to the design basis of 2.5 MW?

  Know: The Betz limit$= 0.5$, size of windmill, and air density is 1.05 kg/m³.

  How: Cross-sectional area of windmill is $\pi d^2/4 = 3.14 \times 100.^2/4 = 7.85 \times 10^3$ m².

  Solve: Using Equation (18.4),

---

[11]http://en.wikipedia.org/wiki/Betz's_law.

$$P_{\text{ideal}} = \tfrac{1}{2}\rho v^3 A = \tfrac{1}{2} \times 1.05 \times 5.00^3 \times 7.85 \times 10^3 \, [\text{kg/m}^3][\text{m/s}]^3[\text{m}^2]$$
$$= 5.15 \times 10^5 \, [\text{N} \cdot \text{m/s}] = 5.15 \times 10^5 \, \text{W} = 0.515 \, \text{MW}$$

The actual power, $P_{\text{actual}} = 0.50 \times 0.515 \, \text{MW} = \mathbf{0.26 \, MW}$.

The design basis for this windmill is 2.50 MW. Since the power varies directly as the cube of the wind speed, **the required wind** is $(2.50/0.26)^{0.333} \times 5.0 = 10.6 \, \text{m/s} = \mathbf{11 \, m/s}$.

Therefore, approximately doubling the wind speed increases the power output by a factor of about 10! However, the calculation does assume that the windmill's operation is otherwise independent of the wind speed, which is unlikely, since the windmill's flow surely depends to some degree on the aerodynamic losses. Unsurprisingly, windmills are located in areas where the wind is both constant and strong.

18.2.5 Hydropower

The Industrial Revolution may be said to have begun using rain runoff from the Pennine Hills in northern England.[12] Water powered many manufacturing industries until the age of steam supplanted it. Instantaneous rainwater runoff was unreliable, but if the water was blocked and released when needed through a channel, it could be put to valuable use. Unlike heat cycles, hydropower uses potential energy and converts it to mechanical energy, a system that is very efficient.

There are two key variables for a successful hydropower system: sufficient potential energy of the upstream water (the water head in Figure 18.11) and the volumetric flow rate of water through the turbine. The flow rate of

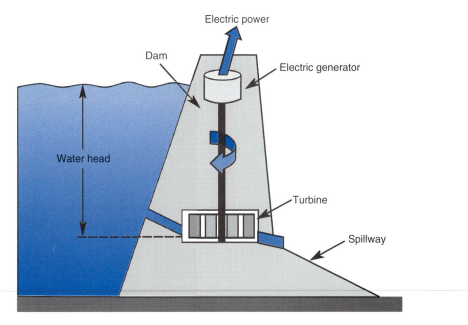

FIGURE 18.11 Hydroelectric power.

[12]You have probably heard of Great Britain's notoriously wet climate!

the water is determined by the geometry of the piping and the internals of the turbines. If the water rate is $\dot{Q}$ (ft^3/s or m^3/s), the power delivered to the turbine is

$$P = \frac{\rho g h \dot{Q}}{g_c} \qquad (18.6)$$

where $g_c = 1$ (dimensionless) in MKS units, and $g_c = 32.2$ lbm · ft/(lbf · s^2) in Engineering English units.

In this formula, power is energy × speed and ρ (Greek rho) is the density of water. The recovery of the potential energy by water turbines is more than 90% efficient. Hoover Dam, shown in Figure 18.12, was built in the 1930s. It has 17 turbines with a combined capacity of 2080 MW; the water head is 590 ft (Example 18.7).

FIGURE 18.12 Hoover Dam, built in the 1930s. *Source:* Bureau of Reclamation, http://usbr.gov/hooverdam.

Example 18.7

A 100% efficient water turbine has a 590. foot water head and produces 161,000 HP. How much water in ft^3/s must flow through it?

Need: The water flow in ft^3/s to 100% efficient turbine system with a head of 590. feet of water. Turbine produces 161,000 HP.

Know: The water density is 62.4 lbm/ft^3, $g = 32.2$ ft/s^2, $g_c = 32.2$ lbm·ft/lbf· s^2, and 1.00 HP = 550. ft·lbf/s.

How: From Equation (18.6), we have $P = \rho g h \dot{Q}/g_c$.

Solve: The water flow rate is

$$\dot{Q} = P g_c / \rho g h$$

$= 161,000 \times 550. \times 32.2/(62.4 \times 32.2 \times 590.)$[HP][(ft·lbf/s)/HP][lbm·ft/lbf·s^2]/{[lbm/ft^3][ft/s^2][ft]}$= \mathbf{2.41 \times 10^3}$ **ft^3/s.**

Hoover Dam has 17 units of about this size so that its total water throughput is about 41,000 ft^3/s. The new Three Gorges Dam in China has electricity production approximately 10 times larger than Hoover Dam and so requires roughly 10 times the water flow through its turbines.

The United States has about 7% of its electric power drawn from hydropower sources. While a decreasing number of sites are still available, hydropower is by far the most successful of green energy sources in the United States.

It should be noted that dams can fail with massive loss of life due to downstream flooding of the break. A serious accident occurred in 2009 at a Russian 2000 MW hydroelectric plant due to a massive electrical failure that produced an explosion with severe loss of life of the plant's operating personnel.

18.3 OTHER GREEN ENERGY SOURCES

Solar energy in its various forms is not an exhaustive list of non-CO_2 power sources. For example, OTEC, or Ocean Thermal Energy Conversion,[13] was suggested for a floating power station that uses the differences in water temperature between deep and shallow zones as a thermal source to make electric power (but at very low efficiency). Another oceanic method relies on wave motion that is mechanically tapped for power. Yet another uses tidal power. The Rance Tidal Power station on the English Channel in the north of France produces 240 MW using tidal water and a dam to capture it. The trapped water is let down through a series of water turbines and thus is similar to standard hydroelectric power. Such systems may work well, but there are few locations where they can be situated.

In northern California, hot steam can be found near the Earth's surface. Using a power plant similar to that for solar thermal power (except that the heat source is geothermal steam), California produces about 1800 MW geothermally. But, in general, there's an obvious difficulty of finding suitable natural geothermal steam sources (Iceland being an exception where 25% of its electrical energy needs to come are met by geothermal power). Given that volcanic areas are most suitable for geothermal energy, it is not surprising that acidic sulfurous gases accompany the production of geothermal power. These gases are smelly and toxic and very corrosive to the materials of construction of the power plant. For these reasons, the gases are collected and disposed of by reinjecting them back underground.

18.4 SUSTAINABLE ENGINEERING

Sustainable engineering is the process of designing devices and systems to use energy and resources at a rate that does not compromise the natural environment (i.e., sustainably). Every engineering discipline is engaged in sustainable design by using initiatives such as life-cycle analysis, pollution prevention, and waste reduction.

In the twenty-first century, engineers need to be aware of potential short- and long-term environmental impacts. Effects beyond the short-term that may not manifest themselves for decades are the main focus of sustainable design. For example, materials previously used by engineers that were thought to be harmless are now known to be hazardous, such as asbestos in floor tiles, pipe insulation, and shingles; lead in paint and pipes; and wall materials that could host toxic molds. These materials illustrate short-term design decisions that became long-term human health hazards.

Some design impacts may not occur until many years in the future. For example, the use of nuclear power to generate electricity may well be a sustainable design issue since the radioactive waste material will take many years for it to decay and become harmless to the environment. How can we design these systems so that they are environmentally safe for all time?

Sustainable engineering often begins with the choice of materials used in a product. For example, exhaust gases from the combustion of traditional hydrocarbon fuels are the major contributors to environmental problems such as global warming and air pollution. The use of renewable fuels is a promising option to reduce the

[13]en.wikipedia.org/wiki/Ocean_thermal_energy_conversion.

environmental impact of fuel combustion. For example, ethanol-gasoline blends contain 10–15% of the renewable fuel ethanol that are commonly in automotive fuel today.

The process of sustainable design involves (a) conducting a life-cycle analysis, (b) prioritizing the most important problems found, and (c) finding the technologies and operations to address those problems. Problems will vary by size, difficulty in treating, and feasibility. The most difficult problems are those that are small but expensive and difficult to solve.

18.4.1 Life-Cycle Analysis

A **life-cycle analysis**[14] (LCA) is a method of predicting the impact of a product, process, or activity on the environment. It analyzes the effects of materials used, manufacturing methods, transportation and distribution systems, customer use, product maintenance, and final disposal. A life-cycle analysis produces a complete cradle-to-grave history of a product as shown in Figure 18.13.

FIGURE 18.13 Product life cycle.

Example 18.8

Describe the life cycle of your family car using words that describe the events involved.

Solution:

1. Raw and recycled materials enter a factory at one end, and our family car is driven out of the other end.
2. It is then shipped to a car dealer who sells it to my family.
3. We purchase the car and drive it to work and on family vacations.
4. It consumes fuel and pollutes the atmosphere, and it occasionally needs repairing.
5. After many years, it breaks down and cannot be fixed anymore.
6. Our car is then sent to a salvage yard, and any useful parts are stripped off.
7. The rest of our car is crushed and then recycled to manufacture a new car or other products.

---

[14]Life-cycle analysis began in the 1960s and is now defined by an International Standard with strict rules about how it is to be applied and relied upon. The standard is available at: AS/NZA ISO 14044:2006.

The rapid adoption of plastics in many applications in recent years is due to their reliability, ease of manufacture, light weight, and affordability. However, with heightened awareness about energy and environmental concerns, plastics made from natural gas and oil are perceived by the public and decision makers as having negative impacts on sustainability.

A life-cycle analysis of plastic packaging was carried out to evaluate the potential environmental effects of these materials.[15] The study analyzed the energy requirements and greenhouse gas emissions of the following six general categories of plastic packaging: caps and closures, flexible beverage containers, rigid containers, shopping bags, shrink wrap, and other flexible plastic packaging. The study concluded that replacing all plastic packaging with nonplastic alternatives for these six types of packaging in 1 year in the United States would

- require 4.5 times as much packaging material by weight, increasing the amount of packaging used in the United States by nearly 55 million tons,
- increase energy use by 80% (equivalent to the energy from 91 oil supertankers), and
- result in 130% more global warming potential (equivalent to adding 15.7 million more cars).

However, there is an ongoing debate about whether paper or plastic is the better choice for the environment. Here's a look at plastic bag use and factors to consider when making the decision.[16]

- Plastic bags were first introduced in 1977 and now account for four out of every five bags handed out at grocery stores.
- Americans make 2.3 trips to the grocery store each week. If people use 5–10 bags each time, that's between 600 and 1200 bags per person per year.
- Stores usually pay less than a penny per plastic shopping bag and 3–4 cents per paper shopping bag.
- Plastic grocery bags consume 40% less energy to produce and generate 80% less solid waste than paper bags.
- Plastic bags can take 5–10 years to decompose. Paper bags take about a month to decompose.
- Paper bags are made from trees, which are a renewable resource. Most plastic bags are made from polyethylene, which is made from petroleum, a nonrenewable resource.
- Two thousand plastic bags weigh 30 pounds, and 2000 paper bags weigh 280 pounds.
- A packed standard-sized paper bag can hold up to 14 items; an average plastic bag often holds 5–10 items.
- When compared to grocery bags made with 30% recycled paper, polyethylene bags use less energy for manufacturing, less oil, and less potable water. In addition, polyethylene plastic grocery bags emit fewer global warming gases, less acid rain emissions, and less solid wastes.

18.4.2 Recycling

Recycling is the process of collecting and processing materials that would otherwise be thrown away as trash and turning them into new products. The benefits of recycling include:

- reducing the amount of waste sent to landfills and incinerators;
- conserving natural resources such as timber, water, and minerals;
- preventing pollution by reducing the need to use new raw materials;

---

[15]http://plastics.americanchemistry.com/Education-Resources/Publications/Impact-of-Plastics-Packaging.pdf.
[16]http://abcnews.go.com/Technology/story?id=97476&page=1.

- reducing greenhouse gas emissions that contribute to global climate change;
- creating new jobs in the recycling and manufacturing industries;
- helping to sustain the environment for future generations.

Recycling is an important part of developing sustainable products. As recycling becomes more socially accept-able, it is important to consider the potential of all the products we use. Even the humble newspaper can be as a valuable resource after its useful life. Of the stuff we threw away in the United States in 2012, about 54% was discarded in landfills, 12% was incinerated for energy, and 34% was recycled or composted.[17]

Recycling materials such as steel, copper, brass, and other metals reduce pollution caused by the excavation and refining of raw materials, and it saves energy and money. It also means that valuable and irreplaceable mate-rials will last longer. For example, it is estimated that in 40 years the world reserve of copper ore will be nearly depleted. Since copper is a vital component in electrical devices, using recycled discarded copper will extent the time needed for engineers to find alternative materials to replace copper.

The challenge of developing a sustainable future depends heavily on the ability of engineers to reduce cost, energy, materials, and natural resources in a structured and environmentally safe manner. In the con-ventional design process, engineers use functional analysis, brainstorming, design matrixes, experiments, computer simulations, risk analysis, cost estimation, and customer feedback. Design constraints, such as bud-get limits, time schedule, federal and state regulations, building codes, existing patents, and public policies, are also included. Sustainable design differs from conventional design in that sustainability issues must be considered in all the design steps. This can be done through "life cycle" thinking that introduces compromises between environmental, economic, social, and technical decisions. These design decisions involve the four life-cycle stages: material extraction and use, production and transportation, customer use, and product disposal.

SUMMARY

Most of what is called **Green Energy** is direct and indirect manifestations of solar energy: **photovoltaic power**, **thermal power**, **solar heating, wind power**, and **hydropower**. Unfortunately, solar power in most forms is limited by the local conditions of clouds and of diurnal and seasonal variations in the Sun's output. Wind power is likely the most important source of green power. Wind power is probably the best Green energy source. Only hydropower has made major inroads into commercial power production.

Photovoltaic power maybe from a free source, but it carries a lot of auxiliary systems; further its efficiency is reduced by the effect of its **band gap** on the useable fraction of the solar spectrum. Direct thermal power for electricity or heat is only making small inroads into commercial use. Wind power is subject to **Betz limit** that reduces the possible electric power output. Hydropower is limited only by finding suitable real estate for col-lection and containment of the required water. All of this is necessary for a sustainable future for the people on earth.

EXERCISES

1. Which value in the table that follows is the best approximation to the efficiency of a diamond-based pho-tovoltaic: 10%, 55%, 75%, or 96%? (a) Why? and (b) Why would diamond be an interesting semiconductor?

---

[17]http://www2.epa.gov/recycle/recycling-basics.

Material	Diamond	Semiconductor X	Semiconductor Y	Semiconductor Z
Band gap, eV	5.5	1.0	2.0	3.0
Efficiency				

2. How long does it take to heat a 1.00 m deep swimming pool by 10.0 °C if the Sun is directly shining over it and there are no heat losses? Assume the water is well mixed. Assume the sunlight totals 1.30 kW/m². The heat capacity of water is $C=4200.$ J/kg/°C, and its density is 1000. kg/m³. Ignore any heat losses from the pool. (**A:** 9.0 hours.)

3. Heat losses from a hot body at T_h in a cold environment at T_c generally follow a rate law in which the losses are proportional to the temperature difference between them $(T_h$-$T_c)$. What is the steady-state temperature of a circulating liquid in a roof-mounted solar heater that receives sunlight at a rate 1.3 kW/m² and loses heat according to $110. \times (T_h - T_c)$ W/m²°C. Assume the outside temperature is 15.0 °C.

4. An industrial thermal plant uses a 15:1 concentrator to catch the Sun's rays. The heat losses from the heated fluid obey the relationship $110. \times (T_h - T_c)$ W/m². What is the daytime temperature of the circulating fluid if the outside temperature is $T_c = 15.0$ °C?

5. The photograph of a modern windmill, Figure 18.9, shows a common three-bladed design. The speed of rotation of these blades is determined from a combination of the wind speed, the size and shape of the blades, the variable blade pitch[18] relative to the wind, and the resistive load imposed by their coupling to the electrical generator. Assume the rotational speed for a 100. m. diameter windmill is 30.0 RPM. What is the tip speed of a blade? Can you also suggest two (or more) limitations on the maximum blade speed?

6. A large manufacturing company is selling a 1025 MW windmill array. In the location chosen, you expect a 10.0 m/s steady wind. If each windmill has a Betz limit of 0.46, how many windmills do you need if each has a diameter of 100. m? The air density is 1.050×10^3 kg/m³.

7. What's the cost for each windmill and its supporting infrastructure in the preceding exercise, given an installed cost of $1200./kW?

8. You work for an electric power utility and send out a RFP[19] for a 1025 MW wind farm and receive several bids, including one particular bid that claims to be able to deliver the specified power with only 400 windmills each 100. m. diameter (and a correspondingly low bid). Your job, as chief engineer for the electric power utility, is to decide if this bid will, in fact, deliver the contracted amount of power, 1025 MW. Assume the wind speed is 10.0 m/s and the air density is 1.05 kg/m³.

9. A dam holds back a lake that is 50. km long and fills a V-shaped channel that is 200. m. deep by 100. m across. How long does it take to empty in a severe time of drought if the outflow is steady at 10.0×10^3 m³/s?

10. You want to build a dam on a large river and extract 5000. MW from it. You can take 10.0×10^3 m³/s of water through the dam's spillways. The turbines you intend to use are 90.0% efficient. Most of the expense goes into the construction costs of the dam, so you would like to locate it where the backup lake will be as shallow as possible. What is this depth?

[18] A small hydraulic motor is used to pump oil to an actuator to rotate the blade pitch. This increases the blade's efficiency to catch the wind.
[19] Request for Proposal. This is how a great deal of engineering work is bid. Either the Government or Corporation X wants to make something and sends out an RFP to qualified companies etc., soliciting their bids. In theory, if you are the lowest bidder, you will receive the contract.

11. The motors and generators in some hydropower systems are reversible. That is, the water turbine can act as a pump and the generator as an electric motor to drive the pump. A power management scheme assumes that you have excess power some of the time and a deficiency at other times. Suppose you have a small dam and a 10.0 MW generator. During the demand period of 6.00 hours, you can produce 10.0 MW, which essentially empties all the water from the impoundment behind the dam. During the night, 12.0 MW of excess power is available from the power grid. How many hours does it take to refill the dam if the pumping and the generator steps are both 100% efficient?

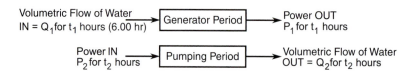

12. In northern France, the Rance tidal power plant captures the inflow from the English Channel[20] behind a 750 m. long dam. The high tide average is 11. m over a year. If the power produced peaks at 240 MW and its capacity factor is 0.40, what's the average energy produced? Assume the turbine is 90.0% efficient. (The *capacity factor* is the actual energy produced divided by the maximum energy if running at 100% rating.)

13. Some British engineers had an idea for a "slightly" tidal electric power concept. The Menai Straits in North Wales is located between the mainland and the small island of Anglesey. The channel between the island and mainland has a strong tide with a 5.0-foot average rise. The British considered damming it at its narrowest point (about 400. m across). The dam was to hold large amounts of an ion exchange resin that would selectively leach natural uranium[21] from the seawater, where it is present at a level of about 3.0 ppb (parts per billion). How much seawater in m^3 must flow through the ion-exchange resin to make 1000. kg of natural uranium?

14. Write out six or more steps you think are in the life cycle of a newspaper. Start with trees and end with recycling and landfills.

15. In their book, *Time to Eat the Dog*: The real guide to sustainable living, the authors Brenda and Robert Vale compare the ecological footprints of pets with those of various lifestyle choices, like owning a SUV. For example, a medium-sized dog eats about 164 kilograms of meat and 95 kilograms of cereal in 1 year. It takes 43.3 m^2 of land to generate 1 kilogram of chicken and 13.4 m^2 to generate a kilogram of cereal per year so that it gives the dog an eco-footprint of about 8400 m^2 (for a big dog like a German shepherd, the figure is 11,000 m^2). An average SUV driven, 6200 miles a year (about 10,000 km), uses 55.1 gigajoules (this includes the energy needed to both build and fuel it). 10,000 m^2 of land can produce approximately 135 gigajoules of energy per year, so the SUV's eco-footprint is about 4100 m^2—less than half that of a medium-sized dog. As well as consuming natural resources, cats and dogs devastate wildlife populations, spread disease, and add to pollution. Is there any good reason to own a dog?

16. You are the chief civil engineer for the county. In your professional capacity, you learn that two of the county's senior politicians each own a piece of land that is ideal to complete a new development. But only one piece of land is needed for the project, and the two politicians are vying to sell their land to the

[20]Francophiles might prefer "La Mance." It translates into English as "The Sleeve" which, to the poetic mind, is the shape of the Channel viewed in an easterly/westerly direction.
[21]The uranium was then to be refined for nuclear power plant use.

development. Understandably perhaps, neither politician has revealed his personal interest in the development to the public. However, you know that neither came by his land recently nor dishonestly and both are asking the same price for their land. There is no difference to the development's costs whichever plan is adopted. None of this has been made public, but a local newspaper reporter has filed court papers under the Freedom of Information Act (FOIA) for all of the county's information about the development. Both politicians ask you to redact data concerning their involvement in the development because it only looks like they are being dishonest while they are, in fact, not. (Use the Engineering Ethics Matrix and the NSPE Code of Ethics for Engineers.)

a. Should you comply with the FOIA request "implicating" the two politicians?

b. Should you comply with the FOIA request but leave out the part played by the two politicians, since neither is being dishonest?

c. Should you approach either politician and suggest one or both should himself leak it to the local press?

d. Should you take the initiative and reveal the land ownership in the regular monthly county meeting and explain the circumstance of the land deal?

17. You sold a house that you claimed is "certifiably" passive. You designed it with lots of windows so that it is warm enough in winter due to solar gain through them. Unfortunately, you suddenly realize that, with so many windows, the interior gets too hot in summer, especially on the south side; indeed, the cooling energy costs for summer could even exceed heating energy costs for winter.

You already signed a contract with the house buyer and may lose money if you have to make too many modifications but you guaranteed only the winter's heating costs. You have several options. Which should you do? (Use the Engineering Ethics Matrix and the NSPE Code of Ethics for Engineers.)

a. Proceed with the original design.

b. Install smaller windows to reduce both the summer's and winter's solar gain, thus reducing the overall energy costs.

c. Use higher-quality windows on the southern exposures to limit summer heating and standard ones on the northern exposure.

d. Add a summer thermal storage system to augment the winter's thermal storage system.

e. Go back to the customer and ask to reopen the negotiation, explaining you cannot build the house for what was previously agreed to.

Hands-On

Introduction to Engineering Design

Source: http://www.mos.org/eie/engineering_design.php.

19.1 INTRODUCTION

This chapter describes the nature of engineering design, introduces modern design philosophy called Design for Six Sigma, suggests some benefits of a hands-on design project, indicates the qualities of a good designer, and explains the need for a systematic approach that consists of an eight-step design process.

19.2 THE NATURE OF ENGINEERING DESIGN

In the course of creating new products, engineering design uses available technology to improve performance, to lower cost, or to reduce risk. For example, if the design of a bridge produces a new structure that is visually stunning with no consideration for its strength, this is design without any engineering. If, on the other hand, the designers of a new concept car use analysis or experiments to evaluate air drag, structural integrity, and manufacturability in addition to style when coming up with a new exterior design, this is engineering design.

Prior to the 1960s, the preceding description of engineering design might have been adequate, but this is no longer the case. The definition has been broadened to include the systematic thought processes and best practices that define the modern engineering design process. This systematic approach, which has become synonymous with engineering design, forms the heart of these design chapters.

Those of you experienced at tinkering with your own inventions in the basement or garage might say that you got along fine without engineering analysis and without training in the systematic approach. Such a view ignores the realities of engineering design. Usually, designers must search for the best possible design under severe conditions of limited time and limited resources (especially cost). For example, in most design situations, there is barely enough time to produce a single prototype. The idea behind an engineering approach, as embodied in engineering analysis and in the systematic approach, is to minimize the number of design iterations required to achieve a successful final design. Fewer design iterations means lower cost and shorter development times.

One of the greatest challenges of engineering design is the breadth of knowledge required of the designer. The diversity of topics covered in earlier chapters provides a hint of what a design might entail; some electro-mechanical designs could conceivably touch on them all. In addition to those topics, there are issues related to manufacturing, economics, aesthetics, ethics, teaming, government regulations, and documentation of the design, to name but a few. In the spirit of these design chapters, we propose that the best way to introduce the multifaceted nature of engineering design is to experience it for yourself through the following hands-on design exercise.

19.3 DESIGN PROBLEMS VS. HOMEWORK PROBLEMS

A design problem is unlike a traditional college homework problem. Homework problems, like those in Parts 1 and 2 ("Lead-On" and "Minds-On," respectively), in this textbook have specific, unique answers; and the student must find those answers using a logical approach such as the need-know-how-solve method. Design problems, on the other hand, do not have unique answers. Several answers (i.e., designs) may satisfy a design problem statement. There may also be a "best" answer based on critical requirements. For example, a design that minimizes manufacturing costs may not produce a very reliable product. If product reliability is a critical requirement of the customer, then minimizing manufacturing costs may not be the "best" design answer.

The challenge of Part 3 of this textbook, which we call "Hands-On," is to transition you from solving well-formulated, single-answer textbook problems to solving what we call *open-ended* engineering design problems. These problems are often complex and sometimes poorly formulated problems. You will find that excellent design work requires considerable creativity beyond that needed to solve textbook problems. In Part 3 of this text, we allow and encourage you to drop your academic inhibitions and explore the joy of engineering creativity.

The problem-solving process we presented in Part 1 of this text (need-know-how-solve) works well for single-answer textbook problems, but it needs to be expanded to encompass open-ended engineering design problems. In recent years, the process of engineering design has been refined to produce more robust and economical designs. The process that most companies embrace today is a multistep process called "Design for Six Sigma."

In Part 3 of this text, we abandon the need-know-how-solve exercise solution method and adopt an eight-step design process based on the Six Sigma design methodology.

19.4 BENEFITS OF A HANDS-ON DESIGN PROJECT

A practicing engineer does not have to be an expert in machining or other basic manufacturing operations. Still, a basic understanding of the challenges involved in manufacturing a product is essential for producing a successful design. The best way to appreciate that fact at an early stage in your career is to manufacture a design yourself.

The lessons to be learned are universal. Don't expect your design to work on the first try. Leave a lot of time for testing. Complicated designs take a lot longer to build and have a lower probability of success, and if you have a choice of manufacturing a part yourself or buying it, buy it. Many such lessons are foretold by the design principles and design for manufacture guidelines of later chapters. The consequences of violating those principles are best understood by experiencing the results of having done so.

Other lessons are to be learned from a hands-on design experience. For electromechanical systems with moving parts, that experience might be the only way to accurately evaluate the design. Also, students gain a sense of accountability by learning that it is not enough for a design to look good on paper. To actually be a good design, it has to lead to an end product that works.

In particular, we recommend that the hands-on design project should be done under a competition format, involving interactions between "machines." It is a natural motivator—the challenge of the task is heightened by having to deal with the unpredictability of your human opponent, and the other designs provide a relative scale against which to assess quality of performance.

19.5 QUALITIES OF A GOOD DESIGNER

These are the qualities of a good designer:

- **Curiosity about how things work:** Seeing other design solutions provides you with a toolbox of ideas that you can draw from when faced with a similar design challenge, so when you come across an unfamiliar device, try to figure out how it works. Take things apart; some companies actually do this and refer to it as *reverse engineering*. Visit a toy store; the products there often demonstrate creative ideas and new technology.
- **Unselfishness:** A key ingredient to effective teaming is suppressing ego and sacrificing personal comfort to serve the best interests of the team.
- **Fearlessness:** It takes a leap of faith to step into the unknown and create something new.
- **Persistence:** Setbacks are inevitable in the course of a design project. Remain resilient and determined in the face of adversity.
- **Adaptability:** Conditions during the design process are constantly evolving. For example, new facts may surface or the rules of the design competition may change in some way. Be prepared to take action in response to those new conditions. In other words, if the ship is sinking, don't go down with the ship—redesign it.

19.6 USING A DESIGN NOTEBOOK

When you are working on a design project and you want to write something down, the design notebook is the place to do it. There is no need for notepads, reams of paper, or sticky notes. The place to record your thoughts is in a permanently bound volume with numbered pages, a cardboard cover, and a label on the front cover identifying its contents. Every college bookstore has them, though they may be called *laboratory notebooks*.

As a starting engineer, now is the best time to start a career-long habit. Just how important the design notebook is can be explained in the case of Dr. Gordon Gould.[1]

> *On November 9, 1957, a Saturday night just given to Sunday, Gould was unable to sleep. He was 37 years old and a graduate student at Columbia University. For the rest of the … weekend, without sleep, Gould wrote down descriptions of his idea, sketched its components, projected its future uses.*
>
> *On Wednesday morning he hustled two blocks to the neighborhood candy store and had the proprietor, a notary, witness and date his notebook. The pages described a way of amplifying light and of using the resulting beam to cut and heat substances and measure distance… Gould dubbed the process light amplification by stimulated emission of radiation, or laser.*

It took the next 30 years to win the patents for his ideas because other scientists had filed for a similar invention. Gould eventually won his patents and received many millions in royalties because he had made a witnessed, clear, and contemporaneous record of his invention.

The lesson is that patents and other matters are frequently settled in court for hundreds of millions of dollars by referring to a notebook that clearly details concepts and results of experiments. You must maintain that notebook in a fashion that will expedite your claim to future inventions and patents.

---

[1]http://inventors.about.com/gi/dynamic/offsite.htm?site=http://www.inc.com/incmagazine/archives/03891051.html.

Another more immediate benefit of using a design notebook is that you will know that everything related to the project is in one place. Finding that key scrap of paper in a pile of books and papers on your desk after working for months on the project can be a rather time-consuming endeavor.

Here are some of the most important guidelines for keeping a design notebook:

- Date and number every page.
- *Never* tear out a page.
- Leave no blank pages between used pages. Draw a slash through any such blank pages.
- Include all your data, descriptions, sketches, calculations, notes, and so forth.
- Put an index on the first page.
- Write everything in real time, that is, do not copy over from scraps of paper in the interests of neatness.
- Write in ink.
- Do not use White Out; cross out instead.
- Paste in computer output, charts, graphs, and photographs.
- Write as though you know someone else will read it.
- Document team meetings by recording the date, results of discussions, and assigned tasks.

19.7 THE NEED FOR A SYSTEMATIC APPROACH

The two main goals of the systematic approach to engineering design are (1) to eliminate personal bias from the process and (2) to maximize the amount of thinking and information gathering done up front, before committing to the final design. The result is fewer costly design changes late in the product development stages.

The engineering design process also provides a blueprint for design of complex systems. For example, you might be able to get along without a formalized design procedure when designing a new paper clip,[2] but when taking on the daunting task of designing a complex system like the International Space Station, brain gridlock can set in. The design process offers a step-by-step procedure for getting started as well as strategies for breaking down complex problems into smaller manageable parts.

19.8 STEPS IN THE ENGINEERING DESIGN PROCESS

A systematic approach to engineering design that uses the elements of the Design for Six Sigma philosophy may be viewed as consisting of eight steps:

1. Define the problem
2. Generate alternative solutions
3. Evaluate and select a solution
4. Detail the design
5. Defend the design
6. Manufacture and test
7. Evaluate the performance
8. Prepare the final design report

These steps are shown in Figure 19.1.

---

[2]But see just how difficult it originally was to perfect the paper clip: Henry Petroski, *The Evolution of Useful Things: How Everyday Artifacts—from Forks and Pins to Paper Clips and Zippers—Came to Be as They Are* (New York: Alfred A. Knopf, 1992).

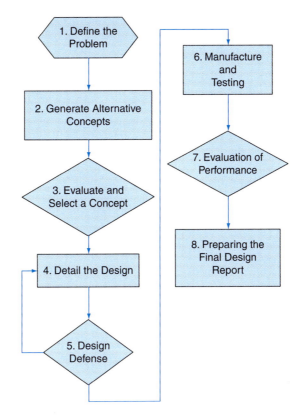

FIGURE 19.1 The design process flowchart.

The next chapter presents two ground rules for engineering design. Subsequent chapters treat each of the preceding eight steps in detail. At the end of each chapter is a suggested milestone for successful completion of the step in the design process described in the chapter. Milestones are crucial for measuring progress toward the eventual goal of a successful design. After the eight steps have been described, this portion of the book concludes with a detailed example of an actual design competition.

19.9 HANDS-ON DESIGN EXERCISE: "THE TOWER"

Your first design objective is to build the **tallest tower** out of materials to be supplied. Since there are both individual and team winners here, this is both an individual and a team competition.

19.9.1 Setup

- Divide the class into about eight teams of three or more students per team.
- Each team should be provided with five sheets of 8.5×11 standard copier paper.
- The following material will be distributed **unevenly**:
  - Two teams each receive one roll of Scotch® tape.
  - Two teams each receive a roll of duct tape.

- ○ Two teams each receive one box of paper clips.
- ○ Two teams each receive one pair of scissors.

19.9.2 Rules

- Each team has just 10 min to build a tower.
- The final tower height measurement must be made by the instructor.
- Teams should indicate to the instructor when they are ready for a measurement.
- If the tower is composed of materials other than the supplied paper, Scotch tape, duct tape, or paper clips, it will be disqualified.
- The tower must be stationary when the measurement is made.
- The tower must be built on a flat surface. The tower cannot lean against or be attached to any other surfaces (wall, table, etc.).
- Any team that intentionally knocks over another team's tower before it has been measured is disqualified.
- After 3 min, **one** individual on each team is offered 8 points to join another team.
- There are no other rules.

19.9.3 Scoring

- One point will be awarded for each inch of tower height. (Heights are to be rounded to the nearest inch.)
- The first team that finishes its tower and is ready to be measured receives a bonus of 10 points.
- The "team" winner is the team with the highest point total.
- The "individual" winner is the student with the highest point total.

19.9.4 After the Exercise

Discuss the importance of the following issues to the outcome of the competition:

- The "quality of teaming" among different teams: Were any of the extra materials (tape, paper clips, or scissors) shared between teams?
- The "quality of teaming" within the team: Was everyone within the team given the opportunity to contribute ideas or did one person dominate the decision making?
- Ethics: Was it ethical to jump to another team, not share materials, or copy the design of another team?
- Manufacturability: How important was it to have the right materials?

EXERCISES

1. What is the major difference between the abstract equations used in a mathematics course and the equations used to describe physical objects in design? (**Ans**. The abstract equations used in mathematics do not have units, but those used in design always have units.)

2. Name three additional learning benefits that you feel come from a hands-on design experience.

3. Two ground rules for design are to use a design notebook and to team effectively. Which of the following would also make a good ground rule?

a) Work alone.
b) Keep the design simple.
c) Manufacture all the parts yourself.
d) Ignore input from less knowledgeable people.

4. There are eight steps in the design process discussed in this chapter (see Figure 19.1). Where does the input from the "customer" enter into the process?

5. Who are the "customers" in a design process? (**Ans.** There are many "customers," including those who purchase the final product. Manufacturing is also a customer, as are sales and marketing.)

6. In the design process shown in Figure 19.1, you are asked to "Generate Alternative Concepts" for the design. Explain why this is a necessary step.

7. In the design process shown in Figure 19.1, you actually begin the design at step 4. Why not start the design earlier and save time?

8. Define the (a) steps, (b) rules, and (c) scoring for a student design exercise that involves using a single wire coat hanger to make a new functional consumer product.

Design teams

Source: http://www.ibope.com.ar/ibope/wp/trabaja-con-nosotros

Great things in business are never done by one person. They're done by a team.

Steve Jobs, cofounder of Apple Computers

20.1 INTRODUCTION

The design process involves a series of creative steps that lead to a solution to a problem. In engineering, the problem often involves designing a new product that accomplishes a task that meets customer needs. An inventor can sometimes carry out these steps alone, but in large companies design problems are solved by teams of trained engineers and technicians, called **design teams**.

While all teams are groups of people, not all groups are teams. Team members must work together toward a common goal, whereas members of a group only share common interests or characteristics. Sometimes it is difficult to tell the difference between a team and a group. For instance, two or more students might meet to discuss homework or watch TV. They may have an agenda and a purpose, but this is a group meeting, not a team meeting. The scope and duration of their meeting is too small to involve the amount of coordination, resources, and effort that teamwork requires.

Engineering students do not always realize that their future success depends on how well they work together in a team. Everyone on a design team comes from a different background. They have different strengths, cultural values, philosophies, and work ethics. Yet they have to join together to form a coherent and productive group. In this chapter, the elements that define a successful design team are discussed. These elements range from project management to team building, team goals, and leadership.

20.2 HOW TO MANAGE A DESIGN TEAM PROJECT

Design team project management is a carefully planned and organized effort to accomplish a specific project (e.g., designing and constructing a robot vehicle to be entered into a competition on a specific date).

Project management includes developing and implementing a plan that defines the project goals, specifying how and by whom the goals will be achieved, identifying needed resources, and developing budgets and time-lines. Project management is usually the responsibility of an individual team leader. When the project team members have been identified, the team leader (or the course instructor) selects one of its members as the project manager. This person has the responsibility for guiding the team design work in a professional, organized, and timely manner. The project manager is also responsible for meeting deadlines and ensuring that the team members are carrying fair workloads. Successful project management involves the following:

- *Understand the projects' goals.* You should be able to state the goals of your project in a single sentence.
- *Engage all the team members.* Subdivide the work using functional decomposition to break the project down into individual work assignments, and make sure that everyone knows what he or she is responsible for to meet the project goals (see Ground Rule Number 2, discussed in the next section).
- *Keep the project moving forward.* Work methodically to meet your benchmarks; don't wait until the last minute and rush to meet a deadline.

20.2.1 Gantt and PERT Charts

Design team leaders need to control the timing of all operations, so they develop *schedules* that show specific tasks to be performed during the design process. They assign tasks to work groups, set timetables for the completion of tasks, and make sure that resources will be available when they are needed. There are a number of scheduling techniques, but the most common are **Gantt and PERT charts**.

A **Gantt**[1] **chart** is a horizontal bar chart that provides a graphical illustration of a schedule that helps to plan, coordinate, and track specific tasks in a project. Gantt charts may be simple versions created on graph paper or more complex automated versions created using project management applications such as Microsoft Project or Excel, as in Figure 20.1.

Gantt chart			Day of Project Work					
Activity	Start day	# of days	1	6	11	16	21	26
Define Problem	1	5	Define Problem					
Brainstorm	5	6	Brainstorm					
Design	10	5			Design			
Construction	14	3			Construction			
Test	16	3				Test		
Improve	19	4				Improve		
Prepare Report	22	4					Prepare Report	

FIGURE 20.1 Gantt chart illustration.

---

[1]Developed as a production control tool in 1917 by Henry L. Gantt, an American engineer and social scientist.

A Gantt chart is constructed with a horizontal axis representing the total time span of the project, broken down into increments (e.g., days, weeks, or months) and a vertical axis representing the tasks that make up the project (e.g., if the project was designing and constructing a robot vehicle, the major tasks might be define the problem, brainstorm possible designs, choose a design, construct the design, and so forth).

Horizontal bars of varying lengths represent the time span for each task. Using the same robot vehicle example, you would put "Define Problem" at the top of the vertical axis and draw a bar on the graph that represents the amount of time you expect to spend on defining the problem, and then enter the other tasks below the first one and representative bars at the points in time when you expect them to be undertaken. The bar spans may overlap, as, for example, you may start construction while refining the design during the same time span.

Gantt charts give a clear illustration of project status, but they do not tell how one task falling behind schedule will affect other tasks (however, the PERT chart described below is designed to do this). More complex Gantt charts show more information, such as the individuals assigned to specific tasks and notes about the procedures. Gantt Charts should be adjusted frequently to reflect the actual status of project tasks when they diverge from the original plan.

Example 20.1

Tom, Jim, and Sue have a difficult team assignment in their freshman design course. They must design and construct a small robot that will collect Ping-Pong balls scattered on a table and deposit them into a box. As the project was not well defined regarding the table's size, robot controls, and collection time limit, they decided to take a week (5 days) to define the project with the course instructor.

Since the assignment needed to be completed in 26 days, they created the Gantt chart below to plan the sequence of events. They realized that some of the tasks could overlap. For example, they decided to begin brainstorming solutions 1 day before the problem definition was finished. Construction could also begin a day before the design was completed, and so forth.

They also began to assign responsibilities to team members. Sue was assigned to clarify the problem definition, all three would contribute to the brainstorming effort, Jim and Sue would focus on the design, Tom would be responsible for construction, and they would all participate in the remaining tasks.

Gantt chart			Day of Project Work					
Activity	**Start day**	**# of days**	1	6	11	16	21	26
Define Problem (Sue)	1	6	Define Problem					
Brainstorm (All)	5	6		Brainstorm				
Design (Jim & Sue)	10	5			Design			
Construction (Tom)	14	3				Construction		
Test (All)	16	3				Test		
Improve (All)	19	4					Improve	
Prepare Report (All)	22	4						Prepare Report

This chart was updated throughout the project and the schedule was adjusted as needed to meet the completion deadline. The Gantt chart allowed continuous progress and kept everyone alert to their responsibilities.

Gantt charts are useful when the design and production processes are fairly simple. For more complex schedules, PERT charts are used.

A **PERT**[2] **chart** is a project management tool used to schedule, organize, and coordinate tasks within a project. It presents a graphic illustration of a project as a network diagram consisting of numbered *nodes* that

---

[2]PERT stands for *Program Evaluation Review Technique*, a methodology developed by the U.S. Navy in the 1950s to manage the Polaris submarine missile program.

represent milestones in the project linked by labeled arrows that represent *tasks* in the project. The direction of the arrows indicates the sequence of tasks. In Figure 20.2, tasks A and E must be completed in sequence. These are called *dependent* tasks. However, task B does not depend on the completion of task A and thus Tasks A and B can be done simultaneously. These tasks are called *independent* tasks.

The *critical path* is the path that takes the longest amount of time. It is determined by adding the times for the tasks in each sequence and determining which path takes the longest time to complete the project. Thus, the critical path determines the total time required for the project. If you want to produce a product quicker, you have to save time on this path. If you only saved the time on any of the other paths, you still wouldn't produce a product any quicker. You gain efficiency only by improving performance on one or more of the activities along the critical path. The amount of time that a noncritical path tasks can be delayed without delaying the entire project is called the *slack time*.

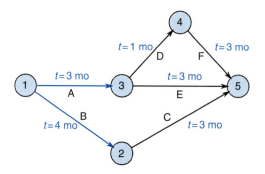

FIGURE 20.2 A PERT network chart for a 7 month project with five milestones (1 through 5) and six activities (A through F). The critical path is tasks A-B-C that require a total of 7 months to complete.

Example 20.2

In the PERT chart below, the letters represent tasks and the numbers represent days to complete the task (i.e., A.2 means task A requires 2 days to complete).

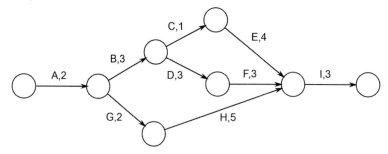

a. Which tasks are on the critical path?
b. What is the slack time for tasks C, D, and G?
c. If task C is delayed by 1 day what effect would this have on the completion date of the project? Why?

Solution

a. Possible paths to completion:

Path 1: A-B-C-E-I $= 2+3+1+4+3 = 13$ days
Path 2: A-B-D-F-I $= 2+3+3+3+3 = 14$ days
Path 3: A-G-H-I $= 2+2+5+3 = 12$ days

Since Path 2 requires the longest time to complete, it is the critical path.

b. Path D-F takes 6 days and Path C-E takes 5 days. The time difference between Path D-F and C-E is 1 day. This is the slack time.

c. If task C is delayed by 1 day it would have no effect because task C has one day's slack time.

20.3 EFFECTIVE TEAMING[3]

Working in teams on a design project is both a joy and a challenge. While there is a sense of security in knowing that others are venturing into the unknown alongside you, the unpredictability of human interactions can be as perplexing as the design itself. To reduce the risk of ineffective teaming, rules of conduct are presented in this section. These are well-accepted best practices based on observations of effective teams.

Several advantages accrue from attacking a design project in teams. First, design requires a wide range of skills and areas of knowledge. No one person is experienced enough to pursue every unfamiliar design challenge in isolation. Teaming provides an opportunity to expand the talents and life experiences brought to bear on the design problem. Second, if done right, teaming serves to keep personal biases in check. Third, more people should mean that more is accomplished in a shorter period of time, although it is puzzling to often see team members standing by politely as one team member does all the work (especially during manufacturing). When best practices are followed, a team is greater than the sum of its parts.

For design projects during the freshman and sophomore years, three students are the ideal team size. Teams of just two students may not experience all of the team dynamics and may not learn as much about teaming. With four student teams it may be too easy for one team member to hide and not contribute. Student design teams at this level may not be assigned a team leader by the instructor. Leadership typically emerges within the team. If a team leader is assigned, the role is *not* to be the boss but rather to organize and facilitate participation by all team members.

Student teams need to understand these basic rules of teaming:

Assign clear roles and work assignments

A few things are best done as a team, such as brainstorming and evaluation of concepts. Most of the time, it will pay off if everyone has his or her own assigned responsibilities and tasks to which each is held accountable by the team. These tasks should be assigned or updated at the end of each team meeting.

Foster good communication between team members[4]

An atmosphere of trust and respect should be maintained, in which team members feel free to express their ideas without retribution. That trust extends to allowing for civilized disagreement, delicately done so as not to suppress ideas or discourage participation. Everyone should participate in the discussions.

---

[3] A funny video about teaming can be found at http://www.bioteams.com/2008/08/29/team_collaboration_funny.html.
[4] An example of poor team communication occurred in 1999 when part of a team of Lockheed Martin engineers used the English system of measurement while the rest of the team used the metric system in designing a Mars orbiter. The unit system mismatch prevented the spacecraft's navigation coordinates from being properly transmitted and the orbiter was lost in space, costing NASA $125 million.

Sometimes, this means reaching out with sensitivity to the shy members of the team. If you succeed, you will have a team operating on all cylinders.

Share leadership responsibilities

If there is a designated team leader, that person should empower the other team members with significant leadership responsibilities. This gives those students a strong sense of ownership in the project. At the same time, team members have to be willing to step forward to assume leadership roles.

Make team decisions by consensus

Teams can make decisions in one of three ways: (1) the team leader makes the decision, (2) discussions continue until everyone agrees, or (3) after discussions are exhausted, the team takes a vote. Those who disagree with the outcome of the vote are then asked if they can put their opinions aside and move forward in the best interests of the team. In a college-level design project, the best ways to make decisions are (2) and (3), which are examples of decision making by consensus.

20.3.1 Team Building

Working as a team implies *group accountability* rather than individual accountability, and results in a collective work product. Team building encourages a "team" approach to working on a project. The advantages of this approach include

- increased flexibility in overall skills and abilities of the team
- more productive than a random student group with individual mindsets
- encourages both individual and collective team development and improvement
- focuses on team goals to accomplish tasks
- improved range of team objectives such as collaboration, communication, and creative or flexible thinking
- realization that a necessary part of team building is negotiation, consensus building, and compromise.

Team building centers around various *exercises* designed to bring a group closer together and create an understanding of individual strengths and weaknesses. It increases camaraderie and team morale, which in turn leads to increased productivity. There are many types of team building exercises that range from simple games to challenges that involve complicated tasks designed to improve team performance. These exercises usually involve interpersonal skills like communication, problem solving, and planning, in contrast to technical skills directly involved with the project. Team building tasks can also stimulate interpersonal skills that will increase team performance. Some examples of team building exercises are shown below.

Communication exercise

Communication exercises are geared toward improving communication skills. The issues teams encounter in these exercises are solved by communicating effectively with each other.

A typical communication exercise is to give everyone in a group a sheet of paper and tell them to close their eyes. Then tell everyone to

1. fold their paper in half
2. fold the lower left corner over the upper right corner
3. turn it 90° to your left
4. fold it in half again
5. turn it 90° to their right
6. rip a half-circle from the middle of the right side.

Now tell everyone to open their eyes and unfold their piece of paper. Even though they all received the same instructions and had the same starting material, nearly everyone will have a different result.

A communication problem occurred because they didn't all start from the same point (some held their paper vertically some held it horizontally), some interpreted ripping a half-circle as removing a big piece, some as a small piece, while others found the instructions vague.

Problem-solving exercise

Problem-solving exercises focus specifically on groups working together to solve difficult problems or make complex decisions. Here is a typical exercise: put a mathematical symbol between 6 and 9 to get a number that is greater than 6 but less than 9 (the answer is a decimal point).

Planning exercise

These exercises focus on aspects of planning and being adaptable to change. A typical example is to plan a surprise party for a mutual friend. The group is to develop a plan to find a time and place for the party without the friend becoming suspicious, invite trusted guests, arrange for food and entertainment, and do so in an organized and effective manner.

20.3.2 Team Leadership

Here are some qualities that a good team leader should have.

Honesty

If you make honest and ethical behavior a key value, your team will trust you.

Ability to Delegate

Delegation involves identifying the strengths of your team members. If they find a task interesting or enjoyable, they will put more effort into it.

Good Communication Skills

Being able to clearly describe what needs to be done is extremely important. If the projects' goals can't be clearly explained to the team, they won't be working toward these same goals.

A Sense of Humor

Team morale influences productivity, so a team leader can inspire a positive attitude by having a sense of humor.

Confidence

A team leader should technically competent and be goal oriented. A team takes cues from its leader, so if the team leader is focused and confident, the team will respond positively.

Commitment

If a leader expects the team to work hard and produce results, she/he needs to lead by example. Once you have gained the respect of your team, they are more likely to deliver the peak amount of quality work possible.

A Positive Attitude

You want to keep your team motivated toward the goals of a project, be enthusiastic about the project.

Creativity

As a team leader, it is important to think outside the box. By utilizing all possible options before making a decision you can more easily reach the project goals.

20.3.3 Team Assessment, Feedback, and Risks

A team can carry out a self-assessment process to determine its effectiveness and improve its performance. To assess itself, a team needs feedback from group members to find out the teams current strengths and weakness. Feedback from the team assessment can be used to identify gaps between the where they are and where they would like to be.

Team assessment can be done in a variety of ways. One of the simplest is to have each team member fill out a questionnaire like the one shown in Table 20.1. Each entry can be given a numerical value, for example, in Table 20.1 we have set Disagree$=0$, Undecided$=1$, Agree$=2$. Then, a total score can be computed for each member of the team by summing all the entries. The best possible score would be 20 ("Agree" is selected for every question). A score of 14 or higher typically means this person is a good team member. A score between 8 and 13 means this person's effectiveness as a team member is sporadic. And a score below 8 indicates that this person has not yet learned how to be an effective team member.

Table 20.1 Team assessment questionnaire

Function		Disagree (enter 0)	Undecided (enter 1)	Agree (enter 2)
1	The team has a clear vision of the project goals			
2	The team has the skills and resources to achieve its goals			
3	Everyone on the team has a clear role in the design, construction and testing			
4	The team has adequate meeting time, space, and resources			
5	Team meetings are attended by all team members			
6	The team members work together effectively			
7	The team members understand the project requirements and timeline			
8	Team members are promptly informed of design changes			
9	Team meetings are run efficiently			
10	Everyone on the team participates at a satisfactory level			
			Total score =	

The major risk of team building is that a team member may become negative and unproductive. This could happen as a result of a teaming event that does not focus on the goals and direction of the design process. For example, if a teaming event seems silly or a waste of time, then it may have a negative impact on team members. This can lead to loss of trust in the leader as well as decrease morale and productivity. Consequently, team building events should be followed by meaningful design project effort.

EXERCISES

1. The balloon tower challenge.
   Equipment: Each group receives 10 small uninflated party balloons and a roll of scotch tape.
   Procedure:
   - Split the class into teams of 3-5 students each
   - Instruct the teams that they need to make the tallest tower possible with the material supplied
   - They must demonstrate a design plan
   - The teams have 20 min to complete the task
   - They must not use any other equipment and the tower must be free standing

   Scoring: Based on the design plan and the final tower height.

2. Arrange to have your team visit a group of working engineers and have them explain how they function as a team. The engineering team should be one that has worked together for some time. Ask them to explain how a project goes from the idea phase through the production phase. After the presentation, have your student team talk about all the ways the working engineers used teamwork to see their idea through. This activity will give your team a chance to see the value of professional teamwork.

3. Gantt chart exercise: Prepare a Gantt chart covering a week of classes that shows how you divide each day into tasks like meals, classes, study time, leisure time, and so forth.

4. Gantt chart exercise: Prepare a Gantt chart for the web site development process shown in the table below.

Item	Task	Duration	Date
Design	Home Page Design	4 days	March 1
	Product Pages Design	1 week	
	Shopping Cart Design—Purchasing	1 week	
	Shopping Cart Design—Checkout	1 week	
Coding	Shopping Cart Coding	2 weeks	
	Contact Us Form Coding	1 day	
Testing	Website Testing	1 day	
	Shopping Cart—Purchasing Testing	1 day	
	Shopping Cart—Payment Testing	1 day	
Delivery	Deliver Website to Client	1 day	
	Receive Feedback & Make Changes	1 week	

5. PERT chart exercise: What is the critical path on the PERT chart shown below.

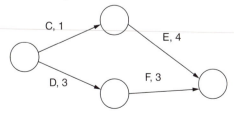

6. PERT chart exercise: Develop a PERT chart that shows how you schedule or organize the tasks you follow when you prepare for school in the morning. Start with getting out of bed and end with arriving at your first class.

7. Communication exercise: Students write directions for making a peanut butter and jelly sandwich for someone who has never made a sandwich before and will be following your instructions. Discuss what the sandwiches might look like after reading your directions.

8. Problem-solving exercise: At a university, all of the students are majoring in engineering, business, or both. 55% of the students are engineering majors and 77% are business majors. If there are a total of 500 students, how many of them are majoring in both engineering & business?

   (*Answer*: If 55% of the students are engineering majors, we know that 45% are not engineering majors. And if 77% of the students do major in business, then 23% are not business majors. 45% + 23% = 68% of the students are not majoring in both engineering and business, therefore 32% are double majors and 500 × 0.32 = 160 students.)

9. Problem-solving exercise: The goal is to build a structure that will prevent an egg from breaking when it is dropped from a height of 8.0 ft. The only materials allowed are 20 straws and 15 in. of ½ in. masking tape.

10. Problem-solving exercise: The goal is to transfer a "radioactive" beach ball from a table in one room onto a table in another room (preferable on another floor of the building) without the beach ball touching ground, any part of the building, or any body part. The only materials allowed are a ball of string, a roll of ½ in. masking tape.

11. Planning exercise: You are with a group of 10 students returning to campus after a bus tour. Your bus is scheduled to leave at 6 pm. While waiting, you and your group go to a coffee shop where one of them was suddenly shot in with a large caliber bullet. While you were helping him, a young girl rushed in to asked for your help with her brother who fell down nearby and broken his leg. Then, a man on in a long-black coat tells you that the person who shot your friend is hidden in the building across the street and is going to shoot all of you. It is now 4 pm and you need to reach campus before 7 pm. Form a plan to deal with this situation.

    (*Answer*: Planning priorities—stop the shooter (call police) and help the person shot (stop blood flow and call an ambulance). Find and help person with broken leg and call another ambulance. Then go to the bus station and catch the 6 pm bus back to campus.)

12. Planning exercise: You are in a cave in the mountains of Colorado. It is winter and a terrible blizzard that will last for a week has covered the ground around the cave with five feet of snow. In the cave you have a metal pot, cigarette lighter, dry wood, warm clothing, knife, two candy bars, snowshoes, and a flashlight. You have no food, water, or telephone. Form a plan to survive.

    (*Answer*: Planning priorities—you need water to survive a week or more in the cave (do not eat snow). Build a fire and melt snow in the metal pot. Ration the candy bars and keep warm. Clear out the opening to the cave and when the blizzard stops put on the snowshoes and head for home.)

13. Ethics exercise: At a team meeting you are asked to report on the progress of your project. A team member asks you to report that his/her part of the project is finished even though it is not. He/she tell you that it will be done in three more days. What do you do? Use the engineering ethics matrix.

Design Step 1: Defining the Problem

21

21.1 INTRODUCTION

The design process begins when somebody, whom we will refer to as the *customer*, expresses a need and so enlists the services of an engineer. The customer can be an individual, an organization, or the consuming public. Most customers are not engineers. It is up to the engineer to translate the customer's need into engineering terms. The result is cast in the form of understanding a need, developing a problem definition, and producing a list of design specifications.

21.2 IDENTIFYING THE NEED

The basis of a good engineering project depends on identifying a viable "**need**" or "**want**" that can be satisfied. Failure to identify, understand, and validate the need, prior to designing, is one of the most frequent causes of failure of the entire design process.

The success of the project then depends on whether or not the customer(s) are satisfied with the result. Ideas for a new engineering project often come from listening to potential customers.

- What do they want?
- Where do they want it?
- When do they want it?
- What price are they willing to pay?

Then you investigate key issues such as

- How large is the market? This is usually determined by conducting a "consumer analysis" through surveys.
- Is anyone else supplying this market? This is usually determined by conducting "competitor analysis" through market studies. If so, can you determine if there are any needs not met?
- Is the market realistic? "I want to go to Mars next week" is not realistic

- What are the financial and legal risks? Research existing patents to see if you risk any violations.
- Finally, can you make a profit by designing and producing a product for this market?

21.3 DEFINING THE PROBLEM

The customer's statement of need does not typically take the form of a problem definition. For example, consider the following statement of need from a fictitious client:

Need: People who work at the Empire State Building are complaining about the long waits at the elevator. This situation must be remedied.

An engineer might translate this need into the following problem definition:

Problem Definition: Design a new elevator for the Empire State Building.

But, is this really a good problem definition? Is the main concern of the management at the Empire State Building to reduce average waiting times or to eliminate the complaints? When turning an expressed need into a problem definition, it is important to eliminate assumptions that unfairly bias the design toward a particular solution. A better, less-biased problem definition might be

Improved Problem Definition: Increase customer satisfaction with the elevators in the Empire State Building.

This would admit such solutions as a mirror on the elevator door or free coffee on the busiest floors.

As another example of an inadequate problem definition, consider the following: *Design a device to eliminate the blind spot in an automobile*. This proposed problem definition also contains an assumption that prematurely limits the designer. The word *device* rules out one solution that achieves the design goal (eliminating the blind spot) by simply repositioning the front and side mirrors.

A third example occurred in a design competition named Blimp Wars (see Figure 21.1). The goal was to *design a system to retrieve Nerf® balls from an artificial tree and return them to the blimp base*. Inclusion of the word *blimp* in the problem definition biased the students toward blimp designs. The alternative of an extendable arm that spans the distance between blimp base and the target balls was not considered.

FIGURE 21.1 Blimp returning to base after retrieving a ball from tree on the left.

21.4 LIST OF DESIGN SPECIFICATIONS

After translating the need into a problem definition, the next step is to prepare a list of design specifications (or requirements) that enforce the *voice of the customer*. The **list of design specifications** includes both "demanded" design characteristics that must be present for the design to be considered acceptable by the customer and "wished for" design characteristics that are desirable but not crucial to the success of the final design. It is the usual practice to classify each specification as either a **demand** (D) or a **wish** (W). Don't confuse the two. If you treat a wish as if it were a demand, your design may become more complicated than is necessary.

Whenever possible, use numbers to express specifications. For example, instead of merely requiring that weight must be low, state, "Weight must be less than 10 pounds." Sometimes use of numbers is impossible. A quality such as "aesthetically pleasing" is difficult to quantify. However, use numbers wherever possible, even if at this early stage they seem like guesses. The numbers can be refined later on as the design begins to take shape.

The specifications should be solution independent to avoid bias. For example, if you are designing a small mobile device, requiring that "the wheels must be made of rubber" will bias the design in two respects: in the use of wheels and in the choice of materials. Such decisions are reserved for later in the design process after careful consideration of alternatives.

Specifications come in the following categories:

- Performance
- Geometry
- Materials
- Energy
- Time
- Cost
- Manufacture
- Standards
- Safety
- Transport
- Ergonomics
- Environment

These categories can also be used as headings by which to organize the list of specifications. Here is an example.

Example 21.1

The following problem definition was posed to three competing design teams.

> *Design and build a RC[1] portable device that will play nine holes of golf at a local golf course with the fewest possible number of strokes.*

The instructor also supplied the following list of demands. It was left to the students to develop a complete list of specifications.

---

[1]A common abbreviation in design for "remote control" or "remote controlled" is "RC."

Demand (D) Specification:

    Must cost less than $600 (not including radio).

    Must be remotely triggered.

    Total number of radio-controlled servos[2] is eight.

    Device cannot be touching the golf ball prior to remote triggering of the shot.

    Entire device must form a single unit.

    Must be portable.

    Design must pass a safety review.

    Ground supports must fit within a 3 ft. circle.

Solution

    The first step was to organize the demands under each heading. Then, using the headings as a guide, additional demand (D) or wish (W) specifications were formulated after talking further with the course instructor (their customer). The results follow:

Performance

    D—Must be remotely triggered.

    D—Device cannot be touching the golf ball prior to remote triggering of the shot.

    D—Driving distance must be adjustable with a range of between 15 and 250 yards.

    D—Putting distance must be adjustable with a range between 0 and 15 yards.

    D—Must operate on inclines of up to 45°.

    W—Must sink 95% of short putts (less than 3 ft.).

    W—Driving accuracy of ±5 yards.

Geometry

    D—Total number of radio-controlled servos is eight.

    D—Entire device must form a single unit.

    D—Ground supports must fit in 3 ft. circle.

Materials

    W—Materials must not degrade under expected range of weather conditions (including rain, snow, $30\ °F < T < 90\ °F$).

Time

    D—Must be designed and manufactured in less than 14 weeks.

Cost

    D—Must cost less than $600 (not including radio).

Manufacture

    D—Must be manufactured using tools available in the machine shop.

    D—Must be manufactured using machining skills available within the team.

    W—Off-the-shelf parts and materials should be readily available.

Standards

    D—Radio must adhere to FAA regulations.

Safety

    D—Design must pass a safety review.

Transport

    D—Must be portable.

    W—Must fit in a car or small truck (for easy transport to the golf course).

---

[2]A servo is a control system to amplify a small signal into a large response. Typically it is an electric motor controlled by a small voltage.

21.5 CLARIFYING THE PROBLEM

Start with what you know. People can be asked in advance to write down what they know about a problem. A brainstorming session is useful in generating the most information.

Decide what information is missing. Information is the key to effective decision making. If you are designing a replacement bridge, do you know how many people will use the bridge? When will they will use the bridge—all the time or mainly in the morning and evening? If that's the case, your problem statement might be "Design a four-lane highway bridge to accommodate 400 cars per hour at peak usage."

Gather information on the problem. You might collect several types of information. It will generally fall into one of the following categories:

- Facts (15% of the people do not use the bridge at all)
- Inference (a significant percentage of drivers bypass the bridge)
- Speculation (many trucks use the bridge)
- Opinion (I think a new bridge will be too expensive)

When you are gathering information, you will probably hear all four types of information, and all can be important. Speculation and opinion can be especially important in gaging public opinion.

For example, if you are trying to reduce the waiting time to pay for products purchased in a large department store you might find that most people have the opinion that the store should open more checkout lanes. Others may speculate that the store manager does not want to pay for more checkout clerks. However, if you observe the checkout line throughout the day and interview the store manager you may find relevant facts that either support or contradict these opinions and speculations.

Define the problem. With the information in front of you, you're ready to write down a "problem statement"—a comprehensive definition of the problem. Define the problem in terms of needs, and not solutions. If you define the problem in terms of possible solutions, you're closing the door to other, possibly more effective solutions.

21.6 DESIGN MILESTONE: CLARIFICATION OF THE TASK

There are two versions of this milestone, depending on the format of the design project. If there is a design competition involved, main responsibility for producing the list of specifications shifts from the students to the instructor, as there is a need for everyone to operate under the same set of constraints. In either case, it is assumed that the instructor provides the problem definition.

For a General Design Project (Version A)

Assignment:

(1) Interview the customer. (In the case of a consumer product, conduct a product survey.)
(2) Prepare a typed list of specifications.

For Design Competitions (Version B)

Assignment:

(1) Review the rules of the competition and ask the instructor for rule clarifications.
(2) Prepare a typed list of design requirements to supplement those already appearing in the official rules of the competition. For example, set performance goals for your machine. You do not have to list requirements already appearing in the rules.

21.6.1 Design Competition Tips

- Probe the rules for holes that will allow for concepts not anticipated by the creators of the competition.
- Avoid any temptation to bias the requirements toward a particular solution or strategy.
- Expect the list of supplemental requirements to be very short if the rules are well defined.

EXERCISES

1. Develop an engineering problem statement for the following need: students need more parking near campus.

2. Develop an engineering problem statement for the following need: shoppers need a faster check-out process in grocery stores.

3. Develop an engineering problem statement for the following need: women need a place to secure their purse in an automobile.

4. Provide a list of desired (D) specifications for the following problem definition. Design and construct a student backpack that will comfortably hold 10 kg of books, a laptop computer, and lunch.

5. Provide a list of desired (D) specifications for the following problem definition. Design and construct an interactive toy for a blind child.

6. Provide a list of desired (D) specifications for the following problem definition. Design and construct a device that will launch peanut butter and jelly sandwiches into a crowd of college students.

7. Make a list of five questions designed to clarify the task of designing a device that will launch peanut butter and jelly sandwiches into a crowd of college students.

8. Make a list of five questions designed to clarify the task of designing a way to secure a woman's purse in an automobile.

9. Make a list of five questions designed to clarify the task of designing a student backpack that will comfortably hold 10 kg of books, a laptop computer, and lunch.

10. Using your own experience, propose a "need" that will help children in some way. Then create an engineering design problem statement that will satisfy that need. Then prepare a list of desired specifications for the design and a list of questions you would ask your target group of children that will clarify the problem statement.

Design Step 2: Generation of Alternative Concepts

22.1 INTRODUCTION

Once the problem statement is in place and the specifications have been listed, it is time to generate alternative concepts. By *concept*, we mean an idea as opposed to a detailed design. The representation of the concept, usually in the form of a sketch, contains enough information to understand how the concept works but not enough information to build it. By *alternative*, we require that the various proposed ideas must be fundamentally different in some way. The differences must go beyond appearance or dimensions. The usual rule of thumb in design courses is to generate at least three fundamentally different concepts.

In this chapter, four aspects of concept generation are discussed: brainstorming, concept sketching, research-based strategies, and functional decomposition.

22.2 BRAINSTORMING

The most common approach for generating ideas is by brainstorming. As the term implies, you rely on your own creativity and memory of past experiences to produce ideas. Usually, team members generate ideas on their own before meeting with the team for a brainstorming session.

Brainstorming is based on one crucial rule: *Criticism of ideas is not allowed.* This enables each team member to put forth ideas without fear of immediate rejection. For example, a professor once recorded the brainstorming session of a small team of students. At one point, a student offered an idea and another student referred to it as "stupid." The voice of the first student was never heard again during the session. Instead of a team of four, it had become a team of three.

It is important to devote some of the brainstorming time searching for bold, unconventional ideas. In the case of a design competition, this could mean searching for holes in the rules that could lead to ideas that the creators of the competition had not anticipated.

Only when brainstorming is complete, should the team eliminate concepts that are not feasible, not legal, or not fundamentally different. After this weeding-out process, at least three concepts should remain. If not, more brainstorming is in order. The following example illustrates this step.

While group brainstorming is effective at generating ideas for solving complex problems, individual brainstorming is effective when you need to solve a simple problem, generate a list of ideas, or focus on a broad issue. Although individual brainstorming misses the benefits of shared experience and expertise, people are usually more creative when they brainstorm on their own rather than in groups. This is because:

- In individual brainstorming, people tend to be more creative at certain times and places. Some are more creative in the morning, some in the evening, some while drinking coffee, some while on a bus, and so forth. This is absent in group brainstorming where people have to meet at a set time and place.
- In group brainstorming, reserved or quiet members are often ignored. This does not happen when they brainstorm on their own.
- Groups sometimes have one or more members with dominating personalities who try to force their ideas onto everyone else. Other members will refrain from expressing their ideas to avoid conflict. People are not likely to share their ideas if they feel intimidated.
- In group settings, some people feel too shy to share what they fear may be viewed as crazy or weird ideas. But it's possible that these ideas can pave the way for more creative designs.

Individual brainstorming puts you in complete control of the creative process. An idea that you may have been hesitant to bring up in a group brainstorming session may blossom during the individual brainstorming process, and it may make the design a success.

22.2.1 Mind Mapping

An effective way for an individual to brainstorm is to write the problem you are trying to solve on a piece of paper and draw a circle around it. Then draw lines out from this circle and label each line with a way to solve the problem. This process is called **Mind Mapping**.

Mind mapping is an effective way of getting information in and out of your brain. It is a creative and logical means of sketching that literally "maps out" your ideas. They have a structure that radiates from the center of a page and use lines, symbols, words, colors, and simple, brain-friendly images. Mind mapping converts a list of information into an organized diagram that works in line with your brain's natural way of doing things.

One simple way to understand a Mind Map is by comparing it to a map of a city. The problem to be solved is in the center of the city, and the main streets leading from the center represent your key thoughts and ideas for a solution. The secondary streets represent your secondary thoughts, and so on, as shown in Figure 22.1. Special images or shapes can represent landmarks of interest or particularly relevant ideas.

The essential characteristics of Mind Mapping are

- The main idea or problem is in the center.
- The main themes *radiate* from the center as "branches."
- Each branch contains a key word drawn or printed on its line.
- Topics of lesser importance are represented as "twigs" of a branch.
- The branches form a connected nodal structure.

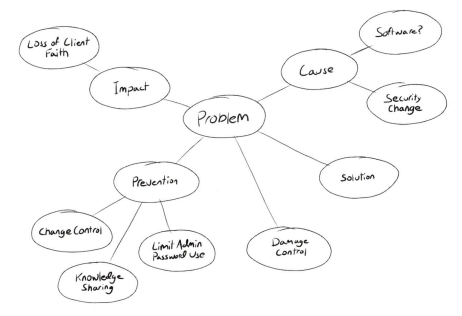

FIGURE 22.1 A simple mind map.

22.2.2 Ideation

Ideation is a creative method of generating, developing, and communicating new ideas. It can be thought of as a form of "structured brainstorming." The **Ideation Process** is a systematic sequence of actions (vs. brainstorming) that have

- Clearly defined event goals (**Why** are we doing this?),
- Measurable event objectives (**What** will we walk away with?),
- Specific event inputs and outputs (Document **deliverables**),
- "Standard" process actions, inputs, and outputs are **adaptable** to accommodate different or changing event goals and objectives.

If you've ever tried to do anything creative, you know that one of the hardest things to do is to begin. Without structure, it is easy to stall after one or two ideas or simply stare at a blank page. One way of drawing ideas out of the brain is to ask questions. For example, if you are designing a container for an egg that will withstand being dropped 20 feet to the ground without breaking, you might ask:

1. How big can the container be?
2. What container shapes are good at absorbing impact?
3. How fast will the contained be traveling at impact?
4. What other technologies are designed to minimize impact?

Just imagine ideation as a funnel. Insights, views, opinions, data, trends, and knowledge are captured at the top. They are filtered in the middle, and the good stuff—new ideas—comes out at the bottom.

Example 22.1

Assuming the alternative concepts in Figure 22.2 are generated as part of an effort to design a new bat for Major League Baseball, in which concepts should be eliminated because they are not feasible, not legal, or not fundamentally different?

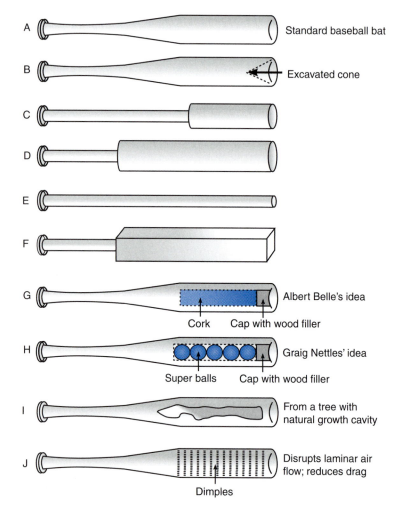

FIGURE 22.2 Alternative concepts for a Major League Baseball bat.

Solution:

A is a standard bat to which the new concepts is to be compared.

Not feasible: **E** because it stands no chance of being competitive; **I** because it is too difficult to find in nature.

Not legal: **F**, **G**, **H**, and **J**.

Not fundamentally different from each other: **C** and **D** because the basic shape is the same; only the dimensions differ.

Therefore, the condensed list of viable alternatives consists of concepts **B** and **C**.

22.3 CONCEPT SKETCHING

For an idea to be considered a feasible alternative concept, it must be represented in the form of a conceptual sketch. The goal in producing a concept drawing is to convey what the design is and how it works in the clearest possible terms. Any lack of clarity, such as failure to represent one of the subfunctions, translates into doubts about the feasibility of the concept when it comes time to evaluate it.

At the same time, however, this is not a detailed design drawing. Dimensions and other details not relevant to understanding the basic nature of how the concept will work are left out.

It is best to proceed through two phases when generating a concept drawing. First, in the creative phase, hand-sketching is done freestyle and quickly, without regard for neatness or visual clarity. A few simple lines, incomprehensible to others, might be enough to remind you of your idea. Sketching is a means for both storing ideas and brainstorming others. The final outcome is a rough sketch of the concept. Second, in the documentation phase, the concept is neatly redrawn and labeled to facilitate communication with team members and project sponsors.

The final outcome is one or more sketches prepared with the following guidelines in mind:

- Can be hand-sketched or computer generated.
- No dimensions. Remember, this is not a detailed drawing.
- Label parts and main features. If the drawing is hand-sketched, handwritten labeling is acceptable.
- Provide multiple views and close-up views if needed to describe how the design works.

The choice of views is up to you. Isometric views, like those shown in Figures 22.3 and 22.4, convey a lot of information in a single picture. Most mechanisms can be described effectively using one or more two-dimensional views, as in Figure 22.5. Despite their apparent informality, the quality of these drawings is crucial to fairly representing the designs during the evaluation process. In some cases, they are the only source of evidence for judging if a design is likely to work.

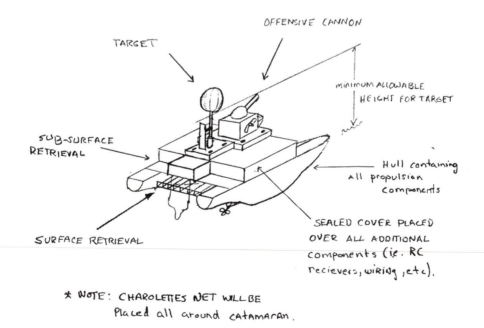

FIGURE 22.3 Concept drawing of a radio-controlled (RC) boat for a design competition (hand-drawn isometric).

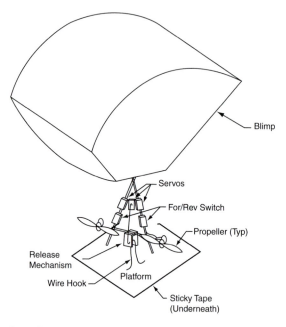

FIGURE 22.4 Concept drawing of a radio-controlled blimp (computer-generated isometric).

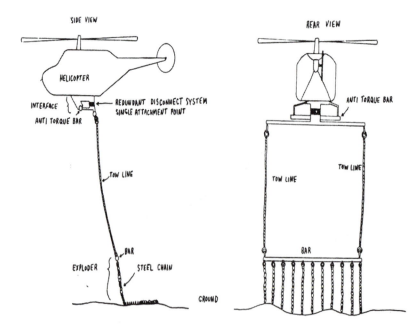

FIGURE 22.5 Concept drawing of an antipersonnel mine clearing system (two hand-drawn views).

22.4 HANDS-ON DESIGN EXERCISE: "THE TUBE"

The design objective is to extract a golf ball from the bottom of a free standing, open-ended mailing tube in the shortest possible time.

22.4.1 Setup

1. Place a mailing tube vertically on the floor and drop a golf ball into the tube.
2. Have a supply of the following materials: string, duct tape, Scotch tape, 8.5×11 standard copier paper, and scissors.

22.4.2 Rules

1. Limited to using the supplied materials.
2. Scissors can be used for manufacturing.
3. Everyone in the group can help in manufacturing, but only one person can extract the ball.
4. Students are not allowed to handle the materials until it is time to test.
5. Must manufacture the design shown on the concept drawing handed to the instructor.
6. Time limit of 3 minutes to manufacture concept and extract ball.
7. Cannot tip over the tube.
8. Cannot touch the outside of the tube with anything.
9. No forces can be applied to the inside of the tube in an effort to hold it vertical; accidental contact with the inside of the tube is okay as long as the tube does not tip over.
10. Violation of any of the preceding rules will lead to immediate disqualification.

22.4.3 Procedure

1. First allow the students 3 minutes to individually brainstorm (encourage them to draw quick sketches of each of their concepts).
2. Then, divide into teams of four students per team.
3. Allow teams 10 minutes to collect ideas, brainstorm as a team, select their best concept, and give a sketch of their best concept to the instructor.
4. Instructor should walk around during brainstorming to remind teams to (a) generate multiple solutions before selecting one and (b) try to involve everyone in the process.
5. Allow teams 2 minutes to assign responsibilities for manufacture and test.
6. One at a time, give each team 3 minutes to manufacture a concept and attempt to extract the golf ball.
7. The team with the shortest retrieval time wins.

22.5 RESEARCH-BASED STRATEGIES FOR PROMOTING CREATIVITY

Some ideas are truly original, but most are drawn from past experience. The following strategies help you look at old designs to generate new ones.

22.5.1 Analogies

One often-used strategy is to look for analogous design situations in other, unrelated fields. To do this, first translate the design objective into an overall function that is general enough to widely apply. For example, you may want to design a system to "climb a vertical wall" or "walk on two legs" or "move efficiently through the water." Nature is filled with solutions to these problems (but because of their complexity, biological solutions usually have to be simplified and adapted before they can be of practical use). If you are designing a system to "throw an object," a survey of ancient artillery could spark ideas.

22.5.2 Reverse Engineering

The basic strategy here is to acquire an existing product that is similar to a design you have in mind, take it apart, figure out how it works, and then either try to improve on it or adapt some of the ideas to your own design. Toy stores are a great place to search for small electromechanical devices that can be reverse engineered.

22.5.3 Literature Search

Web-based search engines are very effective at finding existing design solutions. For high-tech applications, you should also search books and the electronic databases for technical journals (e.g., the Science Citation Index).

22.6 FUNCTIONAL DECOMPOSITION FOR COMPLEX SYSTEMS

When confronted with a complex problem, it is frequently advantageous to break it down into smaller, simpler, more manageable parts. In the case of design, those smaller parts usually correspond to the individual functions (or tasks) that must be performed to achieve the overall design objective. This approach, known as **functional decomposition**, is the basis of the procedure described next for generating concept alternatives.

22.6.1 Step 1: Decompose the Design Objective into a Series of Functions

Start out by decomposing the overall function into four or five subfunctions. Usually, verbs such as *move*, *lift*, and *control* are used in naming the functions. Figure 22.6 shows the functional decomposition of the remote-controlled golf machine of earlier examples. It is given in the form of a tree diagram, which is probably the most common form of representation. If more detail is needed, each subfunction could be further broken down into its respective subfunctions.

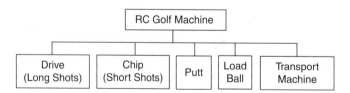

FIGURE 22.6 Functional decomposition for design of a remote-controlled golf machine.

It is very important that the functional decomposition be general enough to avoid biasing the design solution. For example, the separate drive and chip functions in Figure 22.6 may cause the design team to overlook the possibility of using the same device to fulfill both functions. If solution bias is unavoidable, introduce multiple functional decompositions.

When it is not readily apparent what the subfunctions are, it may help to think in terms of the sequence of tasks that must be performed by the design. The "sequential" functional decomposition for a design to assist disabled people into and out of a bathtub is shown in Figure 22.7.

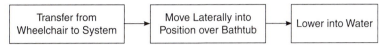

FIGURE 22.7 "Sequential" functional decomposition for a design of a system to aid the disabled into and out of a standard bathtub.

22.6.2 Step 2: Brainstorm Alternative Concepts for Each Function and Assemble the Results in a Classification Scheme

The classification scheme[1] is a two-dimensional matrix organized as shown in Table 22.1. The first column lists the functions resulting from the functional decomposition. The row of boxes next to each function name contains the corresponding design solutions that have been brainstormed. The design solutions are expressed using a combination of words and pictures, so be careful to draw the boxes large enough to accommodate small illustrations.

Table 22.1 Organization of the Classification Scheme

Concepts / Functions	Concept 1	Concept 2	Concept 3	Concept 4
Function A	A1	A2	A3	A4
Function B	B1	B2	B3	B4
Function C	C1	C2	C3	C4
Function D	D1	D2	D3	D4

22.6.3 Step 3: Combine Function Concepts to Form Alternative Design Concepts

Table 22.2 demonstrates how one subfunction concept from each row of the classification scheme is selected to form a total concept. The same subfunction concept can be used with more than one total concept, though keep in

[1]G. Pahl, W. Beitz, *Engineering Design—A Systematic Approach*. (New York: Springer-Verlag, 1988).

Table 22.2 Combining of Compatible Subfunction Concepts

Concepts / Functions	Concept 1	Concept 2	Concept 3	Concept 4
Function A	A1	A2	A3	A4
Function B	B1	B2	B3	B4
Function C	C1	C2	C3	C4
Function D	D1	D2	D3	D4

Total Concept I = A1 + B2 + C2 + D1
Total Concept II = A4 + B2 + C4 + D2

mind that the idea is to generate fundamentally different design concepts. The only other rule when deciding on the best combinations is to be sure that the subfunction concepts being combined are compatible.

22.6.4 Step 4: Sketch Each of the Most Promising Combinations

This is done in accordance with the rules previously presented for concept drawings. Remember that you must end up with drawings for at least three fundamentally different design concepts.

Example 22.2

Use functional decomposition to generate alternative concepts for a proposed radio-controlled blimp, capable of retrieving Nerf balls from an artificial tree and returning them to the blimp base (see Figure 21.1 of Chapter 21).

Solution: The first step is to produce the functional decomposition shown in Figure 22.8.

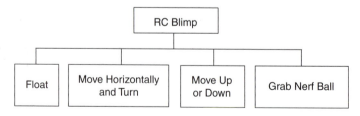

FIGURE 22.8 Functional decomposition for the design of a remote-controlled blimp.

Then, concepts were brainstormed for each of the subfunctions, and the results are assembled in the classification scheme shown in Figure 22.9.

Concepts / Functions	Concept 1	Concept 2	Concept 3	Concept 4
Float	Hot Air	Helium	Hydrogen	
Move Horizontally and Turn	Prop + Rudder	Two Props	Rotating Turret	2 Props at 90°
Move Up or Down	Pivot Props	Vertical Props	Retract String	Heat or Cool Gas
Grab Nerf Ball	Sticky Tape	Rake	Claw	Pin Cushion

FIGURE 22.9 Classification scheme for a remote-controlled blimp.

The total concepts were formed by combining compatible subfunction concepts. The three promising total concepts are

Total Concept I = Helium + 2 Props + Pivot props + Sticky tape
Total Concept II = Helium + Rotating turret + Vertical prop + Rake
Total Concept III = Helium + Prop with rudder + String + Claw

The final step is to represent each alternative design in the form of a concept drawing. The concept drawing for Total Concept I is shown in Figure 22.4.

22.7 DESIGN MILESTONE: GENERATION OF ALTERNATIVES

This milestone assumes that the system to be designed is sufficiently complex (i.e., at least two subfunctions) to warrant the use of functional decomposition.

Assignment

1. For the functional decomposition given in class (or a modification of it that you are at liberty to propose), brainstorm to determine at least five feasible alternatives for each subfunction and assemble the results in a classification scheme.
2. Form three promising design concepts by combining compatible subfunction alternatives from your c' sification scheme.
3. Firm up your three design concepts by sketching them up in the form of concept drawings. Funct' ity (i.e., how it works) should be clearly indicated in the drawings through the use of labeling ar

Grading Criteria

Technical Communication

- Ideas are clearly presented.
- Final concept drawings are neatly rendered.

Technical Content
- All concepts are feasible, legal, and fundamentally different.
- Concepts are presented in sufficient detail.
- Requested number of concepts is generated.

22.7.1 Design Competition Tips

- The goal is to generate three "strong" concepts.
- Search the boundaries of the rules for unusual ideas that could potentially dominate the competition. If you don't, someone else will.
- Include strategy as one of the items to be brainstormed in the classification scheme.
- Redraw your concept sketches to enhance clarity and neatness. The quality of the concept drawing, or lack of it, can do much to sway opinions when it comes time to judge the concepts.

EXERCISES

Use the technique of individual brainstorming for the problems below.

Brainstorming

1. Brainstorm by yourself a list of ways you can get to class each day. Don't be afraid to include methods that don't seem currently plausible (e.g., teleportation) since that may trigger other ideas for you.

2. Brainstorm a list of ways you could use to study for an exam. Include when, what, where, and how long to study.

3. Brainstorm ways you can get an A in this course.

4. Brainstorm ways you could create world peace.

5. Brainstorm ways you think would reduce global warming.

Concept Sketching

6. Sketch a bicycle and label all its components (wheels, frame, etc.).

7. Sketch your cell phone and label its operating functions (on/off button, etc.).

8. Sketch your shoe and label all its parts (heel, sole, etc.).

Sketch your concept of a space ship and label its components.

10. etch a battery-powered wheel chair and label all the parts.

Fun l Deconposition

11 ou a, motely designin an electric footstool. It must have an electric heater, move easily on the floor, and have ur footsool trolld height. Draw a functional decomposition diagram like the one shown in Figure 22.6 for your footstool.

12. You want to design a hot dog stand for a street vendor. Decide what the stand must be able to do (e.g., heat water, store condiments, buns, and wieners, etc.) and prepare a functional decomposition diagram like the one shown in Figure 22.6 for your design.

13. Examine the toaster in your home and determine all its components. Then draw a functional decomposition diagram like the one shown in Figure 22.6 for the toaster.

14. Examine a coffee maker at home or in a department store. Then draw a functional decomposition diagram like the one shown in Figure 22.6 for the coffee maker.

15. You are to design a small robot that will climb a vertical wall. Determine all the subfunctions required of the robot and draw a functional decomposition diagram like the one shown in Figure 22.6 for the robot.

Design Step 3: Evaluation of Alternatives and Selection of a Concept

23.1 INTRODUCTION

Suppose you now generated three concepts that meet the problem definition and fulfill the specifications. Which one should you choose as the basis for your final design? There is no magic formula. However, Professor Nam P. Suh of MIT provided two very helpful design principles for evaluating and improving concepts: **minimize information content** and **maintain the independence of functional requirements**.[1]

This chapter adds three additional considerations for evaluating alternatives: **ease of manufacture**, **robustness**, and **design for adjustability**. It then concludes with a method of pulling together all these ideas: the **decision matrix**.

23.2 MINIMIZE THE INFORMATION CONTENT OF THE DESIGN

When choosing among promising alternatives, the best design is often the one that can be uniquely specified using the least amount of information or, alternatively, can be manufactured with the shortest list of directions. This idea is sometimes inelegantly stated as the *KISS* principle: Keep It Simple, Stupid.[2]

A number of design guidelines naturally follow. A few of the most notable ones are

- Minimize the number of parts.
- Minimize the number of different kinds of parts.
- Buying parts is preferable to manufacturing them yourself.

---

[1]N. P. Suh, *The Principles of Design* (New York: Oxford University Press, 1990).

[2]The term *KISS* (Keep It Simple, Stupid) was coined by the American design engineer Kelly Johnson to indicate that a military aircraft should be repairable with a limited set of tools under combat conditions. Johnson wrote it as 'Keep it Simple Stupid' (without the comma) because it was not meant to imply that an engineer was stupid; just the opposite. A simple solution is better than a complex one even if the solution looks stupid.

23.3 MAINTAIN THE INDEPENDENCE OF FUNCTIONAL REQUIREMENTS

The functions considered in a functional decomposition provide the basis for Suh's second principle. This principle asserts that these functions should be independent of each other in a good design.

A successful application of this principle is illustrated by the decoupled design in Figure 23.1. Independence of the functions "lift" and "move" is maintained by designing physically separate mechanisms for each action (scissors jack for lift, wheeled vehicle for move) and by performing the actions in sequence, rather than at the same time. First, the scissors jack lifts the vehicle, the vehicle then slides horizontally onto the next step, and finally the scissors jack closes upward and is pulled back underneath the vehicle. The coupled design employed four articulated arms, tank-like tracks on each arm, and a complicated motion to both lift and move at the same time. Though both machines performed admirably, the decoupled design has a much higher potential payload and is easier to build, since it requires half as many motors.

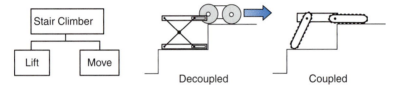

FIGURE 23.1 Two concepts for a stair-climbing machine. The first concept (center) decouples the lift and move. Although it still does the job, but the second concept (right) does not.

The previous example suggests the following design guideline: **Seek a modular design**. A modular design is one in which the design solutions for each function have been physically isolated. The main advantage of a modular design is that the individual modules can be designed, manufactured, and tested in parallel, leading to much shorter product development times.

In looking for opportunities to improve a given design, the situation may arise in which Suh's two principles appear to be in conflict. For example, a design change aimed at increasing the independence of the functional requirements could result in greater complexity. Suh contends that any design change that either increases information or sacrifices the independence of the functional requirements should not be accepted. There always exists a less-coupled design with lower information content.

These two design principles are illustrated in the following example.

Example 23.1

A head-to-head student design competition named Davy Jones's Treasure Trove is based on the following problem definition:

Design a system to retrieve surface (ping-pong balls) and subsurface (1 lb mass) objects from a swimming pool.

This is subject to the following major design requirements:

- It must fit in a 2 ft × 2 ft × 3 ft volume at the start of the competition.
- It must carry a target that disables the boat if struck by opponent.

Concept drawings of two of the student designs are shown in Figures 23.2 and 23.3.

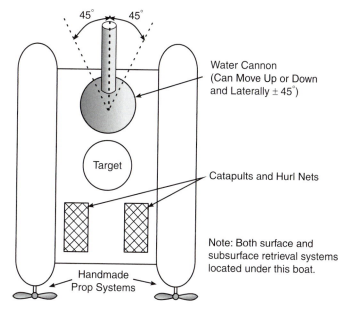

FIGURE 23.2 Water cannon design (top view).

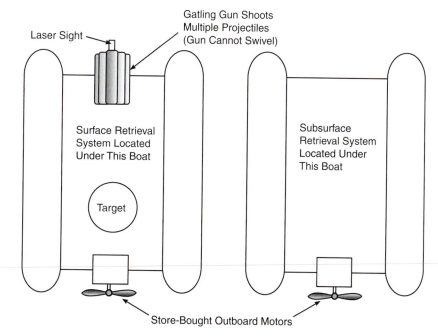

FIGURE 23.3 Twin boat design (top view).

Let's look at these figures and evaluate the two student designs by identifying applications and violations of Suh's design principles. (**Hint**: Only relevant features have been labeled in the figures.)

Solution

Water Cannon Design
- The water cannon might be effective because aiming and steering are independent.
- Catapults will not be as effective because aiming is dependent on steering.
- The boat should be very maneuverable because the two props serve to decouple the move and turn functions, that is, it should be able to turn on a dime.
- Manufacture of the prop systems could be needlessly time consuming.

Twin Boat Design
- The use of two boats will be very effective because it decouples the two retrieval functions. One boat can collect the ping-pong balls, while the other collects the 1 lb masses.
- The use of store-bought propellers saves time.
- The use of two nearly identical hull designs simplifies both design and manufacturing, thus saving more time.
- The boats will not be as maneuverable as the water cannon design because the move and turn functions are not independent, that is, the single-prop design needs to be moving forward in order to turn.
- The Gatling gun will not be as effective as the water cannon because aiming is dependent on steering.

Final Note

The preceding characteristics provide accurate insight into how the boats actually performed. The Twin Boat Design won the competition largely on the strength of its dual retrieval system. Although the water cannon was far more effective than the Gatling gun, the water cannon design was ultimately at the mercy of its handmade props, which took a lot of time to manufacture, left little time for testing, and proved to be unreliable.

23.4 DESIGN FOR EASE OF MANUFACTURE

There are clear advantages to going with a design that is easy to manufacture. If, among competing design teams, yours is the first to complete manufacture of your design, the extra time can be used to test, debug, and optimize performance. For a commercial enterprise, first to market can mean a short-term monopoly in a fiercely competitive marketplace. Often, ease of manufacture goes hand in hand with lower costs. So, given the choice of two concepts, both of which satisfy the design requirements to the same degree, where one is more difficult than the other to manufacture, it makes sense to choose the concept that is easier to manufacture.

At this stage in the design process, evaluation of ease of manufacture should be done at a level of abstraction consistent with the concept drawings. The counting of machining operations and assembly steps is reserved for a later time when the requisite level of detail in the design has been attained. Here, the emphasis is on developing an impression of ease of manufacture as revealed through Suh's design principles.

A student or team with the formidable task of having to build a complex design should be asking the following questions as each concept is evaluated:

Are there a large number of parts?

If there are a lot of parts that need to be made and assembled, it will take a long time to build.

Are there a large number of different kinds of parts?

For parts of comparable complexity, it takes less time to make two of the same part than two different parts because of reduced setup times.

Are there parts with complicated geometry?

These parts take longer to make.

Can some parts be purchased?

This is not always an option in design competitions, but if it is, the time saved and the proven reliability of the prefabricated part usually justify the purchase.

Do you have the skills to make all of the parts?

The safest thing to do is to choose a concept that you know you can build.

Is it a modular design?

We noted earlier that modular components can be manufactured in parallel by subgroups within the design team, thus saving time.

Are there opportunities to simplify manufacture of the design by:

- Reducing the number of parts?
- Reducing the number of different kinds of parts?
- Simplifying the shape of some parts?
- Purchasing some parts?
- Redesigning the parts that are difficult to make?
- Modularizing the design?

23.5 DESIGN FOR ROBUSTNESS

Manufacturing errors, environmental changes, and internal wear can cause unexpected variations in performance. When the designed product is insensitive to these three sources of variability, the design is said to be **robust**. Engineers seek a robust design because performance of such a design can be predicted with greater certainty.

The designer must learn to expect the unexpected. All too often, students conceive of a design while assuming ideal operating conditions. Yet, deviations from those ideal conditions can lead to less than ideal performance, as illustrated by the following example.

Example 23.2

For a design competition, students had to design machines that could accurately throw darts at a dartboard. The machines were powered by large falling masses. Yet, none of the machines was perfectly repeatable. For example, from 8 feet away, the best that the machine in the following figure could do was to keep the darts within a 1-inch circle. What factors contributed to this loss of dart throwing accuracy?

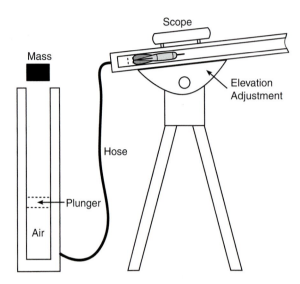

FIGURE 23.4 Concept drawing of a dart throwing machine.

Solution:

Relevant manufacturing errors, environmental changes, and internal wear sites were brainstormed. The following list resulted:

Manufacturing Errors
- Small dimensional differences between darts in the set of three.

Environmental Changes
- Small air currents.
- Inexact repositioning of the plunger.
- Inexact repositioning of the dart within the blow gun.
- Inconsistent releases of the falling mass.

Internal Wear
- Damage to the dart fins.
- Blunting of the dart tip.
- Damage to the dartboard.

When evaluating concepts with respect to robustness, you should be asking yourself the following questions:

Will small manufacturing errors dramatically impair performance?

If parts have to be manufactured perfectly for the design to function properly, you should expect to run into problems. You want a design that will work even when part dimensions are a little off. This is one reason why gear sets are such a popular design choice. Small errors in center distance between mating gears do not change the gear ratio.

Will the design function properly over the full range of environmental conditions?

Environmental conditions subject to variation include applied forces, atmospheric conditions, and roughness of surface terrain. The expected range of relevant environmental conditions should be clearly defined in the list of specifications. If they are not there, now is a good time to include them.

In a head-to-head design competition, have the actions of the opposing teams been anticipated?

Those actions can contribute significantly to the variability of the environmental conditions. For example, an opposing machine can apply forces to your machine or alter the roughness of surface terrain by laying obstacles. Therefore, you want to select a strategy-design combination that performs well irrespective of what the opposing teams may do.

23.6 DESIGN FOR ADJUSTABILITY

In engineering courses, there is usually only enough time and resources to manufacture one design, and that design almost never performs as planned on the first try. Optimizing performance by building several designs is not an option. The only remaining course of action is to design adjustability into the initial implementation.

There are a number of ways to design for adjustability. One way is to design the system with modularity. This can serve to isolate required design changes to a single subsystem. In a mechanical system, dimensional adjustability can be attained by using nonpermanent fastening methods, such as screw joints instead of a permanent method like epoxy.

Design for adjustability can be incorporated into the evaluation process by asking the following questions as each concept is reviewed:

What are the main performance variables?

Usually, only one or two of the most important variables need be considered. Typical performance variables are speed, force, and turning radius.

Can those performance variables be easily adjusted?

Common methods were described previously. Other methods are found by brainstorming and examination of the governing equations.

Example 23.3

A motor-driven moving platform is a common feature of many small-scale vehicle designs. A top view of one such moving platform design follows.

Once manufacture of this design is complete, what adjustments can be made to:

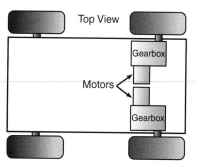

FIGURE 23.5 Moving platform.

   a. Increase the speed of the moving platform?

   b. Increase its pushing force?

Solution:

   a. Alternative methods for increasing the speed are
- Decrease the gear ratio.
- Increase the radius of the tires.
- Increase the voltage from the power supply.
- Switch in motors with higher RPM.

   b. Alternative methods for increasing the pushing force are
- Increase the gear ratio.
- Decrease the radius of the tires.
- Increase the voltage from the power supply.
- Switch to a motor with a higher peak torque.

23.7 HANDS-ON DESIGN EXERCISE: "WASTE BALL"

23.7.1 Scenario

A company that uses radioactive substances for research sometimes has "spills" of spherical radioactive objects. When this happens, the radioactive objects must be transported to a waste container by the emergency team.

23.7.2 Design Objective

Design a method to transfer the radioactive substance (plastic ball) from the site of the spill to a waste container (small refrigerator) at another location.

23.7.3 Setup

- Divide the team, which consists of the entire class, into subfunctional groups of two or three students each. Each subfunctional group is responsible for one leg of the transfer.
- Prior to the class, the instructor has to lay out the course. There must be as many different legs as there are subfunctional groups. The room in which the ball is initially placed and another room that contains the small refrigerator account for two of the legs. Other legs can consist of a corridor, a stairwell, an elevator, or an outdoor excursion. Try to make each challenge a little different to promote development of special-ized designs by the subfunctional groups.
- Distribute the following materials to each subfunctional group:

1 daily newspaper (or equivalent).

1 roll of duct tape.

1 foam plate.

1 plastic cup.

1 pair of scissors (for construction only).

The team also receives three balls of string, which must be shared among the groups.

23.7.4 Rules

1. Since the ball is radioactive, no one can be within 8 feet of the ball.
2. Use only the materials provided.
3. For safety purposes, running is not allowed.
4. If necessary, doors must be safely held open by the teams and closed immediately after waste passes through.
5. If, during transport, the ball accidentally touches something besides the transport container (e.g., floor) a 30-second penalty is imposed and the group carrying the ball must restart at the location where it received the handoff.
6. The team has 3 minutes per group to complete each leg of the transfer.
7. The team with the minimum transit time wins.

23.7.5 After the Exercise

- Assess team performance by comparing times to other sections.
- Discuss the quality of communication between subfunctional groups.
- What were the lessons learned?

23.8 THE DECISION MATRIX

The decision matrix promotes a systematic and exhaustive examination of concept strengths and weaknesses. The entire procedure, from selection of evaluation criteria to filling out the matrix, is designed to remove personal bias from the decision-making process. The results give a numerical measure for ranking alternatives and ultimately selecting the best concept.

23.8.1 Evaluation Criteria

The criteria by which the concepts should be judged are all contained in the list of specifications. To even qualify as a feasible concept, the expectation must be that all the design requirements designated as "demands" will be satisfied. Therefore, the ranking of the feasible concepts ultimately depends on the degree to which they fulfill the design requirements designated as "wishes." However, at the conceptual level, qualities associated with both demands and wishes are included among the evaluation criteria, owing to the uncertainty still associated with estimating their degree of fulfillment.

The design requirements selected to serve as evaluation criteria are usually reworded to indicate the desired quality. For example, instead of weight, cost, and manufacture, the corresponding evaluation criteria become low weight, *low cost,* and *easy to manufacture.*

Evaluation criteria should be independent of each other to ensure a fair weighting of requirements in the decision matrix. For example, low cost and ease of manufacture are redundant and thus double-counted if the cost of labor is a significant fraction of total cost.

The number of evaluation criteria can vary depending on the situation. We suggest a level of detail consistent with the amount of detailed information available about the concept. For most hands-on student projects, five to seven of the most important evaluation criteria should suffice. *Easy to manufacture* and *low cost* are almost always included in this list.

23.8.2 Procedure for Filling out a Decision Matrix

Step (1) Identify the Evaluation Criteria
This step is described in the previous section.

Step (2) Weight the Evaluation Criteria
Weight values are assigned to each evaluation criterion in proportion to its relative importance to the overall success of the design; the larger the weight, the more important is the evaluation criterion. Though not a mathematical necessity, it is usually a good idea to define the weights such that their sum is equal to 1, that is,

$$\sum_{n=1}^{N} W_n = 1 \tag{23.1}$$

in which N is the number of evaluation criteria. This constraint instills the view that weights are being distributed among the criteria and, in so doing, helps avoid redundant criteria.

Step (3) Set Up the Decision Matrix
The organization of the decision matrix is illustrated in Table 23.1. The names of the concepts being evaluated are filled in at the top of each column. Likewise, the evaluation criteria and their assigned weights are written in the leftmost columns of the matrix. Scoring and intermediate calculations are recorded within the subcolumns under each concept and totaled at the bottom of the matrix.

Table 23.1 Organization of the Decision Matrix

Evaluation Criteria	Wt.	Concept 1		Concept 2		Concept 3	
		Val_1	$Wt \times Val_1$	Val_2	$Wt \times Val_2$	Val_3	$Wt \times Val_3$
Criterion 1			|		|		|
Criterion 2			|		|		|
Criterion 3			|		|		|
Criterion 4			|		|		|
Criterion 5			↓		↓		↓
Total	1.0		OV_1		OV_2		OV_3

Step (4) Assign Values to Each Concept
Starting in the first row, each concept is assigned a value between 0 and 10 according to how well it satisfies the evaluation criterion under consideration. The values are assumed to have the following interpretation:

  0 = *Totally useless* concept in regard to this criterion
  5 = *Average* concept in regard to this criterion
  10 = *Perfect* concept in regard to this criterion

and are recorded under the first subcolumn of each concept. This process is repeated for each criterion, going row by row to avoid bias. Usually, assignment of values is based on a qualitative assessment, but if quantitative information is available, they can be assigned in proportion to known parameters.

Step (5) Calculate the Overall Value for each Concept

For each concept-criterion combination, the product of the weight and the value is calculated and recorded in the second subcolumn. After these calculations are completed, the overall value (OV) is computed for each concept using the following expression:

$$OV = \sum_{n=1}^{N} (W_n V_n) \tag{23.2}$$

This is equivalent to summing the second subcolumn under each concept heading. The OVs are recorded at the bottom of the matrix.

Step (6) Interpret the Results

The highest overall value provides an indication of which design is the best. Overall values that are very close in magnitude should be regarded as indicating parity, given the uncertainty that went into assignment of weights and values. The final result is nonbinding. Thus, there is no need to bias the ratings to obtain the hoped-for final result. Rather, the chart should be regarded as a tool aimed at fostering an exhaustive discussion of strengths and weaknesses.

23.8.3 Additional Tips on Using Decision Matrices

1. Every member of the design team should individually fill out a decision matrix prior to engaging in team discussions. This gives everyone a chance to think about the strengths and weaknesses ahead of time and makes it more likely that all will be active participants at the team meeting.
2. Use the matrix to identify and correct weaknesses in a promising design. Give priority to the weaknesses that are most heavily weighted.
3. Feel free to create new alternatives by combining strengths from competing concepts.

Example 23.4

Recalling the golfing machines of earlier examples, the three concepts appearing Figures 23.6 through 23.8 have been proposed as best fulfilling the design requirements. Evaluate these three concepts by using the previously described procedures for filling out a decision matrix.

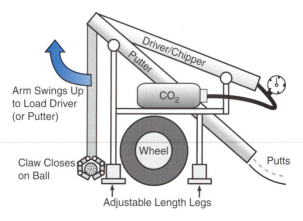

FIGURE 23.6 Concept drawing of the "Cannon."

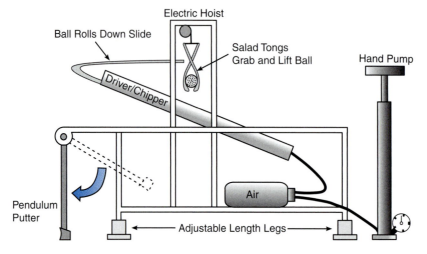

FIGURE 23.7 Concept drawing of the "Original."

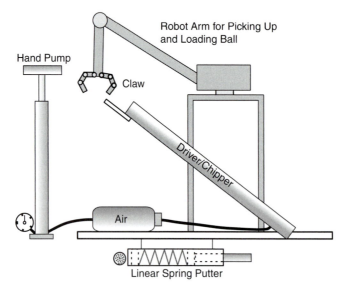

FIGURE 23.8 Concept drawing of the "Robogolfer."

Solution:

The following design requirements were selected to serve as evaluation criteria:

1. Drives well.
2. Putts well.
3. Ball loader is robust (i.e., picks up ball off of all types of terrain).
4. Easy to transport.
5. Easy to manufacture.

Low cost, which usually appears, was not selected because all three concepts met the cost requirement and cost was not involved in the design competition scoring.

With these in hand, the decision matrix can be drawn up and weights assigned to each criterion. The drives and ball loader were considered equally important because one cannot work without the other. The drives/chips were weighted slightly higher than the putts because only 43% of all shots taken by golf professionals are putts. Transport is weighted low because it does not factor into scoring. Ease of manufacture is always important because of its impact on development times. The resulting weights are listed in the decision matrix of Table 23.2.

Table 23.2 Decision Matrix for Golf Machine Concepts

Evaluation Criteria	Wt.	Cannon		Original		Robogolfer	
		Val_1	$Wt \times Val_1$	Val_2	$Wt \times Val_2$	Val_3	$Wt \times Val_3$
Drives well	0.25	9	2.25	8	2.00	8	2.00
Putts well	0.20	4	0.80	4	0.80	8	1.60
Loader is robust	0.25	6	1.50	6	1.50	9	2.25
Easy to transport	0.05	9	0.45	5	0.25	5	0.25
Easy to manufacture	0.25	5	1.25	7	1.75	3	0.75
Total	1.0		6.25		6.30		6.85

Then, proceeding one evaluation criterion at a time, the team analyzes the strengths and weaknesses of each concept in the context of the given criterion and assigns corresponding values to each concept in the decision matrix. The results of the analysis following and the values are recorded in Table 23.2.

1. Drives Well

All three drivers appear to be promising, given the effectiveness of the notorious potato gun. However, since the CO_2 tank comes prepressurized and the hand-pumping is subject to a 60-s time limit on preshot preparation, the Cannon is likely to be firing the ball at higher pressures and so should have a greater range.

2. Putts Well

The greens at the site of the competition are severely sloped and slow. Therefore, the machines must be capable of executing long putts. Of the three machines, the Robogolfer is the most adjustable, since the springs can be easily replaced if the range proves inadequate. On the other hand, the potential energy of gravity powers the other two putters, and it will be difficult to increase starting heights once these machines are built. Therefore, greater risk is associated with these putters.

3. Loader is Robust

A wide range of lies is possible, from severe slopes to sand and divots. The Cannon and the Original address this issue by using legs that are adjustable in length. The robot arm of the Robogolfer is clearly the most flexible design and requires no setup time.

4. Easy to Transport

The rules require that only one student from the team may transport the machine to the next shot location. The Cannon is the easiest to transport because only it has wheels.

5. Easy to Manufacture

The robot arm of the Robogolfer stands out as easily the most complicated system on any machine. As a three-degree-of-freedom mechanism, it requires three independently controlled motors. The Original's loader should be straightforward to manufacture. The salad tongs and the parts for the electric hoist can be easily purchased.

6. Discussion of Results

The Robogolfer is the clear winner on points. But, the challenges involved in designing and manufacturing that robot arm should make you nervous (unless you have a robotics expert on your team). The decision matrix also revealed that the putters for the Cannon and the Original are weak concepts. If they are replaced by the Robogolfer's linear spring putter, the Original ends up with the most points.

The concepts in Figures 23.4 through 23.6 correspond to actual student designs that were designed, manufactured, and tested. The Original team (so named because they were the first team to develop an air cannon) won the design competition. They compensated for their weak putter by chipping the long putts and adding a ramp to make the short putts. The Robogolfer team (who had a robotics expert) took longer to design than the Original and had less time to test. The Cannon had the longest drives and the shortest putts.

23.9 DESIGN MILESTONE: EVALUATION OF ALTERNATIVES

Successful completion of this milestone requires three strong design concepts, an open mind, and a lot of careful thought.

Assignment

1. Decide on five to seven evaluation criteria that will be used with a decision matrix to evaluate the three concepts from the previous milestone.
2. Assign weights to the evaluation criteria.
3. Fill out a decision matrix. One row at a time, discuss the strengths and weaknesses of all of the concepts in the context of the given criterion and assign values by consensus before moving on to the next criterion.
4. Analyze the results of the decision matrix. Use the matrix to look for weaknesses and attempt to correct them by combining ideas from different concepts.
5. Select the best concept.
6. Document your evaluation process as per Example 23.4.

Grading Criteria

- Are weights and values accurate and fully justified?
- Were the results of the decision matrix interpreted thoughtfully when searching for and selecting the best concept?
- Were all three concepts strong designs?
- Is the documentation typed and clearly written?

23.9.1 Design Competition Tips

- There is no need to rig the results of the decision matrix to come out to the concept you want as the results are nonbinding.
- Do not blindly obey the results of your decision matrix; the selection of evaluation criteria may have been flawed to begin with.
- Engage everyone in the decision-making process.
- Do not shy away from bold designs just because they are different from everyone else's. Those differences could lead to victory at the final competition.

EXERCISES
Minimize Design Information Content

1. A company received a complaint from a customer that she bought their product, but the box it was supposed to be in was empty. The company then installed an X-ray machine on the assembly line and staffed it with two workers to make sure that no more empty boxes escaped detection. How would you solve this problem using the KISS principle? (**Ans.**: Install a fan on the assembly line that would blow off any empty boxes.)
2. NASA spent $12 million to develop an ink pen that would write in zero gravity and in a wide range of temperatures. How would you solve this problem using the KISS principle? (**Ans.**: Use a pencil.)
3. To test the design ability of an engineering student, a professor filled a bathtub with water and then offered the student a tea spoon, a cup, and a bucket to empty the tub. How would you use the KISS principle to empty the tub? (**Ans.**: Pull the plug.)
4. Internet web pages are often confusing. Use the internet to find a web page you find too complex and suggest ways to KISS simplify it.
5. Shakespeare wrote "This life, which had been the tomb of his virtue and of his honor, is but a walking shadow; a poor player, that struts and frets his hour upon the stage, and then is heard no more: it is a tale told by an idiot, full of sound and fury, signifying nothing." Use the KISS principle to simplify this quote.
6. Shakespeare also wrote "Shall I compare thee to a summer's day? Thou art more lovely and more temperate. Rough winds do shake the darling buds of May, and summer's lease hath all too short a date. Sometime too hot the eye of heaven shines, and often is his gold complexion dimmed and every fair from fair sometime declines, by chance, or nature's changing course, untrimmed. But thy eternal summer shall not fade nor lose possession of that fair thou ownest. Nor shall Death brag thou wanderest in his shade, when in eternal lines to time thou grownst; So long as men can breathe or eyes can see, so long lives this, and this gives life to thee." Use the KISS principle to simplify this message to his beloved (hint: "Hey babe, how yuh doin" is not a good answer).

Functional Requirement Independence

7. Explain whether or not the steering and speed of a large ship are functionally independent? (Ans. they are not because the radius of the turn depends on the speed of the ship.)
8. Explain whether or not the steering and speed of a bicycle functionally independent?
9. Is there a component in a system that you think should be functionally dependent on another component (think safety)?
10. How could you uncouple the water cannon design steering and catapult aiming shown in Figure 23.2?
11. How could you uncouple the twin boat design move the turn functions shown in Figure 23.3?

Design for Ease of Manufacturability and Robustness

12. What are the seven questions you should ask when designing for ease of manufacture?
13. The Boeing 777 uses 3 million parts that come from 500 suppliers around the world. Explain why this would be an example of ease of manufacturability.
14. What are the three questions you should ask when designing for robustness?
15. End users often use products for things that the designers did not anticipate. How would you increase the robustness of a ceramic coffee cup so that someone could use it as hammer?
16. What adjustments can be made to the motor-driven moving platform in Example 23.3 to decrease the cost?

Decision Matrix

17. Develop a decision matrix for the two student designs in Example 23.1 using the following criteria: (1) It must fit in a 2 ft × 2 ft × 3 ft volume at the start of the competition, (2) it must carry a target that disables the boat if struck by opponent, (3) the steering and motion must be independent, and (4) it is easy to manufacture.
18. Develop a new decision matrix for Example 23.4 (Table 23.2) by adding "Low cost" to the list of criteria and change the weighting factors as follows: drives well—0.15, putts well—0.20, loader is robust—0.15, easy to transport—0.05, easy to manufacture—0.25, and low cost—0.2. Determine your own values for each of the three golf machine concepts.

Design Step 4: Detailed Design

24.1 INTRODUCTION

The goal of this step in the design process is to specify the details of the design so that it can be manufactured. Those details are typically the dimensions and material composition of parts, as well as the methods used to join them. The decisions made during detailed design are guided by the following four methods to reduce the "risk" that additional design changes will be needed later on.

1. Analysis
2. Experiments
3. Models
4. Detailed drawings

24.2 ANALYSIS

Analysis refers to the application of mathematical models to predict performance. The role of analysis in freshman design projects is limited because the analytical capabilities engineering students are just starting to develop. Calculation of power requirements is the first step in motor or gear box selection for student design projects. If the mechanical power required for a given application is known, then the power ratings for various motors can be reviewed to determine the best motor for use in the design project.

24.2.1 Calculating Mechanical Power

Direct current (DC) brushless motors are the commonly used in student design projects due to their availability, price, and performance. Hobby websites offer many DC motors with different specifications, and a method of selecting the appropriate motor is described below. To begin understanding DC motors, Figure 24.1 shows the torque vs. speed curve of a typical DC motor with a constant applied voltage.

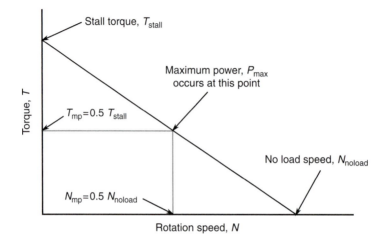

FIGURE 24.1 A typical motor curve for a small direct current (DC) motor.

Each point on the motor curve shown in Figure 24.1 represents a different equilibrium state of the motor. For example, if the motor shaft is allowed to freely spin, it has a rotational speed equal to the "no-load speed," denoted by N_{noload}. Conversely, if you grab the spinning motor shaft between your fingers and gradually increase the pressure, the shaft stops spinning when the torque you apply reaches a value equal to the "stall torque," T_{stall}. Therefore, to determine the actual motor speed, N_{motor}, we need to know N_{noload} and T_{stall} of the given motor, as well as the actual torque, T_{motor}, acting on the motor shaft.

In DC motors, electrical power, P_{elect}, is converted to mechanical power P_{mech}. However, there are power losses within the motor, so $P_{mech} = P_{elect} - P_{loss}$. For rotational motion, the mechanical power output is the product of shaft torque multiplied by its rotational speed, or

$$P_{mech} = T \times \omega \tag{24.1}$$

where T is the shaft torque and ω is the shaft rotational speed in *radians per second*. If the shaft rotational speed is given as N in revolutions per minute (RPM), then you must convert it to ω in radians per second by multiplying N by $2\pi/60$ (radians per revolution divided by seconds per minute), or ω(in rad/s) $= N$(in RPM) $\times (2\pi/60)$, then the shaft power can be written as

Example 24.1

You need to determine the power required to drive small vehicle for a student head-to-head competition. You have determined that the torque on the driving wheels shaft is $T = 3.0$ in-lbf. Your design calls for the maximum wheel rotational speed to be $N = 50$. RPM. What motor power will be required to meet these conditions?

Solution

The motor power comes from Equation (24.2):

$$P_{mech} = T \times \omega = T \times N \times (2\pi/60)$$
$$= (3.0 \text{ in} - \text{lbf})(1 \text{ m}/39.37 \text{ in})(1 \text{ N}/0.225 \text{ lbf})(50 \text{ RPM})(2\pi/60)$$
$$= 1.77 \text{ Nm/s} = 1.77 \text{ J/s} = 1.8 \text{ W (to two significant figures)}$$

So considering the losses in the motor, you would need a 2.0 watt motor.

$$P_{mech} = T \times N \times (2\pi/60) \tag{24.2}$$

Brushless DC motors have a motor velocity constant called Kv, which is the motor's no-load speed in RPM per volt of electricity applied to the motor (or RPM/volt).[1] The motor's no-load shaft speed can then be calculated from[2]

$$N_{noload} = Kv \times (\text{Applied Voltage}) \tag{24.3a}$$

Maximum power output of a small DC motor is important in the design of small vehicles and robots because you need to match a motor's best performance with its usage. The rotational speed at which maximum power is achieved is one-half of the no-load speed, or

$$N_{\text{RPM at max power}} = \frac{Kv \times (\text{Applied Voltage})}{2} = \frac{N_{noload}}{2} \tag{24.3b}$$

Thus the maximum power is achieved at a torque and rotational speed that are half of the motor's maximum capability (see Figure 24.1).

Example 24.2

An unloaded DC motor has a Kv of 5710 RPM/V and is powered by 11.1 V battery.

(a) What is its no-load speed?

(b) What will be its rotational speed when it is producing its maximum power?

---

[1]Kv is *not* the abbreviation for "kilovolt."
[2]This calculation ignores the electrical and mechanical losses within the motor, and the actual no-load motor speed will be somewhat less than this value.

Solution

(a) Using Equation (24.3a), it will run at a no-load speed of

$$N_{noload} = Kv \times (\text{Applied Voltage}) = 5710 \text{ RPM/V} \times 11.1 \text{ V} = 63,381 \text{ RPM} = 63,400 \text{ RPM}$$

(to three significant figures)

(b) Its speed at its maximum power output is determined by Equation (24.3b) as

$$N_{max\ power} = [Kv \times (\text{Applied Voltage})]/2 = 63,381/2 = 31,690 \text{ RPM} = 31,700 \text{ RPM}$$

(to three significant figures)

24.2.2 Determining Gear Ratios

One calculation that is useful for small robotic devices is the determination of the optimal gear ratio. In developing this mathematical model, we draw from the equations on gearing and gear ratios in Chapter 14 (Mechanical Engineering).

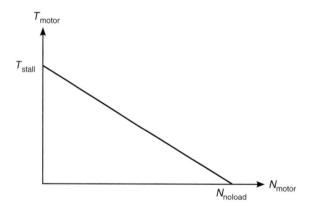

FIGURE 24.2 Schematic of the drivetrain for a robotic moving platform.

Assume that we want to determine the best overall gear ratio for the drivetrain of Figure 24.2 in which the overall gear ratio corresponds to the gear ratio of the gearbox.

$$\text{GR} = \frac{N_{motor}}{N_{axle}} = \frac{\omega_{motor}}{\omega_{axle}} \tag{24.4}$$

N_{motor} and N_{axle} are the rotational speed of the motor shaft and the axle in RPM, and ω_{motor} and ω_{axle} are the rotational speeds in radians per second (rad/s). The linear speed of the robotic moving platform (V_{rmp}) is related to the rotational speed of the axle through the following relationship:

$$V_{rmp} = \omega_{axle} R_{tire} \tag{24.5}$$

where ω_{axle} is in radians/s and R_{tire} is the radius of the driven tire. Changing units from radians per second to RPM in Equation (24.5), we obtain

$$V_{rmp} = \frac{2\pi N_{axle} R_{tire}}{60} \tag{24.6}$$

Rearranging Equation (24.6) to obtain an expression for N_{axle} and substituting the result into Equation (24.4) lead to

$$GR = \frac{2\pi \, N_{motor} R_{tire}}{60 \, V_{rmp}} \tag{24.7}$$

where the length units of R_{tire} and V_{mp} must be the same, and the time units of V_{rmp} are seconds. Equation (24.7) can be used to calculate the overall gear ratio required to achieve a desired speed V_{rmp}, if we know the rotational speed of the motor, N_{motor}.

With small DC motors, the determination of the motors actual operating speed under load, N_{motor}, is not always a straightforward matter. To begin with, N_{motor} is linearly dependent on the actual torque on the motor shaft, as illustrated in Figure 24.1. If the motor shaft is allowed to freely spin, it has a robotic speed equal to the "no-load speed," denoted by N_{noload}. Conversely, if you grab the motor shaft between your fingers and gradually increase the pressure, the shaft stops spinning when the torque you apply reaches a value equal to the "stall torque," T_{stall}. Therefore, to determine the value of N_{motor} to be used in Equation (24.7), we need to know N_{noload} and T_{stall} of the given motor, as well as the actual torque T_{motor} acting on the motor shaft.[3]

Given these constraints, if you cannot actually measure T_{motor} required for your application, we recommend that you calculate gear ratio using N_{noload} (i.e., assume T_{motor} is close to zero for the application). This requires that the vehicle be very small and light and that it moves only on a level plane without pushing against anything. Then Equation (24.7) reduces to

$$GR = \frac{2\pi \, N_{noload} R_{tire}}{60 \, V_{rmp}} \tag{24.8}$$

Since frictional losses are hard to avoid, you can expect the moving platform to run at a speed that is somewhat smaller than the target value. If there are times during operation when T_{motor} is not negligible, such as when climbing a hill or pushing against an opponent, Equation (24.8) is no longer valid. Then you should try to measure the actual torque required on your shaft axle to complete the task at hand.

When the value of N_{noload} is not provided for your motor, you may (at your own risk) assume an average value of 9000 RPM, given that N_{noload} for most small DC motors is in the range of 6000–12,000 RPM. This will not work if your DC motor is a gearhead motor, which already has a built-in gearbox. These spin at much lower rates, so your only recourse, if you want to use Equation (24.8), is to measure its N_{noload}.

Example 24.3

We want to design a robotic moving platform with a top speed of 6.00 inches/s on a flat surface. The motor to be used has specifications of $T_{stall} = 0.210$ oz-in and $N_{noload} = 11,600$ RPM, and we have 2.00 in. diameter tires. Determine the overall gear ratio required to achieve the desired speed.

Solution: Substituting into Equation (24.8), we get

$$GR = \frac{2\pi \, N_{noload} R_{tire}}{60 \, V_{rmp}} = \frac{(2\pi \text{ radians/rev}) \times (11,600 \text{ rev/min}) \times (1.00 \text{ in})}{(60 \text{ s/min}) \times (6.00 \text{ in/s})} = 202.$$

Note that V_{rmp} and R_{tire} are expressed using the same length units (inches).

---

[3] The motor's stall torque can be measured by attaching a pulley to the output shaft and winding a string around the pulley. With the motor running and winding up the string, slowly add weights to the free end of the string until the motor can no longer lift the weights. The stall torque is then the product of the weight and the pulley radius. Do not stall the motor for a long period of time as that will damage the motor.

24.2.3 Common Mechanical Linkages

A *mechanical linkage* is a series of rigid *links* connected with *joints* to form a closed chain. A linkage with two or more links movable with respect to a fixed link is called a *mechanism*. Mechanical linkages are designed to perform certain tasks, such as steering a RC vehicle or making a robot climb stairs.

Linkages have different functions. The functions are classified depending on the primary goal of the mechanism. The simplest closed-loop linkage is the four-bar linkage. It has three moving links, one fixed link and four pin joints.

The following six common four-bar linkages are shown in Figure 24.3.

a. A *reverse-motion linkage* makes an object or force move in opposite directions. If the fixed pivot is centered between the moving pivots, output link movement is the same as the input link movement, but in the opposite direction. By varying the position of the fixed pivot, the output link can produce a mechanical advantage.

b. A *push–pull linkage* makes the output link to move in the same direction as the input link. It can be rotated through a full 360°.

c. A *parallel-motion linkage* makes links move in the same direction, but at a fixed distance apart. The moving and fixed pivots on the opposing links must be equidistant for this linkage to work correctly. This linkage can also be rotated through 360°.

d. A *bell-crank linkage* changes the direction of a link by 90°. If the pivots are at the midpoints of the cranks, link movements will be equal. However, if those distances vary, a mechanical advantage will occur.

e. A *slider-crank linkage* is used to convert linear motion into rotary motion (and vice versa). When the slider is at its maximum distance from the axis of the crankshaft, it is a *top dead center* (TDC); when the slider is at its minimum distance from the axis of the crankshaft, it is at *bottom dead center* (BDC).

f. In a *crank and rocker linkage*, one link rotates continuously while another link rocks back and forth. The windshield wiper on a car uses this mechanism.

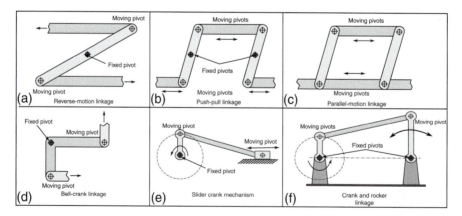

FIGURE 24.3 Common mechanical linkages.

24.3 MECHANISM CONTROL
24.3.1 Simple Control

Simple controls involve the use of some type of switch or sensor to activate an output or change the motion of a robot. For example, a robot might use a *photoelectric sensor* to track a light source or it may use a *limit switch* to determine when it has run into a wall. A simple *mercury tilt switch* can be used to indicate an incline or decline in the robot's path, and an *infrared sensor* can be used to find a heat source. Simple controls are usually used when you have wires running from the robot to a hand-held control device.

The material in the chapter on Electrical Engineering in this text can be used to design simple electromechanical control circuits.

Example 24.4

A student design competition requires a robot to turn on a light whenever it runs into a solid object. Figure 24.4 shows how one team designed their bumper, and Figure 24.5 illustrates the simple control circuit.

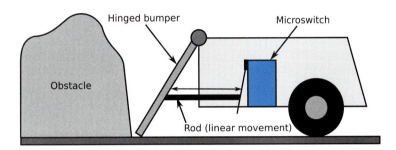

FIGURE 24.4 Schematic of the bumper design.

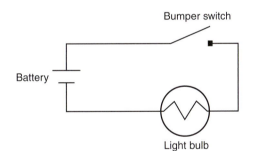

FIGURE 24.5 Bumper control circuit wiring diagram.

24.3.2 Remote Control (RC)

Remote control devices typically use infrared or radio frequency signals to communicate with the device.[4] They can control a wide variety of robotic functions such as speed, turning, lights, and so forth. A large number of relatively inexpensive remote-controlled toys are on the market today, including cars, boats, and helicopters.

[4]In 1898, Nikola Tesla developed and patented a "Method of an Apparatus for Controlling Mechanism of Moving Vehicle or Vehicles" and publicly demonstrated his method by operating a remote-controlled boat. In 1939, a low-frequency, battery-operated radio transmitter became the first wireless remote control for consumer electronic devices, and in 1950, the Zenith Corporation invented the first television remote control.

They can be disassembled, and their parts used in building remote-controlled vehicles or robots for student projects as shown in Figure 24.6. Their parts can also be used to construct autonomously controlled robots.

FIGURE 24.6 A remote-controlled model car.

24.3.3 Autonomous Control

Systems that automatically respond to unanticipated events during their operation are called "autonomous." A typical autonomous controller designed for student use is made by *Arduino©*. It is an open-source electronics system based on easy-to-use hardware and software.[5] Arduino can be used to develop interactive robots, taking inputs from a variety of switches and sensors, to control a variety of outputs such as lights, motors, and other physical devices. Arduino projects can be stand-alone, or they can communicate with software running on an imbedded computer.

24.4 EXPERIMENTS

Physical **experiments** are a particularly effective way to reduce risk when working with small robotic systems. Because of the small scale, materials needed for the experiments can be scavenged or at least obtained at low cost, and realistic forces can easily be applied. Also, physical experiments are often more accurate than idealized mathematical models at this scale.

Since the actual design has not been built yet, the subfunction being investigated may have to be idealized for the purposes of the experiment. For example, you might use cheaper materials or use your hands to create the motion. The errors introduced by these approximations are tolerable if they are much smaller than the changes in performance being observed.

Knowing when to use experiments requires a keen awareness of the sources of risk[6] in a design. This is no time for overconfidence; you can safely assume that *if something can go wrong, it will*. Therefore, it is vital that you be able to distinguish between the aspects of the design that you are sure about and those aspects that you are not so sure about. The latter are candidates for physical experiments.

The steps for formulating an experimental plan are as follows:

[5]There are numerous student project microcontrollers available on the Internet, for example, Parallax Basic Stamp, Netmedia's BX-24, Phidgets, MIT's Handyboard, and many others.
[6]Risk is the possibility that something bad or unpleasant (such as having to redesign your robot) will happen.

1. Identify aspects of the design and its performance about which you are uncertain.
2. Associate the aspects in step 1 with one or more physical variables that can be varied by means of simple experiments.
3. Carry out the experiments that will do the most to reduce risk of redesign within the available time frame.
4. If possible, document the results in the form of graphs or tables.

Example 24.5

A concept for a robotic design competition named Dueling Duffers has been proposed and is shown in Figure 24.7. The object of this head-to-head competition is to be the first to deposit up to 10 golf balls into a hole in the center of the tabletop playing field shown in Figure 24.8.

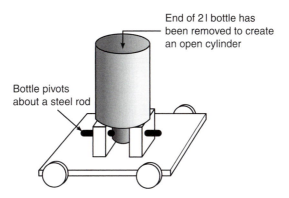

FIGURE 24.7 Proposed concept for the "Dueling Duffers" design competition.

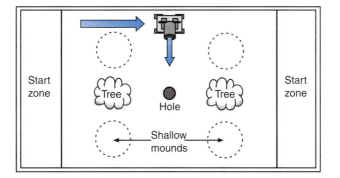

FIGURE 24.8 Playing field for the "Dueling Duffers" design competition.

These are the main features of the proposed design:

- It holds 10 golf balls in the top half of a 2 l bottle.
- It uses the side rail to steer.
- When it is even with the hole, it dumps the golf balls in the general direction of the hole.

The path of the vehicle and the direction in which the balls are dumped are indicated in Figure 24.8. For this example, you are asked to (a) identify the main sources of risk and (b) propose experiments to address those sources of risk.

Solution

Sources of Risk

The sources of risk are presented here in the form of questions, the answers to which are currently unknown. These are the same type of questions that will be asked by the jury at the oral design defense.

1. Will all 10 balls fit in the top half of the 2 l bottle?
2. What is the optimal height from which to dump the balls?
3. Will all 10 balls drop into the hole when dumped in this manner?
4. Is it important to have the vehicle perfectly positioned before dumping the balls?
5. Will the machine travel slowly enough that its position can be easily controlled?

Proposed Experiments

The experimental setup in Figure 24.9 can be used to address each of the first four sources of risk just listed. The last item in the list is best handled by computing the overall gear ratio using Equation (24.8).

The bottle, the steel rod, and the two blocks of wood are all easily obtained. Assume that the actual playing field is available for testing. The bottom of the plastic bottle will need to be cut off, and the steel rod will need to be inserted through the top of the bottle. There are no other parts that need to be joined. You can use your hands to keep the rod in place above the blocks, while taking care to let the bottle and balls fall under their own weight.

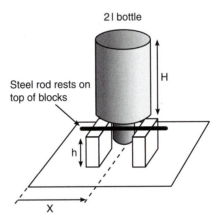

FIGURE 24.9 Experimental setup for establishing key dimensions and reducing risk.

The numbers of the experiments described next correspond to the preceding numbered sources of risk.

1. The control variable is H (refer to Figure 24.9). Put 10 golf balls into the bottle and measure the minimum H required to hold 10 balls.
2. The control variable h determines how fast the balls roll when they pass the hole. Increase h by inserting books under the blocks; decrease h by sawing the ends of the blocks. For each value of h, dump the balls three times and record the number of balls that drop into the hole. Document the results by plotting h versus the average number of balls that dropped.
3. Here, the concern is less with the speed and more with the distribution of the balls as they pass the hole. Control variables might be H or the manner in which the balls are packed within the bottle (e.g., a vertical divider could be used to keep the balls on the side of the bottle facing the hole). Again, use three trials for each value of the control variable and plot the control variable versus the average number of balls that dropped.
4. The control variable is X. Perform three trials for each value of X and plot X versus the average number of balls that drop. The shape of this graph provides the answer to question 4.
5. Calculate the gear ratio using Equation (24.8).

24.5 MODELS

Models are scaled replicas constructed out of inexpensive, readily available materials. In the case of small electromechanical devices, they are often constructed out of cardboard or foam board. Models are used to check geometric compatibility, establish key dimensions of moving parts, and visualize the overall motion.

Typical examples are shown in Figures 24.10 and 24.11. The foam board model of the stair-climbing device in Figure 24.10 is used to prove the feasibility of the design. Several other stair-climbing concepts, mainly wheel and track designs, were regarded as feasible until models proved otherwise. In Figure 24.11, a dart-throwing mechanical linkage is modeled in 3D using cardboard. All the links move, including the grip mechanism that releases the dart when the arm strikes a stopper.

FIGURE 24.10 Model of a stair-climbing device.

FIGURE 24.11 Model of a dart-throwing machine.

24.6 DETAILED DRAWINGS

By definition, a **detailed drawing** contains all the information required to manufacture the design. The drawings should be so complete that if you handed them off to someone unfamiliar with the design, that person would be able to build it. The various elements of engineering drawing, dimensioning, and sketching are discussed in Chapter 2. Table 2.1 illustrates the five common types of drawing projections.

The usual practice is to specify dimensions on multiple orthogonal views of the design. An isometric view is also sometimes provided to assist with visualization. In all, six orthogonal views are possible: front, back, left, right, top, and bottom. Three views, however, are most common. Figure 24.12 shows a detailed drawing with five orthogonal views.

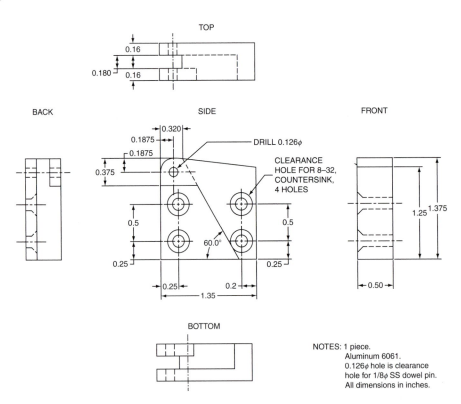

FIGURE 24.12 Five views of the right toe of an animatronic eastern grey squirrel.

Additional information such as material specification, part type, and assembly directions are conveyed through written notes on the drawings. Close-up views can be employed to clarify small features.

While practicing engineers generate drawings like Figure 24.12 using computer-aided design (CAD) software, first-year engineering students may or may not be taking a CAD course (or had taken one in high school). Therefore, we recommend that the usual standards for preparation of detailed drawings should be relaxed somewhat and replaced by the following set of guidelines:

- Drawings can be neatly hand-drawn using a ruler and compass.
- Drawings must be drawn to scale, though not necessarily full scale.

- Drawings of at least two orthogonal views of the design should be prepared. An isometric view is not required, but close-up views should be used to clarify small features.
- Show hidden lines only when they enhance clarity. These are dashed lines that are used to show edges not visible from the viewer's perspective.
- It is acceptable to show only essential dimensions, that is, key dimensions that either have a direct impact on performance or are needed to demonstrate that geometry constraints are satisfied.
- Use notes or labels to indicate material specification, part type, and assembly directions.
- It is acceptable if the manufacturing details are incomplete. Students who lack the experience to fully specify them have to discover those details by trial and error during building.
- If an electric circuit was designed, show it in the form of a neatly hand-drawn but fully specified circuit diagram.

An example of a detailed drawing prepared in accordance with these guidelines is shown in Figure 24.13.

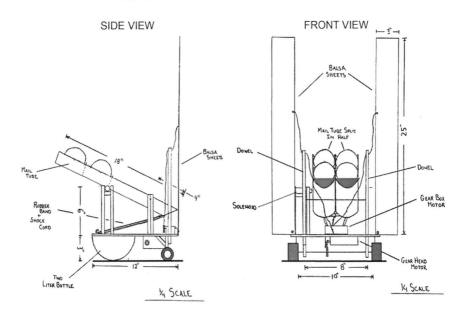

FIGURE 24.13 Hand-drawn detailed drawing of a competition vehicle.

Although the freedom to leave out some dimensions and manufacturing details has been allowed, keep in mind that missing details amount to higher risk in the minds of those being asked to provide resources to the project. Thus, a design with fewer missing details may be viewed by the instructor as having lower risk.

24.7 DESIGN MILESTONE: DETAILED DESIGN

This milestone is all about reducing risk, not only in your own mind but in the minds of the jury at the upcoming oral design defense.

Assignment
1. Use analysis, experiments, and models to help establish dimensions and proof of concept.
2. Prepare detailed drawings of the design concept you selected.

Documentation
- Write the analysis details in the usual format, stating all assumptions.
- For each experiment, (a) state the purpose of the experiment, (b) describe the experimental procedure, (c) present results, and (d) state conclusions.
- Summarize useful information yielded by models and turn in the models constructed.
- Attach hand-drawn detailed drawings.

Grading Criteria
- From examination of the detailed drawings, does the design have a chance of working?
- Have opportunities to reduce the level of risk (through analysis, experiments, and models) been fully exploited?
- Is there enough information in the detailed drawings to manufacture the design?
- What is the overall quality of the detailed drawings?

24.7.1 Design Competition Tips

- Time spent now on analysis, experiments, models, and detailed drawings pays off later in fewer design changes during manufacturing and testing.
- Of the four methods used in reducing design risk (analysis, experiments, models, and detailed drawings), good detailed drawings reap the most rewards in a freshman design project.

EXERCISES
Analysis

1. If the rotational speed of a shaft is 250. RPM, what is its robotic speed in radians per second?

2. What is the mechanical power output of a shaft rotating at 1500. RPM that has a torque of 200. Nm?

3. If a small DC motor has a power output of 3.0 watts at 700. RPM, what is its torque?

4. A small 8.00 watt DC motor on a robot is geared down to produce an output torque of 350. Nm. What is the rotational speed of the output shaft?

5. If the no-load motor speed in Example 24.3 was 3600. RPM, what would the overall gear ratio required to allow the moving platform to have a top speed of 4.0 inches per second?

6. If the overall gear ratio in Example 24.3 was 50, what would be the top speed of the moving platform?

7. What is the translational kinetic energy of the moving platform in Example 24.3 if it weighs 5.0 newtons[7]?

8. If the motor in Example 24.3 was replaced with a similar motor that had a torque of 3.42 ounce-inches and a no-load speed of 1759 RPM, what overall gear ratio is required to have the moving platform to reach a top speed of 6.89 inches per second?

---

[7]Note that the unit named after Isaac Newton is not capitalized when it is spelled out (newton) but is capitalized when it is abbreviated (N). See Section 2.2 of Chapter 2 for more information on units.

9. If the moving platform in Example 24.3 had a constant acceleration of 0.056 ft/s², how long would it take for the platform to reach its top speed of 6.0 inches per second?

Experiments

10. The Dueling Duffers concept shown in Figure 24.7 requires the bottle to be tipped over far enough to spill all the golf balls out of it. How would you develop and experiment to determine the how far the bottle needs to be tipped so that all the golf balls escape?

11. In the Dueling Duffers concept shown in Figure 24.9, it was decided to tip the bottle by pushing on it half way up its surface at H/2. How would you develop and experiment to determine the force required at this height to tip the bottle far enough to spill the golf balls?

12. During the development of the Dueling Duffers concept shown in Figure 24.7, it was decided to glue the bottle to the steel pivot rod and use a gear motor to turn the rod to tip the bottle and spill the golf balls. How would you develop and experiment to determine how much torque would be required by the motor?

13. In Example 23.2 of the previous chapter, the concept of a dart-throwing machine was presented. How would you develop an experiment to determine how much mass the plunger should have to propel the dart into a target 8 feet away?

14. In the three concept drawings shown in Example 23.4 of the previous chapter, how would you develop an experiment to determine how much air or CO_2 pressure is required to drive the golf ball 250. yards?

Models and Drawings

15. Make a hand-drawn detailed sketch of the Dueling Duffers concept shown in Figure 24.7. Choose your own dimensions for the components.

16. Make a hand-drawn detailed sketch of the stair-climbing device shown in Figure 24.10. Choose your own dimensions for the components.

Design Step 5: Design Defense

25.1 INTRODUCTION

Engineers must convince customers that a design is worth expenditures of money and the time of skilled people. In a student design project, this process of convincing customers is simulated by an *oral design defense*. The goal of this oral presentation is to win the confidence of the project sponsors, henceforth referred to as the *jury*.

25.2 HOW TO PREPARE AN ORAL DEFENSE

In assessing a team's chances for future success, the jury is searching for answers to the following questions:

- Did the team adhere to the systematic approach?
- How does the final concept work?
- What level of risk is associated with this design?
- Do the students appear to be teaming effectively?

The jury's concerns suggest some strategies that should be effective. First, the organization of the presentation should parallel the steps in the design process, as shown in Table 25.1. This is your way of saying that you followed a systematic approach. Second, you should try to get the jury to understand how your final concept works as quickly as possible. This frees up more time during questions for alleviating concerns about the design. Third, you should anticipate that the jury will ask questions about potential sources of risk and prepare evidence in advance to quell those concerns. This evidence should be in the form of high-quality detailed drawings, results of calculations and experiments, and models. If you built models, bring them; if you conducted experiments, try to bring some evidence that, indeed, you did them. The strategies that serve to reduce risk, and their counterparts that don't, are summarized in Table 25.2.

Table 25.1 Suggested Organization of the Oral Design Defense

Organization	Slides
Title	1
Outline of presentation	1
Problem definition	1
Important design requirements	1
Alternative concepts not selected	2
Final concept ■ Describe main features ■ Explain why you selected it	1–2

Detailed design
■ Show main drawings
  – Explain how the design works
  – Explain how you will construct it
	2

■ Zero in on special features with close-up views ■ Present results of analyses, experiments, and models	1–2 1–2
Summary ■ Summarize strengths of the design ■ Quantify performance expectations (e.g., top speed) ■ Describe strategy at the final competition	1
Total slides	12–15

Table 25.2 Strategies That Either Reduce or Amplify Risk

Risk reducers

1. High-quality concept drawings and detailed drawings.
2. Calculations, experiments, and models that establish proof of concept.
3. Explanation of manufacturing details.
4. Quality visual aids.

Risk amplifiers that could delay manufacture

1. Poorly detailed drawings.
2. One or more subfunctions obviously will not work.
3. No thought given to manufacturing.

Meanwhile, the jury is also evaluating your teaming. It will base its impressions on the quality of your design and oral presentation (see Table 25.3 for some tips on delivery and visual aids). There are other telltale signs. For example, did everyone contribute equally to the presentation? Was everyone involved in answering questions? Did team members refer to themselves as "we" or "I" when citing accomplishments?

When answering questions, be forthright and honest. Failure to do so will lead to an unending chain of questions. If your response is an opinion and not a fact, state so because one erroneous answer can damage your credibility and thus elevate the risk associated with your design.

Table 25.3 Oral Presentation Tips

Tips on delivery

- Do not read sentences directly off the slides.
- Look at the audience.
- Stand next to the screen when speaking.
- Practice the presentation.
- Be positive and dynamic.

Tips on visual aids

- Avoid using too many words.
- Use a font size that can be seen easily.
- Include concept or detailed drawings.
- Put a heading on each slide.
- Show results of calculations or experiments.
- View the projected images in advance to make sure all the words and pictures are visible to the audience.

25.3 DESIGN MILESTONE: ORAL DESIGN DEFENSE

To qualify to receive parts and materials for the manufacturing phase of a hands-on design project, a majority of the jurors must be convinced that your design will work. In the event such a consensus is not achieved, teams are asked to revise their designs and resubmit at a later date.

Assignment
1. Prepare the visual aids for the oral design defense. You must use PowerPoint or an equivalent software package. Relevant drawings should be scanned.
2. Practice the presentation.
3. Deliver the oral presentation to a jury of evaluators.

Typical format
- Eight minutes for the oral presentation; four minutes for questions.
- All team members participate in the presentation and in responding to questions.

Grading Criteria
- What level of risk is associated with the design?
- What was the quality of the drawings and the other visual aids?

25.3.1 Design Competition Tips

- Grading tends to be proportional to the quality of the drawings.
- Evidence of the use of calculations, experiments, and models do much to reduce the level of risk in the minds of the jurors. Visual representations of the results are more effective than just saying you did them.

EXERCISES

Preparing an Oral Defense

1. A team of students prepared an oral defense for their design project contained only the following items: a statement of the problem and the final detailed design. What else should they have included?

2. When you are called upon in class to make an oral defense for your design project, you suddenly realize that you do not have your presentation slides with you. What can you do to make a successful oral defense without them?

3. While making their oral defense, a student group begins by telling a joke and showing a cartoon slide. This approach is often effective for certain types of presentations; do you think it is appropriate for a design project oral defense? Why or why not?

4. When you are making your oral defense your instructor does not want you to use notes of any kind (3 by 5 cards, etc.). What is the best way to move through your defense without reading your notes? (*Ans.*: use your slides to prompt your memory.)

5. One of the tips in Table 25.2 is to look at the audience. Some students are shy and find this very difficult to do. How can you look like you are looking at the audience without actually looking at them? (*Ans.*: look at the back wall of the room.)

Design Step 6: Manufacturing and Testing

26.1 INTRODUCTION

Manufacturing begins once the detailed design has been approved and ends when the machine is placed in the starting zone of the final competition. In between, the machine undergoes numerous modifications. Few new designs work on the first try. Manufacturing and testing tend to take much longer than expected—probably three to five times as long. This chapter begins with a summary of good manufacturing and testing strategies and then moves on to describe materials, joining methods, and hand tools.

26.2 MANUFACTURING AND TESTING STRATEGIES

You can employ strategies during manufacturing to minimize the time it takes to get the first prototype ready for initial tests. The extra time freed up for testing and design refinements can prove decisive in a design competition.

The following time-saving **manufacturing strategies** have been observed in successful teams:

- **Talk to a machinist.** Professional machinists are considered partners in the design process. What they lack in knowledge of the engineering design process and analysis, they make up for in manufacturing and practical experience. As a practicing engineer, you will be required to consult with a machinist before finalizing your detailed design.
- **Don't delay in getting started.** Take the leap of faith and begin manufacturing as soon as possible. Only then does the team gain a realistic sense of the manufacturing timeline.
- **Divide up responsibilities so that team members can work in parallel on different subassemblies.** Otherwise, you may find the entire team standing around waiting for the same glue joint to dry.

- **Keep detailed drawings up to date.** If they are not up to date, only one person knows what the actual design looks like. That one person ends up doing most of the manufacturing, while the other team members watch. With accurate drawings, team members can work in parallel.
- **Set and enforce intermediate deadlines.** Manufacturing can span several weeks. The instructor sets the big deadlines through the milestones; the teams should set the little ones in between.

Testing is as important as manufacturing in preparing for a design competition. For example, three design teams were assigned the task of designing machines capable of playing 18 holes of miniature golf at a local course. The first team was stocked with experienced machinists, so it built its machine out of thick steel parts. The second team had an expert welder, so it welded together its machine out of steel beams and plates. The third team chose to make its machine out of wood. Three very different manufacturing skill sets, yet all three machines failed at the final competition for the same reason,—not enough testing. The first team did not test its machine on synthetic grass before the competition. Its steel machine was heavy, which created large friction forces between its tank-line treads and the synthetic grass. When it attempted to turn, the treads broke, immobilizing the machine. The second team did not have time to test their steering mechanism due to last-minute modifications. As a result, it could not consistently maneuver into position for putts within the time constraint. The third team completed manufacture and testing of its moving platform 2 weeks before the final competition. Over the next 2 weeks, while the putting mechanism was being made, it did not test the moving platform again until about 30 min before the start of the competition. The machine never moved.

The lessons learned by these three teams apply to all design competitions. They are summarized in the following testing tips:

- Always leave a lot of time for testing.
- When conducting tests, try to simulate as closely as possible the conditions at the final competition. If these conditions are not known, test under a variety of conditions to insure robustness.

26.3 MATERIALS

When possible, the designs should be made of wood to facilitate manufacture and keep costs down. Manufacturing can be simplified still further by constraining the designs to be small (less than 1 ft^3) and lightly loaded. Under these conditions, balsa can be used as the main structural material.

Recommended *materials* for a small, lightly loaded electromechanical device are listed in Table 26.1, along with their relative attributes. Balsa is listed as easy to use with hand tools because it can be cut and shaped easily

Table 26.1 List of Recommended Materials for a Small Electromechanical Design Project

Material	Relative Strength	Relative Stiffness	Ease of Manufacture	Useful Forms
Balsa	1	1	Very easy	Sheets, beams, blocks
Woods	5	4	Easy	Plywood sheet, wood dowels
Plastic	5	1	Hard	Prefabricated gears
Steel	50	80	Medium	Thin rods
Rubber	0.5	Very small	Very easy	Long strands

Note: *The strength (resistance to breaking) and stiffness (resistance to deformation) are normalized with respect to values for balsa wood.*

with a sharp knife. On the other hand, plastic tends to deform rather than shear cleanly under the action of cutting tools, and so it is listed as hard to work with.

When selecting a material from Table 26.1, the strength and stiffness requirements of the given part also need to be considered. For example, if there are concerns about a part breaking under load, strength considerations override ease of manufacture, leading to the use of plywood instead of balsa. If a small-diameter axle requires high stiffness so that gears can remain engaged, then a steel rod may be the best choice.

26.4 JOINING METHODS

For small-scale balsa and wood structures, the preferred *joining methods* for parts are adhesives, wood screws, and machine screws with nuts. Typical joint configurations employing these methods are shown in Figure 26.1. Use of tapes, especially duct tape, is frowned upon for their nonpermanence and poor aesthetics. Nails are not particularly compatible with balsa because of the large impact forces involved and the possibility of wood splitting.

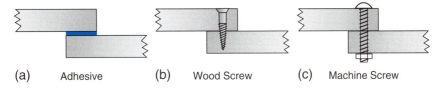

(a)   Adhesive   (b)   Wood Screw   (c)   Machine Screw

FIGURE 26.1 Different joining methods.

Adhesives, in particular hot glue, are the method of choice when balsa is the predominant structural material. Hot glue cures quickly, and though essentially permanent for wood-to-wood bonds, metal-to-wood bonds can be adjusted or broken by heating. It is so general purpose that it can be used to mount a motor to a plywood base. The main drawback of hot glue is its low strength, but this is usually not an issue for lightly loaded balsa structures. When it does become an issue—for example, when the surface area available for the glue joint is very small—one of the higher strength adhesives listed in Table 26.2 may be substituted.

Table 26.2 Common Adhesives

Adhesive	Typical Uses	Setting Time*	Curing Time†	How to Apply	Relative Strength
Wood glue	Wood, paper	8 h	24 h	Apply in liquid form direct from bottle.	13
Epoxy	Wood, metal	5 min to 12 h	3 h to 3 days	Comes in two tubes; mix equal amounts and apply with stick.	14
Hot glue	Almost anything	1 min	2 min	Place glue sticks in heated gun and apply with gun.	1

*Setting time = Time to harden.
†Curing time = Time to reach maximum strength.

On balance, wood screws are a less popular alternative for balsa-to-balsa joints. To reduce the chances of wood splitting, a pilot hole equal in diameter to the screw without the threads should be drilled prior to inserting the screw. This material removal plus the persistent possibility of wood splitting reduces the effective strength of the balsa members. In situations where adjustability is needed, the benefits of easy screw removal may override these costs.

Wood-to-wood joints are a different matter. Wood is much stronger than balsa, so a higher strength joint is justified. In such cases, use of screws is often preferable to the long curing times of the higher-strength adhesives.

26.5 USEFUL HAND TOOLS

Small-scale electromechanical devices can be crafted using common *hand tools*, provided balsa or wood is the primary structural material. In this section, we offer a short compendium of those tools for your easy reference. Our goal in presenting them is to make you aware of the alternative manufacturing operations at your disposal. We do not go into much detail on how to use them. Instead, we advise you to observe your classmates, talk to a machinist, or just give it a try.

For each tool, we will (1) show a picture of it (so that you can find it), (2) name it (so that you can ask for it), (3) describe its use, and (4) provide additional comments on usage.

26.5.1 Tools for Measuring

FIGURE 26.2 Tape measure/ruler.

Name: Tape measure; steel ruler (Figure 26.2).
Use: For measuring length.
Comments: A steel ruler can also be used as a straight edge when making straight cuts in balsa with the X-acto knife.

26.5.2 Tools for Cutting and Shaping

Name: X-acto knife (Hunt Manufacturing Co.) (Figure 26.3).

FIGURE 26.3 X-acto™ knife.

Use: For cutting and shaping balsa.
Comments: Blades come in different shapes and are replaceable.

FIGURE 26.4 Coping saw.

Name: Coping saw (Figure 26.4).
Use: For cutting curves in wood.
Comments: The blade can be mounted in the frame to cut on either the push or pull stroke. Unscrewing the handle relieves tension on the blade so that it can be removed, rotated up to 360°, or inserted through a small drilled hole to cut out a larger hole.

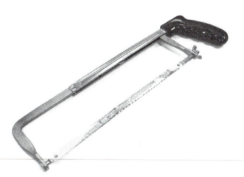

FIGURE 26.5 Hacksaw.

Name: Hacksaw (Figure 26.5).
Use: For making straight cuts in metal.

Comments: It cuts on the push stroke when the blade is mounted correctly. Though designed for metal, it can also be used with wood. Blades are detachable and can be rotated to cut in four directions.

Name: Dremel (Robert Bosch Tool Corporation) (Figure 26.6).
Use: For shaping and drilling holes in wood.
Comments: This is an electric hand drill without the handle. Because of its shape, forces can be applied more easily when grinding or sanding.

26.5.3 Tools for Drilling Holes

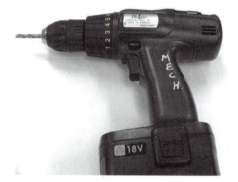

FIGURE 26.7 Cordless hand drill.

Name: Cordless hand drill (Figure 26.7).
Use: For drilling holes in wood and plastic.
Comments: Release trigger lock and pull trigger to start drilling. Power switch controls the direction of spin.

26.5.4 Tools for Joining Parts

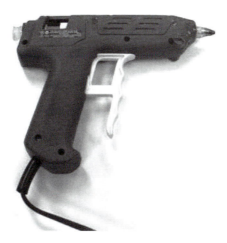

FIGURE 26.8 Hot glue gun.

Name: Hot glue gun (Figure 26.8).
Use: For applying hot glue.
Comments: Insert a glue stick in the back, wait 3–5 min for the gun to heat up and then apply the glue by pulling the trigger. Handle with care, as the glue can reach temperatures of 400°F.

FIGURE 26.9 Spring clamps.

Name: Spring clamps (Figure 26.9).
Use: For holding parts together when waiting for an adhesive to set.
Comments: If the grips do not open wide enough for your application, C-clamps are a good alternative.

FIGURE 26.10 Screwdrivers.

Name: Screwdriver (Figure 26.10).
Use: For inserting and removing screws.
Comments: The longer and thicker the handle, the easier it is to apply torque. The screwdriver in Figure 26.10 has a removable tip. You can pull out the tip, rotate it 180°, and reinsert it to switch from a regular screwdriver ($\ominus$) to a Phillips-head screwdriver ($\otimes$).

FIGURE 26.11 Adjustable wrench.

Name: Adjustable wrench (Figure 26.11).[1]
Use: For holding nuts when tightening a screw.
Comments: Thumb adjustment changes the distance between jaws. Unlike pliers, there is no gripping force, so parts being held must have flat surfaces.

FIGURE 26.12 Claw hammer.

---

[1]In the United Kingdom, this is called a *spanner*.

Name: Claw hammer (Figure 26.12).
Use: For driving or removing nails.
Comments: Nails should not be used to join balsa parts.

26.5.5 Tools for Wiring

FIGURE 26.13 Soldering iron.

Name: Soldering iron (Figure 26.13).
Use: For attaching or connecting wires with solder.
Procedure: (1) Create a mechanical connection (e.g., by twisting wires together), (2) let iron heat up, (3) apply solder to tip of iron (called *tinning*), (4) heat wires with iron, (5) bring solder in contact with heated wires until solder melts and flows, and (6) let wires cool.

FIGURE 26.14 Long-nosed pliers.

Name: Long-nosed pliers (Figure 26.14).
Use: For bending wires, grasping parts, cutting wires, and reaching into tight places.
Comments: Clamping forces are highest at the wire cutters and smallest at the tip of the nose.

FIGURE 26.15 Wire cutters.

Name: Wire cutters (Figure 26.15).
Use: For cutting or stripping wires.
Comments: You may not need this tool if your long-nosed pliers have a good wire cutter.

FIGURE 26.16 Wire strippers.

Name: Wire strippers (Figure 26.16).
Use: For stripping insulation from wires, cutting wires, and shearing small diameter bolts.
Comments: To strip insulation, lay wires across jaw in notch of same diameter as metal wire and then pull wire while holding jaws compressed.

26.6 DESIGN MILESTONE: DESIGN FOR MANUFACTURE - ASSESSMENT I

The **Design for Manufacture** (DFM) assessments are so named because the teams that gave serious consideration to DFM principles when first formulating their designs have the best chance of doing well on these milestones.

This first DFM assessment should occur about halfway through the first manufacturing iteration. For a design project of the scale proposed in these chapters, that's about 1 week after receiving the box of parts.

Assignment

Make as much progress as possible toward completion of manufacturing and testing of your design.

Grading

Teams should bring their materials and a detailed drawing of their machine to the next design studio. Grading will be based on the amount of progress made.

100 points = Manufacturing is more than 50% completed
80 points = Manufacturing is between 25% and 50% completed
60 points = Some progress made (but less than 25%)
0 points = No progress

26.6.1 Design Competition Tips

- Team members should work in parallel on different aspects of the design to accelerate progress.
- From here on, the design project is a race to begin testing. Teams with the most testing time tend to win competitions.

26.7 DESIGN MILESTONE: DESIGN FOR MANUFACTURE - ASSESSMENT II[2]

By the time this milestone is reached, teams should have completed their first manufacturing iteration and begun testing. For a freshman project, this could be as soon as 1 week after the previous DFM milestone.

The instructor evaluates progress by conducting one (ideal) or more well-defined performance tests to determine whether or not the various subfunctions work. Evaluation of how well they work is reserved for the next milestone. Subfunctions to be evaluated must be carefully selected, so as not to be biased toward a particular design or strategy.

Assignment

Get all subfunctions working by the assigned due date.

Performance Test

To be defined by the instructor.

Grading

Team grades are based on the level of functionality demonstrated in the best of three trials and computed as follows:

If manufacturing is 95% completed, then

$$\text{Grade} = B + \sum_{i=1}^{I} (W_i s_i)$$

where

Grade = Assigned grade on a scale of 100 points
B = Base grade (typically $B = 70$ points)
I = Total number of subfunctions being evaluated

---

[2]The instructor must tailor this milestone to the design project.

W_i = Number of grade points associated with the ith subfunction and defined such that

$$100 = B + \sum_{i=1}^{I} W_i$$

$s_i = 1$ (if the subfunction worked)
$s_i = 0$ (if the subfunction did not work)

If manufacturing is less than 95% completed, then

$$\text{Grade} = fB$$

where

Grade = Assigned grade on a scale of 100 points
   B = Same base grade as previously
   f = Fraction of manufacturing that was completed

EXERCISES

Manufacturing and Testing Strategies

1. In Chapter 20, we discussed Gantt charts as a way if planning your design activities. It can also be used to develop manufacturing and testing strategies. Suppose you are in a design team with three other students (Sara, Jim, and Leslie) and are beginning to construct (i.e., manufacture) your design. Develop a Gantt chart that will show the responsibilities and deadlines for each student (including yourself) for the following activities: (1) talk to a machinist for advise on making unique components of your design, (2) search for components on the Internet or in hardware stores (wheels, bolts, etc.), (3) collect receipts for purchases, (4) keep drawings up to date, and (5) assemble the components to make a working model of the design.

2. Now that you have built your design, it's time to test it. You only have 2 weeks left in the semester, and during your first test, you determine that your prototype does not function properly. How do you salvage your design before the semester ends?

Materials and Joining Methods

3. When you begin testing the prototype of your design, you find that an important component you made from balsa wood breaks under load. You need to fabricate a new version of the component and replace it quickly. What material do you choose for its replacement?

4. Your team's final design calls for gluing two pieces of metal together. During testing, the glue doesn't hold and the prototype fails to function properly. How do you fix this problem quickly and securely?

5. One of your team members wants to assemble the main components of the design with bolts. It turns out that the bolts add a lot of weight to the prototype and are difficult to keep from coming loose during testing. What would you recommend using instead of bolts at this late date and why?

Design Step 7: Performance Evaluation

Source: http://www.aerospacenews.com/content/view/356/1/.

27.1 INTRODUCTION

Once the design has been manufactured, it is time to evaluate its performance. If machines are required to interact with each other, the performance should be measured in two stages. First, comes individual performance testing. The manufactured device is tested alone, under controlled conditions, to verify that it is capable of doing what the problem definition requires. Then comes the final competition. The device is tested against other machines in a series of head-to-head matches to determine the best overall design. The student grade resulting from performance evaluation typically constitutes up to 50% of the project grade.

27.2 INDIVIDUAL PERFORMANCE TESTING

Performance of a given machine in head-to-head matches may vary with the opponent. For example, an offensive or defensive strategy that works well against one opponent may not work against another. So, the only way to test all of the machines under the same set of conditions is to test them in isolation, with no opponent.

The basic approach is to measure one or more quantities, referred to as "**metrics**," that are good predictors of success in the head-to-head matches. Typical metrics are time, speed, pushing force, or number of points scored against a stationary obstacle representing the opponent. Some basic rules when selecting metrics are: (1) they should be easily measurable, (2) they should be continuously variable to maximize information content, and (3) they should not be biased toward particular design solutions.

The number of different physical tests required to measure all of the metrics depends on the choice of metrics. Ideally, you want to be able to design a single test that measures all of the metrics. Sometimes, each metric requires its own test.

The performance grade is the overall measure of performance expressed on a scale of 100. We recommend that it be computed as a weighted sum of the metric values as expressed in the design milestone at the end of this chapter.

The specifics of the performance tests are often not revealed to the students until a week before they take place. This is to prevent students from tailoring their machines to the performance tests instead of to winning the final competition, which is the real design objective.

27.3 THE FINAL COMPETITION

The final competition pits the machines against each other, usually in a series of head-to-head matches that may involve direct interactions between the machines. This is the ultimate test of the machines, as it is the only way to accurately evaluate the effectiveness of offensive and defensive strategies, robustness, durability, and the wisdom of past design choices.

However, it is probably best that the results of head-to-head competition not be linked to the performance grade, since each machine faces a different set of challenges. For example, one machine may not match up well against a particular opponent, or a prefabricated part may fail unexpectedly due to an accidental collision.

27.4 DESIGN MILESTONE: INDIVIDUAL PERFORMANCE TESTING[1]

Grade wise, this is the most important milestone. The testing regimen enforced by this milestone and the previous one prepare the machines for the final competition.

Assignment
Optimize performance of your machine in preparation for individual performance testing.

Performance Test
To be defined by the instructor.

Grading
Team grades are based on the quality of performance demonstrated in the best of three trials and are computed as follows:

$$\text{Grade} = B + \sum_{i=1}^{I} W_i \left(\frac{m_i}{M_i}\right)^n - \sum_{j=1}^{J} P_j$$

where

    Grade = Assigned grade on a scale of 100 points

---

[1]The instructor must tailor this milestone to the design project.

B = Base grade (typically, $B = 70$ points)
I = Total number of metrics
J = Total number of penalties assessed for rules violations
W_i = Number of grade points associated with the ith metric and defined such that

$$100 = B + \sum_{i=1}^{I} W_i$$

m_i = Measured value of the ith metric
M_i = Best value of m_i recorded by any team in the class
$n = 1$ (if performance is directly proportional to m_i)
$n = -1$ (if performance is inversely proportional to m_i)
P_j = Number of grade points associated with the jth penalty

If the grade calculated is less than or equal to the base grade (B), then

$$\text{Grade} = f \times B$$

where

Grade = Assigned grade on a scale of 100 points
B = Base grade (typically, $B = 70$ points)
f = Fraction of manufacturing that was completed

EXERCISES

1. What metrics are normally used to predict the performance of your entry in a head-to-head competition?

2. What additional metrics beyond those listed for exercise 1 would you use to predict your design's performance in a head-to-head RC model boat competition?

3. How would you measure the pushing force of your entry in a head-to-head competition?

4. If you had designed and constructed an entry in a head-to-head RC walking robot competition, what performance tests would you carry out before the competition?

5. You are to design and construct a small autonomous vehicle that will seek out a lit candle and blow it out. What performance tests would you implement before the competition?

Design Step 8: Design Report

28.1 INTRODUCTION

The design report documents the final design. It enables someone unfamiliar with a design to figure out how it works, evaluate it, and reproduce it. It is the final step in the design process. This chapter summarizes the organization of the report and provides some writing guidelines.

28.2 ORGANIZATION OF THE REPORT

Like the oral design defense, the organization of a design report follows the steps of the design process (see Table 28.1). The report begins with a concise statement of the "Problem Definition," in the student's own words.

The "Design Requirements" section does not need to repeat the competition rules. Instead, it should begin with a short paragraph describing competition strategy. Then, list all performance requirements, for example, "Must have a top speed of 1 ft/s"; "Must deposit at least six ping-pong balls"; and "Must be able to steer."

Almost all the content of the "Conceptual Design" section should be available from previous milestones. Section 3.1 of Table 28.1 presents the sketches of your alternative design concepts and briefly describes

Table 28.1 Suggested Organization of the Design Report

Organization	Pages
Title and authors	1
Table of contents (with page numbers)	1
List of individual contributions to the report	1
1. Problem definition	0.5
2. Design requirements	1
3. Conceptual design	
3.1. Alternative concepts	2
3.2. Evaluation of alternatives	1
3.3. Selection of a concept	0.5
4. Detailed design	
4.1. Main features and how they work	3
4.2. Results of analysis, experiments, and models	1
4.3. Manufacturing details	1
5. Performance evaluation	1
6. Lessons learned	1-2
Total	15-16

how each one works. Section 3.2 of Table 28.1 presents your decision matrix and uses the matrix as a vehicle to discuss the strengths and weaknesses of each concept. Section 3.3 of Table 28.1 indicates the concept you selected and gives your rationale.

The "Detailed Design" section describes the design that appeared at the final performance evaluation. New detailed drawings have to be prepared. Place these new drawings in Section 4.1 of Table 28.1, along with text describing the operation and main features of the final design. Describe the overall design first, and then zero in on the details of special features. If possible, include digital photographs of your machine. To create Section 4.2, retrieve the results presented at the oral design defense and add text. Section 4.3 is primarily a summary of the joining methods used.

The results of the performance tests are summarized in the "Performance Evaluation" section. Describe how your machine fared, both during individual performance testing and at the final competition, being as quantitative about it as you can. Also, compare performance predictions to actual results.

The "Lessons Learned" section is an opportunity to reflect back on the design experience. Write it in the form of three paragraphs, with each paragraph dedicated to answering one of the following questions: (1) How would you redesign your machine to improve performance? (2) What general lessons did you learn about the design process? and (3) What general lessons did you learn about teaming?

28.3 WRITING GUIDELINES

Use double spacing to leave room for instructor comments. Use the section and subsection headings of Table 28.1 and boldface them so that they stand out. Finally, figures should be embedded within the text (rather than placed at the end of the report) for ease of reference. In addition,

1. Be concise.
2. Begin each paragraph with a topic sentence that expresses the theme or conclusion of the entire paragraph. The reader should be able to overview the entire report by reading just the topic sentences.
3. Generate high-quality concept drawings and detailed drawings to pass the "flip test." The first thing the instructor does before reading the report is to flip through the pages to examine the figures. The figures are the instructor's first impression of the report.
4. Give each figure (i.e., drawing, graph, etc.) a figure number and a self-explanatory figure caption, for example, *Fig. 3: Decision matrix*. Figures should be numbered consecutively and referenced from the text using the figure numbers. For example, "The results of the comparison are summarized in the decision matrix of Figure 3."
5. Use a spelling checker. Spelling mistakes cause the reader to question technical correctness.
6. Employ page numbers both in the text and in the table of contents.

28.4 TECHNICAL WRITING IS "IMPERSONAL"

Technical writing is always done using an impersonal writing style. This style is limited to using the passive voice, third person pronouns, and impersonal "things" rather than people as subjects of sentences.

In a sentence with an action verb, the subject of the sentence performs the action indicated by the verb, and the sentence is in the *active voice*. If the subject of a sentence is *acted upon* by the verb, the subject is *passive*, and the sentence is in the *passive voice*. So, in a sentence using active voice, the subject of the sentence *performs* the action expressed in the verb. In the passive voice, the subject *receives* the action of the verb.

In technical writing, we do not use the active voice. The passive voice is always used in writing technical and laboratory reports because the person writing is not important (passive) but the process or principle being described is important. Examples of appropriate passive and inappropriate active voice sentences in technical writing are shown in Table 28.2.

Table 28.2 Examples of Passive and Active Voice Writing

Use Passive Voice	Do Not Use Active Voice
The angle was observed to be ...	I observed the angle to be...
It is suggested ...	We suggest...
A graph was used to ...	We used a graph ...

Technical writing also only uses third person pronouns. Examples of the correct third person and incorrect first person pronouns are shown in Table 28.3.

Table 28.3 Examples of Third Person and First Person Writing	
Use Third Person	**Do Not Use First Person**
It was found that…	I found…
It was assumed that…	I assumed that…

Finally, impersonal items rather than people are subjects of sentences in technical writing. Examples using impersonal rather than personal (people) as sentence subjects are shown in Table 28.4.

Table 28.4 Examples of Impersonal and Personal Sentence Subjects	
Use Impersonal Sentence Subjects	**Do _Not_ Use People as Sentence Subjects**
Analysis of the data indicates…	During the tests I noticed that…
This report presents…	In this report I show…

28.5 DESIGN MILESTONE: DESIGN REPORT

This is a time for both documentation and reflection.

Assignment
Prepare a design report in accordance with the guidelines of this chapter.

Grading Criteria
- Is the report complete?
- Does it pass the flip test?
- Is the writing style solid?
- Is it clear how the design works?
- Could someone unfamiliar with your machine manufacture it from the drawings and information in the report?

EXERCISES

1. While writing a design report, a team member copies material from the Internet and pastes it into his/her section of the report. He/she feels that this is important information relevant to the operation of your design. How do you deal with this situation?

2. In reviewing your team's design report, you come across this statement in the section on "Conceptual design": "We brainstormed three alternate solutions and then we used the design matrix to decide in which one we should use for our project." How would you restate this sentence to make it sound more professional?

3. One of your team members didn't contribute very much to the design but spent a lot of time constructing the competition entry. When writing design report section on "Manufacturing details," he/she wrote the

following sentence: "I made the base out of cardboard and then I glued the motor and gear wheel thing on it and I hoped it would work out ok." How would you restate this sentence to make it sound more professional?

4. Sometimes you can be too concise in writing your design report. Consider the concise statement "we made it out of wood." How would you change this to still be concise and yet give the reader more relevant information?

5. A team member wants to put several cartoons in the design report to entertain the instructor who will be reading it. The cartoons have nothing to do with the design process, but they do inject some humor into the performance evaluation. List several reasons why you think this is acceptable or not.

Examples of Design Competitions

Source: Mars Challenge at Union College.

29.1 INTRODUCTION

In this chapter, we present a typical design competition along with one team's solutions to the first four milestones. Design competitions like the one described in this chapter have proven to be very successful with first- and second-year engineering students. The rules, tabletop playing field, and list of parts provided may be used as a template in defining similar competitions. This chapter has two design competitions for the student to study: "A Bridge Too Far" and the "Mars Meteorite Retriever Challenge."

29.2 DESIGN COMPETITION EXAMPLE 1: A BRIDGE TOO FAR

A design project begins on the day that the instructor distributes the rules governing the design competition. This package of rules typically consists of a statement of the design objective, a list of design requirements, a drawing of the playing field, and a list of parts and materials.

In the design competition named A Bridge Too Far, the objective is *to design and build a device to outscore your opponent in a series of head-to-head matches*. The playing field is shown in Figure 29.1. A team receives +1 point for every ball resting in its scoring pit at the end of play. In all, there are 17 scoring balls. Six of the balls start out in the possession of the two teams; the remaining 11 balls have starting positions on the playing field as indicated in Figure 29.1. Other key rules are:

- The device must fit within a 1-foot by 1-foot by 2-foot high volume.
- Parts and materials are limited to those listed in Table 29.1.
- Each device can have up to three independent tethered controls.
- There is a 1 minute setup time.
- One game lasts 30 seconds.
- If there is a tie, both teams lose.

The parts and materials of Table 29.1 are not supplied to the teams until they successfully defend their designs at the oral design defense. The complete set of rules is given in Section 29.4 of this chapter.

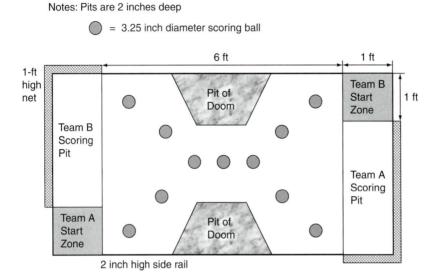

FIGURE 29.1 A Bridge Too Far tabletop playing field.

Quantity	Item
	Table 29.1 The List of Parts and Materials for "A Bridge Too Far"
1	Motor with adjustable gear box
1	Gearhead motor
1	Electric relay (for use with gearhead motor)
1	Electric diode (for use with gearhead motor)
1	Electric pull solenoid
1	Set of six gears
1	3/8 in. by 36 in. wooden dowel
2	$1/16$ in. by 3 in. by 36 in. balsa wood sheets
2	$3/8$ in. by $3/8$ in. by 36 in. balsa wood beams
1	¼ in. by 12 in. by 12 in. plywood sheet
10	Large craft sticks (tongue depressors)
1	$1/8$ in. diameter by 20 in. long steel rod
1	10 in. piece of string

Table 29.1 The List of Parts and Materials for "A Bridge Too Far"—cont'd

Quantity	Item
2	1 in. long metal hinges
1	¼ in. diameter by 10 in. long elastic cord with hooks
1	12 in. long rubber band strip
4	2 in. diameter wheels with hubs (for $1/8$ in. diameter shaft)
1	2 in. diameter mailing tube
1	Two liter plastic soda bottle (provided by the team)

29.3 DESIGN MILESTONE SOLUTIONS FOR "A BRIDGE TOO FAR"

29.3.1 Design Milestone 1: Clarification of the Task

In the case of a design competition, clarification of the task is mostly the responsibility of the instructor. There were just two things left for the design team to do to complete this milestone.

First, the team directed the following questions to the rules committee:

1. Can a machine score from the Pit of Doom?
2. Can you drop obstacles to interfere with the other machine?
3. Can you score points by driving a machine loaded with balls into a scoring pit?
4. Can a machine attach itself to the other machine?
5. Can a machine expand beyond the 1-foot by 1-foot by 2-foot high dimensions once the game begins?

The answer to all these questions is "yes," since nothing is against the proposed actions in the rules. The motivation behind the questions is to fully understand the design constraints and probe for omissions that could lead to a design advantage.

Second, the team compiled a list of performance requirements that were specific to their design and their competition strategy.

D = Must score at least four points.
W = Must score the first point within 10 s.
D = Must hinder the opponent's ability to score.

The list is short to avoid solution bias. Later, after a design has been selected, other performance requirements can be added.

29.3.2 Design Milestone 2: Generation of Alternative Concepts

The functional decomposition settled on is shown in Figure 29.2. Consideration of offensive strategy is done through the subfunction "deliver balls."

The results of brainstorming each subfunction were collected in the classification scheme of Figure 29.3.

Compatible subfunction concepts were combined to form the four promising designs shown in the concept drawings of Figures 29.4–29.7.

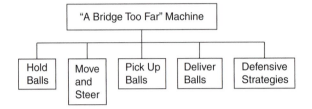

FIGURE 29.2 Functional decomposition for "A Bridge Too Far."

Concepts / Functions	Concept 1	Concept 2	Concept 3	Concept 4	Concept 5
Hold balls	Place on machine	Keep under machine	Push balls in front	Container to put balls in	Drag behind machine
Move and steer	2 Motors	2 Motors	1 Motor steers	No steering	Don't move
Pick up balls	Don't pick up balls	Wedge	Scoop	Plow	Claw
Deliver balls	Push them into pit	Throw them into pit	Drive into pit with balls	Roll them down a ramp	Push opponent into pit
Defensive strategies	Block bridge	Ignore opponent	Throw extra balls into Pit of Doom	Pick up opponent	Throw obstacles

FIGURE 29.3 Classification scheme for "A Bridge Too Far."

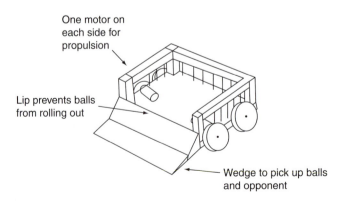

One motor on each side for propulsion

Lip prevents balls from rolling out

Wedge to pick up balls and opponent

FIGURE 29.4 Concept drawing of the "Wedge."

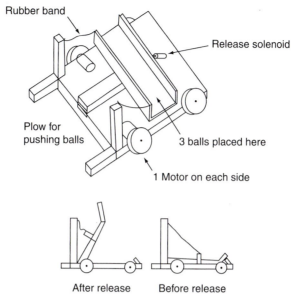

Rubber band

Release solenoid

Plow for pushing balls

3 balls placed here

1 Motor on each side

After release   Before release

FIGURE 29.5 Concept drawing of the "Catapult."

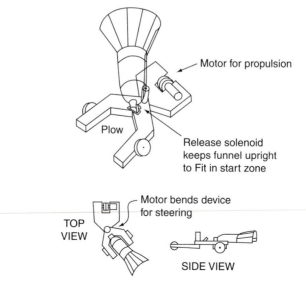

Motor for propulsion

Plow

Release solenoid keeps funnel upright to Fit in start zone

Motor bends device for steering

TOP VIEW

SIDE VIEW

FIGURE 29.6 Concept drawing of the "Snake."

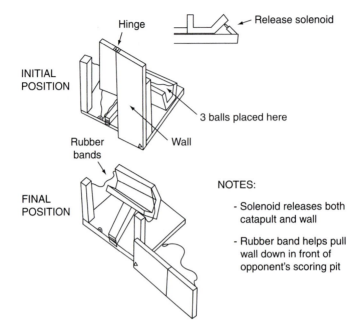

FIGURE 29.7 Concept drawing of the "The Wall."

29.3.3 Design Milestone 3: Evaluation of Alternative Concepts

The four alternative concepts developed in the previous milestone are illustrated in Figure 29.8 for easy reference.

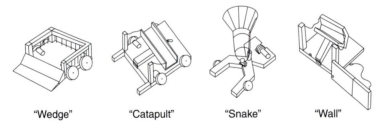

"Wedge"   "Catapult"   "Snake"   "Wall"

FIGURE 29.8 Alternative design concepts from "A Bridge Too Far."

The selection of evaluation criteria was tricky, because a high-scoring machine that consistently gathers 9 of the 17 balls should be ranked on the same level as a defensive machine that consistently wins by the score of 1-0. The evaluation criteria and their respective weights are shown in Table 29.2. Both of the first two criteria are needed to describe scoring potential, because it is not enough that you can transport a lot of balls; you must be able to score with them with an opponent in your way. Their combined weighting of 0.4 is slightly higher than the weight of 0.3 assigned to defensive capabilities (third criterion), since you have to score to have a chance of winning. "Easy to manufacture" has its usual importance, but "low cost" was not included because it is not a factor in the competition. "Easy to control" appears because of the potential challenges involved in maneuvering across the bridge. The discussion of concept strengths and weaknesses is organized by evaluation criterion.

Table 29.2 Decision Matrix for "A Bridge Too Far"

Evaluation Criteria		Wt	Wedge		Catapult		Snake		Wall	
			Val_1	$Wt \times Val_1$	Val_2	$Wt \times Val_2$	Val_3	$Wt \times Val_3$	Val_4	$Wt \times Val_4$
1	Large payload	0.2	6	1.2	5	1.0	8	1.6	2	0.4
2	Robust scoring	0.2	5	1.0	9	1.8	3	0.6	7	1.4
3	Disrupts opponent	0.3	7	2.1	5	1.5	4	1.2	8	2.4
4	Easy to manufacture	0.2	7	1.4	5	1.0	5	1.0	7	1.4
5	Easy to control	0.1	7	0.7	6	0.6	3	0.3	9	0.9
	Total	1.0		6.4		5.9		4.7		6.5

(1) Has a Large Payload

The "Wall" is rated low because it can score with only the three original balls. For the other three concepts, ratings are proportional to carrying capacity. Since the funnel design of the snake expands to be much larger than the start zone, it has a much larger carrying capacity than the other two.

(2) Robust Scoring Capability

The two designs that launch the balls score highest here, because it is much harder to block a ball that is thrown than one that is pushed. Also, these machines can score without having to cross the bridge.

(3) Can Disrupt Opponent

The "Wall" is an obvious favorite for this category, since it is the only one that actively blocks the opponent's goal. The "Wedge" is designed to be able to pick up the opponent and drop it in the appropriate goal to gain more points, so it too can theoretically disrupt the opponent well.

(4) Easy to Manufacture

The "Wedge" is rated higher than the "Catapult," because it has fewer functions to build. The "Wall" also does well in this category, because it does not involve any motors or corresponding drivetrains. In fact, it would be rated even higher were it not for the difficulties anticipated with sequencing the release of the wall and the catapult arm.

(5) Easy to Control

The "Wall" simply requires flipping one switch, which is almost as easy as doing nothing. The "Snake" is slightly alien to most things people will have controlled and, as a result, would be hard to drive.

29.3.3.1 Discussion of results

All the machines demonstrated strength in some areas. Observed weaknesses cannot be corrected without diminishing the attributes that made them strong. For example, the "Snake's" funnel, which gives it the largest payload, also makes it vulnerable to being pushed around by the opponent.

The "Wall" is the highest rated design and also the boldest, in that it dares to remain stationary. Students tend to shy away from designs like this one, because they are different. This design was not selected because it was felt that its defensive capabilities were overrated. The design may not be able to resist pushing by an opponent at the end of the wall.

The "Wedge" was finally selected as the best design. It is interesting that it is rated high, even though it is not a clear winner in any of the categories. It compares very closely with the "catapult." The simplicity, lifting potential, and large payload of the "Wedge" were the decisive factors.

29.3.4 Design Milestone 4: Detailed Design

As you may recall, the "Wedge" (shown on the left in Figure 29.9) was selected as the final concept. The original wedge design was dual purpose: (1) to lift and push the opponent's machine and (2) to allow balls to roll up into the holding bin above the moving platform.

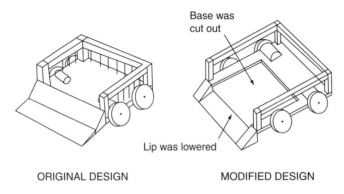

FIGURE 29.9 Design modifications in response to experiments.

29.3.4.1 Experiments

There were concerns that the forward speed of the vehicle might not be sufficient to allow the balls to roll up high enough into the bin. Experiments (see Figure 29.10) were conducted to check this out. The results showed that the concerns were justified. Acceptable dimensions for the wedge and holding bin were established, based on the results of the experiments. The modified concept is shown on the right in Figure 29.9.

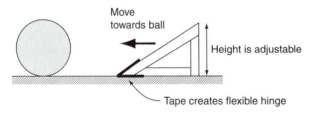

FIGURE 29.10 Experiment used to determine optimal wedge geometry.

29.3.4.2 Analysis

Two identical gearhead motors were available for creating the drivetrains for the left and right sides of the vehicle. The no-load angular speed (N_{noload}) at 24 V was listed in the catalog as 145 rpm. To maintain a speed of 1.00 ft/s or 12.0 in./s on the flat with tire diameters of 2.00 in., the overall gear ratio needs to be (see Equation (21.5) of Chapter 21):

$$GR = \frac{2\pi N_{noload} R_{tire}}{60 V_{mp}} = \frac{2\pi(145)(1.00)}{60(12.0)} = 1.26$$

which we round up to 2, based on the set of gears provided. With the pushing requirement of this design, it might seem that the actual gear ratio needs to be much higher. However, this particular motor is quite powerful; the motor shaft could not be visibly slowed by manually gripping the ends of the shaft. Given the power of the motor and the low weights of the vehicles, this gear ratio was deemed to be reliable, even with the pushing requirement. The resulting drivetrain geometry is shown in Figure 29.11.

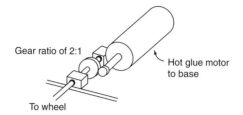

FIGURE 29.11 Close-up view of the drivetrain.

29.3.4.3 Detailed drawing

Two views of the final design are shown in the hand-drawn detailed drawing of Figure 29.12. Some manufacturing details are provided.

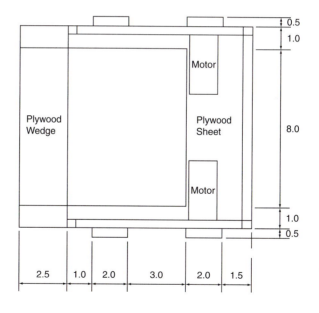

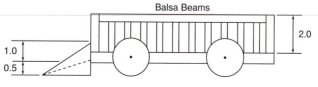

NOTE: All dimensions in inches

FIGURE 29.12 Detailed drawing (two views).

29.4 OFFICIAL RULES FOR THE "A BRIDGE TOO FAR" DESIGN COMPETITION

29.4.1 Objective

Design and build a device to outscore your opponent in a series of head-to-head matches.

29.4.2 Constraints

1. The parts and materials that may be used in the construction of the device are limited to those listed in Table 29.1. Each team is provided with a box containing all legal parts and materials.
2. The devices must be constructed entirely by the members of the team (i.e., if you lack the required expertise or tools to manufacture a certain part of the device, redesign it).
3. Each device can be loaded with up to three scoring balls. If a team chooses not to load all three balls, the discarded balls are removed from the playing field.
4. When placed on the table at the start of the game, each machine must fit completely within its assigned 1-foot by 1-foot by 2-foot high start zone.
5. An external power source is made available to each device for use during the matches. It consists of
    *   Overhead wires that connect to the device. (Note: these wires are considered to be part of the playing field.)
    *   A control box with two forward-reverse-off switches and one on-off switch.

29.4.3 The Game

1. Just prior to the game, there is a 1-min setup time, during which each team should:
    *   Place its device in the starting zone.
    *   Attach and check the electrical connections.
    *   Load its device with up to three scoring balls.
2. Except for manipulation of the electrical control boxes located at one end of the playing field, no human interaction with the device (or playing field) is allowed once the game begins.
3. The game begins when indicated by the referee and ends 30 seconds later when the power is switched off.
4. The game, and all scoring, ends as soon as one of the following occurs:
    *   All movement stops as a result of power being switched off.
    *   Five seconds have elapsed since power was switched off.

29.4.4 Scoring

1. At the end of the game, each team receives 1 point for every ball in its scoring pit, irrespective of which team caused the ball to fall into the pit. The team that scores the most points wins.
2. In all, there are 17 scoring balls; six of the balls start out in the possession of the two teams, and the remaining 11 balls have starting positions on the playing field as indicated in Figure 29.1.
3. A ball is counted as lying in a scoring pit if an imaginary vertical line through the center of the ball lies within the boundary of the scoring pit. The referee can ascertain the status of a scoring ball at the end of the game by removing other balls from the scoring pit; if the ball in question then falls into the pit, it counts as lying in the scoring pit.
4. In the event of a tie,
    *   If neither team has scored, both teams lose.
    *   If both teams have scored, the winner is the machine that has advanced farthest down the field as based on final positions. If this criterion proves indecisive, both teams advance to the next round.

29.4.5 Other Rules

1. If the tethers (i.e., electrical connections) should entangle as a result of the machines passing each other, time stops and both machines are returned to their respective starting zones. Play then continues at the signal of the referee. This situation can occur, because each machine is tethered to the nearest of two overhead rods running parallel to the length of the table. To avoid entanglement, keep to the right of the opposing machine when passing.
2. Devices that permanently damage the playing field or the balls are disqualified.
3. Any attempt to intentionally inflict permanent damage on an opponent's device results in immediate disqualification. However, devices should be designed to hold up under expected levels of nonaggressive contact. For example, some pushing should be expected, but that pushing should not occur at significant ramming speeds.
4. Any device deemed unsafe is not allowed to participate in the matches.
5. Implementation of any strategies that are not directly addressed in the rules, but that are clearly against the spirit of the rules (e.g., intentionally interfering with the person at the controls) leads to disqualification.
6. The rules committee has the final word on any interpretation of the rules.

29.5 DESIGN COMPETITION EXAMPLE 2: MARS METEORITE RETRIEVER CHALLENGE

Meteorites from Mars occasionally land on Earth. Those that land on Antarctica are particularly easy to spot because of their color contrast with ice. NASA is developing automated rovers to retrieve these meteorites. Your job is to implement such a rover on a simulated Antarctica: a wooden board 96 in. long and 48 in. wide. The meteorite is simulated by a small object containing a light source. NASA identified the most promising "landing zone," simulated by a rectangle on the board 18 in. by 24 in. The meteorite can be anywhere within that landing zone.

Your *autonomous* vehicle begins its trip at a robot depot, simulated by a 12 in. by 12 in. square on the board. Your challenge is to locate the meteorite, travel to it, pick it up, carry it without touching the ground at any point (to avoid contamination), and deposit the meteorite at the meteorite lab (another 12 in. by 12 in. square), see Figures 29.13 and 29.14. Two completed student designs are shown in Figure 29.15. The vehicle starts at the robot depot and the meteorite is placed randomly within the landing zone.

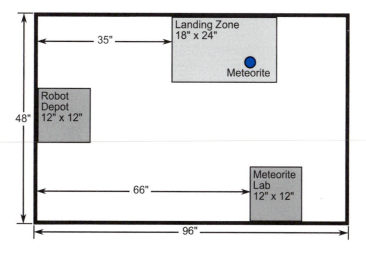

FIGURE 29.13 Layout of simulated Antarctica.

FIGURE 29.14 The actual Mars meteor retriever challenge game board.

FIGURE 29.15 Student vehicles.

29.6 SOME DESIGN MILESTONES FOR THE "MARS METEORITE RETRIEVER CHALLENGE"

29.6.1 Design Milestone 3: Performance Testing

29.6.1.1 Grading

Team grades are in proportion to level of functionality attained in the best of two tests, as defined by the following increasing levels of performance: The vehicle

- Is completely manufactured, 60%.
- Moves when powered, 70%.
- Moves with program control, 75%.
- Navigates to the loading dock, 80%.
- Navigates to the loading dock and touches meteorite under program control, 85%.
- Lifts the meteorite from the table surface and leaves the loading dock completely, 95%.
- Lifts the meteorite from the table surface and deposits the light source in the laboratory area, 100%.

29.6.2 Design Milestone 4: Design Report

29.6.2.1 Purpose

The purpose of the report is to provide a concise, accurate, and informative record of your design efforts. The report must be no more than 12 pages long, including the cover page, table of contents, and drawings.

29.6.2.2 Organization

1. Cover page.
2. Table of contents.
3. Introduction: State the design objective and cite some of the design constraints.
4. Strategy: Describe your competition strategy.
5. Conceptual design:
   - Discussion of alternatives with clearly labeled concept drawings. Discuss the strengths and weaknesses of each.
   - Include your functional decomposition, classification scheme, decision matrix, and the results.
6. Detailed design:
   - How it works (i.e., details of the operation of each subfunction).
   - Include detailed drawings of the final design and explain how it is put together.
   - Include the results of any calculations or experiments.
7. Performance evaluation: Summarize the results of performance tests.
8. Conclusions and recommendations: Discuss the strengths and weaknesses of your design and your execution, and conclude with a discussion of lessons learned.

29.6.2.3 Grading

Technical communications, 50%
- The report is neat, concise, well organized, and includes all the items just listed.
- The report is free from spelling and grammatical errors.
- The report is presented attractively.

Technical content, 50%
- The report shows clear evidence that you understand and have followed the design methodology learned in the design studio.
- The report describes how the design works clearly enough that a person who has never seen the device could produce a working machine from the descriptions and drawings in the report.

29.6.2.4 Oral presentation

During the competition each team gives a 2-min presentation to the judges. This presentation must include a detailed sketch or picture of the final design and a brief description of the design features, such as the steering mechanism, pickup mechanism, and the like that make the team's design unique. You should also provide a brief description of a problem encountered and the way in which the problem was solved using teamwork. If you did not encounter any difficulties in your design, discuss how teamwork made this possible.

Each team must practice its presentation and time it to ensure that it does not exceed 2 min. Your presentation should consist of two PowerPoint slides: a title page that includes the team members' names and team name (if you have no team name, select one) and a slide with a detailed drawing of the final design.

29.7 OFFICIAL RULES FOR THE "MARS METEORITE RETRIEVER CHALLENGE" DESIGN COMPETITION

29.7.1 Objective

The objective of the competition is to construct an autonomous vehicle that can retrieve a (simulated) Mars meteorite from the (simulated) icy wastes of Antarctica.

29.7.2 Constraints

1. Parts and materials that may be used in the construction of the device are limited to those listed in Table 29.3. Each team will be provided with a box containing all legal parts and materials.
2. The devices must be constructed entirely by the members of the team (i.e., if you lack the required expertise or tools to manufacture a certain part of the device, redesign it).
3. When placed on the table at the start of the game, each machine must fit completely within its assigned 1-cubic foot start zone.

Table 29.3 List of Parts and Materials for the "Mars Meteorite Retriever Challenge"

Quantity	Item
1	$7/8$ in. by $11 7/8$ in. by $1/4$ in. plywood
2	$3/8$ in. by $3/8$ in. by 36 in. balsa wood beam
1	$3/8$ in. by 36 in. hardwood dowel
6	Large craft sticks (tongue depressors)
1	$3/32$ in. by 3 in. by 36 in. balsa sheet
2	Parallax continuous rotation servos
2	Boe-Bot wheels and tires
1	Tamiza 4-speed crank axle gearbox and motor
1	Plastic ball caster
2	20 in. long 6-32 threaded rod with eight nuts and washers
2	Battery holders for four AA batteries
1	Parallax microcontroller board with a BASIC Stamp processor
1	Box of paper clips
2	Light sensors
2	Microswitches

29.7.3 Scoring

The overall challenge is to accumulate points for design quality and vehicle performance. The teams with the most points will be designated "superteams." Design quality points (up to 30) are awarded by the judges for creativity, construction, and presentation. Vehicle performance points (up to 70) are awarded for accomplishing the tasks of the challenge in the demonstration.

1. Teams give a 2-min presentation about their design for the design quality judging.
2. Teams demonstrate their vehicles. Vehicles start from a designated robot depot.
3. At the judges' "go" signal, the teams immediately operate any necessary switches and release their vehicles. No team member may subsequently make either direct or indirect contact with the vehicle or playing field until the "stop" signal is given.
4. Such contact, even if accidental, results in immediate disqualification.
5. The vehicle operates until it fully completes the task or until 3 min have elapsed. If 3 min have elapsed, the judges give a "stop" signal, and the team terminates the vehicle's operation.
6. Points are awarded for the demonstration of tasks as follows:
   - 5 points, the vehicle completely departs the robot station.
   - 5 points, the vehicle at least touches the landing zone.
   - 10 points, the vehicle touches the meteorite.
   - 10 points, the vehicle picks up the meteorite.
   - 10 points, the vehicle moves the meteorite outside the landing area (5 points for moving it, but letting it touch the surface).
   - 10 points, the vehicle delivers the meteorite to the meteorite lab.
   - 10 points, the vehicle deposits the meteorite completely within the meteorite lab.
   - 10 points, the vehicle navigates completely outside the meteorite lab.

29.7.4 Rules

1. Vehicles must be a single unit.
2. Vehicles must at all times fit within a cube 12 in. on a side
3. No vehicle can damage or be attached to the playing field. All vehicles must move in a way that does no harm to the field.
4. All vehicles using batteries must include in their circuits a length of fuse wire between one pole of each battery pack and the next connected component. No vehicle may use more than eight AA batteries or any other kind of battery.
5. The microcontroller board must be mounted in a way that can be removed after the competition and returned undamaged.
6. Imaginative strategies in the spirit of the game are encouraged.
7. However, any strategies determined by the judges to be contrary to the spirit of the game are excluded. Contestants have the responsibility of clearing with the judges before the competition any strategies that might possibly violate this rule.

29.7.5 Additional Supplies

A circuit design and parts that allow the Tamiya motor to be reversed are provided to those who would like to implement them. Teams must purchase AA batteries for the microcontroller. Teams may purchase simple and inexpensive connectors, both mechanical (e.g., screws and nails) and electrical (e.g., simple contact switches). The simplicity and inexpensive nature of such additional supplies must be cleared with your instructor and made known to and available to other contestants.

Teams that, in the opinion of the judges, are attempting to win the competition by exclusive use of expensive or otherwise difficult-to-access outside technology (i.e., components not made by the team itself from approved parts) are disqualified.

EXERCISES

1. Other examples of first-year engineering student design competitions can be found on the Internet. An excellent list of well over 100 first-year competitions from a wide variety of universities can be found at the *Directory of Projects for First-year Engineering Students* (http://www.discovery-press.com/discovery-press/studyengr/projects.asp).

2. The Rube Goldberg design projects are especially interesting for first-year engineering students (see: http://www.rubegoldberg.com).

3. TryEngineering offers a variety of lesson plans that allow teachers and students to apply engineering principles in the classroom (http://www.tryengineering.org/lesson-plans).

Closing Remarks on the Important Role of Design Projects

Source: © iStockphoto.com/Malcolm Romain

If you ask professors or students why they do design projects in engineering courses, you can expect to hear responses like this:

- "They are motivational tools."
- "They apply the analytical methods taught in courses."
- "They help develop written and oral communication skills."
- "They teach teaming."

Indeed, these are all valuable outcomes of a design project, but each one can be achieved by some other means. The answer must lie elsewhere.

Part of the answer is found in the view that every engineering endeavor is ultimately about finding or designing a solution to an expressed need. The analytical methods, the teaming skills, and the rest are tools for achieving that goal; that is, they are the means to the end, not the end itself. Each design project offers a rare opportunity for students who spend most of their time deeply immersed in learning analytical methods to see the big picture.

The rest of the answer has to do with the real purpose behind these design chapters. Engineering design is at its core an unbiased and structured methodology for dissecting and solving complex problems. It is the way engineers should and must think. In contrast to analytical methods that are each limited to their own special class of problems, design methodology has universal applicability—to design, to research, to all fields of study. Design projects are the best way we know to exercise and develop this most fundamental of all engineering methods.

Hands-on design projects come closest to fully realizing these goals. They complete the design process; for as we have seen, it does not end with the detailed design; there will be design modifications to be made during manufacturing, testing, and the final performance evaluation. Students learn the importance of design for manufacture principles by experiencing the results of having failed to heed them. They also gain a sense of accountability by learning that it is not enough for a design to look good on paper—it has to *work*.

Index

Note: Page numbers followed by *b* indicate boxes, *f* indicate figures and *t* indicate tables.